W9-CRC-170

Cauldrons in
the Cosmos

Theoretical Astrophysics
David N. Schramm, series editor

Cauldrons in the Cosmos

NUCLEAR ASTROPHYSICS

Claus E. Rolfs and
William S. Rodney

With a Foreword by
William A. Fowler

The University of Chicago Press
Chicago and London

CLAUS E. ROLFS is professor of physics at the University of
Münster. WILLIAM S. RODNEY is the former program
director for nuclear physics at the National Science
Foundation. Both have published many journal articles in
the field of astrophysics.

The University of Chicago Press, Chicago 60637
The University of Chicago Press, Ltd., London

© 1988 by The University of Chicago
All rights reserved. Published 1988
Printed in the United States of America
97 96 95 94 93 92 91 90 89 88   5 4 3 2 1

Library of Congress Cataloging in Publication Data
Rolfs, Claus E.
    Cauldrons in the cosmos.

    Bibliography: p.
    Includes index.
    1. Nuclear astrophysics.   I. Rodney, William S.
II. Title.
QB464.R65   1988   523.01   87-16786
ISBN 0-226-72456-5
ISBN 0-226-72457-3 (pbk.)

Contents

Foreword

Cauldrons in the Cosmos is a potent witches' brew distilled from the ferment in the authors' hearts and minds, stimulated by their joy and delight in the nature of the universe which they, and we, inhabit. The title betrays their willingness to eschew the traditional low-key parlance of science textbooks, but the contents will reward the reader with a deep understanding of the field of nuclear astrophysics—the union of nuclear physics and astronomy.

In the minds of some there exists the patronizing belief that nuclear physics is a mature science. The same is not believed about nuclear astrophysics, which has been an active branch of astrophysics for over forty years but is now in the midst of an exciting revival in experimental and theoretical research around the world. University and government laboratories lead the way. The ultimate goal is to understand how nuclear processes generate the energy of stars over their lifetimes and, in doing so, synthesize heavier elements from the primordial hydrogen and helium produced in the big bang which led to the expanding universe.

Progress toward that goal has been recognized and rewarded, but much more will come in the next decade than during the past four decades. *Cauldrons in the Cosmos* covers quite adequately what we have learned in those four decades, and is still able to devote a substantial number of pages to the fundamental questions still to be solved. I stress the word *fundamental*. Indeed there are details to be attended to, but they are overshadowed by serious difficulties in the most basic concepts of nuclear astrophysics. On square one the solar neutrino problem is still with us (chap. 10), indicating that we do not even understand how our own star really works. On square two we still cannot show in the laboratory and in theoretical calculations why the ratio of oxygen to carbon in the sun and similar stars is close to two-to-one (see chap. 7). We humans are mostly (90%) oxygen and carbon. We understand in a general way the chemistry and biology involved, but we certainly do not understand the nuclear astrophysics which produced the oxygen and carbon in our bodies.

The authors have made important and significant contributions to the field of nuclear astrophysics in both experiment and theory. They have written an impartial record of the work of many nuclear physicists. They have studied in great detail the astronomical and astrophysical environments in which energy generation and element synthesis take place. Their book will serve for many years as an important reference source for astronomers and physicists alike. Moreover, the work will serve as an excellent textbook for advanced undergraduate and graduate students.

WILLIAM A. FOWLER

Preface

Nature offers no greater splendor than a star-filled sky on a clear moonless night. Silent, timeless, jeweled with the constellations of ancient myth and legend, the night sky has invoked wonder throughout the ages. It is a wonder that lets our imagination roam far from the confines of earth and the present into boundless space and cosmic time. It is human not just to accept things as they appear but rather to ask questions and search for answers. Astronomy and astrophysics, born of that wonder and those questions, are sustained by two of the most fundamental traits of human nature: the need to explore and the need to understand. Curiosity drives us to explore our surroundings; we want to see the unseen and to know what lies beyond the horizon. Over the centuries the urge to explore has carried the human mind outward into the realm of the galaxies and inward into the microscopic domain of elementary particles. Through the interplay of exploration, discovery, and analysis, the key to understanding, answers to questions about the universe have been sought since the earliest of times. Yet, the search has never, since its earliest beginnings, been more vigorous or exciting than it is today.

Astronomy is concerned with nature on its largest scale, the physical universe beyond the earth. Its purpose is to extract information on celestial objects from electromagnetic and other types of radiation that falls on the earth. Such studies have revealed a scene of ever-increasing magnitude and grandeur. Astronomy began with the geocentric view of the early Greeks, in which the earth was the central figure. The sun, moon, and planets revolved around the earth within a rotating sphere of fixed stars. Next, the heliocentric view, dating formally from the time of Copernicus, put the planetary system on a more nearly correct basis and brought to rest the sphere of the stars. There emerged then the idea that the stars are other suns, many of them perhaps also attended by planetary systems. In this century the picture of the universe has been unfolding further with spectacular rapidity. Our Galaxy, the Milky Way, having the sun in its suburbs, now stands out clearly as a majestically rotating spiral structure in the foreground of the vast celestial scene. The formerly mysterious spirals and associated nebulae lying beyond the Milky Way are established as galaxies. It is also recognized that in the universe at large violent processes are the norm rather than the exception. The discoveries of the past twenty years have radically changed our concepts of the origin and evolution of stars, galaxies, and the universe itself.

Astrophysics—the union of astronomy and physics—applies physical laws observed and studied on earth to the laboratory of space. The enunciation of the law of gravitation by Newton inaugurated a dynamical interpretation of the motion in the solar system, and it was soon recognized that this interpretation was not limited to the solar system but rather represented a key to the understanding of many features of the entire universe. Looking outward and building on the observations of astronomy is astrophysics, one of the frontiers of physics, which is research on the vast space in the cosmos, seeking to understand the structures in the universe. Looking inward, the frontier is the study of atoms, nuclei, and elementary particles, the building blocks of matter. Although these two research fields, one infinite and the other infinitesimal, appear to be basically disconnected, it was in understanding the microscopic structures and interactions of matter that the major link was provided that was to furnish insight into questions and problems of astrophysics and the mysteries of the origin and history of the universe. There is still much to be learned, much to be discarded, and much to be revised. We have excellent astrophysical and cosmological theories, but theories, in all of science, are only guides to understanding. They are not truth itself. They must therefore be continuously revised if they are to lead us in the right direction. They must be constantly confronted with experimental and observational data to ensure that they do not evolve in some entirely meaningless direction. The present picture of the physical universe is incomplete, and doubtless it will always be incomplete, but there is hope that in the light of new explorations it will improve. In future years we may see greater grandeur and simplicity added to our view of the universe.

Investigations during the last fifty years have shown that we are connected to distant space and time not only by our imagination but also through a common cosmic heritage: the chemical elements that make up our bodies. These elements were created in the hot interiors of remote and long-vanished stars over many billions of years. Their fuels finally spent, these giant stars met death in cataclysmic explosions, scattering afar the atoms of heavy elements synthesized deep within their cores. Eventually this material, as well as material lost during the red giant stages, collected into clouds of gas in interstellar space; these, in turn, slowly collapsed, giving birth to new generations of stars, thus leading to a continuing cycle of evolution. In this scenario, the sun and its complement of planets were formed nearly 5 billion years ago.

Drawing upon the material gathered from the debris of its stellar ancestors, the planet earth provided the conditions that eventually made life possible. Thus, like every object in the solar system, each living creature on earth embodies atoms from distant corners of our Galaxy and from a past thousands of times more remote than the beginning of human evolution. Thus, in a sense, each of us has been inside a star and truly and literally consists of stardust; in a sense, each of us has been in the vast empty space between the stars; and—since the universe has a beginning—each of us was there. Every molecule in our bodies contains matter that once was subjected to the tremendous temperatures and pressures at the center of a star. This is where the iron in our blood cells originated, the oxygen we breathe, the carbon and nitrogen in our tissues, and the calcium in our bones. All were formed predominantly in fusion reactions of smaller atoms in the interior of stars. The smaller atoms themselves (i.e., hydrogen and helium) were created prior to star formation in the very early universe.

The hypothesis that the energy which powers the sun comes from thermonuclear reactions appears to be due mainly to Eddington, following shortly after the work of Aston, showing that an enormous amount of energy is stored in nuclei, and the discovery of nuclear reactions by Rutherford, showing that this nuclear energy can be liberated and that one element can be transmuted into another. Eddington had these discoveries of nuclear physics in mind when, in an address to his colleagues at a meeting in Cardiff, Wales, in 1920 he said, "What is possible in the Cavendish Laboratory may not be too difficult in the sun." This basic concept for understanding the nature and history of the sun (and other stars) was substantiated in the 1920s by the work of Gamow, Atkinson, and Houtermans and further elucidated in the 1930s by von Weizsäcker, Bethe, and Critchfield. From the hypothesis that stars generate most of their energy through thermonuclear reactions, it followed that nuclei are at the same time being transmuted in stars.

The explosive growth in the physics of nuclei during the early 1940s resulted in a wealth of empirical data on nuclear reaction rates. When cross sections for neutron-induced reactions on a variety of heavy nuclei became available, Hughes and Alpher, Bethe, and Gamow immediately noted that there is an approximately inverse relationship between neutron-capture cross sections and the relative abundances of elements found in the solar system. Thus, for the first time, a clear connection between nuclear reactions and element synthesis was established, at least for the heavy elements. From subsequent improvements of data on elemental abundances, in particular by Suess and Urey in 1956, it became increasingly clear that all the elements were formed in accordance with their basic nuclear properties. Thus, our world bears clear signs of being the collective ashes of what Suess and Urey called a "cosmic nuclear fire." In 1957, Burbidge, Burbidge, Fowler, and Hoyle, and independently Cameron, integrated all of the new ideas and information on element formation into a coherent picture, referred to as the theory of nucleosynthesis of the elements and their isotopes. The present picture is that all elements from carbon to uranium were produced entirely within stars.

There is persuasive evidence that the "cooking" of the elements, i.e., nucleosynthesis, is a continuing process. Perhaps the most dramatic example of this was the discovery in 1952 of lines from the element technetium in stellar spectra. Since the longest-lived isotope of this element has a half-life of only 4 million years, the element must have been made relatively recently. A more recent example (1982) is the discovery in γ-ray astronomy of significant amounts of ^{26}Al, which has an even shorter half-life of 0.7 million years.

The recognition of our cosmic heritage is thus a relatively recent achievement. The detailed understanding of this heritage combines astrophysics and nuclear physics and forms what is called nuclear astrophysics. Included in this scientific field are the structure and evolution of stars; the generation of energy in stars; the synthesis of elements in stars; the nuclear debris from the beginnings of the universe; the structure and formation of neutron stars, pulsars, and black holes; the origin of cosmic rays and their interactions with interstellar gas; the chemical evolution of galaxies; the history of the planets and the moon; and neutrino and γ-ray astronomy. In a sense, nuclear astrophysics involves much of astrophysics, because there are few important events in astronomy, cosmogony, and cosmology that have not left nuclear clues. We may gather these clues, study the properties of the atomic nuclei, and, if we are lucky, figure out what has happened. There remain puzzles and problems, which challenge the basic ideas underlying nucleosynthesis in stars and elsewhere. Thus, the ultimate goal of the field has not been attained, and much work is needed on all its aspects (experiment, theory, and observation) before the picture is complete.

Laboratory nuclear astrophysics is often a frustrating science. The desired cross sections are among the smallest measured in the nuclear laboratory, often requiring long data-collection times with painstaking attention to background. From a purely nuclear point of view, the reactions studied are often of comparatively little interest. It is their application to astrophysics that provides the major intellectual motivation. However, on many occasions evaluation of the collected data has provided unexpected intellectual rewards in nuclear physics itself. The grand concept of elemental nucleosynthesis will not be truly established until we attain a deeper and more precise understanding of the many nuclear processes operating in astrophysical environments.

This textbook is basically an introduction to nuclear astrophysics. It is divided into ten chapters, each of which may be regarded as an introduction to the chapter's subject matter. The first two chapters give a limited survey of discoveries and of our understanding of objects in the universe and thus set the general scene for the rest of the book. Chapters 3 and 4 describe the characteristics of stellar reaction rates, followed in chapter 5 by experimental equipment and techniques used in the laboratory to obtain such information. This latter chapter is addressed to those students who are interested in seeing how such experiments are actually carried out and what the basic problems and requirements are. The major burning phases in stars are the subject of the succeeding chapters 6–8, followed by nucleosynthesis via neutrons in chapter 9. Here we review the results of laboratory measurements of nuclear cross sections and their correlations with observed elemental and isotopic abundances, and derive from this some remarkable quantitative information about specific processes that occurred in the creation of the elements. These chapters are written in such a way that they can be understood without reference to the sizable chapter 5. Finally, miscellaneous topics of current interest are described in chapter 10. The notation and units used in this book are defined and described in the Appendix.

The book assumes some knowledge of quantum mechanics and elementary nuclear physics. The material in this book has been used as the basis for a two-semester graduate course given at the Universität Münster and the Ohio State University at Columbus. Because of the rapid progress in astrophysics, it cannot be hoped that a technical book such as this will remain up to date for long. Consequently, the purpose of this book can only be to make the general ideas and the present status of understanding more easily available to all those who may wish to take the next step in this exciting part of today's astrophysics. In citing source material, we have often referenced review articles which summarize the state of the art on a particular subject. These articles are based on the work of many investigators, who may not be individually cited. Thus, the reference list is far from complete and does not do justice to the many dedicated scientists who have contributed to this fascinating field.

In writing this book we have had generous help from many friends who either have cleared up questions, have provided us with data, or have read through parts of the manuscript. We should particularly like to thank S. M. Austin, K. Brand, A. G. W. Cameron, B. Cleff, M. Cohen, M. El Eid, F. Käppeler, R. W. Kavanagh, J. M. Lambert, K. Rohlfs, E. E. Salpeter, M. Schmidt, W. H. Schulte, J. Schweitzer, R. Taylor, F. K. Thielemann, T. A. Tombrello, H. P. Trautvetter, G. Wallerstein, D. Wilkinson, and S. E. Woosley. We are especially grateful to W. A. Fowler and R. N. Boyd for thoroughly reading and commenting on the entire manuscript. Valuable criticism and constructive suggestions have come from many students, in particular P. B. Corn, who have read through the preliminary manuscript as part of lecture courses given on nuclear astrophysics. We also should like to thank Dawn Froehlich and Phyllis Hurley, who have cheerfully typed several chapters of the manuscript, and W. Hassenmeier, who did most of the drawings.

Some of the joint work of the normally widely separated authors has taken place at one or the other's home institution. We are grateful for the support we received from the National Science Foundation, Georgetown University in Washington, D.C., and the Westfälische Wilhelms-Universität in Münster. Further, we gratefully acknowledge support by the Stifterverband für die Deutsche Wissenschaft, the Deutsche Großforschungseinrichtungen, the Deutsche Forschungsgemeinschaft, and the Fulbright Foundation. We also express our appreciation to P. Dano, H. Pelster, and P. Treado for gracious hospitality. Finally, we are grateful for the hospitality shown during extended visits to the Kellogg Radiation Laboratory of the California Institute of Technology, where many ideas were born.

1

Astronomy—
Observing the Universe

Historically, any science in its earliest stage consists of unsystematized observations, frequently without specific objectives. Astronomy, the science of the stars and the oldest and perhaps most comprehensive of all sciences, followed such a pattern. People of ancient times, attentive observers of the skies, were attracted by the splendor of the heavens, as we are today, and by its mystery. The early observations, stimulated by man's natural curiosity, were carried out with only the unaided eye. These observations revealed that, while all objects in the heavens appeared to move, a few of them appeared to move more rapidly, while others appeared to be nearly fixed. As is natural, people tried to understand this relatively ordered motion, and, since from experience everything appeared to move around the observer, it was assumed that the motion would be most easily understood with the observer at the center. Indeed, gazing upward at the sky on a clear night, it is difficult to avoid the impression that the sky is a great hollow spherical shell with the earth at the center. After watching the sky for several hours, it becomes clear that the celestial sphere is slowly changing its orientation.

The early Greeks were attentive observers and regarded the sky literally as a celestial sphere of very great size with the earth standing motionless in the center. They imagined the sphere as the boundary of the universe with the stars fixed in the inner surface. It was thought to rotate daily about an axis that passed through the earth and in turning caused the stars to rise and set. The Greeks were also aware that the sun gradually changed its position on the celestial sphere in an independent motion, quite apart from the daily apparent rotation of the celestial sphere. They also noted that other objects moved among the stars. The moon and each of the five planets visible to the unaided eye (Mercury, Venus, Mars, Jupiter, and Saturn) change position from day to day. The Greeks therefore distinguished between what they called the "fixed stars," the stars that appeared to maintain fixed patterns among themselves on the celestial sphere (and identified with the mystic stellar "constellations"), and the "wandering stars" or "planets," which moved around on the celestial sphere. It was observed that as the planets moved among the constellations of stars, they followed complex paths relative to the background stars.

This view of the universe remained essentially unchanged for about 2000 years, although alternative views had been suggested from time to time. Astronomers from the time of antiquity to the time of Galileo (1564–1642) devoted most of their energy to observing carefully the positions and motions of the heavenly bodies (notably Brahe, 1546–1601) or to constructing models or schemes with which they could accurately calculate planetary positions at any time in the future. An important incentive to the early cultivation of astronomy was its usefulness in ordinary pursuits, such as serving as a basis for time scales, developing and refining the calendar, which is important in agriculture, and providing a guide for travelers on land and sea. The Chinese astronomers had a working calendar at least as early as 1300 B.C. They also kept rather accurate records of special events in the sky such as comets and meteor showers.

The Greek astronomers recognized from the moon's phases that it must be spherical. Also, the lunar eclipses showed that the earth is not a disk but is also of spherical shape. They arrived at nearly correct values for the diameter of the earth and the moon as well as the moon's distance from the earth. These studies culminated in the description by Ptolemy (about A.D. 140), the most important aspect of which was a geometrical representation of the solar system that predicted the motions of the planets with considerable accuracy and accounted successfully for the observations available at that time. His geocentric scheme of cosmology, with some modifications, was accepted absolutely throughout the Middle Ages until it finally gave way to the heliocentric theory in the seventeenth century. An astronomical clock based on the Ptolemaic system can be seen in the cathedral at Münster, West Germany. It was built about 1540, just prior to the introduction of the Copernican system.

In addition to the planets and the stars, which appear to the naked eye as pointlike objects in the sky, there are other extended areas of the sky shining faintly. The most prominent of these was known in ancient times and was called the Milky Way (Fig. 1.1). Another such extended luminous area just barely visible to the naked eye is the Andromeda "Nebula" (Fig. 1.15), labeled M31 in Messier's catalog or NGC 224 in the *New General Catalogue*. It is the nearest galaxy to our own and the only other spiral galaxy visible to the naked eye.

To the casual observer, the fixed stars appear to be changeless. There are, however, dozens of stars whose variable brightness is clearly visible to the unaided eye. The Chinese astronomers reported stars which they called "guest stars," stars that are normally too faint to be seen but suddenly flare up to become visible for a few weeks or months. Such stars are now called novae and supernovae. These observations suggest that violent events are occurring

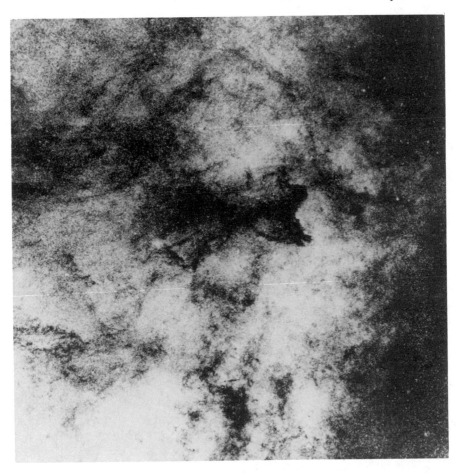

FIGURE 1.1. Stellar clouds of the Milky Way in the constellations of Sagittarius and Scorpius. The stars are so numerous that they blend together and give a milky appearance of diffuse light. The center of the Galaxy (Greek: *galactein* = milk) is thought to be in this direction of the sky. (Palomar Observatory photograph.)

in the universe. An improved understanding of such events as well as of the extended luminous areas in the sky had to await the invention of the telescope.

1.1. Observational Techniques

1.1.1. Optical Astronomy

It is not certain when the principle of the telescope was first conceived. It is known that in 1609 Galileo constructed such an instrument, and he deserves the credit for having been the first to make significant astronomical observations with it. The most important part of a telescope (Kui60, Woo82a) is the objective lens or primary mirror (Fig. 1.2), which collects radiation over a large area and focuses it to form an image of the object. The brightness of an image of a point source (star) increases in direct proportion to the area of the telescope objective. This light-gathering power of a telescope allows one to observe objects too faint to be seen with the unaided eye and thus to extend viewing farther into space of objects with the same intrinsic brightness. Another function of a telescope is to separate sources that would otherwise be indistinguishable. This feature is referred to as the resolving power (or resolution) of a telescope. The resolution is limited because of diffraction by the objective (or mirror) of a telescope. The theoretical resolving power is approximated by the relation

$$\alpha_{\text{theo}} \simeq 1.22\lambda/D \text{ radians}$$

$$\simeq 2.52 \times 10^5 \lambda/D \text{ seconds of arc} , \qquad (1.1)$$

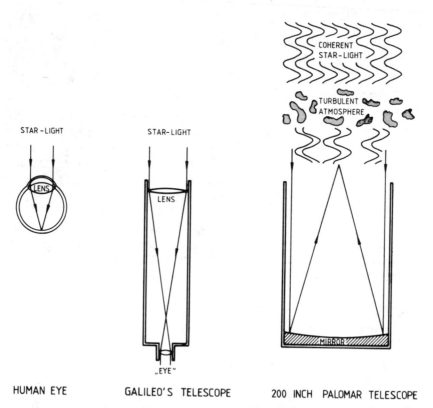

FIGURE 1.2. The major functions of a telescope are to collect radiation over a large area and focus it (light-gathering power) and to resolve separate sources of light (resolving power). Both functions improve with increasing diameter of the aperture. There is, however, a limiting resolution of about 1″ due to the turbulence in the earth's atmosphere.

where λ is the wavelength of the radiation and D is the diameter of the aperture, both in the same units (Appendix). Thus, in addition to making it possible to see fainter objects, a larger aperture telescope has better resolving power.

Presently the largest optical telescope in the United States is the 200 inch telescope on Mount Palomar, which is a millionfold more powerful as a "light bucket" than the human eye. (Note that a 240 inch telescope has been built in the USSR, but its efficiency is lower.) The 200 inch telescope outperforms Galileo's best telescope by a factor of 40,000 (Table 1.1). Although this telescope should have a theoretical resolving power of $0''.025$ (Appendix) at a wavelength of 5000 Å, small turbulent cells in the atmosphere limit the resolution to about $1''$ for long-lasting photographic exposures. This resolving power is roughly the angle subtended by a small coin at a distance of 4 km or by 2 km at the distance of the moon. Recently, however, techniques have been developed that make it possible to achieve the full theoretical resolution under special circumstances.

TABLE 1.1 Comparison of Astronomical Instruments

Instrument	Aperture D (mm)	Relative Light-Gathering Power	Resolving Power[a] α	
			Theoretical[b]	Practical
Eye	5	1	$25''$	$200''$[c]
Galileo's telescope	25	25	$5''$	$10''$[d]
Palomar telescope	5000	1,000,000	$0''.025$	$1''$[e]

[a] For $\lambda = 5000$ Å $(1 \text{ Å} = 10^{-8} \text{ cm})$.
[b] Equation (1.1).
[c] Due to effects of the eye's retina.
[d] Due to quality of available optics.
[e] Due to turbulent cells in the earth's atmosphere.

The invention of the telescope opened a vast new astronomical world. Galileo remarked that with his rudimentary instrument he could see stars so numerous as to be beyond belief. To better observe the faintest objects, astronomers have continued to build larger and more ingeniously designed telescopes.

Until late in the nineteenth century, however, the detector at the focus of the telescope was still the eye. Although the eye works well over a remarkably wide range of brightness, it cannot store light for more than a few tenths of a second. No matter how long you look at the night sky, you will not see stars fainter than some limiting brightness. With the advent of photography came revolutionary results, and photography quickly became the chief detection method in astronomy. A photographic emulsion is not more sensitive than the eye, but it has the advantage that it can build up a picture of a faint object by accumulating light over a long period of time, thus significantly extending the range of brightness detectable. With the addition of a prism or grating, the recorded spectra of celestial bodies provide information on the chemical composition of the light source. Modern photographic emulsions are sensitive to a wider range of wavelengths than is the eye—from the ultraviolet (UV) region to the near-infrared (IR). The eye, which can discern subtle differences in light intensity, is a poor judge of absolute brightness. Conversely, brightness can be measured photographically to considerable accuracy, provided that the transformation from photographic density to intensity can be calibrated. There is still a limiting faintness beyond which an object cannot be detected even photographically, because for very long exposures the ever-present background light from the night sky eventually saturates the entire emulsion. In recent years new light-detection techniques involving photomultipliers, image tubes, and microelectronic technology have come into use, as has the use of computers for data analysis. These technologies significantly improved the capability of the telescope to detect the fainter cosmic objects. In science it is a general rule that the availability of improved experimental techniques strongly influences the direction research will take.

A considerable obstruction to optical astronomy is the earth's atmosphere. Weather conditions limit the observing times; the atmosphere absorbs large parts of the electromagnetic spectrum (Fig. 1.3); and the unsteady air leads to star twinkling, which limits the resolving power of the telescope. To increase

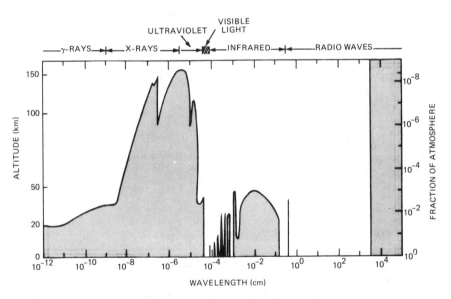

FIGURE 1.3. Atmospheric absorption of electromagnetic radiation is strong in all regions except for the windows in the visible, the near-infrared, and a broad range of radio wavelengths. The upper boundary of the shaded areas specifies the altitude at which the intensity of incident radiation is reduced to one-half its original value.

further the effectiveness of optical telescopes, astronomers today are thinking not only of larger and more expensive optical telescopes but also of ways to improve the present ones. In particular, improvements in detection techniques have resulted in studies of the sky using radiation in the infrared region, opening up a new branch of astronomy—infrared astronomy (Faz76).

1.1.2. Radio Astronomy

With the advent of radio technology in the 1930s, another window in the atmosphere became available (Fig. 1.3), and the new field of radio astronomy was born. Because of their long wavelengths ($\lambda \simeq 1$ mm to 50 m), radio waves pass more readily through the clouds of the earth's atmosphere and the interstellar gas and dust. For the same reason, radio reception is not greatly affected by weather conditions or atmospheric turbulences. Indeed, radio observations are as effective by day as by night, and the resolving power of a radio telescope is close to its theoretical value.

The most common form of radio telescopes (Kui60) is the steerable paraboloid, where the radiation is collected at the focal point of the paraboloid. Equation (1.1) for the resolving power shows that discrimination between closely spaced points is much more difficult with an ordinary radio telescope than with an optical telescope. For example, to match the best optical resolving power of 1″ with a radio telescope operating at a wavelength of 1 m, the aperture of the paraboloidal antenna would have to be 250 km, an apparently unsurmountable engineering obstacle. Present radio telescopes have apertures in the range of 10–100 m (Fig. 1.4). The largest existing radio telescope is a 305 m fixed bowl located in a natural depression at Arecibo (Puerto Rico). It has a resolving power of 160″ at $\lambda = 0.2$ m, or about one-tenth of the angular diameter of the full moon. The 100 m telescope at Effelsberg (Germany) achieves its maximum resolving power of about 18″ at $\lambda \simeq 4$ mm. Radio telescopes generally are accepted as being fundamentally poorer in angular resolution compared with optical telescopes.

New instruments, called interferometers (Kui60), were introduced in the 1920s when Michelson used such an instrument to measure the diameter of a few bright nearby stars. In this interferometer (Fig. 1.5) light from a distant source is intercepted by two mirrors (placed at a distance D from each other). The two beams are reflected to a common point, where they are combined. If the path of one beam is slightly different from that of the other, the light waves in one beam will be out of phase with the waves in the other. When the beams are combined, the two waves will interfere both constructively and destructively, creating interference fringes. If such interference patterns are obtained for several distances D between the two mirrors and if they are subjected to Fourier transformation, one obtains the image of the celestial object directly. The angular resolution of such an interferometer is given by the ratio λ/D and is thus essentially identical to the resolution one would get from a single giant telescope of diameter D.

FIGURE 1.4. The Westerbork telescope (Netherlands) consists of 12 parabolic antennas, each 25 m in diameter, arranged in a line 1.6 km long. If the individual telescopes are linked together and used as an interferometer, the resolution is equivalent to that of a single reflector 1.6 km in diameter.

OPTICAL INTERFEROMETER

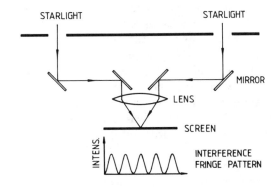

STARLIGHT   STARLIGHT

MIRROR

LENS

SCREEN

INTENS.

INTERFERENCE
FRINGE PATTERN

RADIO INTERFEROMETER

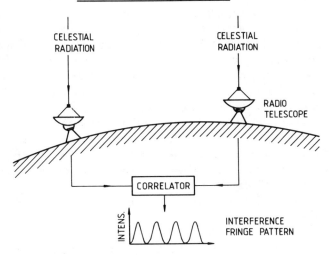

CELESTIAL
RADIATION

CELESTIAL
RADIATION

RADIO
TELESCOPE

CORRELATOR

INTENS.

INTERFERENCE
FRINGE PATTERN

FIGURE 1.5. An optical interferometer is an instrument that combines light from two separated points originating in a common source. The waves from the two points interact to produce a pattern of interference fringes. The radio interferometer is, in a sense, an exact analogue of an optical interferometer.

In the late 1940s radio astronomers began using this interferometric technique. As with an optical interferometer, in a radio interferometer (Figs. 1.4 and 1.5) the advancing wave front from a celestial object falls on two (or more) relatively small separate radio telescopes. The signals are carried to a common point and combined in an adding correlator. The superposition must be done in a way that preserves the phase relations of arriving radio waves. In conventional radio interferometers the telescopes are separated by a few kilometers. Adding coherent signals from each telescope was done by employing cables to link the telescopes. By the 1950s a resolution of the order of 60″ was obtained (Kui60). The most advanced of these interferometers is the Very Large Array (VLA) near Socorro, New Mexico, a collection of 27 movable radio telescopes. A significant improvement in resolution was obtained by replacing the cables with microwave links. This technique was employed for baselines of about 100 km, and a resolution of the order of 1″ was obtained. Further large improvements in the resolution of such radio-linked interferometers are not practical, since microwave links are limited by propagation irregularities of the intervening atmosphere. A fundamental innovation was to eliminate the direct connection between interferometer elements by separately recording the signals at each radio telescope and later comparing and correlating the two recordings. This technique needed stable atomic frequency standards (atomic clocks for relative phase synchronization) and high-speed tape recorders. It allowed radio interferometers with baselines extending across continents and even across oceans, the baselines thus being comparable to the diameter of the earth. This technique is called very long baseline interferometry (VLBI). The California to Australia baseline, for example, is 11,000 km long, more than 80% of the earth's diameter. The skillful application of this technology coupled with the evolution of computer technology permits resolution approaching 0″.0001, which corresponds to about the span of a human hand at the distance of the moon. This constitutes an improvement of some 4 orders of magnitude over the resolution of earth-based optical telescopes (Fig. 1.6). The high angular resolution obtained with such networks of radio telescopes is required for studies of very distant extragalactic objects such as radio galaxies and quasars, the most luminous and distant objects in the universe (sec. 1.2.6).

1.1.3 Space Astronomy

Because the earth's atmosphere blocks out most of the electromagnetic radiation emitted by celestial objects, until recently astronomers could only guess about the spectrum of celestial radiation in the blocked regions. Now the ability to put telescopes and other instruments effectively above the earth's atmosphere (balloons, rockets, satellites, space laboratories) is providing astronomers with an opportunity to observe the sky over the full range of wavelengths.

The utility of telescopes in space goes beyond their capability of using radiation that cannot penetrate the atmosphere. The absence of the troublesome

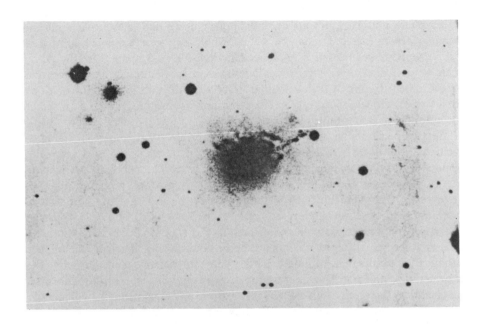

FIGURE 1.6. The elliptical galaxy NGC 1275 at a distance of 300 million light-years (Appendix) is a powerful source of radio waves. In the top photograph, taken with the Mount Palomar telescope, it appears as the fuzzy spot in the center whose diameter is 80″. In the bottom figure is the same object as it appears in a radio image made by VLBI. Here the resolution is 0″.0004, revealing details 2500 times smaller than any that can be observed optically. The bottom figure is only 0″.023 across.

atmosphere and background light will push down the achievable limit for angular resolution and increase the allowable exposure times as compared with ground-based telescopes. A space telescope with a 3 m aperture should be able to detect an object 100 times fainter than the faintest object detectable from the earth, and with an angular resolution of 0″.05. Such data on faint objects are critical for settling major questions in cosmology such as whether or not the universe is finite (chap. 2).

Several new subdivisions of space astronomy have been created, each concerned with its own region of the electromagnetic spectrum and requiring unique observing techniques and instrumentation. The subdivisions include X-ray and γ-ray astronomy, ultraviolet (UV) astronomy, and infrared (IR) astronomy, although much important work in the last category can still be accomplished with ground-based instruments. A fourth division, particle astronomy, is concerned with the detection of cosmic rays, which are actually nuclear species detected via mass spectrometry and nuclear techniques (chap. 5). Cosmic rays originate in the sun, the Galaxy, and probably the extragalactic universe. Considering the huge variety of objects in the universe, it is almost a certainty that the results from space astronomy will greatly change our ideas of the nature of the universe. Improvement in detection techniques for neutrinos (chap. 10) may soon provide an additional branch of observational astronomy; further, in the future perhaps better detectors may lead to routine detection of gravity waves (Web70).

1.2. Observed Structures in the Cosmos

In the sixteenth century Copernicus (1473–1543) inaugurated a new era in astronomy by discarding the ancient theory of the central motionless earth. As had been suggested earlier by some Greeks such as Aristarchus (287 B.C.), Copernicus proposed the theory that the planets revolve around the sun rather than the earth and that the earth is simply one of these planets. The rising and setting of the stars were now ascribed to the daily rotation of the earth on its axis. The new theory placed the sun and its family of planets sharply apart from the stars. No longer required to rotate around the earth, the sphere of the stars could be imagined to be much larger than before. This altered condition and the sun's new status as the dominant member of the solar system prepared the way for the thought that the stars are remote suns at various distances from us. When this theory, referred to as the heliocentric or Copernican system, was published in 1543, there was not a single item of unambiguous observational evidence in its favor, and it came at a high cost. For example, it threw the earth into dizzying flight around the sun somehow bringing the moon along with it. The greater simplicity in representing celestial motions was the only argument Copernicus could offer in its defense.

Before the invention of the telescope, the unaided eye could see about 5000 stars counting all those visible in the different seasons. In the fall of 1609

Galileo turned his telescope to the heavens and startled Europe with his remarkable discoveries. He found a myriad of new stars too faint to be seen with the naked eye. Some nebulous blurs were resolved into collections of stars, and the Milky Way was revealed to be the confluence of a multitude of individual stars. Galileo also discovered that there were satellites (moons) circling Jupiter, phases of Venus like those of the moon, movement of spots across the sun's disk, and mountains on the moon. These discoveries strongly supported the Copernican system, which not only predicted the phases of Venus but also explained many other observations. If the earth was a planet, the other planets might be earthlike as well, and so indeed the moon, with its mountains, turned out to be. The Copernican system arranged the planets naturally by period. Similarly, the satellites of Jupiter were found to be arranged sequentially by period as in a miniature solar system. Galileo's explanation that the movement of the spots on the sun's disk is caused by the sun's rotation provided an argument by analogy for the earth's rotation as well. With the gradual acceptance of the Copernican system, the stars being regarded as remote suns at different distances from the earth and in motion in various directions, the way was prepared for exploration of the stars themselves, leading to the more comprehensive view of the universe that man has today.

1.2.1. The Solar System

The solar system consists of the sun and the many smaller bodies revolving around it. These smaller bodies include the planets with their satellites, the comets, the asteroids, and the meteorites. It was not until the advent of the telescope that the planets Uranus, Neptune, and Pluto were discovered, completing the list of the nine known principal planets (Table 1.2). The planets are relatively small, dark globes shining only as they reflect the sunlight, while the stars are remote suns emitting their own light. Because of the large differences in density (Table 1.2), some planets must have different compositions from others. The Jovian planets, which approach stellar masses, must be composed mainly of hydrogen, while the terrestrial planets have lost most of their initial hydrogen and helium content because the thermal velocity of a given particle may exceed the escape velocity. The study of the inner structure of the planets is much more complicated than the study of the solar (stellar) interior. The stars are composed of hot gases which follow relatively simple laws for pressure, temperature, and density (chap. 2), while the planets are cool objects, either liquids or solids, to which physical laws cannot be applied in a simple manner.

The motion of the principal planets around the sun conforms to the following regularities. The orbits are nearly circular and nearly in the same plane. Most of them revolve and rotate in about the same angular direction. Upon assuming the correctness of the Copernican system, the observed motion of the planets led to the celebrated laws of Kepler (1571–1630). After describing

TABLE 1.2 Properties of the Planets

Planet	Mean Distance from Sun[a]	Orbital Period (y)	Equatorial Radius[b]	Mass[c]	Mean Density[d] (g cm^{-3})	Rotational Period (days)
Terrestrial:						
Mercury	0.387	0.241	0.38	0.054	5.4	59
Venus	0.723	0.615	0.95	0.81	5.1	243
Earth	1.0	1.0	1.0	1.0	5.5	1
Mars	1.52	1.88	0.53	0.11	4.0	1.03
Jovian:						
Jupiter	5.20	11.9	11.2	318	1.3	0.41
Saturn	9.54	19.5	9.5	95	0.7	0.43
Uranus	19.2	84.0	3.7	15	1.6	0.45
Neptune	30.1	165	3.9	17	1.7	0.66
Pluto	39.5	248	0.47	?	?	(6.4)

[a] In units of the mean earth-sun distance, $a_\oplus = 149.6 \times 10^6$ km = 1 astronomical unit (AU).
[b] In units of the earth radius, $R_\oplus = 6378$ km.
[c] In units of the earth mass, $M_\oplus = 5.98 \times 10^{24}$ kg.
[d] From the relative density, one may conclude that the first four planets are earthlike, while the last five, known as the Jovian planets, are more like Jupiter.

the paths of the planets the major scientific question was what caused them to follow Kepler's laws. In this regard Galileo's chief contribution to our knowledge of planetary motion was his pioneering work on the motion of massive bodies in general. His conclusions prepared the way for a new viewpoint in astronomy. Interest was beginning to shift from the kinematics to the dynamics of the solar system—from the courses of the planets to the forces controlling them. The observations of Galileo and others concerning the motion of bodies were consolidated by Newton (1642–1727) into three statements known as Newton's laws of motion. Using these laws of motion and mathematical reasoning, Newton succeeded in explaining Kepler's geometrical description of the planetary system by a single comprehensive physical law, the law of gravitation. Every particle of matter in the universe attracts every other particle with a force that varies directly as the product of their masses and inversely as the square of the distance between them:

$$F = Gm_1 m_2/r^2 ,$$

with the gravitational constant $G = 6.670 \times 10^{-8}$ cm^3 s^{-2} g^{-1} (for a description of the unit system used see the Appendix). Note that the attraction between 1 g masses 1 cm apart is only one-fifteen millionth of a dyne. Although the gravitational force is very feeble between ordinary bodies, it becomes large when one considers the huge mass of celestial bodies.

The sun, a typical star, is a sphere of very hot gas, having a visible surface of radius $R_\odot = 6.96 \times 10^{10}$ cm = 696,000 km, a mass $M_\odot = 1.99 \times 10^{33}$ g, and a mean density $\rho_\odot = 1.4$ g cm^{-3}. The sun rotates in the same sense as the rotation and revolution of the earth with a period of 25 days at the equator. The gradual movement of sun spots across the disk of the sun is a well-known effect of this rotation. The sunspots are of great interest. They occur in a cycle of about 11 years and are related to the magnetic fields in the sun's atmosphere. What appears to be a changeless celestial object with a smooth uniform surface is seen, upon scrutiny, to be a highly structured, mottled sphere with complex motions and activities (Noy76). The sun has an effective surface temperature $T_s = 5800$ K and radiates energy into space at a prodigious but controlled rate. The energy output, or stellar luminosity L, is $L_\odot = 3.83 \times 10^{33}$ ergs s^{-1} = 2.39 $\times 10^{39}$ MeV s^{-1}. On the basis of biological evidence and other information, the luminosity of the sun appears to have been essentially constant during its entire history. With the use of advanced detection techniques of astronomy, the sun has been shown to be a source not only of visible light but also of radiation over essentially the entire electromagnetic spectrum. In addition, the sun is the source of a solar wind consisting of ions of a number of atomic species, but predominantly of hydrogen. It also emits neutrinos arising from nuclear reactions in the interior (chaps. 6 and 10). The details of the interior of the sun are known only indirectly from theoretical studies (chap. 2) and to some extent through neutrino observations (chap. 10).

Once the general characteristics of the solar system became known, the quest began to explain its origin. Any theory not only must explain the regularities of the system but also must provide for any major irregularities. Many theories have been advanced since the first one by Descartes in 1644 (Cam62, War75, Wet81, and chap. 10). All the theories encounter problems, notably the quantity and distribution of angular momentum (sun, 2%; Jupiter, 60%; the three other giant planets, 38%).

The solar system is the only known system of its kind, although others may very well exist, since such a system associated with the nearest star would not be seen even with the largest telescope. Our solar system viewed from the closest star with a similar telescope would appear only as a bright star. Recent observations using infrared telescopes indicate that other planetary systems may indeed exist.

1.2.2. Normal Stars and Clusters of Stars

The sun is one of a multitude of stars whose properties are similar. Some stars are much larger than the sun and others much smaller ($M/M_\odot \simeq 0.10$–100). Most stars have masses comparable to or smaller than that of our sun. Very massive stars with $M/M_\odot \geq 10$ are comparatively rare, and supermassive stars with $M/M_\odot \geq 100$ have not yet clearly been identified. Blue stars are hotter than the sun, which is a yellow star, and red stars are cooler. Stellar surface temperatures are found to be in the range 3000–50,000 K. All stars radiate huge amounts of energy. They are the power plants and, as we shall see later, the cooking pots of the universe—the cauldrons in the cosmos.

Because of the vast space between the stars, it is convenient to express interstellar distances in units of light-years (1 light-year = 1 ly = 9.46×10^{17} cm; see the Appendix). The sun's nearest stellar neighbor, the bright multiple-star system consisting of the double star Alpha Centauri and Proxima Centauri, is about 4 ly away. Since interstellar distances are so great, the chances of stellar collisions are almost zero, and indeed such collisions have never been observed. The distances of the nearest stars can be measured directly by trigonometric methods using stellar parallaxes. The techniques are similar to those a surveyor uses to measure distances here on earth. This direct method can be used for measuring distances up to about 300 ly. For larger distances, indirect methods requiring detailed calibration must be used.

As a result of trying to determine stellar parallaxes, Halley in 1718 demonstrated that the stars are not "fixed." He showed that certain bright stars had moved a distance of about the moon's apparent diameter from the places assigned them in Ptolemy's ancient catalog. The stars are moving in various directions relative to one another. Although these movements through space are often swift, their angular motions seem very slow because of the great distances involved.

The brightness of a star is measured by determining the radiative energy arriving from it per unit of time and per unit area of the detector. It is one of a star's basic observable parameters. If the star's distance is known, the absolute stellar luminosity L (total energy radiated per unit of time) can be determined. A star's luminosity is expressed commonly in terms of the solar luminosity $L_\odot$. It turns out that stars differ very greatly in luminosity, with the ratio $L/L_\odot$ ranging from 10^{-4} to 10^6 for masses $M/M_\odot$ in the range 0.1–100.

From its spectrum one can infer the chemical composition of a star (sec. 1.3.1 and Fig. 1.23). In general, hydrogen (H) is the most abundant element, followed by helium (He) and the heavier elements in much smaller amounts. In astronomy carbon and all elements heavier than carbon conventionally are called metals. Stars that are relatively high in metallic content are classified as Population I stars and are regarded as young stars (chap. 2). Alternatively, Population II stars are very low in metallic content and are considered to be older stars. The sun is a Population I star.

Interstellar space is not completely empty. In the vicinity of the sun (as in many other regions) it contains about 3%–5% as much gas as there is in the stars themselves. The gas is accompanied by smaller amounts of dust. Some clouds of this interstellar material are made luminous by the radiations of neighboring stars and constitute the most obvious bright and dark nebulae. These clouds are being studied effectively by radio and infrared astronomy. Other clouds are essentially black. The dust clouds can be detected by the dimming and reddening of stars behind them; they are responsible for the dark rifts that cause most of the variety in the Milky Way (Fig. 1.1). The interstellar material has a chemical composition similar to the Population I stars themselves.

Stars frequently appear in pairs bound by their mutual gravitational attraction. They are called binary stars. The coupling is sometimes shown decisively by their mutual revolution, but it is often indicated only by their common proper motion and radial velocity. Pairs of stars that can be resolved by telescopes as two stars are called visual binaries. Pairs of stars identified by their regular variation in radial velocity and their eclipsing features (sec. 1.2.3) are called spectroscopic and eclipsing binaries, respectively. The study of binary stars provided the first data on the masses of stars and the mass-luminosity relation (sec. 1.3.3). Binary stars are now known to be very common objects in the sky and may be the rule rather than the exception. Indeed, there are astronomers who hold that Jupiter is a quasi-star, the solar system thus being nearly a binary system. From the irregular motion of a visible star the presence of a much fainter or invisible companion can be discerned, as was done by Bessel in Königsberg in the middle of the nineteenth century, constituting an "astronomy of the invisible," in which unseen celestial objects are detected by their gravitational interaction with visible objects.

Clusters of stars are physically related groups of stars held together, at least temporarily, by their mutual gravitational attraction. The common motion of the members of a cluster through the star field suggests that they have a common origin in the condensing and fragmenting of a large cosmic gas cloud (chap. 2). Because the stars of a particular cluster are practically at the same distance from us, they may be compared with one another. Although their ages are about the same, the members have different masses. Since the more massive ones have evolved faster (chap. 2), studies of stars in clusters can provide information on the course of stellar evolution. They are used to test the theory of stellar evolution. Star clusters are associated with our Galaxy as well as with other galaxies. They are of three types (Dic68): (1) galactic (or open) clusters, so named because those in our Galaxy are near its principal plane and contain up to several thousands of stars; (2) globular clusters, which are more compact and spherical in form; and (3) associations containing blue (hot) stars, which are located in the spiral arms. There are about 500 globular clusters in our Galaxy. They are larger, more populous, and more luminous than the galactic clusters, and are not confined to the vicinity of the plane of the Milky Way. The globular cluster M13 in Hercules (Fig. 1.7) is estimated to contain 0.5×10^6 stars of average mass equal to about half that of the sun. Around the center the density distribution of stars may be 100 times as great as the average for the cluster and 50,000 times that of the stars in the sun's neighborhood. These great clusters are presumably as old as the galaxies themselves. They are massive enough to remain stable even if perturbed by tidal forces from the Galaxy and may provide important clues to the early history of the galaxies (chap. 2). The observation of variable stars in such globular clusters provided information for determining the distances of the clusters from earth. The results of such studies (by Shapley in 1917) established the size and shape of our Galaxy and prepared the way for the recognition of

FIGURE 1.7. The globular cluster M13 (or NGC 6205) in the constellation Hercules belongs to the Milky Way and is 33,000 ly away from the sun. It contains 0.5×10^6 Population II stars of the same age (about $10\text{–}15 \times 10^9$ years). (Palomar Observatory photograph.)

exterior galaxies (sec. 1.2.5). It is interesting to note that Shapley at that time opposed the idea that the spiral nebulae were extragalactic.

1.2.3. Unusual Stars

Most stars expend their supply of nuclear fuel at a remarkably constant rate for hundreds of millions or billions of years, then die out in a number of ways (chap. 2) which fit into a general scheme of classifications (sec. 1.3.3). A minority of stars deviate from the usual pattern in that they vary in brightness and in other respects as well. They are called variable stars. These variable stars are monitored with great interest by astronomers who use them to gain a better understanding of the nature and the evolution of stars (chap. 2). The variable stars can be divided into three major classes: eclipsing, eruptive, and pulsating.

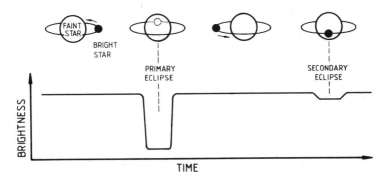

FIGURE 1.8. Periodic dimming of a bright star is evidence that the star is a member of an eclipsing binary system (Wil81). Such evidence has established that nova systems and X-ray stars are close binaries.

1.2.3.1. Eclipsing Stars

Variable stars of one class vary in brightness because they are eclipsing binary stars. Telescopically, these eclipsing binaries often appear as single stars, their binary character being shown by periodic change in their brightness (Fig. 1.8). In contrast, the other classes of variable stars vary in brightness as a result of intrinsic characteristics. The variation in brightness (luminosity) as a function of time is referred to as the light curve of the stellar object.

1.2.3.2. Eruptive Stars

The eruptive stars show rapid irregular increases in brightness. The most famous among the eruptive stars are the novae (Fig. 1.9). The word *nova* literally means "new" and was used by astronomers in the past to identify stars that increase in brightness from relative obscurity to high visibility. Usually they remain visible to the unaided eye for not more than a few weeks, grad-

FIGURE 1.9. Nova DQ Hercules (*right*) was photographed in 1934 at the Yerkes Observatory of the University of Chicago. At the outburst the star increased in brightness by a factor of 10^5 times its normal brightness (*center of picture at left*).

ually fading thereafter. About 10 such novae, which suddenly increase their brightness by a factor between 10^4 and 10^6, are detected every year in our region of the Milky Way. For a brief period such a star may rival the brightest stars in the sky. At the time of a nova outburst an outer layer of the star is ejected in the form of a shell of gas with velocities between 300 and 3000 km s^{-1}. The mass released in the outburst is only a small part of the star's total mass ($\Delta M/M \simeq 10^{-3}$). The energy released is about 10^{45} ergs. Some novae have been observed to undergo more than one outburst and thus are called recurrent novae. Evidence, now rapidly accumulating, indicates that the nova phenomenon is directly related to the existence of very close double-star systems (eclipsing binaries), separated generally by little more than the diameter of a typical star. A popular theory concerning this phenomenon (chap. 2) involves the transfer of mass from one star to the companion with the possibility that the exchange triggers an outburst on the companion (Wil81). Thus, in spite of their name, novae are not new stars but rather are old binaries making violent adjustments in their lives and releasing large amounts of mass and energy in the process. They are therefore not part of the evolution of normal (single) stars (chap. 2).

Perhaps the most spectacular of the cataclysms of nature is the supernova (Zwi74). A supernova is an eruptive star which, in contrast to an ordinary nova, increases in brightness to many millions, perhaps billions, times its former brightness, dominating its region of the sky. A supernova represents the brief final event in the evolution of a very massive star ($M/M_\odot \geq 10$), its explosive death taking only about one second (chap. 8). The amount of light emitted in a supernova explosion is, at its maximum, comparable to that emitted by our entire Galaxy, the Milky Way. When supernovae explode, they blow gaseous material into space amounting to several solar masses. The energy released in a supernova is of the order of 10^{51} ergs. The matter is ejected explosively at velocities ranging from 1000 to 20,000 km s^{-1}. Supernova remnants continue to emit energy at the rate of 10^{36} ergs s^{-1}. The envelope of a supernova expands rapidly, persisting for centuries as a discrete source of X-rays and radio waves. The Crab Nebula in Taurus is the remnant of a supernova that exploded in A.D. 1054 (Fig. 1.10). It was seen in the daytime as Venus is, and was visible during the day for 23 days and observed

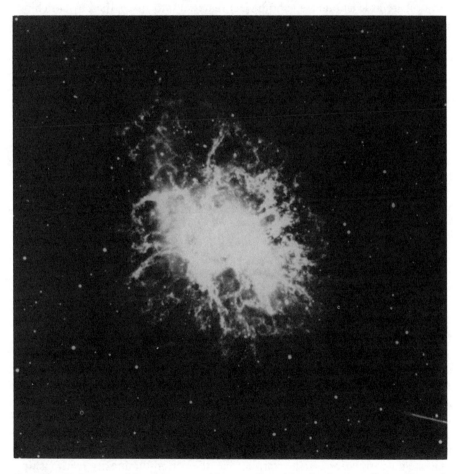

FIGURE 1.10. Because of its appearance, this expanding object in the constellation Taurus is named the Crab Nebula. It is some 5000 ly away and has a diameter of about 6 ly. On 4 July 1054 Chinese astronomers observed a gigantic supernova explosion in Taurus at the position where we now find the Crab Nebula. There is a stellar cinder left behind, a pulsar, which is a rapidly spinning neutron star. (Palomar Observatory photograph.)

at night for a total of 21 months. At its peak the brightness was billions of times that of our sun. The Crab Nebula harbors a fascinating stellar pulsing object called a pulsar, as do other remnants of supernovae (sec. 1.2.3.3). Supernovae have been divided by astronomers into two major classes (Tam77): Type I exhibits the presence of heavy elements and very little H, and the light emitted varies with time according to a rather definite pattern, while Type II contains primarily H and has a more variable light curve (Bar73, Kir73). In a typical galaxy there are about 40 novae every year, but there are only about two or three supernovae per century (Tam74). No supernova has been seen in our Galaxy since 1604. In spite of their many differences, novae and supernovae share at least one trait: both eject material into space, thereby enriching the interstellar matter with heavy elements—the metals, which they have synthesized in the generation of their emitted energy.

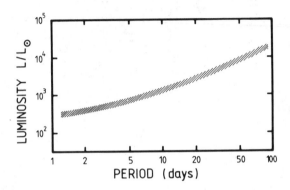

FIGURE 1.11. Schematic diagram of the period-luminosity relation of Cepheid stars (Type II). Cepheids are fundamental tools for distance measurements.

1.2.3.3. Pulsating Stars

Many giant and supergiant stars are variable in brightness, with periodic light curves. They are called pulsating stars because they are alternately contracting and expanding, in turn becoming hotter and cooler. The physical movement of the stellar surface can be inferred from the Doppler shift of their spectral lines. Among these pulsating stars are included the historically very important Cepheid variables named for the prototype Delta Cephei, a variable star discovered in 1784. The period of pulsation for these Cepheid variables was found to be correlated with the absolute luminosity of the star. It is important that such a correlation, often referred to as the period-luminosity relationship, could be established (Fig. 1.11), since the distance of a number of Cepheids could be determined by independent means. Thus, if a star is identified as a Cepheid variable with a given pulsating period, its absolute luminosity can be deduced. This value plus the observed brightness of the star is sufficient to determine its distance. Thus it is possible to use Cepheids as "standard candles" to establish the distance of more remote galaxies (Kra59). Pulsating stars such as the RR Lyrae variables are most commonly located in the horizontal branch of the red giant region of the Hertzsprung-Russell (H-R) diagram (Fig. 1.24), a region of instability. Thus, these pulsations represent episodes in their normal evolution (chap. 2).

In 1967 a group of radio astronomers at the University of Cambridge (Hew67, Hew70, Wad75) announced that a strange new class of radio-emitting objects had been found outside the solar system. The strange thing about them was that they emitted sharp—intense—rapid and extremely regular pulses as regularly spaced as a broadcast time service. As a result they were quickly named "pulsars." Although the name misleadingly connotes a pulsating (expanding and contracting) object, it has become the accepted term. Excitement rose when it was found that the pulses were coming from an object no larger than a planet situated relatively close to us (about 400 ly) among the nearer stars in our Galaxy. Were the pulses some kind of message from another civilization? This possibility was entertained simply for the lack of any obvious natural explanation for signals that seemed so artificial. The credibility of such a possibility declined when similar pulses were discovered coming from other directions in space and when the absence of any planetary motion associated with the sources was noted. The "tick" period of a pulsar ranges from a few milliseconds to about a few seconds. The few milliseconds combined with the speed of light limits such objects to a size less than a few thousand kilometers, or comparable to the size of a planet. This follows because a large body cannot emit a pulse of radiation in a time shorter than the time required for light to travel across it. Suppose, for example, the sun could be instantaneously switched off. First we would see a dark spot at the center of the sun, since this part is nearest to us. The dark area would then enlarge until the sun was a bright ring. Finally, even its outer edge would disappear. The entire sequence would take about 4 s. Therefore, if the sun were to flash on and off like a pulsar, the flashes could not have a duration shorter than that. Because of the very faint luminosity of the pulsars ($L/L_\odot \le 10^{-4}$), they had to be either small or cold or both. The luminosity and the immense stability of the time-keeping mechanism imply that some object with the mass of a star is involved.

Some stellar behavior had to be found to explain the precise clocklike timing of pulses. In astronomy periodic behavior has been found in three settings. A single star can expand and contract regularly (pulsating stars); a pair of stars can orbit around each other, causing periodic eclipses (eclipsing stars); and a spinning star can be seen by a distant observer to vary regularly in brightness if its surface is not uniformly bright. From the observed periods of the pulsars it was concluded that they could be neither pulsating stars nor orbiting stars, leaving only two types of spinning stars, white dwarfs and neutron stars, which are small enough. White dwarfs, commonly observed

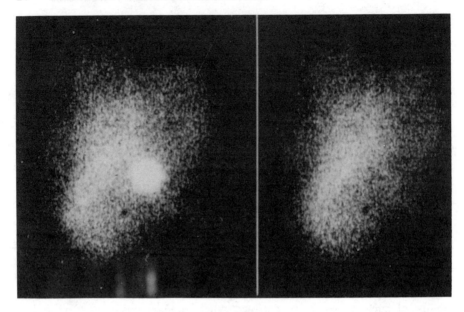

FIGURE 1.12. The pulsar in the Crab Nebula, part of the remnant of a supernova described by Chinese observers in A.D. 1054, caused the bright central spot in this image recorded by the *Einstein Observatory* X-ray telescope. (NASA photograph.)

objects in the sky, are stars that have collapsed after exhausting their nuclear fuel. Their mass, roughly equal to a solar mass, is packed into a volume the size of a planet (chap. 2). The existence of neutron stars was postulated on theoretical grounds as early as the 1930s (chap. 2), but before they were actually observed they seemed almost too strange to be credible. They would have roughly the same mass as white dwarfs, but their diameter would be on the order of only several tens of kilometers. Because their atoms are literally crushed out of existence by intense gravitational forces, they have extremely high central densities ($\simeq 10^{15}$ g cm^{-3}), comparable to the saturation density of nuclear matter. The limit on the rate at which a stable star can rotate is set by the critical speed at which the centrifugal force equals the gravitational force at the surface in the equatorial region. The period is approximately inversely proportional to the square root of the mean density. Thus, a white dwarf would rotate with a period of a few seconds at most, which is too long for most pulsars, while a neutron star would rotate with a period as short as a few milliseconds. Astronomers and astrophysicists now generally agree on the nature of pulsars: these objects are neutron stars in rapid rotation (Gol68, Pac68). According to theory, the outburst that accounts for one kind of supernova originates in an explosion in a shell intermediate in position between the star's center and its surface (chap. 8). The explosion blows the outer part of the star out in a nebula (like the Crab Nebula) and implodes the central regions into an extremely dense remnant (i.e., neutron star). Thus, a pulsar at the center of the Crab Nebula, a supernova remnant, is consistent with theory. Calculations indicated that neutron stars might emit detectable X-rays, a fact subsequently verified with X-ray telescopes (Fig. 1.12). It was also noted that the implosion producing the neutron star contracts the magnetic field of the original star into a field of fantastically high strength (more than 10^{12} gauss). As the neutron star rotates, high-energy charged particles (electrons and nuclei) from its surface spew out into the magnetic field, perhaps from the poles of the star. The interaction of the particles with the magnetic field would cause the loss of angular momentum. The pulses from a pulsar should slowly lengthen in period and eventually die out (Gol68). These predictions turned out to be correct (Ric69). It was also found that the loss of angular momentum, i.e., the loss of rotational energy, roughly balances the total luminosity of the Crab Nebula. At present there is no widely accepted explanation for the origin of the radio pulses themselves. Most theoretical models have adopted the lighthouse geometrical picture (Fig. 1.13), with the beam of light generated by charged particles streaming from the magnetic poles.

1.2.3.4. Planetary Nebulae

Many stars are known to be surrounded by an extended atmosphere—an expanding gas shell—even though no outbursts have been observed. They are recognized by their bright-line spectra and the Doppler shift of their radiation.

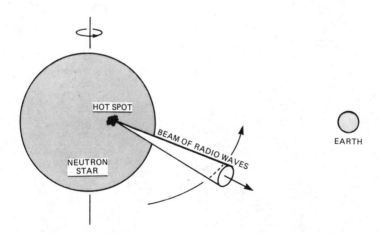

FIGURE 1.13. A rapidly spinning neutron star emits a highly directional radio beam around the sky once per rotation. It appears to have some type of "searchlight beam" fixed to its surface, possibly at a hot spot. In some cases two beams or "pulses" are observed. An observer receives a pulse of radio waves each time the star's radio beam points toward the earth ("lighthouse" effect). The name pulsar reflects the original idea that the stars were pulsating. Pulsars also have been observed at other wavelengths (see, e.g., Fig. 1.12).

The most important class of stars with extended atmospheres are the planetary nebulae. They are large shells of gas ejected from and expanding about certain extremely hot stars, which are about as massive as the sun. They derive their name from the fact that a few bear a superficial telescopic resemblance to planets. Actually they are thousands of times larger than the entire solar system and have nothing whatever to do with planets. Planetary nebulae often appear large enough to be resolved with a telescope. The most famous example of a planetary nebula is the Ring Nebula in Lyra (Fig. 1.14). Planetary nebulae differ from the other kinds of stars that eject gas shells in that an appreciable amount of material is ejected into the shell ($\Delta M/M = 2\%-20\%$) and that typically the shells move away from their parent stars at speeds of

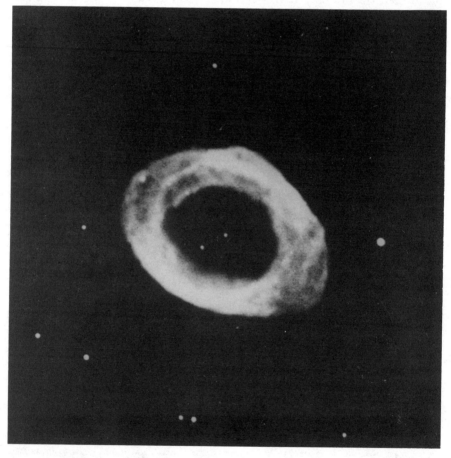

FIGURE 1.14. The Ring Nebula (M57 or NGC 6720) in Lyra is an example of a planetary nebula. It has a small, hot star in its center surrounded by a shell of glowing gas, which is moving away from the central star and was presumably ejected by it. The off-center star within the ring is not part of the nebula. (Palomar Observatory photograph.)

20–30 km s^{-1}. Considering that planetary nebulae are temporary phenomena existing only for a relatively short time, they are actually very common. Indeed, an appreciable fraction of all stars must sometime evolve through the planetary nebula phase (chap. 2).

The gas shells of planetary nebulae shine by the process of fluorescence. They absorb UV radiation from their central stars and reemit this energy as visible light; thus the UV luminosities of the central stars must be very high. Since the luminosities are very high, the stars must be very hot. Nearly all these stars are hotter than 20,000 K, and some have surface temperatures well in excess of 100,000 K, making them the hottest known stars. Despite their high temperatures, the central stars of planetary nebulae do not have exceedingly high luminosities ($L/L_\odot \simeq 1$). They must, therefore, be stars of small size with very high densities. Some in fact appear to have the dimensions of white dwarfs. It does not follow, however, that the stars were small and hot when the gas shells were ejected. Analysis shows that it is more likely that the nebulae were ejected from their parent stars when the latter were in the red giant phase of their evolution (chap. 2).

1.2.3.5. X-Ray Stars

The celestial zoo of abnormal stellar systems was extended with the discoveries of X-ray astronomy. A number of objects that are intense emitters of X-rays have been detected. They are a new class of stellar objects called X-ray stars because their X-ray radiation is overwhelmingly greater than their visible and radio wave radiation. Their luminosity in the visible region is a factor of 10^2 larger than that of the sun, while in the X-ray region they pour out 10^9 more energy than the sun. Hence, X-ray stars are among the most luminous pointlike objects in our Galaxy. Refined observations show that there are at least two classes of X-ray stars (binaries).

Class 1 is characterized by highly periodic short-term variations in the X-ray intensity with a regular pulsation period of a few seconds and a much shorter pulse duration. Arguments similar to those applied in the identification of pulsars lead to the conclusion that the X-ray source must be an extremely small stellar object, a compact star such as a neutron star with a mass of 1–2 $M_\odot$. Long-term studies of the pulsation period reveal that the rotation of the X-ray star is speeding up rather than slowing down. This surprising observation implies that the energy for the X-ray emission cannot come from rotational energy as it does with pulsars. Another feature of these X-ray stars is that they appear to be located in the spiral arms of our Galaxy, the place where many young stars and large clouds of gas and dust are found. Observations show that some of these sources are definitely eclipsing double stars with periods of a few days. Other sources with similar characteristics may also belong in the same category. The companions, discovered subsequently with optical telescopes, were found to be blue supergiants with masses as large as $M/M_\odot = 15$–20.

Class 2 X-ray stars exhibit no regular pulsations but are characterized by short "bursts" of X-rays emitted by sources in globular clusters and sources near the center of our Galaxy. The bursts are not isolated, singular events, but neither do they keep a fixed schedule. Instead, they repeat at irregular intervals ranging from several hours to a few days. They appear to signal explosive events that release prodigious amounts of X-ray energy. A typical burst reaches its maximum intensity in a few seconds or less; the source then fades to its steady, preburst level of X-radiation in about a minute. In this brief period some 10^{39} ergs of X-ray energy are emitted.

A popular scenario identifies both classes of X-ray stars (chap. 2) with close binary systems in which the X-ray–emitting object is a neutron star accreting matter from the normal companion star that is still consuming its nuclear fuel. The accretion of matter onto the neutron star is the source of the enormous energy output; it is gravitational energy released by the accretion process. Class 1 describes a system with a youthful neutron star, which is spinning rapidly, leading to regular pulsations in X-ray intensity. Class 2 contains a relatively old neutron star whose rotation has slowed down, and hence no regular pulsation occurs. In the class 1 system Cyg X-1, a compact X-ray–emitting object was found with a mass greater than 6 $M_\odot$. Since this mass is too large for a neutron star (chap. 2), there is strong evidence that this compact object is a black hole.

1.2.4. Our Galaxy—the Milky Way

The faint hazy band of light easily visible on a moonless night and completely encircling the sky has the descriptive name "Milky Way" (Fig. 1.1). Galileo solved the first mystery of the Milky Way when he turned his telescope on it and saw that it was really a myriad of faint stars. Later telescopic photographs of the Milky Way revealed that it consists of literally millions of stars, glowing gas, and dark clouds. The Milky Way is thus a vast belt in the sky glowing by the combined light of a huge number of stars.

Some general features of our Galaxy were pointed out quite early by Herschel. The concentration of stars in the direction of the Milky Way shows that the main body of our Galaxy is like a flat disk. The near-equality in the number of the stars on both sides of the Milky Way (north and south of the galactic equator) indicates that the sun is not far from the principal plane. Thus, looking toward the equator, one looks the long way through the Galaxy and therefore sees many of its stars. They appear as the glowing band of the Milky Way.

The modern era of galactic structure studies began in 1917 with Shapley's research on globular clusters. Shapley showed that the Milky Way is finite and that the sun is far from its center. An earlier discovery that the mysterious spirals and associated "nebulae" were receding from the sun at fantastic velocities led Hubble to study in detail the biggest such object, the spiral galaxy M31 in Andromeda (Fig. 1.15). Hubble's discovery in 1924 of Cepheid variable

FIGURE 1.15. Shown is the Andromeda "Nebula," the spiral galaxy M31 or NGC 224, which is the twin of our Milky Way. Its distance is about 1.5×10^6 ly and its diameter about 0.1×10^6 ly. It contains about 10^{11} stars. The stars which can be seen scattered over the figure are foreground stars of our own Galaxy. (Lick Observatory photograph.)

stars in M31 conclusively showed that M31 was in fact an independent assemblage of stars—called a galaxy—thus placing our own galactic system in better perspective. When referring to our own galactic system, it is common to write "Galaxy," the capitalization distinguishing it from the multitude of other galaxies (sec. 1.2.5). Many astronomers using a variety of telescopes have verified the general characteristics, and continuing studies are providing an increasingly clear picture of the Galaxy (Bok74). The principal features in the modern view of the Galaxy are as follows (Fig. 1.16). The Galaxy is an assemblage of about 100 billion stars, one of which is the sun. They are bound together by their mutual gravitational attraction. The central region is a spheroidal concentration of stars 10,000 ly or more in diameter situated in the direction of the constellation Sagittarius (Fig. 1.1). Here the stars are much more densely packed than they are in the sun's vicinity, solely as a consequence of gravity. Refined studies of this central "nucleus" of stars are being

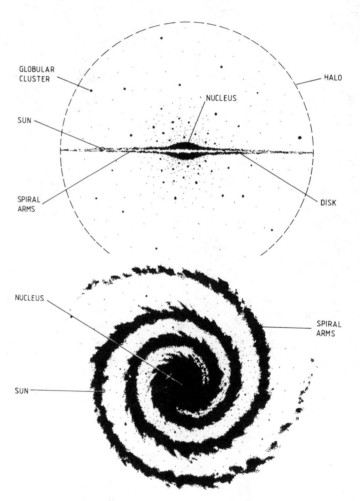

FIGURE 1.16. Schematic representation of our Galaxy seen from the edge (*top*) and along the axis (*bottom*). Most objects are confined to a relatively flat disk containing a central nucleus of stars and spiral arms, while the globular clusters are contained in a spheroidal region. A "haze" of individual stars and the clusters form the galactic halo, a region whose volume exceeds many times that of the main disk.

carried out using infrared and radio telescopes. There is growing evidence that at the very heart of the central region there is an ultracompact object, possibly a massive black hole, embedded in a dense swirling mass of stars, gas, and dust. Although the black hole cannot be observed directly, its existence is suggested by the behavior of the surrounding material (Geb79). The spheroidal central region is surrounded by a flat disk of stars about 100,000 ly in diameter. The disk (thickness $\simeq 2000$ ly) also contains much gas and dust. The sun, located near the principal plane of the Galaxy, is about two-thirds of the distance from the center to the edge of the disk ($d_\odot \simeq 28,000$ ly). After Hubble in 1924 demonstrated the existence of other galaxies, the possibility was advanced that our Galaxy might have a spiral structure similar to that of some other galaxies (Fig. 1.15). This possibility became established in the early 1950s as a result of studies using radio waves, which are not affected by the intervening dust in the Galaxy. The spiral arms (Fig. 1.16) contain a considerable amount of gas and dust as well as stars. Associated with the gas and dust clouds are many young stars (Population I), a few of which are very hot and luminous. It is in the interstellar gas and dust clouds of the spiral arms that star formation is believed to be still taking place (chap. 2). Stars in the central region, where little gas and dust remain, belong mainly to Population II. In addition to individual stars and clouds of interstellar matter, the Galaxy contains many star clusters. The globular clusters are scattered in a roughly spherical distribution about the disk of the Galaxy. They are part of a more or less spherical "halo" or "corona" surrounding the main body of the Galaxy.

As might be inferred from its flattened shape, the Galaxy is rotating like a gigantic pinwheel around an axis through its center. In the rotation the sun is moving in the whirl of the highly flattened disk in the direction of Cygnus at a speed of about 220 km s^{-1}. The period of the rotation in the sun's neighborhood is of the order of 250 million years. On the basis of this and the size and shape of the Galaxy, the mass of the Galaxy can be calculated. The calculation reveals a galactic mass of about 10^{12} $M_\odot$; thus, if the mass of the sun is taken as average, the Galaxy contains some thousands of billions of stars. It turns out that only about 10% of the mass required by this analysis can be accounted for by detectable stars, gas, and dust. This discrepancy is known as the problem of the "missing mass." One suggestion that has been made is that invisible stars, perhaps infrared stars, may be quite numerous and might account for a significant part of the total mass. However, infrared astronomy has shown that, in spite of the numerous new stars that have been observed, infrared stars account for only a very small part of the total mass. Thus the nature of the missing mass (dark matter) remains a mystery (Bok81, Rub83).

1.2.5. Galaxies and Clusters of Galaxies

Since early times, faintly glowing spots in the heavens have been called nebulae. Later, some of these proved to be remote star clusters. Others seemed outside the region of the Milky Way and came to be known as extragalactic

FIGURE 1.17. Shown are (a) the elliptical galaxy NGC 147 in Cassiopeia, (b) the spiral galaxy NGC 5457 in Ursa Major, and (c) the barred spiral galaxy NGC 1300 in Eridanus. (Palomar Observatory photograph.)

a

nebulae. The brightest of these nebulae is the great nebula in Andromeda, now called the Andromeda galaxy (Fig. 1.15), which appears to the naked eye as an elongated hazy spot. With the introduction of long-exposure photography, the existence of many diffuse patches of light, not at all like stars, was disclosed. These extragalactic nebulae, many of them beautiful spirals, are seen in all directions and in great profusion. As early as the eighteenth century Herschel and Kant suggested that these nebulae were actually "island universes," huge aggregations of stars lying far beyond the Milky Way. The validity of this hypothesis was not confirmed until 1927, when Hubble succeeded in measuring details of a number of spiral nebulae.

Separate stars in the spiral arms of the Andromeda galaxy (Fig. 1.15) were first observed by Hubble, who also discovered the halo of globular clusters surrounding the spiral. Resolved stars in the central region and in the disk outside the conspicuous arms were first observed by Baade, and similar observations were made for other nebulae. Hubble found Cepheid variables in some of the nearer nebulae, and from their period he was able to deduce their absolute luminosity. Using this information, he could estimate their distances (sec. 1.2.3). Hubble also established that the nearest spiral nebulae were vast systems of stars situated a million or more light-years outside our own Galaxy (Hub20, Hub25). Examination of these individual galaxies shows that they are by no means alike (Fig. 1.17). Hubble subsequently developed a scheme for classifying galaxies according to their morphology, ranging from systems that are amorphous, reddish, and elliptical (containing many red stars and little gas and dust) to systems that are highly flattened disks with a complex spiral structure containing many blue stars and lanes of gas and dust, with the entire system rapidly rotating (San61, Fer82). The rich variety of galactic forms is illustrated in Fig. 1.18. Regular galaxies contain all the features and objects found in the Milky Way. Except for its larger dimensions (180,000 ly in diameter), the neighboring Andromeda spiral galaxy (Fig. 1.15) seems to resemble very closely the structure of our own Galaxy. Clearly, the structural

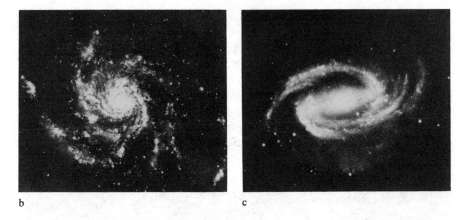

b   c

features of Andromeda or any galaxy can be observed better from outside. What has been learned about Andromeda has served to guide investigations of the Milky Way itself.

Millions of galaxies extend in all directions in space as far as the largest telescopes can explore, and many of these are spirals. Galaxies represent the major structures of the physical universe. The masses of galaxies turn out to be typically 10^{11} $M_\odot$, ranging from 10^8 to 10^{12} $M_\odot$. The diameters of the larger spirals such as our own Galaxy are about 100,000 ly. The spectra of galaxies

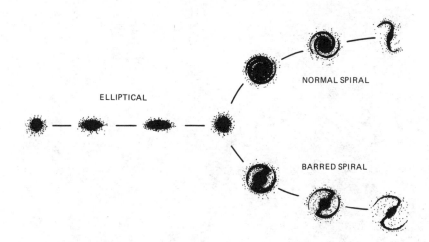

FIGURE 1.18. The separate series of regular galaxies having rotational symmetry were joined by Hubble, in the 1930s, into a continuous sequence from the compact spherical form at one end to the most open spirals at the other. Going from left to right in the figure might suggest the streaming of material from opposite sides of the flattened elliptical system. As a system evolves, there is a gradual building up of the spiral arms at the expense of the central regions. Whether or not galaxies actually evolve along these lines is an open question.

are composites, as would be expected for assemblages of stars. The Doppler effect alters the spectral lines owing to the different radial velocities of the individual stars and the rotation of galaxies. In addition to their rotation, galaxies may also be moving toward or away from our Galaxy, leading to an additional Doppler shift. Observations show that the more distant galaxies all exhibit redshifts and thus are moving away from us. It is found observationally by comparing the apparent galactic luminosities that the greater the redshift, the farther the galaxy. The latter feature provides the principal observational basis for the spectacular theory of the expanding universe (sec. 1.3.4).

FIGURE 1.19. The giant elliptical galaxy M87 in the constellation Virgo, one of the brightest galaxies known, shows a luminous jet 6000 ly long issuing from its center. The jet may not be as one-sided as it appears in this figure. There could well be an oppositely directed jet, which does not happen to emit much visible light in our direction. (Lick Observatory photograph.)

The apparently rather rare galaxies falling outside the scheme of Figure 1.18 were regarded by Hubble as "irregular." Recently, astronomers have become very interested in these irregular galaxies, and it is now thought that the peculiarities which they exhibit are not quite so rare as Hubble supposed. Even apparently normal galaxies, when examined closely, often show peculiarities (Fig. 1.19). The galaxy M87 appears as a normal globular form belonging to the elliptical class. Yet on a closer look M87 has a peculiar jetlike feature which can be seen emerging from the center on one side only, which can by no means be considered "normal." The light from this jet comes not from stars in the galaxy but from very high speed electrons interacting with a magnetic field. The electrons also emit other forms of radiation, especially radio waves. Confirming this, M87 is now known to be a radio galaxy (sec. 1.2.6).

There appears to be a fundamental tendency in nature for all things of a given class to clump together forming units of a new, higher order. Nucleons and electrons clump together to form atoms, atoms to form molecules, atoms and molecules to form stars and planets, and so on up the chain of complexity to galaxies and clusters of galaxies. The clusters of galaxies, the last firmly established level at the top of the hierarchy, provide a laboratory several million light-years across for studying the interaction of gas, stars, and galaxies on a grand scale. Studies suggest that every galaxy in the universe belongs to a cluster and that there might not be any isolated galaxies, although some astronomers believe that genuine single-field galaxies do exist. Some 10% of all galaxies are observed to belong to rich clusters, each of which consists of thousands of galaxies. In comparison, our own Galaxy is a member of a very small system known as the Local Group, consisting of no more than two dozen galaxies, most of them much smaller than ours. The Local Group is about 3 million ly in diameter, with the Andromeda galaxy being some 2 million ly distant. Studies of several rich clusters have shown that most of the thousands of galaxies within them are travelling through space at thousands of kilometers per second. The high velocity of these galaxies and the density of their distribution in space imply that they are bound together by gravitational forces much greater than can be accounted for by the observable mass. This problem of the "missing mass" (sec. 1.2.4) was introduced into the study of clusters of galaxies by Zwicky (Rub83).

Recent observations at X-ray and radio wavelengths have disclosed that the space between the galaxies in rich clusters is filled by hot gas at a temperature of some 10 million K and that in certain giant elliptical galaxies found at the center of the clusters there have been titanic explosions that have ejected vast clouds of high-energy particles into this hot intergalactic gas. What causes these explosions? What is the origin of the hot gas? Where is the extra mass that is needed to keep the speeding galaxies from flying apart (nearly a factor of 10 or more mass is needed)? No clear answer is possible. The emerging picture is that in rich clusters the interactions of gas, stars, and galaxies result in a gravitationally induced maelstrom generated by the dense concentration

FIGURE 1.20. Core of the cluster of galaxies in the constellation Coma Berenices, a small part of a vast supercluster. At least 300 moderately bright elliptical galaxies can be counted, each a giant collection of tens of billions of stars at a distance of some 300 million ly. The only prominent object in the picture that is not a galaxy is a bright blue star, slightly above and to the right in the center (*object with spikes*), which is a member of our Galaxy. (Palomar Observatory photograph.)

of galaxies at the center of the cluster. For example, the giant galaxy M87 in Virgo (Fig. 1.19) may have in its center a black hole with a mass equal to that of five billion suns. It is also possible that this galaxy is about 30 times more massive than previous estimates. The assignment of very large masses to giant elliptical galaxies such as M87 is important in accounting for the missing mass in clusters of galaxies.

In recent years large-scale surveys of selected regions of the sky have provided evidence for the existence of enormous superclusters of galaxies, organized structures composed of multitudes of clusters of galaxies spread over 100 million ly (Fig. 1.20). They represent the largest structures in the universe. On a scale larger than that of superclusters, the universe appears to be uniform. Superclusters offer an insight into evolutionary history that is not obtainable with smaller systems, since on smaller scales the original distribution of matter is smeared out by evolutionary "mixing." Astronomers hope that an understanding of the largest structures in the universe will clarify the processes that give rise to structures of all dimensions, ranging downward from galaxies to stars and planets.

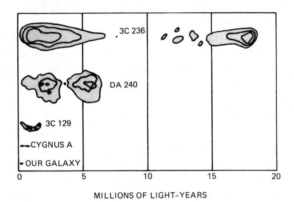

MILLIONS OF LIGHT-YEARS

FIGURE 1.21. Four radio galaxies are compared in size with our Galaxy, which is 0.1 million ly in diameter. They range from Cygnus A, 0.6 million ly across, to DA 240 and 3C 236, which are giant radio galaxies. The galaxy DA 240 has large, nearly circular components surrounding a central strong source.

1.2.6. Radio Galaxies and Quasars

Radio astronomy has revealed the existence of pointlike sources, many of which coincide with visible galaxies. Visually, these galaxies do not appear to be very unusual, but they are clearly strong radio wave emitters and are therefore called radio galaxies (Bre76). Many radio galaxies, when studied with high-resolution telescopes (VLBI), turn out to be double sources of radio emission with the optical source situated nearly in the middle (Fig. 1.21). This double structure appears to be the preferred morphology for strong radio sources outside our Galaxy. The galaxy Cygnus A (Fig. 1.21), one of the most powerful of all radio galaxies, has two outer components (jets) expelled from the parent galaxy like two streams of water from a double-ended hose. They span a distance of 0.5 million ly and are displaced symmetrically on each side of the visible galaxy. Since radio components move away from their parent galaxy at speeds of a few thousand kilometers per second, these objects are about 10^9 years old. The radio sources always cover a much larger area than optical sources, and in some radio galaxies three or even more sources of radio emission have been detected (Figs. 1.6 and 1.21). The optical source is not centrally located in such systems, implying perhaps a nonsymmetrical violent origin for the present configuration. The long jets (outer components) are always emitting synchrotron radiation, that is, radiation emitted by high-energy charged particles spiraling in a weak magnetic field. Radio galaxies are generally giant elliptical galaxies emitting energy in the radio spectrum at a rate of about 10^{43} ergs s^{-1}. The question of the origin of this energy is one of the most challenging in modern astrophysics. Radio galaxies also exhibit powerful outbursts. For relatively nearby radio galaxies the energy is about 10^{52} ergs per outburst, which is of the same order as the energy released in a powerful supernova explosion.

Seyfert (Sey43) called attention to a class of objects that looked like spiral galaxies but were distinguished by a very sharp bright source in the nucleus. These Seyfert galaxies, as they are now called, were found to be radio sources. Although most subsequently discovered Seyfert galaxies are now known to be strong radio emitters, some are not.

The successful identification of the Seyfert galaxies and other cosmic objects led to a diligent effort in the 1960s to identify all observed radio sources with an optical counterpart. Because the radio telescopes then available could not detect radio emission from normal stars, the coincidence of a radio source with an optical source was used by astronomers to differentiate radio sources from stars. As a result, the majority of the structures first identified were strong radio sources of pointlike (starlike) size and were called "quasi-stellar radio sources" or "quasars." In one suggested scenario a quasar is a radio galaxy with a particularly active nucleus so inordinately bright that it overwhelms the ordinary starlight. Seen from a great distance, such an object would appear as an isolated brilliant point of light, as quasars are observed to be. When photographed directly, the quasars appear starlike but with a resolvable fuzz surrounding the nearer objects (Fig. 1.22).

It was discovered, however, that a much larger population of quasars were weak radio emitters and therefore had escaped detection by radio telescopes. Sandage found that quasars emitted much more UV radiation than ordinary stars and therefore could be identified on this basis. He also showed that quasars that are quiet at radio wavelengths greatly outnumber those having strong radio emission. Another method of identifying quasars depends on the

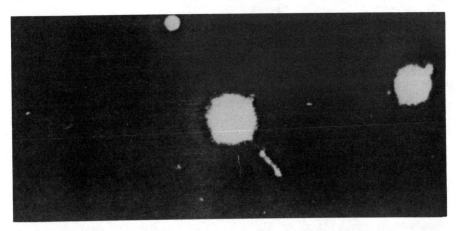

FIGURE 1.22. Quasar 3C 273, the first identified "quasi-stellar radio source," observed in 1963 by M. Schmidt (Sch63, Gre64). A long exposure shows a curious one-sided jet (rather similar to that of the galaxy M87; Fig. 1.19), which was subsequently also seen by VLBI mapping. (Kitt Peak National Observatory photograph.)

large redshifts of spectral lines (sec. 1.3.4). In fact these redshifts are so large ($Z \leq 3.53$; Wam73) that identification of such lines (Sch63) took several years (Fig. 1.22). For example, the Lyman-alpha emission line of hydrogen shifts from the far-UV region [$\lambda(0) = 1216$ Å] for a stationary atom to the green part of the visible [$\lambda(v) = 5508$ Å] for an atom with relative velocity v for which $Z = 3.53$ (sec 1.3.4). Some 1500 quasars have been observed, each of which can be 1000 times brighter than an entire galaxy of 100 billion stars. They are the most powerful emitters of electromagnetic radiation in the universe. Since quasars are known to vary significantly in optical and radio emission in less than a year, their vast output of energy must be generated within a volume no larger than 1 ly in diameter. Some quasar outbursts release about 10^{58} ergs of energy in several days. Quasars have also been discovered by their X-ray emission. The *Einstein Observatory* has in recent years recorded X-ray emission from every known quasar. The summed X-ray contribution of ancient quasars may be sufficient to account for the integrated background flux of X-radiation from outside our Galaxy.

The spectra of quasars also disclose that the emitting atoms are more highly ionized than the atoms in nebulae around hot young stars in our Galaxy. Moreover, in many instances the emission lines are wide, indicating that some of the gas surrounding the quasar is moving at velocities as high as 10,000 km s^{-1}. The physical conditions deduced from the intensities of the various lines (H, C, N, O) show that the gas is hotter than the gas in normal nebulae and that the central source in the quasar does not radiate at all like a normal star. The cosmic objects quasars, galaxies, and stars can then be identified best by general features in their spectra: quasars have emission lines which are all highly redshifted; normal galaxies are characterized by absorption lines and have cosmological redshifts; and galactic stars exhibit no cosmological redshifts.

The curious one-sided jet observed for the quasar 3C 273 (Fig. 1.22) was a surprising result, since the large-scale radio structure of extragalactic objects is generally highly symmetrical, consisting of two large radio lobes that straddle the optical object. VLBI mapping of several quasars and galaxies with highly energetic nuclei has shown that such asymmetric structures appear to be a common feature of such objects. These studies also revealed that these active nuclei steadily eject blobs of matter at nearly the speed of light in one direction. This rapid motion may hold the key to the one-sided structure. According to the special theory of relativity, if a radiating body is moving at nearly the speed of light, the radiation is beamed into a narrow cone in the direction of motion. It seems likely that the active nuclei in these celestial objects are actually ejecting two jets in opposite directions, with only the jet beamed toward us being observed. This is an attractive idea because it serves to reconcile the large-scale symmetry observed in many extragalactic objects with the small-scale asymmetry observed in many of these nuclei. At the point where the jets plow into the intergalactic medium and are slowed down, the radiation is no longer beamed but travels in all directions, so that both large blobs are observed (Fig. 1.21).

The explanation of the nature of the energy source in quasars and radio galaxies as well as of the mechanism of matter ejection from them remains controversial (Bur70). First, the source must produce a vast amount of energy, equivalent to the conversion of the total mass of many millions of suns into energy, in a region much smaller than a light-year across and perhaps not much larger than the solar system. Second, it must focus the ejected matter into a rather narrow cone. Third, it must remain stable and keep the jet stable while it ejects an amount of mass equal to the mass of a small galaxy. Finally, it must be able to eject matter at a speed close to the speed of light. It appears that only one kind of object is known that is theoretically capable of satisfying all these requirements: a spinning, supermassive black hole located at the center of a large galaxy. The object would have to have the mass of about a billion suns, and it would have to be spinning so that gyroscopic effects would keep it stable. The accurate and persistent alignment of the jets can be explained if one assumes that they are being extruded along the black hole's axis of spin. The radiated power would be generated by the release of gravitational energy as infalling stars are torn apart and swallowed by the giant black hole. The center of a quasar would then be in essence a stellar graveyard. As plausible as this speculation appears, it is likely that the problems of the quasars and radio galaxies will remain one of the great enigmas of astrophysics for some time. Improved observations will help to place explanations of these structures on a firmer ground.

1.2.7. The Universe

Within the observable part of the universe there are a few tens of billions of galaxies, where every galaxy contains many billions of stars. The total number of stars in the universe is therefore about 10^{22}, and the total mass also about $10^{22}\ M_{\odot}$. In spite of this enormous amount of matter in the form of stars, because of the large distances between them, the major feature of the universe is its emptiness. Within a typical galaxy, the stars are separated by several light-years. In a model in which stars are represented by raindrops, their separation would be about 100 km. Therefore stars within a galaxy fill only about 10^{-25} of the available space. Scaling such considerations to the universe, analysis indicates (Sha71) that the density of the visible matter is only about 2×10^{-31} g cm^{-3}. This density is equivalent to a gas pressure of about 10^{-23} torr or to about 0.1 H atoms per cubic meter. The universe is therefore a space of ultrahigh vacuum, within which there is on the average " nothing."

1.3. Selected General Properties of the Universe

The evolution of the universe became plausible only after extensive observations and comparisons of the properties of the objects within it. These observations were compared as to their relationships in time, space, and magnitude. For example, the elemental abundances could be referred to as universal or cosmic abundances only after they were determined for a diverse set of cosmic objects. Similarly, the mass-luminosity relationship and the Hertzsprung-Russell diagram were derived from observational data on a very large number of stars in all parts of the Galaxy. In like manner, the extremely important Hubble's law was derived from observations on a large number of distant galaxies. While Hubble's observations implied a spectacular origin of the universe, it was the discovery of the primordial photon remnant, the 2.76 K microwave background radiation, that provided the capstone to the big-bang hypothesis. This allowed astrophysicists and cosmologists to refine greatly their models of the early universe. Also pushing back the curtain of time are the observations on quasars, which appear to be at the very horizon. Finally, one takes all the observations and with great patience, some skill, and a good deal of imagination constructs the history of the physical universe. A discussion of some properties of the universe and their implications for its history is what this subchapter is about.

1.3.1 Observed Abundances of the Elements

The idea that there might be some systematics in the abundances of the elements as a function of mass has a lengthy history, going back at least to 1889, when Clarke (Cla89) reported an unsuccessful attempt to find some regularity. Such attempts were doomed to failure as long as the only data available were the abundances of the elements in the earth's crust. For example, the elements oxygen and silicon are much more abundant than hydrogen and helium in terrestrial samples, while hydrogen and helium are the most abundant elements in the sun. Other elemental abundances are also misleading. One reason for these differences lies in the geological and chemical fractionation and differentiation processes to which the material of the earth has been exposed during the 4.5 billion years since its formation. Furthermore, because of the escape of H and He from a body the size of a proto-earth, H and He are grossly underrepresented on earth. Conversely, when the relative abundances of the isotopes of a given element were examined, they were found to be nearly identical in all samples. This follows since—except for hydrogen—isotope abundances are influenced very slightly by chemical and other fractionation processes. The abundances of the isotopes as a function of mass number were found to be sufficiently regular in some groups of elements that certain empirical relationships could be established (Har17).

The situation improved when the abundances of the elements in meteorites (carbonaceous chondrites) were studied. These abundances were measured extensively in the 1920s and 1930s (Gol37). Additional information on the elemental abundances were obtained from the spectra of stars (in particular the sun) and from interstellar matter as well as from cosmic rays. While an accurate determination of the elemental abundances depends on extraterrestrial sources, the determination of isotopic abundances can be done largely using more easily available terrestrial samples. Attempts to put together a "cosmic" abundance table for the elements and isotopes combining meteoritic, terrestrial, and astronomical sources of data were begun by Brown (Bro49) and Suess and Urey (Sue56) and continuously refined as better data became available (Cam82, And82). The resulting abundance distribution of the elements is illustrated in Figure 1.23, which shows maxima for H, He, Fe, and Pb. A plot of the nuclidic abundances as a function of mass number (Fig. 1.23) exhibits many more pronounced features. Among the striking features are the gap between He and C, in which the abundances are very low (the elements Li, Be, and B), the steady decrease in abundances from the C-O region down through Ca to the very low valley at Sc, followed by the majestic Fe peak, succeeded in turn by a rolling landscape with a variety of hills and valleys. Because this abundance distribution is found to be nearly the same for all Population I stellar objects and interstellar gas, the term "cosmic or universal abundance" is used in referring to it. It represents the relative proportion of the elements and their isotopes at least for the matter of the visible universe. The data show that about 60%–80% of the matter by weight is H and 96%–99% is H plus He. The remaining few percent is made up of the other, heavier chemical elements (the "metals").

The abundance table of Suess and Urey has had enormous influence on studies of the origin of the elements and in the development of the science of nuclear astrophysics. The abundances of elements in the sun and similar Population I stars indicate that matter evolved from rather simple forms to much more complex ones. The stellar evolutionary process begins with H and He atoms from the big bang, which combine to form more complex atoms via nuclear reactions that transform one element into another. The entire process is referred to as nucleosynthesis.

Gamow and associates proposed in 1948 (Alp48a) that the elements and their isotopes were built up from protons and neutrons all in the course of one

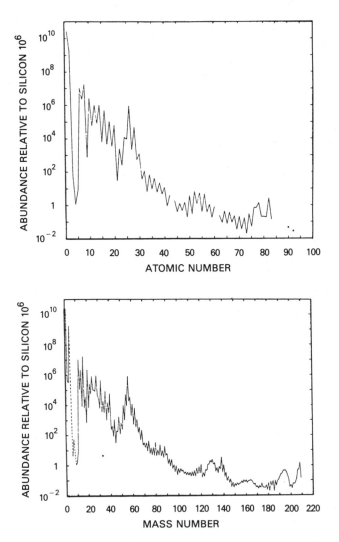

FIGURE 1.23. Shown are the elemental and nuclidic (isotopic) abundances in the solar system (Cam82). The abundances are normalized in such a way that Si has the value 10^6. Note the enormous abundance range displayed—a factor of about 10^{12} from the most abundant element (H) to the least abundant elements shown.

half-hour following the primordial big bang (chap. 2). Although the observed abundance curve (Fig. 1.23) agreed in general with their theoretical curve based on successive neutron captures, other investigators with better data pointed out that the process could not bridge the gaps of the mass instabilities at $A = 5$ and $A = 8$ (chap. 2).

With the abundance table of Suess and Urey it was possible for the first time to subdivide the table into groups of nuclides so that the abundances within any one of the subgroups could be attributed primarily to specific nuclear reactions. From this point it was but a small step to look to the stellar environment in which these processes of nucleosynthesis operate. This research led to the extensive analysis of nucleosynthetic processes which gave rise in 1957 to the fundamental paper by Burbidge, Burbidge, Fowler, and Hoyle (B^2FH). A parallel, independent analysis by Cameron (Cam57) came to many of the same conclusions.

The theory proposed by B^2FH and Cameron suggests that the heavy elements (the metals) have been and still are being synthesized in the interiors of evolving stars. The observed abundance curve of the elements (Fig. 1.23) offers one of the most powerful clues to the history and evolution of stars. Since stars play a critical role in the origin of matter and hence also set the stage for the

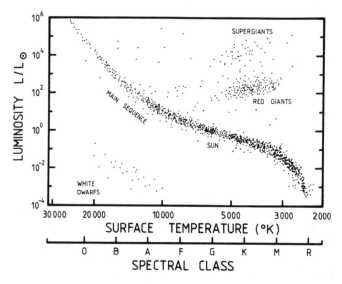

FIGURE 1.24. The Hertzsprung-Russell (H-R) diagram is shown schematically. Most of the stars (including the sun) are grouped along a band called the main sequence. As one goes from the upper left-hand corner to the lower right-hand corner along the main sequence, the temperature, mass, size, and luminosity decrease, while the mean density increases. The diagram represents a snapshot in the history of the stars' lives. The wide variety of stellar masses and the resulting variety of evolutionary tracks reflect the diversity of stellar objects. Note that only a small fraction of low-mass main-sequence stars is shown.

origin of life, considerations of the birth, life, and death of stars are of considerable human interest (chap. 2).

1.3.2. The Hertzsprung-Russell Diagram

The effective surface temperatures T_s of stars (2000–50,000 K) vary over a
much smaller range than the luminosities (10^{-4}–$10^6\ L_\odot$). Hertzsprung and
Russell compared the surface temperatures (or color, or spectral class) and
luminosities of stars within several clusters by plotting their luminosities
against their temperatures (Rus14). These investigations led to an extremely
important discovery concerning the relation between these quantities
(Fig. 1.24), called the Hertzsprung-Russell (H-R) diagram in honor of both
astronomers.

The most significant feature of the H-R diagram is that the stars are not
uniformly distributed all over this diagram but rather cluster in certain areas.
The majority of stars are aligned along a narrow sequence running from the
upper left-hand part (blue, hot, highly luminous) of the diagram to the lower
right-hand part (red, cool, less luminous). This band of points is called the
"main sequence." A substantial number of stars lie above the main sequence
in the upper right-hand region (red, cool, highly luminous). These stars are
called "red giants." At the top of the diagram are stars of even higher luminosity, called "red supergiants." Finally, there are stars in the lower left-hand
corner (white, hot, low luminosity) known as "white dwarfs."

Stars differ not only in surface temperature and luminosity but also in size,
as indicated by the above names. For the assumption of blackbody radiation,
the surface temperature, radius, and luminosity of a star are related to each
other by the Stefan law:

$$L = 4\pi R^2 \sigma T_s^4 \ ,$$

with $\sigma = 5.67 \times 10^{-5}$ ergs $K^{-4}\ s^{-1}\ cm^{-2}$. In terms of the sun's properties this
relation leads to

$$(L/L_\odot) = (R/R_\odot)^2 (T/T_\odot)^4 \ .$$

With $L/L_\odot = 10^6$ and $T/T_\odot = 4$ for the stars in the upper left-hand corner,
one finds $R/R_\odot = 60$. Likewise, for the red and cool supergiants on the upper
right-hand region of the H-R diagram, where $L/L_\odot = 10^4$ and $T/T_\odot = \frac{1}{2}$, one
obtains $R/R_\odot = 400$. They are truly "giant" stars. Their density ratio is
$\rho/\rho_\odot \leq 10^{-6}$ for $M/M_\odot \leq 50$. Thus these stars are characterized by extremely
low mean densities. The red, cool main-sequence stars at the lower right-hand
corner have $L/L_\odot = 1/2300$ and $T/T_\odot = \frac{1}{2}$, hence $R/R_\odot = 0.1$. Using the
mass-luminosity relation of such main-sequence stars (sec. 1.3.3), it follows that
$M/M_\odot = 0.1$ (consistent with measured masses) and hence $\rho/\rho_\odot = 200$.
Finally, the white dwarf stars in the lower left-hand corner of the H-R diagram
have extremely high densities. With $T/T_\odot = 2$ and $L/L_\odot = 1/200$, one finds
$R/R_\odot = 0.02$, which leads to $\rho/\rho_\odot = 8 \times 10^4$ for a mean value of $M/M_\odot =$

0.4. The theory of white dwarfs (chap. 2) predicts that the masses, radii, and densities of white dwarfs should be in the range $M/M_\odot = 0.1$–1.4, $R/R_\oplus = 4$–0.5 ($R/R_\odot = 0.04$–0.005), and $\rho/\rho_\oplus(H_2O) = 5 \times 10^4$–$10^6$, respectively. White dwarfs are believed to be one of the most common final states in stellar evolution.

An important application of the H-R diagram is in the determination of stellar distances. If the spectral class or color of a main-sequence star is known, its absolute luminosity is determined from its position in the H-R diagram (Fig. 1.24), which leads to the distance of the star by comparison with its observed brightness. This indirect method is called the method of spectroscopic or photometric parallaxes.

The H-R diagram for the main-sequence stars leads to a relation between luminosity and surface temperature

$$L \propto T_s^{5.5} \, .$$

This relation can be derived theoretically for stars whose internal structure is similar to that of the sun (chap. 2). The existence of the main sequence, where at least 95% of all stars are found, implies that the intrinsic structure of all stars is governed by the same physical laws as the sun. These stars must also have nearly the same chemical composition. The H-R diagram was therefore recognized as a powerful tool for the study of the intrinsic structure of stars as well as stellar evolution.

1.3.3 Mass-Luminosity Relation of Main-Sequence Stars

The data shown in the H-R diagram (Fig. 1.24) all refer to two stellar characteristics, the luminosity and the surface temperature. These two characteristics are important because they can be measured. Hence they provide the clues and checks for the theory of stellar evolution. But they are not the most basic characteristics. Rather, they are determined from moment to moment by the conditions inside the star, which are in turn dependent on the most basic properties of a star—its mass and chemical composition.

Studies of binary stars with well-separated components have provided fairly accurate knowledge of the masses of individual stars. When the masses and luminosities of those stars for which both of these quantities are well known are compared, it is generally found (Fig. 1.25) that the most massive stars of the main sequence are also the most luminous. This well-defined characteristic is known as the mass-luminosity relation and is, together with the H-R diagram, one of the most powerful tools for the investigation of stellar interiors and evolution. Most stars fall along a narrow sequence running from the lower left-hand corner (low mass, low luminosity) of the figure to the upper right-hand corner (high mass, high luminosity). The relation between the mass and the luminosity of a star is not accidental or mysterious but results from the fundamental laws that govern the internal structure of stars (chap. 2). It is estimated that about 90% of all stars obey the relation $L \propto M^{3.5}$. These main-sequence stars have a relative mass range of $M/M_\odot = 0.1$–50 and a relative luminosity range of $L/L_\odot = 10^{-2}$–10^6.

Although a star with $M/M_\odot = 10$ initially possesses a much larger reservoir of nuclear fuel than the sun, it burns this fuel more than 1,000 times faster, resulting in a much shorter life and a much quicker run through the evolutionary stages. The mass of a star is one of the basic characteristics determining the star's evolution along the route, and the age of a star is the changing parameter. As a star grows older, it will run through the H-R diagram on an evolutionary track fixed from the outset by its initial mass, its chemical composition, and the mass lost along the way.

A valuable clue to the possible evolutionary tracks is provided by the stars of an old globular cluster. Such stars are assumed to have formed essentially simultaneously and thus have similar ages and chemical compositions, differing from one another only in their masses. The globular cluster M3 (Fig. 7.2) shows its faint stars (low mass) still on the main sequence, while the brighter stars (higher mass) have moved to the red giant region. The turnoff point of the brighter stars from the main sequence is a phenomenon common to all clusters. The position of the turnoff point in the H-R diagram is different from cluster to cluster and depends on the age and the chemical composition of the cluster. There is also a horizontal branch in the red giant region (between the main sequence and the red giants) within which there is an instability strip, where variable stars of the RR Lyrae type (periods of 0.3–0.7 days) are found.

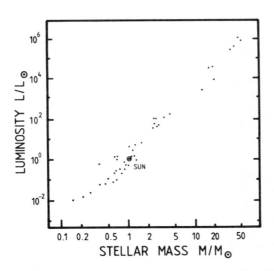

FIGURE 1.25. The mass-luminosity relation of main-sequence stars ($L \propto M^{3.5}$) and the H-R diagram show that the quantities luminosity, surface temperature, and mass of a star on the main sequence are related to one another. These relationships represent one of the basic tools for the understanding of stellar structure and evolution.

1.3.4. The Expansion of the Universe and Hubble's Law

The spectra of the stars are generated by their surface chemical elements, each of which emits a characteristic pattern of lines whose wavelengths are precisely known from laboratory measurements. When a galaxy (and hence its stars) is moving away from an observer, the wavelength of each spectral line increases as a result of the Doppler effect, so that all the lines appear to be displaced toward longer wavelengths. For visible radiation this means toward the red end of the spectrum. This displacement is called a redshift. By measuring its magnitude, the velocity of recession can be calculated. When an object is moving toward an observer, the wavelengths of the spectral lines are decreased by the Doppler effect, and the lines appear to be displaced toward the blue end of the spectrum, an effect called a blueshift. The value of the red- or blueshift, Z, is obtained by subtracting the rest (laboratory) wavelength $\lambda(0)$ of a galaxy's stellar spectral line from the observed wavelength $\lambda(v)$ and dividing the difference by the rest wavelength:

$$Z = [\lambda(v) - \lambda(0)]/\lambda(0) = \Delta\lambda/\lambda .$$

The Doppler effect predicts this shift Z to be related to the velocity v of the object relative to the observer via the relation

$$Z = [(1 + v/c)/(1 - v/c)]^{1/2} - 1 \simeq v/c \qquad \text{for } v/c \ll 1 .$$

Thus, an observed shift Z can be used to calculate the velocity of the object relative to the observer (see, however, Wei72 for cosmological applications).

The optical spectra of all distant galaxies and clusters of galaxies, as measured by Slipher and Hubble and other astronomers, exhibited redshifts. Since it was rather certain that the relatively small redshifts seen in the spectrum lines of the galaxies were due to the Doppler effect, one could conclude that the observable galaxies are in orderly motion and that the motion is expansion, with all galaxies rapidly receding from us. This systematic expansion does not operate within the galaxies themselves or within the groups and clusters of galaxies, but only over larger "cosmological" distances.

Establishing the distances of galaxies and showing them to be in recession were only part of Hubble's achievements. In 1929 Hubble (Hub29) announced a profound discovery of another remarkable property of the recessional motion, that is, that the redshift of a galaxy is directly proportional to its distance. Hubble's extensive work established that the ratio of velocity to distance is constant:

$$H = v/d = \text{constant} .$$

The reciprocal H^{-1} has the dimension of time and is called the Hubble time. This linear relation persisted when the inquiry was later extended to more distant galaxies. Small departures from this relation are commonly observed (Fig. 1.26) because most individual galaxies are members of groups or clusters of galaxies and have random motions with a component along the line of sight between them and the earth. Since the motions are essentially random, within any large sample of galaxies they cancel one another. The principle that the ratio of distance to velocity is constant has since become known as Hubble's law. There is no general agreement on the absolute value for the Hubble time H^{-1} because of uncertainties as to the distance of galaxies. The measurements have been repeatedly revised (Hod81) since Hubble's first estimate of about 2 billion years (Hub36). Present estimates of the Hubble time range from about 9 to 23 billion years (Fow85).

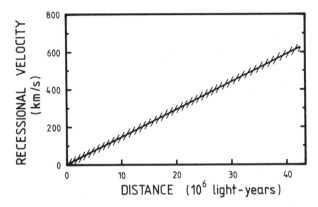

FIGURE 1.26. Schematic representation of Hubble's law, which was established by measuring the ratio of velocity to distance for many galaxies ($H = v/d$). One estimate of the ratio (San84) is about 15 km s^{-1} per million light-years (or, equivalently, 50 km s^{-1} Mpc^{-1}). This law is predicted by all models of an expanding universe, in the limit $v \ll c$. Note that the Hubble "constant" H does vary with time (chap. 2).

Another characteristic of the cosmic expansion is its isotropy: it is the same in all directions. No matter in what direction in the sky a galaxy is found, its recessional velocity is related to its distance by the same proportionality (Fig. 1.27). This observation seems to suggest that the universe is remarkably symmetrical and, what is even more extraordinary, that the earth happens to be at its very center. There is, of course, another plausible explanation, which can be understood most readily by considering a simple two-dimensional analogue of an expanding universe. Imagine a spherical balloon with small dots pasted on its surface, each dot representing a galaxy. As the balloon is inflated, the distance between any two dots (always measured on the surface of the sphere) increases with a speed proportional to the distance between them. No matter which dot is designated as the center, all the other dots recede from it uniformly in all directions. Thus each dot observes the same expansion and no one of them has the privileged position. Such an expansion has no center, or conversely, every point is its center. Similarly, in the real universe galaxies which happen to lie at the corners of a cube in one epoch will remain at the corners of a cube, albeit a larger one, as the universe expands. Thus progressive redshifts in the spectra of galaxies argue for an expanding and isotropic universe. Explaining these redshift observations is a fundamental requirement for any theory of the universe.

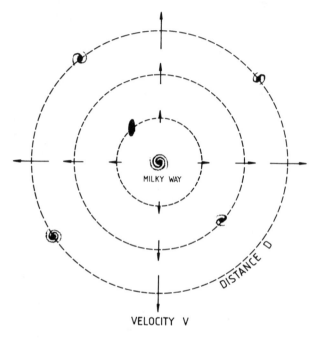

FIGURE 1.27. Pictorial representation of Hubble's law, which states that the velocity v with which a galaxy recedes is proportional to its distance d from the observer. This cosmic expansion seems to place the observer at the center of the universe from which all distant galaxies are fleeing. However, since the galaxies are also systematically receding from one another the cosmic expansion has no center or, more precisely, every observer (galaxy) is its center. Although the space between the galaxies expands, the size of each galaxy remains the same.

If the recessional velocity of every galaxy remained unchanged through all time, any galaxy now receding from us was once arbitrarily close, and the time that has elapsed since then is equal to the ratio between the galaxy's distance and its velocity. Since the ratio is the same for all galaxies, all of them must have been crowded together at the same ancient time. In other words, at some unique time in the past all the matter in the universe was compressed to an arbitrarily high density everywhere! This time is approximately given by the Hubble time of about 16 ± 7 billion years—the age of the universe (Dav76, Fow85). This simple calculation has led to the cosmological hypothesis that the world began with an unimaginably great primordial explosion called the "big bang" involving all the matter and energy in the universe.

The light of some clusters of galaxies set out toward us about 5 billion years ago, so we can be sure that some galaxies are even older than that. Estimates of the age of stars suggest that our Galaxy, and others like it, are unlikely to be much less than 10 billion years old (Fow85). Hence we are presented with a remarkable coincidence that most galaxies appear to be about as old as the universe. This implies that galaxies must have been formed when conditions in the observable universe were much different from those now prevailing. It seems clear, then, that the formation of galaxies cannot be treated apart from cosmological considerations (chap. 2).

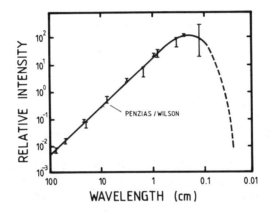

FIGURE 1.28. The first measurement of the intensity of the cosmic background radiation was made in 1965 by Penzias and Wilson (Pen65) at a wavelength of 7.5 cm. The observed intensity and the results obtained subsequently at other wavelengths follow the energy spectrum of a blackbody with a temperature of 2.76 K (*solid line*). The existence of such a background radiation with a temperature of about 5 K had been predicted in the late 1940s by Gamow and coworkers (chap. 2).

1.3.5. The Universal Background Radiation

In 1965 the radio astronomers Penzias and Wilson (Pen65) discovered another basic and most important cosmological phenomenon which, like the recession of the galaxies, is a truly fundamental feature of the universe. A historical account of this discovery is given elsewhere (Wei77). It is the low-energy cosmic radio radiation that apparently fills the universe and bathes the earth from all directions in space. This microwave radiation has since been shown to be highly isotropic, varying by less than one part in 10,000 in temperature over the entire area of the sky. Intensity measurements at several wavelengths show (Fig. 1.28) that this background radiation is consistent with that of a blackbody (Planck's radiation law) at a temperature of 2.76 K. Radiation of this type cannot have been generated by any of the known astronomical objects. Galaxies and quasars certainly emit radio waves, but they do not generate an adequate intensity in this particular wavelength range and, in any case, their radiation would be directional. The microwave radiation is therefore believed to have originated at an epoch of the universe much more remote than any time observable by the usual astronomical methods. This microwave radiation is the most ancient signal ever detected. It is therefore a background in front of which all astronomical objects lie. The current view is that this microwave radiation was there at the beginning of the universe. In the first few seconds of the history of the universe, the radiation had a temperature of about 10 billion degrees. The present-day low temperature of only 2.76 K has come about because in the expansion of the universe the radiation has continuously cooled from its initially very hot state. Using the Stefan-Boltzmann law, the number of photons per cubic centimeter can be calculated by means of the relation

$$N_\gamma = (\pi/13)(kT/\hbar c)^3 = 20.25T^3 \,,$$

with T in K. The microwave radiation corresponds, then, to a photon gas filling the universe with a density of about 430 photons cm^{-3}.

From the data on this background radiation, theorists were able to calculate a new fundamental quantity: the ratio of the number of photons in the universe to the number of nucleons (the protons and neutrons). The ratio is about 10^9 photons to one nucleon and is proportional to the average thermal entropy associated with each nucleon; we will subsequently refer to this as the astrophysical entropy. Entropy, usually represented by the letter S, is defined by the equation $S = k \ln W$, where k is Boltzmann's constant and W is the number of states possible for the system. To put it in another way, entropy is a measure of randomness or disorder. A crowd of people in a marketplace has a much higher entropy than a regiment consisting of the same number of soldiers and moving in step. A crystal lattice is an example of a low-entropy system because its atoms are arranged in a highly ordered way. If the atoms have spin $\frac{1}{2}$, there are two possible spin orientations (up or down) according to quantum theory, and the entropy per atom (in units of k) is then 0.69. The ratio of the density of photons to the density of nucleons averaged over a large volume of the universe is a measure of the average entropy because photons constitute the most disordered states of thermal energy, and nucleons, atoms, molecules, and cells constitute the most ordered state. Hence the relative abundances of these two extreme states are a measure of the average entropy. According to the second law of thermodynamics, the total entropy of the universe increases continuously as time goes on. This means that at the time of the big bang (chap. 2) the astrophysical entropy was less than 10^9. An entropy of 10^9 is quite large compared with the entropy of about 1 exhibited by systems in the terrestrial environment, and thus the universe as a whole is a relatively "hot" place.

1.3.6. Quasars as Probes of the Distant Universe

If the redshifts of quasars are interpreted the same way as those of galaxies (by the Doppler effect), the observed quasar PKS 2000−330 with a redshift of $Z = 3.78$ is evidently receding at 92% of the speed of light. On the same scale, stars within galaxies have velocities of 0.1% of the speed of light, and nearby galaxies move away from us at no more than 1% the speed of light. If the quasar redshifts are interpreted according to Hubble's law, they cannot be nearby objects that happen to be moving at high velocity but must be extremely distant objects. On the basis of the current scale of the universal expansion ($H^{-1} \simeq 16 \times 10^9$ y), a galaxy with a redshift of $Z = 0.01$ (or $v = 3000$ km s^{-1}) will be about 130 million ly away. This is already a considerable distance. The nearest galaxy, the Andromeda galaxy, is only about 2 million ly away. Quasars are vastly more distant in space and time. The concept of an expanding universe implies also that an object with a large redshift is observed as it existed long ago. The light now reaching us from a quasar with a redshift of $Z = 3$ was emitted some 13 billion years ago. Quasar redshifts therefore supply some clue to the structure and character of the early universe (Osm82).

In the late 1960s M. Schmidt (Sch71) discovered a remarkable property of distant quasars of given luminosity: they are far more numerous at large distances than they are in our vicinity (Fig. 1.29). At a redshift of $Z = 2$, or about 12 billion years ago, their density was 1000 times greater than it is now. Evidently, whatever process gave rise to quasars must have been extremely active when the universe was young and has subsided to practically nothing today.

At the time of Schmidt's work on the spatial density of quasars, the largest redshift known was $Z = 2.88$. Enormous as this value was compared with what had been expected only a few years earlier, Schmidt's result suggested that quasars with much greater redshifts should be plentiful. Naturally astronomers kept searching for quasars of ever greater redshifts. In spite of enormous efforts and improved detection sensitivities, all quasars detected had redshifts of $Z \leq 3.78$. This work indicated that the density of quasars in space continues to increase only up to redshifts of $Z = 3.8$ (Fig. 1.29). The quasar redshift limit of $Z = 3.8$ implies that for at least one type of cosmic object astronomers were seeing things at the horizon of the universe (chap. 2). It also implies an abrupt change in the properties of the universe. One of the simplest explanations is that quasars suddenly formed in a great burst of activity about 14 billion years ago, which would certainly have been a remarkable occurrence in the evolution of the universe. Such a concept, of course, would greatly influence ideas about the fundamental nature of quasars. Alternatively, an absorbing screen of dust or some other kind of material may be present at a redshift of $Z = 3.8$, blocking our view of the most distant quasars. For the universe to be transparent on one side of the limit and opaque on the other side would be equally remarkable. In either case, further work on the subject is likely to bring important results in the future.

The above discussion depends on the assumption that the observed large redshifts of quasars can be interpreted, as in the case of galaxies, as due to the Doppler effect. The essential question is whether these objects are at the extreme distances implied by the redshift or whether the redshift is due to some other factor, such as intense gravitation, scattering from cosmic dust, or something not yet known. It is therefore well to remember that, plausible as the above picture of quasars may be, there is some chance that it is entirely wrong and some chance that it is wrong in important particulars (sec. 1.2.6). In any case, it is hard to image the prevailing concept of quasars being over-turned without their becoming even more amazing than they already are.

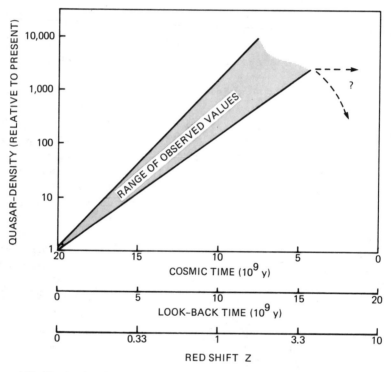

FIGURE 1.29. The density of quasars of given luminosity 15 billion years ago was more than 1000 times greater than it is today (Sch71, Osm82). Some 1500 quasars are now known; however, the absence of quasars with a redshift larger than $Z = 3.8$ implies that astronomers have probed to the earliest epoch of quasar formation.

2

Astrophysics—
Explaining the Universe

At the turn of the century, astronomers viewed the stars as a bewildering collection of objects without any discernible set of systematic interrelations. To them, the sky was filled with a widely diverse set of bodies whose characteristics could be measured and described. Subsequent astronomical observations, as presented in chapter 1, implied systematic and general properties of the universe and its objects, but these characteristics were not clearly understood. Although some of the observed phenomena remain mysterious and new phenomena continue to appear, the apparent chaos in the sky is being resolved into some order. This resolution is derived from the union of astronomy and physics, called astrophysics, a branch of astronomy that applies the theories and laws of physics to the universe and the multitude of objects therein (Shu82). One might consider that astrophysics began in 1812 when Fraunhofer (1787–1826) discovered the dark lines in the spectrum of the sun.

In this chapter we will try to organize the results of the astronomical observations into a coherent picture of the universe. Any theoretical model must explain the universe on a grand scale as well as predict the observations made on the galaxies and stars in our part of it. The models developed are based on results from many different areas of astrophysics and cosmology. They incorporate the physical processes known to govern the galaxies, i.e., the dynamic processes that determine the galaxy's overall size and shape as well as stellar formation, evolution, mass loss, and death. The basic concept required for understanding the nature and history of stars came from the discovery that thermonuclear reactions provide the energy to sustain their luminosity over their long lifetimes. This discovery led to a clearly formulated theory of stellar evolution which touches every aspect of astronomy and is beginning to provide the structural elements for a coherent picture of the universe.

There is also evidence that unusual processes are at work in the universe. For example, from the pulsars and quasars one deduces that matter in these objects is in a condition unlike that found anywhere else and that perhaps sources of energy are being tapped that are not now fully understood. Perhaps new basic laws of physics will be revealed in the "laboratory of space."

2.1. Big-Bang Cosmology

Although cosmological inquiry is ancient, it is only in this century that an understanding has emerged of how the universe began and what its ultimate fate may be (Bon52, Wei72, Wei77, Dol81, Clo83). The crucial perception came in the 1920s when Hubble and others demonstrated that the spiral nebulae are not local objects but independent systems of stars much like our own Galaxy. This demonstration indicated that the universe is much larger than had previously been imagined. Thus, detailed observations of a large variety of simple and large-scale phenomena are essential to the understanding of the universe or the development of cosmology as a truly scientific field.

2.1.1. Standard Cosmological Models

Since Hubble's original discovery, increasingly precise observations have shown not only that the cosmic expansion is isotropic but also that all other large-scale features of the universe are independent of direction. For example, the distribution of galaxies on the celestial sphere, in particular the clusters and superclusters, seems to be quite uniform, as does the distribution of quasars. The most compelling evidence for isotropy is the universal background radiation which bathes the earth, coming at us from every direction with no discernible difference in intensity. This is now established to better than 0.01%. The observation of such remarkable isotropy led to the adoption of a "cosmological principle," a powerful generalization, which theorists use to build models of the universe (Wei72, Wei77). It states simply that the evolving universe appears isotropic to all observers participating in the expansion—everywhere and at all times. In other words, our Galaxy is indeed at the center of the universe, but so is every other galaxy. There are, of course, local anomalies; but averaged over significant portions of the universe, the principle holds.

If the motions of the galaxies are reversed and extrapolated back into the past as far as possible, eventually a state is reached in which all the observable galaxies are crushed together at infinite density, an inferno of particles and radiation, a singular state of affairs representing the beginning of the so-called big bang or big-bang universe. (Fri22, Lem27, Gam48, Gam53). It marks the origin of the universe and all that is in it. If, as some speculate (Wei72, Wei77, Dol81), the recessional velocities have changed in the past, it does not alter the fact of the big bang but merely changes the time since its occurrence. We arrive, therefore, at the most remarkable conclusion that the universe has a finite history and that it all began some time between 9 and 23 billion years ago (chap. 9).

That the universe began with a big bang is an inevitable conclusion assuming that the laws of physics are correct and complete. It is conceivable, however, that within the laws of nature there are effects which are negligible on the scale of the terrestrial laboratory, or even on the scale of the solar system, but which might dominate the behavior of the universe as a whole. One theory for the universe requiring such new laws of nature is the "steady-state" cosmology, according to which the universe is unchanging and infinitely old and the density of matter remains constant (Bon48, Bon52, Hoy49). In order to explain the cosmic expansion, the steady-state theory postulates the continual creation of matter in the voids at the very small rate needed to keep the density constant. This model generated considerable speculation and controversy. Essential to this model is the creation of matter from nothing, something that has never been observed. This is not surprising, however, since the required creation rate is only one nucleon every 50 years in a cubic kilometer. This estimate of the creation rate is easily obtained by calculating the rate of mass change with time in a given volume of space and setting this equal to zero:

$$\frac{dM}{dt} = -4\pi \rho r^2 \frac{dr}{dt} + \frac{4}{3}\pi \xi r^3 = 0 \, ,$$

where the first term is the mass outflow. This is related to the Hubble constant through $dr/dt = v = Hr$, with $H \simeq 15$ km s^{-1} per 10^6 ly $= 50$ km s^{-1} Mpc^{-1} (chap. 1). The second term represents the postulated creation, the constant ξ being the rate of mass creation. This equation yields $\xi = 3\rho H$, which, for $\rho = 2 \times 10^{-31}$ g cm^{-3}, gives $\xi \simeq 1 \times 10^{-48}$ g cm^{-3} s^{-1}, or one nucleon created every 50 years in a cubic kilometer.

The steady-state model requires every part of the universe to look the same independent of time. Observations on quasars (and radio galaxies), however, lead to the conclusion that in earlier times there were more quasars than there are now (Fig. 1.29), which argues strongly against the steady-state model. Also, in the steady-state model of the universe, it is particularly difficult to account for the background radiation. This radiation field has the spectral characteristics of thermal radiation emitted by a blackbody at a temperature of 2.76 K (Fig. 1.28). It seems to be satisfactorily explained as a relic of an epoch when the universe was very hot and very dense (see below). A steady-state universe cannot have had such conditions because in that model all conditions, by definition, are unchanged. The most ancient of all astronomical observations, namely, that the sky grows dark when the sun goes down, also argues against the steady-state model. To see the significance of this observation, note that if absorption is neglected, the apparent luminosity of a star of absolute luminosity L at a distance r will be $L/4\pi r^2$. If the number density of such stars is a constant n, then the number of stars at distances between r and $r + dr$ is $4\pi nr^2 dr$, so the total radiant energy density due to all stars is

$$\rho_s = \int_0^\infty (L/4\pi r^2) 4\pi n r^2 dr = Ln \int_0^\infty dr .$$

The integral diverges in an infinite universe, leading to an infinite energy density of starlight, and thus the night sky should never get dark. This is known as Olbers's paradox. In big-bang models Olbers's paradox is avoided (Har74) owing to the finite age of the universe, i.e., the luminous emissions from stars are much too feeble to fill—in their lifetime—the vast empty spaces between stars with radiation of any significant amounts. In these models the background radiation is also a natural consequence of conditions in the early universe.

2.1.2. Basic Physics and Dynamics of the Standard Big-Bang Model

The present state of the universe is characterized by four large-scale phenomena: the expansion of the universe (Hubble's law), the 2.76 K background radiation, the ratio of 10^{-9} for the number of nucleons (baryons) per photon, and the universal abundance by weight of about 25% ^{4}He and 75% ^{1}H (sec. 2.2). Working backward from the present state of the universe to gain some knowledge of the initial conditions within the big-bang model is not at all arbitrary. To get a handle on the conundrums of cosmology, we need to start by describing the dynamical history of the universe. The gross features should be clear at the onset. The gravitational force of matter pulls everything in the universe toward every other thing. It halts expansion and speeds contraction. To have a basis for such elaborations, we need first to discuss the physical laws characterizing the evolution of the universe.

The energy density of the photon gas is described by the Stefan-Boltzmann law:

$$u_\gamma = (\pi^2/15)[(kT)^4/(\hbar c)^3] = 4.72 \times 10^{-3} T^4 \text{ eV cm}^{-3} ,$$

with T in K. The equivalent mass density is obtained by dividing u_γ by c^2:

$$\rho_\gamma = 8.40 \times 10^{-36} T^4 \text{ g cm}^{-3} .$$

For $T = 2.76$ K one finds $\rho_\gamma = 4.9 \times 10^{-34}$ g cm^{-3}, which is negligibly small compared with the matter density of $\rho_0 \simeq 2 \times 10^{-31}$ g cm^{-3} (chap. 1). Thus, the largest fraction of energy of the present universe is stored in the form of matter, and such epochs of the universe are called matter-dominated. An alternative way to arrive at this conclusion is to consider energy densities: from Wien's law ($T\lambda \propto T/E = $ constant), one obtains for $T = 2.76$ K a mean photon energy of 8×10^{-4} eV, and since there are 10^9 photons per nucleon, the mean radiation energy is 8×10^5 eV, which is small compared with the 10^9 eV energy equivalent of the nucleon mass.

As the universe expands, it is like a container with expanding walls which is filled with particles. Because of the expansion, the Doppler effect increases the wavelengths of the photons (particles) in proportion to the size and gravitation (general relativity) of the universe, $\lambda \propto R$, where the size R can be interpreted as the distance between two typical galaxies. This interpretation seems to restrict the discussions to a small spherical volume of the universe. However, since the expansion is universal (cosmological principle), the laws derived for any small volume must apply to the universe as a whole (Cal65). By spherical symmetry the matter outside this small volume exerts no force, and for a limited volume the forces are small and the expansion velocities nonrelativistic. Thus, to a good approximation one can use Newtonian mechanics. A remarkable feature of a blackbody spectrum is that, when it is viewed in a frame of reference moving with respect to the emitter, it retains the characteristic blackbody shape, but with a different characteristic temperature (Wei77). Thus Wien's law yields

$$T \propto 1/\lambda \propto 1/R \,,$$

i.e., the temperature drops (increases) linearly with increasing (decreasing) size of the universe. For example, a factor of 1500 decrease in size increases the radiation temperature from its present 2.76 K to 4140 K ($kT = 0.36$ eV), which gives a radiation mass density of $\rho_y = 2.47 \times 10^{-21}$ g cm^{-3}. Because of the wave nature of massive particles the above proportionality $T \propto 1/R$ is also valid for them. Taken together with $\rho_m \propto 1/R^3$, this leads to a temperature dependence of the matter density given by

$$\rho_m = \rho_0(T/T_0)^3 = 4.76 \times 10^{-33}T^3 \text{ g cm}^{-3} \,,$$

where for the present universe $\rho_0 \simeq 2 \times 10^{-31}$ g cm^{-3} and $T_0 = 2.76$ K. At $T = 4140$ K, one finds $\rho_m = 6.7 \times 10^{-22}$ g cm^{-3} and thus $\rho_y \gg \rho_m$. Such an epoch is called radiation-dominated, i.e., the universe contained mainly radiation, with a relatively insignificant amount of matter.

As the size of the universe is reduced further, the temperature increases accordingly, and the radiation density becomes higher and higher. For example, a reduction in size by a factor 10^{10} increases the temperature to $T = 2.76 \times 10^{10}$ K, which is above the temperature threshold $T_{\text{thres}} = 6 \times 10^9$ K for photons capable of producing electron-positron pairs ($kT_{\text{thres}} = mc^2 = 0.511$ MeV). At about this temperature radiation domination ends and matter again plays an important role. However, matter at this temperature is quite different from the cold matter of the present universe (Zel65, Chi66). At still higher temperatures (smaller volume), muons and antimuons and other particles plus their antiparticles can be created. Since at these temperatures the particles and their antiparticles were in equilibrium with the photons, their number densities were roughly equal to that of the photons. At temperatures $T \gg T_{\text{thres}}$ matter behaves like photons (Wei77) and hence $\rho \propto N_i T^4$, where N_i is the number of particles and antiparticles in thermal equilibrium with photons and other particles ($N_i \simeq N_y$).

Next we examine the time dependence of the parameters of the universe (Wei77). Consider a sphere of radius $R(t)$ enclosing a portion of the universe, and suppose that at time t a typical galaxy of mass m is at the surface of the sphere. The mass contained in this sphere is given by the volume and the density $\rho(t)$:

$$M = \tfrac{4}{3}\pi R^3(t)\rho(t) .$$

It can be shown that the galaxy is gravitationally influenced only by the matter within the sphere. The total energy of the galaxy is then given by the sum of the potential energy and the kinetic energy:

$$E = -[mMG/R(t)] + \tfrac{1}{2}mv^2(t)$$

$$= mR^2(t)[\tfrac{1}{2}H^2(t) - \tfrac{4}{3}\pi\rho(t)G] ,$$

where we have used Hubble's law, $v(t) = H(t)R(t)$, and $H(t)$ and $\rho(t)$ are the values of the Hubble "constant" and the cosmic mass density, respectively, at time t. Since the total energy E must remain constant at all times and since $\rho(t) \propto 1/R^3(t)$, the potential energy goes to infinity for $R(t) \to 0$, and thus the two terms in the brackets must cancel in this limit:

$$\tfrac{1}{2}H^2(t) \simeq \tfrac{4}{3}\pi\rho(t)G .$$

This relationship defines the characteristic expansion time, which is just the inverse of the Hubble constant:

$$t_{\exp}(t) \equiv 1/H(t) = [3/8\pi\rho(t)G]^{1/2} .$$

This result is also valid for a relativistic universe if the density $\rho(t)$ includes the normal matter density as well as the "mass" equivalent of radiation, E/c^2. For example, at $\rho(t) = 10^6$ g cm^{-3} one finds $t_{\exp} = 1.4$ s. From the expression for the expansion time $t_{\exp}$ we see that the gravitational force is dominant on the largest scale and that it controls the dynamics of the universe.

In the matter-dominated epoch the total mass within an expanding sphere of radius $R(t)$ stays constant (conservation of baryon number), and thus the matter density is given by $\rho_m(t) \propto 1/R^3(t)$. In contrast, the radiation-dominated era has a density given by $\rho_\gamma(t) \propto T^4(t)$, and since $T(t) \propto 1/R(t)$, one arrives at the relation $\rho_\gamma(t) \propto 1/R^4(t)$. In general, one can write $\rho(t) \propto 1/R^n(t)$, with $n = 3$ and $n = 4$ for the matter- and radiation-dominated eras, respectively. Using the above relation $H(t) \propto \rho(t)^{1/2}$ and Hubble's law, $v(t) = dR(t)/dt = H(t)R(t)$, we arrive at the differential equation

$$dR(t)/dt \propto R(t)^{(1-n)/2} ,$$

which has as a solution

$$t_1 - t_2 = \left(\frac{2}{n}\right)\left[\frac{1}{H(t_1)} - \frac{1}{H(t_2)}\right] = \left(\frac{2}{n}\right)\left(\frac{3}{8\pi G}\right)^{1/2}\{[\rho(t_1)]^{-1/2} - [\rho(t_2)]^{-1/2}\} .$$

For example, in the radiation-dominated era ($n = 4$) a temperature decrease from $T_1 = 10^9$ K to $T_2 = 10^7$ K corresponds to a density change from $\rho(t_1) =$

8.4 g cm^{-3} to $\rho(t_2) = 8.4 \times 10^{-8}$ g cm^{-3}, giving the time necessary for this cooling as $t_1 - t_2 = 26.7$ days. The above result can be generalized to give the time t needed to decrease to a density $\rho(t)$ as

$$t \simeq (2/n)[3/8\pi G\rho(t)]^{1/2} ,$$

which is equal to $\frac{1}{2}t_{\text{exp}}$ for the radiation-dominated era and to $\frac{2}{3}t_{\text{exp}}$ for the matter-dominated era. For example, at $T = 4000$ K the radiation density is $\rho(t) = 2.1 \times 10^{-21}$ g cm^{-3}, and the time needed for the universe to drop to this temperature is 470,000 y, which is one-half the expansion time t_{exp}.

If a system is in thermal equilibrium, its properties are uniquely determined by the values of the conserved quantities. Since the universe has gone through such a state of thermal equilibrium, a description of the early universe requires knowledge of the values of the conserved quantities during the expansion of the universe. One such conserved quantity is energy, which for a system in thermal equilibrium may be replaced by an equivalent quantity, the temperature. Since the universe is composed predominantly of radiation and of particles and antiparticles in equal numbers (see, however, sec. 2.1.6), one needs to know only the temperature to develop the equilibrium properties of the system. At high temperatures particles and antiparticles are created from photons and subsequently annihilated. There are then only three conserved quantities: the electrical charge Q, the baryon number N_B, and the lepton number N_L (see below). Since the values of these quantities are conserved in a volume expanding with the universe, their densities vary as R^{-3}. Since the number density of photons, N_γ, also varies as R^{-3}, the ratios of these quantities, N_B/N_γ and N_L/N_γ, are independent of R and thus of the size of the universe.

Although particles and antiparticles with opposite charges can be created in processes, charge conservation requires that the net amount of charge not be changed in the process. Since the universe as a whole appears to be electrically neutral (Dol81), the net charge of the universe must be zero; hence $Q/N_\gamma = 0$.

Elementary particles, such as protons and neutrons, which feel the strong nuclear force are called baryons. Each baryon is given a baryon number $+1$, and each antibaryons a baryon number -1, and it is found as a "bookkeeping rule" that the sum of baryon numbers minus the sum of anti-baryon numbers remains constant in nuclear processes:

$$\sum |\text{baryons}| - \sum |\text{antibaryons}| = N_B = \text{constant} ,$$

where N_B is called the total baryon number. For example, charge and energy conservation together would allow for the decay of the proton into a positron and a neutral meson (pion): $p \rightarrow e^+ + \pi^0$. However, the left-hand side of this process has baryon number $+1$ and the right-hand side baryon number zero. Thus, the process is forbidden by conservation of baryon number, and the proton should be a stable particle (the lower limit on the proton lifetime is found to be about 10^{32} y). The presence of matter in the universe clearly indicates that the baryon number of the universe is not zero but finite and positive. Its ratio to the photon density has the value $N_B/N_\gamma \simeq 10^{-9}$. Note that the condition $N_B > 0$ means that the universe is not symmetrical with respect to matter and antimatter.

Light elementary particles, such as electrons, muons, tauons, and neutrinos and their antiparticles, are called leptons. They do not experience the strong nuclear force. It is also found that processes involving leptons obey a lepton number conservation law:

$$\sum |\text{leptons}| - \sum |\text{antileptons}| = N_L = \text{constant} ,$$

with N_L as the total lepton number. Since matter is electrically neutral, the number of electrons must be equal to the number of protons (baryons): $N_e = N_B$. If electrons are the only types of leptons in the present universe, we have $N_L/N_\gamma \simeq 10^{-9}$. There could, of course, exist neutrinos and antineutrinos. However, if these have been produced in equal numbers, their contribution to N_L is zero. It is usually assumed that $N_L/N_\gamma \ll 1$, i.e., all types of leptons are nondegenerate (sec. 2.4.10).

In summary, with matter and radiation in thermal equilibrium, the conserved quantities charge, baryon number, and lepton number of the universe are all very small relative to the photon number. These and other assumptions discussed above define a particular model of the big bang called the "standard" big-bang model, which is usually used in calculating such things as the primeval nucleosynthesis (sec. 2.2).

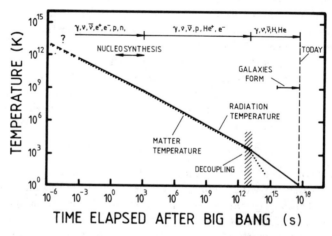

FIGURE 2.1. Shown is an outline of events in the universe since the big bang. Events within the first 10^{-3} s are not well known, primarily because they are dominated by interactions of exotic particles and photons at energies beyond those we can currently investigate (sec. 2.1.6).

2.1.3. Evolution of the Early Universe

The initial state in the big-bang models is one of high temperature and density sometimes called the "cosmic or primeval fireball." Figure 2.1 charts the history of the universe from the time when its overall temperature was 10^{12} K. Before that time the universe was full of short-lived, exotic particles and antiparticles at tremendous density, temperature, and pressure (sec. 2.1.5). These particles were in thermal equilibrium with a radiation field of high-energy photons and other particles. The particles could, for example, interact to produce photons; and, conversely, photons could interact to produce particles. The higher the energy of the photons and particles, the more massive and peculiar the particles they could create. By the time the temperature of the expanding universe dropped to 10^{12} K, the only particles participating in the creation-annihilation process were neutrinos, antineutrinos, electrons, positrons, and photons. Some trace amounts of protons and neutrons were left ($N_B/N_\gamma \simeq 10^{-9}$), but essentially all the other heavy particles and antiparticles had been annihilated.

At 10^{12} K the temperature of the universe is still high enough to impose statistical equilibrium among all particles present. The condition of statistical equilibrium ensures that for all time the properties of the universe are independent of its history (Dol81). As a consequence, the number densities of the particles present are not determined by the details of history but are the result of equilibrium between creation and annihilation processes. The evolution of the constituents begins in this environment. Thus, the density of the universe is related to the densities of all its constituents:

$$\rho = \rho_\gamma + \rho_e + \rho_\nu + (\rho_m) \simeq \rho_\gamma + (7/4)\rho_\gamma + (7/4)\rho_\gamma = (9/2)\rho_\gamma = N_* \rho_\gamma ,$$

where ρ_γ is the photon density, $\rho_m \ll \rho_\gamma$, and N_* is a total weighting factor. The weighting factor for each constituent (Wei77) represents the effective number of particle states (degrees of freedom), including their number of antiparticles, their number of spin orientations and their statistical properties (Fermi or Bose statistics). Note that the total weighting factor N_* is temperature-dependent: the higher the temperature, the more particles with $kT > mc^2$ are present to contribute to the density:

$$N_*(kT < 1 \text{ MeV}) \simeq 1 , \qquad N_*(kT = 10^{15} \text{ GeV}) \simeq 100 .$$

For $T = 10^9$ K one finds $\rho = 3.78 \times 10^{13}$ g cm^{-3}, which is nearly the density of nuclear matter. For this density, the expansion time of the universe is $t_{\exp} = 1.1 \times 10^{-4}$ s, and the time elapsed to reach this density is $t = 0.5 \times 10^{-4}$ s.

Because the mass difference between neutrons and protons is relatively small [$\Delta mc^2 = (m_n - m_p)c^2 = 1.293$ MeV], and because of the presence of energetic leptons (for $T = 10^{12}$ K, $E_l = 85$ MeV $\gg \Delta mc^2$), neutrons and protons are changing into each other via the reactions

$$\bar{v}_e + p \rightleftarrows e^+ + n; \qquad v_e + n \rightleftarrows e^- + p \,,$$

where in the conventional notation the antineutrino is denoted by $\bar{v}$. In this epoch, the normal neutron decay $n \rightarrow p + e^- + \bar{v}_e$ is negligible because of its long half-life of 10.6 minutes. Since $N_L/N_\gamma \ll 1$ and $Q/N_\gamma = 0$, there are as many neutrinos as antineutrinos and as many electrons as positrons. The neutron-to-proton abundance ratio in thermal equilibrium is given by the Boltzmann factor (Saha equation):

$$n/p \simeq \exp\left(-\Delta mc^2/kT\right) \,.$$

For $T = 10^{12}$ K we find $n/p \simeq 1$, i.e., protons and neutrons are present in almost equal amounts. Although the neutrons and protons interact to form deuterium via the reaction $p + n \rightarrow d + \gamma$, the low binding energy of 2.22 MeV for deuterium combined with the high density of high-energy photons leads to immediate photodisintegration of deuterium. Thus, the synthesis of heavier and more stable nuclei such as ^{4}He is blocked by the fragility of deuterium. At this stage there are about 10^9 photons per baryon, as there are today. Thus, nucleosynthesis becomes possible only when the energy of the photons falls below the deuterium binding energy.

When the temperature dropped to 10^{11} K, the density, although still high by present standards, was low enough ($\rho = 3.8 \times 10^9$ g cm^{-3}) that neutrinos and antineutrinos ceased interacting with other particles and with photons and fell out of equilibrium, or "decoupled." From that time on ($t_{\text{exp}} = 0.022$ s), virtually no "new" neutrino pairs were produced and no "old" pairs annihilated. The relic neutrinos and antineutrinos behaved as free particles expanding with the universe, their wavelengths increasing with the size of the universe. They ceased to play an active role, but continue to contribute to the density of the universe. They should, as with the microwave photons, still be around today, but at a somewhat lower temperature (about 2 K). There should be about as many relic neutrinos and antineutrinos in the neutrino background as there are photons in the microwave background radiation, that is, about 100 neutrinos of each kind per cubic centimeter. It is unfortunate that at present there appears to be no way of detecting them (chap. 10).

When the temperature of the universe reached 1×10^9 K ($kT = 0.086$ MeV), most of the photons no longer had enough energy to create electron-positron pairs. This disrupted the equilibrium between photons and pairs of particles, since, while positrons and electrons continued to combine to produce photons, they could no longer be replaced by the reverse reaction. All positrons were eventually consumed, leaving a small excess of electrons ($N_e/N_\gamma \simeq 10^{-9}$). As a consequence, the density of the universe is now characterized by a smaller effective weighting factor (Wei77) leading to $\rho = 1.45\rho_\gamma$, which yields $\rho(T = 10^9$ K$) = 12.2$ g cm^{-3}, giving $t_{\text{exp}} = 380$ s and $t = 190$ s. When the temperature dropped to 10^9 K, it became increasingly easier to convert neutrons into protons than to convert protons into neutrons, and the ratio n/p fell. Calculations of their relative abundances depend now on a precise knowledge of the transition rates in the weak-interaction processes (rates $\propto T^5 \propto 1/t^{5/2}$), since the rates become too small to sustain the n/p ratio at its equilibrium value and the abundance ratio n/p effectively "freezes out" of equilibrium at a certain temperature T_*. To a first approximation (Dol81),

$$(n/p)_* \simeq \exp\left(-\Delta mc^2/kT_*\right) = \exp\left(-15.01/T_{9*}\right),$$

where the freeze-out temperature is determined by the competition between the weak-interaction rate and the expansion rate of the universe. For an equality of both rates, the freeze-out temperature is found to be $T_{9*} = 7.5$ (in units of 10^9 K), i.e., the freeze-out occurs long before deuterium stops being photodisintegrated at $T_9 \simeq 1.0$. It follows that an increased expansion rate leads to a higher freeze-out temperature T_* and thus to a larger ratio $(n/p)_*$. The calculations must also include the normal neutron-decay process. One finds at $T_{9*} = 7.5$, abundances of 13% n and 87% p, and this ratio is fixed from there on. At $T_9 = 1.0$ the photons had a low enough energy to allow heavier nuclei to be built up without being immediately photodisintegrated (chap. 3). At this time the protons and neutrons underwent a series of nuclear reactions resulting predominantly in the formation of ^{4}He nuclei (sec. 2.2) with no free neutrons surviving. Since neutrons constitute one-half of the ^{4}He nuclei and since practically all neutrons (and a similar number of protons) were bound in ^{4}He nuclei, the fraction by weight of ^{4}He became simply twice the fraction of neutrons in the baryon matter, giving a ^{4}He abundance of about 26%. This predicted ^{4}He abundance agrees very well with the observed abundance throughout the universe (sec. 2.2). The universe at the time of ^{4}He formation was composed of excess protons and helium nuclei, which, together with the electrons, made up an ionized gas (plasma) in thermal equilibrium with the photon sea. The enigmatic neutrinos were also still around. The origin of the excess of electrons and protons over positrons and antiprotons is a fundamental question of cosmology (sec. 2.1.6).

The temperature of matter and radiation continued to decrease steadily as the universe expanded. In the time interval between 10^2 and 10^{13} s after the big bang, the density of high-energy photons dropped, and photons and particles no longer interacted as freely. As a result, the large-scale dynamics of the universe changed, and the temperature dropped a little faster than before (Fig. 2.1).

When the temperature dropped below 5000 K ($kT = 0.43$ eV, $\rho = 7.6 \times 10^{-21}$ g cm^{-3}, $t_{\exp} = 500{,}000$ y), the ions and electrons of the ionized gas combined, resulting in a neutral gas. Note that the combination occurred below a temperature corresponding to the ionization energy of hydrogen ($kT = 13.6$ eV), owing to the effects of the energy distribution of the gas (Planck distribution) and the population of excited states according to the Boltzmann factor (chap. 3). This period (shaded area in Fig. 2.1) is usually called the "moment of decoupling," after which there were essentially no longer any free charged particles to scatter the photons. At about this time, the universe switched from being radiation-dominated to being matter-dominated, and the temperature of the radiation dropped faster than before, namely, as $T \propto 1/t^{2/3}$ (Fig. 2.1). The neutral gas followed the ideal gas law and cooled with a time dependence of $T \propto 1/t$. We see, then, that the matter and radiation temperatures are radically different after decoupling and up to the present, since the reaction rates between matter and radiation are not rapid enough compared with the expansion time scale to maintain equilibrium distributions, and departures from equilibrium occur. We are today in such a nonequilibrium state. Departures from equilibrium are extremely important, in the epoch of big-bang nucleosynthesis (freezing of the neutron/proton ratio) as well as in other epochs of the early universe (sec. 2.1.6).

With the neutralization of charge the previously opaque universe became transparent, allowing radiation to travel unscattered through space, preserving an image of the plasma from which the photons were last scattered. The temperature of the plasma was about 4500 K, corresponding roughly to that of the surface of the sun. Originally emitted as visible and infrared radiation with a peak wavelength of about 0.7 μm, this background radiation has been redshifted by a factor of 1500 (Wei72). We now observe its peak wavelength at about 1 mm (Fig. 1.28). In a frame moving with the plasma at the time of decoupling, the characteristic temperature of the radiation is about 4500 K. In our frame of reference, it is 2.76 K. Originating some 5×10^5 years after the big bang, this background radiation represents a veil beyond which we cannot see farther back in time using electromagnetic radiation. However, there are other relics of earlier epochs such as the abundances of light nuclides (discussed in more detail in sec. 2.2), the densities of matter relative to photons, and the excess of matter over antimatter.

In the late 1940s Gamow, Alpher, and Herman (Gam48, Alp49) predicted the existence of a thermal background radiation, as discussed above, entirely on the basis of the big-bang model. Because the radiation was emitted by a hot source in thermal equilibrium, its intensity should vary with wavelength as that of an ideal thermal radiator, or blackbody. Gamow and coworkers estimated its present temperature to be about 5 K. They also predicted that the radiation surviving from the primeval fireball should be isotropic; that is, it should have the same intensity in every direction of the sky. The general agreement between predictions and later observations (chap. 1) is the most compelling evidence for the big-bang cosmology and eliminates a number of competing cosmologies (Wei72, Clo83). Actually, earlier observations in 1941 of the excitation of the first excited rotational band of interstellar CN molecules by McKellar (McK41) required the existence of such a background radiation. However, as is frequently true in such cases, McKellar and Gamow were not aware of each other's work.

With the expansion of the universe the tremendous energy density of radiation has been greatly reduced. It was not until this reduction occurred that matter could clump into the observed structures of the universe such as galaxies and stars (Jeans criterion, sec. 2.5.1). It is interesting to note that matter itself originated in the first few minutes of the universe (sec. 2.1.6).

2.1.4. Versions of the Big-Bang Universe

We have seen that in the beginning the universe was in a state of infinite density. This occurred at 1 Hubble time—16 ± 7 billion years ago (Fow85). It is then not meaningful to ask what happened before the big bang or where the big bang took place. The point universe (the cosmic fireball) was not an object isolated in space: it was the entire universe; the big bang happened everywhere. It was not simply an explosion of a clump of matter into an otherwise vast and empty space. There was no primordial clump of matter, no center to the explosion, and no outer edge to the distribution of matter. The big bang was not an explosion of matter within space but was an explosion of space itself. Since within this picture the dynamics and structure of the universe are largely beyond the scope of Newtonian physics, it is necessary to use Einstein's general theory of relativity (Ein16, Wei72). The general theory of relativity coupled with the cosmological principle produces a relativistic model of the universe where the presence of matter is manifested by uniform curvature of space (Wei72). This model predicts an expanding universe that started at some unique time in the past. The model is generally accepted and is referred to as the standard big-bang model of the universe (Wag73). There are three possible versions of the standard model, depending on the average density of the universe, which are discussed below.

In the standard model of the evolving universe, all matter was thrust apart at the moment of the big bang, eventually condensing to form galaxies (Fig. 2.1), which continued to move in free flight. In truly free flight they should continue in uniform motion forever. Actually, the galaxies continue to interact as they fly apart, via the gravitational force, which has a significant effect on the expansion. For objects leaving the surface of the earth, the escape velocity is determined by the mass of the earth. It is similar for a galaxy because, on the surface of an arbitrary sphere in space, the escape velocity is determined by the total mass within that sphere. Since the universe is assumed to be homogeneous, the relevant quantity can be shown to be the average density of matter in the universe. If the density is smaller than some critical value, the gravitational deceleration is too small to halt the cosmic expansion; and the galaxies will recede forever (although ever more slowly). This is the "open" universe (Fig. 2.2). If the density is greater than the critical density, gravitational deceleration will prevail. The expansion will slow to a stop, and contraction will begin—ending in a final cataclysm, which might be called the "big crunch." Such an universe is called "closed." If the density just equals the critical density (a marginally bound universe), we have the "flat" universe, in which the universe forever expands, but at an ever slower rate. Its space curvature is zero—hence the term "flat." It is a misnomer in a sense but has become

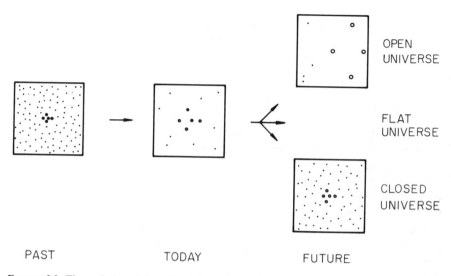

OPEN
UNIVERSE

FLAT
UNIVERSE

CLOSED
UNIVERSE

PAST   TODAY   FUTURE

FIGURE 2.2. The evolution of the universe is described through mathematical models of its behavior. In one class of models the universe starts with an infinite density and expands continuously and indefinitely, albeit at an ever slower rate. In the other class, the universe expands to a maximum size, then contracts, eventually again reaching infinite density. The alternatives, called open and closed universes, are illustrated here for an arbitrary volume of space (Got74). The borderline between the alternatives is called the flat universe.

the accepted "terminology." It does not mean "flat" in the sense of a "flat" earth. Note that small perturbations will cause the flat universe to behave either as open or as closed (sec. 2.1.7).

Using the expression $v^2 = 2GM/R$ for the escape velocity, $\rho = 3M/4\pi R^3$ for the average density, and $H_0 = v/R = 50$ km s^{-1} Mpc^{-1} from Hubble's law (chap. 1), the value of the critical density (Wei72, Dav76, Wei77, and sec. 2.1.2) is found to be

$$\rho_c = (3H_0^2/8\pi G) \simeq 5 \times 10^{-30} \text{ g cm}^{-3},$$

which is equivalent to about three hydrogen atoms per cubic meter. While that seems to be an exceedingly small density, it should be remembered that on the average the universe is quite empty (chap. 1). Estimates of the mass of a great many visible (luminous) galaxies indicate that ρ/ρ_c is approximately 0.04 and thus that the universe is open. However, there are many investigations in progress searching for the possible "missing" mass (secs. 1.2.4 and 1.2.5). This missing mass, which, if it exists, must be in the form of nonluminous matter (massive neutrinos or other invisible forms) and relativistic to a high degree (in order not to clump gravitationally; Fow85a), could be sufficient to close the universe. In this case the density of the invisible (dark) matter would have to exceed the density of the luminous matter by about a factor of 20. This possibility would carry the Copernican principle to its extreme: first the earth was not at the center of the solar system; then the sun was not at the center of the Galaxy; and finally the Galaxy was not at the center of the universe. Now it appears possible that the stuff we are made of may not even be the dominant kind of matter in the universe.

The origin of the universe was thus a kind of springboard from which the universe launched itself. It gave matter the composition that would serve later as an energy source for the stars. Rather like a clock, the universe started tightly wound, and like a clock it is running down as the galaxies fly apart and the nuclear evolutionary processes within the stars approach completion. If the universe is open, the run-down will be followed by a state of total stagnation. If the universe is closed, it will eventually fall back on itself, going out of existence just as mysteriously as it came into being. Does the world we are aware of constitute the whole universe? Some astronomers believe that the world of our experience sprang into existence some 16 billion years ago in a sudden transition from another existence out of another world. It is possible for this to happen if the universe is closed and if in the collapsing universe— before the density becomes infinite—an unknown mechanism caused the universe to "bounce" and begin expanding once again. In this case the universe is cyclic. Recent developments such as the inflationary universe are described in section 2.1.7.

2.1.5. The Beginning of the Standard Universe

We have seen that the observations of an expanding universe and of the 2.76 K background radiation imply a hot and dense state of the universe at earlier times. Another signature of this state is the nucleosynthesis of the light nuclides ^{2}H, ^{3}He, ^{4}He, and ^{7}Li taking place at temperatures around 10^9 K and times between 10^{-3} s and a few minutes after the big bang. The currently observed abundances of these elements are in such good agreement with the calculated values of primeval nucleosynthesis (secs. 2.1.3 and 2.2) that one is encouraged to consider the state of the universe at times still earlier than 10^{-3} s and at still higher temperatures and densities (Fig. 2.1). If we extrapolate the expansion of the universe backward, we find a time when the size of the universe is zero. Since this is the natural starting point for the universe, one calls it $t = 0$. Since the expansion rate at $t = 0$ is singular [in the standard model of a radiation-dominated universe $\dot{R}/R \equiv H(t) \propto \rho(t)^{1/2} \propto 1/R^2(t) \propto t^{-1}$], the universe starts with a bang. This singular point is, of course, a poor place to specify any parameters describing the initial state of the universe.

An object of mass M has a Schwarzschild radius $R_S = 2GM/c^2$, inside which the object becomes a black hole (sec. 2.4.13), a ghastly beast beyond the realm of known physical laws. If this radius is the same as the object's Compton wavelength $\hbar/Mc$ (a size where its quantum behavior becomes evident), we arrive at a mass (neglecting numerical factors)

$$M_{Pl} = (\hbar c/G)^{1/2} = 2.18 \times 10^{-5} \text{ g},$$

which is called the Planck mass. This mass corresponds to a Planck energy of $M_{Pl}c^2 = 1.22 \times 10^{19}$ GeV or an equivalent Planck temperature of $T_{Pl} = M_{Pl}c^2/k = 1.4 \times 10^{32}$ K. Using the uncertainty relation $\Delta E \Delta t \simeq \hbar$, one arrives at the corresponding Planck time,

$$t_{Pl} = (G\hbar/c^5)^{1/2} = 5.3 \times 10^{-44} \text{ s} ,$$

and the associated Planck length,

$$l_{Pl} = ct_{Pl} = (G\hbar/c^3)^{1/2} = 3.3 \times 10^{-33} \text{ cm} .$$

Note that these Planck units contain only the fundamental constants $\hbar$, G, and c (as Planck himself emphasized) and thus yield a starting point for a discussion of the very early universe. At times around and earlier than the Planck time, the space time metric must be quantized, and our understanding to the effects of quantum gravity is very uncertain.

In order to avoid these uncertainties, the discussions are usually restricted to energies of less than 1% of the Planck energy, that is, 10^{17} GeV. In the standard scenario the radiation-dominated universe cooled to a temperature of 10^{30} K (or an energy of 10^{17} GeV) just about 10^{-40} s after the initial singularity, the big bang, i.e., at this point in time the universe had persisted for a large number (about 10^3) of the fundamental time unit, the Planck time. This very hot and very dense epoch of the universe provided an environment in which all kinds of unusual particles were produced copiously and, through frequent interactions, quickly came into thermal equilibrium with one another. As discussed earlier (secs. 2.1.2 and 2.1.3), such a thermal equilibrium permits, in the cosmological context, an enormous simplification and a unified treatment of all particles present: the physical quantities of interest, such as the number density and energy density, effectively depend only on the temperature, and it is not necessary to investigate the detailed reactions that lead to equilibrium or to know the prior evolution of the universe. As the universe expanded and cooled during the subsequent evolution, the interaction rates began to lag behind the expansion rate, and at an early time thermal equilibrium could no longer be maintained. Such departures occurred frequently during the evolution of the universe. As a consequence, crucial differences developed among the various particle species as, one by one, they dropped out of equilibrium, leaving behind "relics" of those early epochs. How many relics survive and how they affect the subsequent evolution of the universe depends on masses, lifetimes, and interaction strengths of the constituent particles. Many of these particles may not have been discovered because their masses are too high to be produced by present accelerators, which are limited to energies below about 10^4 GeV (chap. 5). In the very early universe particle energies were not constrained by accelerator technologies (or budgets). Thus, Zel'dovich (Zel70) viewed the universe as "a hot laboratory for the nuclear and particle physicist," and other physicists called it the ultimate laboratory or "the poor man's accelerator."

Clearly, a description of the evolving universe in the time interval from about 10^{-40} s to about 10^{-3} s involves particle physics, and thus there is a natural interplay between particle physics and cosmology. The interdisciplinary work involving these two fields is sometimes referred to as "particle astrophysics." In order to take advantage of this natural interplay, one must find "footprints," observable today from the very early universe, such as stable relic particles. Fortunately, such footprints do exist, as discussed below.

2.1.6. Matter-Antimatter Asymmetry and the Origin of Baryons

Perhaps the most curious features of the present universe are the observations that it appears to be composed almost entirely of matter (antimatter/matter $\leq 10^{-6}$; Ste76) and that this matter-dominated universe is characterized by a remarkable ratio of 10^{-9} baryons per photon. These are the observable signals, the beacons from the early cauldron, that show what occurred in those hidden first milliseconds of the very early universe. In contemplating these features, some fascinating questions arise. Why does the universe prefer matter over antimatter (matter-antimatter asymmetry)? Where did the matter of the universe come from? And why is the proportion of matter to radiation characterized by the ratio $N_B/N_\gamma \simeq 10^{-9}$; that is, why is the number of baryons swamped by the number of photons? For many years there seemed to be no natural explanation for these cosmological puzzles. These footprints of the very early universe seemed to some physicists to be some kind of initial conditions of the big bang; they always have been characteristic features of the universe, and they were facts as inexplicable as the existence of the universe itself. For others this assumption appeared too arbitrary: if nature plays favorites in preferring one kind of substance over the other, there ought to be some reason, and thus these features could not merely be initial conditions (Kol83).

The very existence of antimatter—the principle that for every kind of particle in nature there is an antiparticle of equal and opposite charge and identical mass—arises from fundamental symmetries within relativity and quantum theory, as shown by Dirac in 1928. Accelerator physics now verifies this concept every day: particle-antiparticle pairs are produced copiously every time an energetic beam interacts with a target. Indeed, it has long been assumed that the laws of nature express no preference for matter over antimatter. From observations of particle reactions physicists also concluded that the number of baryons in a system minus the number of antibaryons was a conserved quantity (sec. 2.1.2) and thus was absolutely fixed for all time (baryon conservation law; Wil80, Clo83). No interactions or transformations of the particles could ever change this quantity. If this were true for the universe as a whole, the preponderance of matter over antimatter in the universe today (its positive baryon number density) always existed and must have been built into the initial structure of the big bang.

Paradoxically, despite the vast quantities of matter in the universe and the virtual absence of antimatter, the violation of the matter-antimatter symmetry is quite tiny, about 10^{-9}. To see how astrophysicists arrive at this number, let us look again at the evolution of the very early universe. According to the standard model of the big bang, the very early universe was hot enough for particles, antiparticles, and photons to be in thermal equilibrium, and their enormous number densities (abundances) were nearly equal. No sooner would a baryon annihilate an antibaryon than an identical pair would be regenerated in another reaction, i.e., matter and antimatter were continuously created and destroyed. But these reactions could not be quite in balance. The arguments for a tiny imbalance are as follows: to a first approximation the number of photons in the universe has remained constant since the very early universe. Their density has simply fallen with cosmic expansion by the same factor as has the density of matter. We see these photons today as the 2.76 K microwave background radiation with about 400 photons cm^{-3}. Counting photons now is nearly the same thing as counting baryons and antibaryons then. Note that the baryon number of a system represents the difference in number densities of baryons and antibaryons. It follows that this baryon number divided by the number of photons, N_B/N_γ, is a constant, independent of time or rate of expanion of the universe. The numerical value of that ratio observed today, $N_B/N_\gamma \simeq 10^{-9}$, therefore represents the fractional discrepancy between matter and antimatter just after the big bang: for every billion antibaryons there were one billion and one baryons.

As the temperature of the universe fell below 10^{13} K (threshold temperature for nucleons) some 10^{-6} s after the big bang, the rate of pair production began to fall behind the rate of annihilation. Baryons and antibaryons started consuming one another; very quickly, all that remained was a hurricane of photons and that one leftover baryon out of one billion. It was this tiny remnant of matter that one day would give rise to the earth and all the planets, stars and galaxies, that is, our present universe. It is this thin sliver of asymmetry that astrophysicists have sought so long to explain. If the baryon number is absolutely conserved and if the universe started with a perfect baryon-antibaryon symmetry, then it would remain symmetrical at all times, contrary to the remarkable absence of the antimatter anywhere in the present universe. Because of departures from thermal equilibrium near freeze-out temperature (sec. 2.1.2), the baryon-antibaryon annihilation would today be not complete but nearly so, with a baryon-to-photon ratio of only about 10^{-18} (Smo77, Ste85), which is about 9 orders of magnitude smaller than that observed. In this case there would not have been sufficient matter in the cosmos for the formation of galaxies, stars, and planets. Obviously, these assumptions do not describe our universe. Since the matter-antimatter asymmetry already existed at about 10^{-6} s, its source must clearly be sought at much earlier times. It is to these early epochs of the universe that new theories of particle physics have shed light on these cosmological puzzles.

Enormous progress has been made over the last decade in particle physics, and the following picture has emerged (Wei80, Gla80, Gol80, Wil80, Clo83).

All of the hadrons (baryons and mesons) are built from more fundamental pointlike constituents, the "quarks." For example, three (two) quarks are needed to make a baryon (meson). The leptons are themselves fundamental pointlike particles. Renormalizable "local" or "gauge" theories describe the interactions of these fundamental particles, the quarks and leptons. In these theories there are also gauge particles (vector bosons), the photon being the most familiar, which mediate particle interactions, and Higgs particles (spinless X bosons), which have not been observed but are postulated to generate masses for the other particles. The simplest gauge theory, called quantum electrodynamics, or QED, describes the electromagnetic interactions. The Glashow-Weinberg-Salam gauge theory has unified the weak and electromagnetic interactions (called electroweak force) and is in good agreement with the experiments done to date. The color gauge theory, called quantum chromodynamics, or QCD, has been very successful thus far in describing the strong interactions. Spurred by the success of these gauge theories, "grand unified theories" (GUTs for short) have now been proposed to unify strong, weak, and electromagnetic interactions. Supersymmetric theories (super-GUTs and various "string" theories), which attempt to unify all the interactions including gravity, are also being investigated.

The evolution of such gauge theories toward more and more unification of the forces of nature is not arbitrary. At the relatively low energies accessible in terrestrial laboratories, the forces appear quite different. The nuclear force is strong, but of such short range (about 10^{-13} cm) that its effects are feeble at atomic dimensions. If we were unable to probe distances of 10^{-13} cm, we would be unaware of this force. However, such distances can be probed by collisions of particles at energies of several tens of MeV. According to the electroweak gauge theory, the weak force is as strong as the electromagnetic force but is of extremely short range. To see its full strength, one must probe distances of the order of 10^{-17} cm requiring particle energies of about 100 GeV and higher. Such particle energies have become available in recent years, and the results of experiments at these energies could be successfully explained by the electroweak gauge theory. It turns out that under these conditions weak effects are as probable as electromagnetic effects, while when studied from afar they appear to be more feeble. This perspective raised the question of whether there are powerful forces at work on very short distance scales or, equivalently, at very high energies, far beyond the reach of present accelerator technologies, and whose effects are thereby hidden from us. The grand unified theories suggest that this is indeed the case: at energies of about 10^{15} GeV (or distances of about 10^{-29} cm) the weak, electromagnetic, and strong forces become equally strong, and thus these fundamental forces of nature coalesce and can be described as different manifestations of one underlying force. This unification is hidden, and separate identities have developed at the cold low energies available on earth. We are not likely ever to be able to create energies of 10^{15} GeV in the terrestrial laboratory and to see the full glory of unity of the forces. However, in the very early universe such energies were available, and collisions at energies of 10^{15} GeV were abundant in that epoch. It appears, then, according to the GUT, that in the high temperatures of the early big bang there was a profound unity among the natural forces, and this innate symmetry possessed by the universe was obscured ("broken") as it cooled and expanded to the present day. The influence of temperature on "symmetry breaking" is familiar in many contexts. One is the transition from liquids to crystalline solids. At high temperatures liquids are isotropic, all directions being equivalent. As the temperature drops, the liquid solidifies and may form crystals. These are not isotropic; the original full symmetry has been lost even though well-defined symmetries, less than total symmetry, remain. Magnets yield a second example: at high temperatures the atomic spins are isotropically randomly ordered and no macroscopic magnetism is observed. At lower temperatures, north and south poles occur.

All the grand unified theories have a common feature, namely, that baryon and lepton numbers are no longer absolutely conserved. The necessity of baryon and lepton nonconservation in GUTs is qualitatively easy to understand. The unification of the fundamental forces in GUTs is achieved in part by treating quarks and leptons as equals and allowing transformations from one to the other, i.e., a quark into a lepton, and so on. In a wide class of these theories the difference of the baryon and lepton numbers, $N_B - N_L$, is conserved, so that each change of baryon number is accompanied by an equal change of lepton number. These baryon (lepton) nonconserving transformations are believed to arise from interactions mediated by gauge particles, such as the superheavy X bosons with masses of the order of 10^{15} GeV. For example, in the familiar world of strong, weak, and electromagnetic forces, a quark—no matter how much it may be knocked around otherwise—always remains a quark; let it interact with a X boson, however, and it becomes an antiquark or even a lepton. It is this feature of GUTs which may resolve the cosmic puzzles of the excess of matter over antimatter in the universe and of the remarkable ratio $N_B/N_\gamma \simeq 10^{-9}$. Clearly, this amplifies the natural interconnections of cosmology and particle physics, the study of the very large and the very small.

Many different scenarios have been suggested for the actual details of baryosynthesis (Dim78, Wei79, Tou79, Ell79, Lak84). One popular model begins with a baryon-symmetrical universe. It is noted that the processes that violate baryon conservation were enormously accelerated at the high temperatures and densities of the very early universe. Even if the universe had started out with some kind of quark-antiquark imbalance, the copiously produced X bosons would quickly have brought the constituents into symmetry. Before 10^{-35} s, the universe was in thermal equilibrium: the density of X bosons equaled the density of anti-X ($\bar{X}$) bosons, quark density equaled antiquark density, and lepton density equaled antilepton density. The net baryon (lepton) number of the universe was zero ($N_B = N_L = 0$). As the universe expanded and cooled, the production of very massive constituents became disfavored and soon ceased totally. When the temperature reached a level where X and $\bar{X}$ bosons were decaying faster than they were regenerated ($E \simeq 10^{14}$ GeV, $t \simeq 10^{-33}$ s), thermal equilibrium was no longer possible, and the processes controlled by the X bosons started to freeze out. Any X and $\bar{X}$ bosons that had not mutually annihilated decayed into pairs of quarks or quarks and leptons. Suppose that the X boson decays into a quark pair (qq) a fraction a of the time and into antiquark and antilepton ($\bar{q}\bar{l}$) a fraction $1 - a$ of the time:

$$X \rightarrow a(qq) + (1 - a)(\bar{q}\bar{l}) \,.$$

Correspondingly, the $\bar{X}$ boson decays into an antiquark pair ($\bar{q}\bar{q}$) and a quark-lepton pair (ql) the fractions b and $1 - b$ of the time, respectively:

$$\bar{X} \rightarrow b(\bar{q}\bar{q}) + (1 - b)(ql) \,.$$

The decay of the X and $\bar{X}$ bosons violates both baryon (i.e., quark) and lepton number conservation; however, if the rates of their decays are equal ($a = b$), the number of quarks (N_q) equals the number of antiquarks ($N_{\bar{q}}$), and the number of leptons (N_l) equals the number of antileptons ($N_{\bar{l}}$). In this case there is matter-antimatter symmetry with a zero net baryon and lepton number ($N_B = N_L = 0$), and, in spite of baryon nonconservation, the constituents would have stayed symmetrical. There is no fundamental requirement in GUTs that $a = b$. Suppose that the fraction of X bosons that decay into quark pairs is greater than the fraction of $\bar{X}$ bosons that decay into antiquarks: for $a > b$ it follows that $N_q > N_{\bar{q}}$ and $N_l > N_{\bar{l}}$. Even though we have an equal number of X and $\bar{X}$ bosons to begin with, we can end up in this case with an excess of quarks over antiquarks and of leptons over antileptons and thus an excess of matter over antimatter.

Not only can this symmetry breaking in the decay of the X bosons happen in principle, but there is evidence that it does indeed happen in other systems. In 1964 a Princeton group (Chr64) discovered that such matter-antimatter asymmetry does occur, at least in the decay of the K meson. Starting with an equal mixture of K^0 and $\bar{K}^0$ mesons, their decays

$$K^0 \rightarrow e^+ + \pi^- + \nu_e \,,$$

$$\bar{K}^0 \rightarrow e^- + \pi^+ + \bar{\nu}_e$$

are matter-antimatter correspondents ($e^+ \equiv \bar{e}^-$, $\pi^- \equiv \bar{\pi}^+$). Yet the two sets of products are not equally produced: the K^0 decay is about 0.7% more frequent than the $\bar{K}^0$ decay. This asymmetry between matter and antimatter leads to the possibility that an analogous asymmetry exists for an equal mixture of X and $\bar{X}$. Indeed, most of the GUTs allow X and $\bar{X}$ bosons to decay in slightly different patterns, and in the end, when all X bosons had disappeared, quarks and leptons would outnumber antiquarks and antileptons by some small margin, leading to a small excess of matter over antimatter. Later, all the antiquarks were annihilated with quarks and all the antileptons were annihilated with leptons, winding up in a high flux of photons, with $N_\gamma \simeq N_{\bar{q}} + N_{\bar{l}}$. The excess quarks and leptons that remained formed our universe. The excess quarks coalesced into baryons (such as protons and neutrons) a few instants later ($N_B \propto N_q - N_{\bar{q}}$, $N_L \propto N_l - N_{\bar{l}}$). Thus, most of the antimatter did not survive the first millionths of a second. In this simplified model the ratio of baryons to photons is given by

$$N_B/N_\gamma \simeq (a - b)N_x ,$$

where N_x represents the relative abundance of X bosons at their freeze-out temperature. In the grand unified schemes $N_x \simeq 1\%$, and since the observed ratio N_B/N_γ is 10^{-9}, the GUTs must place the difference in the decay rates $(a - b)$ in the vicinity of 10^{-7}. We see, then, the dropping out of the X bosons plus a tiny asymmetry in their decay patterns as one of the possible scenarios giving rise to one left over baryon out of a billion and hence to all the matter of the observable universe; the present matter is only a small remnant (relic) of what was originally produced.

Despite the undeniable elegance and naturalness of GUTs in explaining the long-standing cosmological puzzles and in accounting quantitatively for baryosynthesis, these theories are still not complete and remain unproved. Some of the basic ideas describing GUTs may be tested in terrestrial laboratories. Processes common in the hot big bang are not entirely absent under cooler conditions but are merely slowed down and are thus rare in occurrence. One prediction of GUTs is that the proton is unstable. If an excess of protons can result from $X\bar{X}$ decay, i.e., protons can be "created," then protons can "decay." The proton, which is thought to be a bound system of three quarks, could decay if a constituent quark underwent a quark-lepton transformation as mentioned above. Most GUTs predict the half-life of the proton to be around 10^{29}–10^{33} y, which is just at the edge of detectability. Several experimental investigations are currently underway to search for the predicted decay (Gol80, Wei81, Per84, Par85a), already yielding a current limit of $\geq 10^{32}$ y. It is interesting to note that only some fifty years ago baryon number conservation was invented to exclude proton decay (Wey29, Stu39, Wig49). Another prediction of GUTs is a finite electric dipole moment for the neutron of around 3×10^{-28} (e cm). This has a direct bearing on baryosynthesis in the very early universe. The current upper limit is 1.6×10^{-24} (e cm), and it might be at least a decade before the electric dipole experiments are in the predicted range (Ell81). Both experiments will eventually either provide confirmation of the scheme sketched here or suggest new theories. Such developments of GUTs will presumably determine whether processes which generate a baryon and lepton asymmetry become a priori inputs of big-bang cosmologies. As knowledge of the fundamental particles and their interactions increases, and as the determination of cosmological observables improves (or new observables are discovered), the close relationship between cosmology and particle physics promises to continue to be exciting and fruitful.

2.1.7. The Inflationary Universe

The standard big-bang cosmology has been very successful in explaining many features of the evolution of the early universe: the universal expansion, the microwave background radiation, and the primordial nucleosynthesis. All these characteristics of the universe are related to events that presumably took place some seconds or minutes after the big bang. Recent developments in particle physics, in particular the development of GUTs, make it possible to describe the universe during its first small fraction of a second. Armed with GUTs, cosmologists are now attempting to understand the history of the universe back to 10^{-40} s after time zero. Such attempts already have shed light on the mystery of baryosynthesis (sec. 2.1.6). However, when the standard big-bang model is extended to these earlier times, some problems arise. One set of problems has to do with the special conditions imposed on the model as the universe emerged from the big bang (Dic79, Gut81).

One of these problems is the difficulty in explaining the large-scale uniformity of the observed universe. The large-scale uniformity is most evident in the microwave background radiation, which is known to be uniform (isotropic) in space and temperature to better than 0·003%. In order to understand the origin of this problem, recall that in the singular expansion of the standard model the early universe is radiation-dominated and expands adiabatically according to the relations (secs. 2.1.2 and 2.1.3):

$$\rho = N_* \rho_\gamma = 8.40 \times 10^{-36} N_* T^4 \text{ g cm}^{-3} ,$$

$$t = (3/32\pi G\rho)^{1/2} = 2.3 \times 10^{20}/(N_*^{1/2} T^2) \text{ s} .$$

For example, in the very early universe with $kT \simeq 10^{14}$ GeV and $N_* \simeq 100$, we find $t \simeq 10^{-35}$ s, and at the moment of decoupling, with $kT \simeq 1$ eV and $N_* \simeq 1, t \simeq 10^{12}$ s. With $RT = $ constant, we find

$$R(t) \propto t^{1/2} ,$$

and the scale factor $R(t)$, and thus the physical "size" of the universe, increases with the square root of time:

$$R(t_1)/R(t_2) = (t_1/t_2)^{1/2} .$$

The most distant objects observed lie at a distance of roughly 10^{10} ly. If we take this distance as the nominal size of the present universe, its size in the very early universe $(t_1 = 10^{-35}$ s) has decreased to the value $R(t_1 = 10^{-35}$ s$) \simeq 10$ cm. At the moment of decoupling, the size is $R(t_1 = 10^{12}$ s$) \simeq 2 \times 10^{25}$ cm [matter-dominated universe with $R(t) \propto t^{2/3}$]. Since no information or physical process can propagate faster than a light signal, at any given time t there is a maximum distance $L_h \simeq ct$, known as the horizon distance, that a light signal could have traveled since the time the big bang started (Wei72, Gut81). Consider two points with a spatial separation of L_h near the beginning of time. Prior to the time it takes for light to travel between them, events at either of the two points will not affect events at the other. Thus, regions of the universe with a spatial separation greater than L_h (at the time radiation decoupled from matter) will not know of each other's existence until a time t has elapsed. The particle horizon corresponds to that region of space wherein it is possible to have seen a particle. Note that once seen, always seen. For example, at the moment of decoupling we have $L_h \simeq 3 \times 10^{22}$ cm and $R \simeq 2 \times 10^{25}$ cm, and thus, in the standard model, the sources of the microwave background radiation observed from opposite directions in the sky were separated from each other by about 670 times the horizon distance when the radiation was emitted. The corresponding ratio of physical volume to horizon volume is $(2R/L_h)^3 \simeq 2 \times 10^9$, i.e., there are about 2 billion causally disconnected regions in the universe at this time. Since these regions could not have communicated and causally influenced each other at any time in the past, it is difficult to see how they could have evolved conditions so nearly identical. The puzzle of explaining why the universe appears to be so homogeneous and isotropic over distances that are large compared with the horizon distance is known as the "horizon problem" (Rin56). At the very early universe we have $R \simeq 10$ cm and $L_h \simeq 3 \times 10^{-25}$ cm, and thus there are $(2R/L_h)^3 \simeq 10^{78}$ causally disconnected regions. The problem arises because, in approaching the beginning of the universe ($t \rightarrow 0$), the physical size of the radiation-dominated universe decreases much slower than the horizon length, since $R \propto t^{1/2}$ and $L_h \propto t$. Hence, the horizon encloses smaller and smaller parts of the universe the farther back we look. Although starting from an initially small volume, the region of the universe in causal contact is initially zero and grows as the rate of expansion decreases and time elapses for signals to propagate. As a consequence, particles say always "hello" and never "farewell" in a radiation-matter universe with particle horizons. The horizon problem is not a genuine inconsistency of the standard model; if the large-scale uniformity is assumed as an initial condition of the standard model, the universe will evolve uniformly. Thus, the problem is that one of the most salient features of the observed universe—its large-scale isotropy and homogeneity—cannot be explained by the standard model; it must be assumed as an initial condition.

Even with the assumption of large-scale uniformity, the standard big-bang model requires yet another assumption to explain the nonuniformity observed on smaller scales. On scales less than several hundred million light-years the universe is clearly inhomogeneous: there is the clumping of matter into galaxies, clusters of galaxies, and superclusters of clusters (chap. 1) in an otherwise homogeneous and isotropic universe, as indicated by the universal blackbody radiation. The existence of these clumps of matter implies (Pee80) the presence of small deviations from homogeneity and isotropy of the matter in the early universe, i.e., a spectrum of primordial matter inhomogeneities must be assumed as part of the initial conditions. In the gravitational instability theory of galaxy formation (sec. 2.3), primordial density fluctuations begin to grow by their self-gravitational attraction after the matter and radiation decouple. Density fluctuations of the order of $\Delta\rho/\rho \simeq 10^{-3}$ on a mass scale of the order of $10^{12}\ M_\odot$ are required at the decoupling time to ensure that the present structure "grows up" via the Jeans instability criterion (sec. 2.5.1). The spatial distribution and mass spectrum of galaxies today reflect the nature of the primordial fluctuations. The fact that the spectrum of inhomogeneities has no explanation is a drawback in itself, but the problem becomes even more pronounced when the model is extended back to 10^{-40} s after the big bang. The incipient clumps of matter develop rapidly with time as a result of their gravitational self-attraction, and so a model that begins at a very early time must begin with very small inhomogeneities. To begin at 10^{-40} s the matter must start in a peculiar state of extraordinary but not quite perfect uniformity ($\Delta\rho/\rho \simeq 10^{-12}$). A normal gas in thermal equilibrium would be far too inhomogeneous, owing to the random motion of particles. This peculiarity of the initial state of matter required by the standard model is called the "smoothness problem." The puzzle is why the fluctuations, though there, are so very small.

Another subtle problem of the standard model concerns the energy (mass) density of the universe. We have seen in section 2.1.2 that the total energy E of a typical galaxy in terms of its potential and kinetic energies is given by

$$E = mR^2(t)[\tfrac{1}{2}H^2(t) - \tfrac{4}{3}\pi\rho(t)G]\ ,$$

which with the substitution $2E/m = -k$ leads to an Einstein field equation with a cosmological constant $\Lambda = 0$ (known as the Friedmann model):

$$H^2(t) = (8\pi/3)\rho(t)G - k/R^2(t)\ ,$$

where the term $k/R^2(t)$ is called the curvature term, with the curvature index k having values of 0, ± 1. In the case $E = 0$ ($k = 0$), we arrive at the flat Einstein–de Sitter universe with

$$\rho_c = (3/8\pi G)H^2(t)\ ,$$

i.e., this universe is expanding forever but at an ever slower rate asymptotically approaching zero [$\rho(t) = \rho_c$]. Note that for any given value of the Hubble "constant," there is a critical density ρ_c which gives a precisely flat universe. The closed universe (positive curvature, or $k = +1$) has a negative total energy E, which implies that the universe is "bound" and will eventually collapse [$\rho(t) > \rho_c$]. Finally, the open universe (negative curvature or $k = -1$) has positive energy E, which implies that the universe is "free" and will expand forever [$\rho(t) < \rho_c$]. Today one can state conservatively that $\rho/\rho_c = \Omega$ lies in the range

$$0.1 \leq \Omega_{now} \leq 2$$

(although more and more astronomers say $\Omega_{now} \leq 0.2$). No one is surprised by how narrowly this range brackets $\Omega = 1$ (i.e., a flat universe). The quantity Ω is time-dependent, with

$$\Omega(t) = 1 + k/[R^2(t)H^2(t)] ,$$

which leads at the very early universe (radiation-dominated universe at time t_i with Ω_i) to the relation[1]

$$\Omega(t) = 1 + (\Omega_i - 1)(t/t_i) .$$

In order for $\Omega(t)$ to lie in the above range at the present time ($\Omega_{now} \simeq 1$ at $t \simeq 10^{10}$ y), the Ω_i value at some time in the past, say at $t_i = 10^{-35}$ s, must also have been very near to 1, and it had to be extremely fine-tuned:

$$\Omega_i = 1 \pm t_i/t \simeq 1 \pm 10^{-53} .$$

We see then that a value near $\Omega = 1$ (flat universe) represents in the standard model a state of unstable equilibrium. If Ω was ever exactly equal to 1 (case $k = 0$ or $\Omega_i = 1$), it would remain exactly equal to 1 forever. If Ω differed slightly from 1 an instant after the big bang (i.e., much more than 10^{-53}), however, the deviation from 1 would grow rapidly with time. For irregularities leading to $\Omega_i = 1 + \xi$ ($\xi \ll 1$), the universe would have very quickly recollapsed, forming black holes (in a few Planck times). For $\Omega_i = 1 - \xi$ ($k = -1$) the universe becomes curvature-dominated: $H(t) \simeq 1/R(t)$, i.e., $1/T(t) \propto R(t) \propto t$, and regions of space expand so rapidly as to now appear empty and devoid of structure. Clearly, this does not describe the universe around us. If the universe had become curvature-dominated at the Planck time (10^{-43} s), today with $T = 2.76$ K it should be only about 10^{-11} s old. This paradox is called the "oldness problem." Given the $\Omega = 1$ instability, it is surprising that Ω is measured today as being between 0.1 and 2. In order for Ω to be in this rather narrow range today, its value a second after the big bang had to equal 1 to within one part in 10^{17}. The standard model offers no explanation of why Ω began so extremely close to 1 but simply assumes this fact as an initial condition. This shortcoming of the standard model is called the "flatness problem" (Dic79).

To sum up, although the standard big-bang model has proved to be a remarkably simple and reliable framework for understanding the evolution of the universe, the same model leads to several cosmological conundrums by requiring that very special conditions be imposed on the very early universe.

In order to solve some of these cosmological puzzles, a new model of the very early universe has recently been invented, known as the "inflationary universe" (Bro78, Bro79a, Gut81, Gut84, Ste84, Lak84). It coincides with the standard big-bang model for all times after the first 10^{-30} s. For this first

---

[1] The following arguments for $\Omega = 1$ at early times have been questioned (Fow85, Fow85a), since $\rho \propto R^{-3}(t)$, and thus $\Omega(t)$, asymptotically approach unity by definition for $R(t) \to 0$.

fraction of a second, however, the story is dramatically different. According to the inflationary model, the universe evolves after the Planck time in the usual way; however, as the universe expands and cools, there is a brief period thereafter ($\Delta t = t_B - t_A$) during which the energy density of the universe remains nearly unchanged. Owing to this constant energy density ρ_{vac} (being related to a finite cosmological constant $\Lambda \neq 0$), the Einstein field equation becomes qualitatively different:

$$(\dot{R}/R)^2 = H^2(t) = (8\pi/3)G\rho_{vac} = H_*^2 = \text{constant} ,$$

where we have omitted the curvature term and the ρ term because they are negligible at very early times. The time evolution of the scale factor $R(t)$ is then given by

$$R(t) = R(t_A) \exp\left[H_*(t - t_A)\right] ,$$

i.e., the size of the universe grows exponentially with time. With the relation $R(t) \propto 1/T(t)$ it also follows that the temperature decreases exponentially with time:

$$T(t) = T(t_A) \exp\left[-H_*(t - t_A)\right] .$$

Thus, in this brief period Δt the universe had an extraordinarily rapid expansion ("inflation") and a tremendous drop in temperature ("supercooling"). These features represent the key elements of the inflationary model of the universe and differ drastically from the assumption in the standard model that the universe has been adiabatically expanding since the big bang with $R(t) \propto t^{1/2}$ during the radiation-dominated period and $R(t) \propto t^{2/3}$ during the matter-dominated period, both of which lead to a "weak" expansion.

Before the inflation begins, the physical size of the universe is smaller than the horizon distance:

$$\text{Physical size} \leq \text{horizon distance} \simeq 10^{-25} \text{ cm} ,$$

and the universe had time to homogenize and reach thermal equilibrium. Thus, in the inflationary model the observed universe evolves from a region that is much smaller in size (by a factor of 10^{26} or more) than the corresponding region in the standard model. The small homogeneous region is then inflated during the inflationary era by the accelerated expansion to become large enough to encompass the entire volume of the observable universe now seen by us:

$$\text{Physical size} \simeq 10 \text{ cm} = \exp\left(H_* \Delta t\right) \times \text{horizon distance} .$$

This evolution requires the condition

$$\exp\left(H_* \Delta t\right) \geq 10^{26} \qquad \text{or} \qquad H_* \Delta t \geq 60$$

and poses a restriction on the minimum duration Δt of the inflationary era. It is assumed that after this colossal expansion the universe continues to expand

and cool as described by the standard big-bang model. If the above condition can be met by the inflationary model, the horizon problem is avoided in a straightforward way. For example, the sources of the microwave background radiation arriving today from all directions in the sky were once in close contact; they had time to reach a common temperature before the inflationary era began. Thus, this time when they meet, they are not strangers, having met before the inflation. Whereas signals say "hello" and never "farewell" in the standard big-bang model, there are both "hellos" and "farewells" in the inflationary universe.

The flatness and oldness problems are also evaded in a simple and natural way. At the beginning and at the end of the inflationary era we have

$$\Omega(t_A) = 1 + k/R^2(t_A)H^2(t_A) \,,$$

$$\Omega(t_B) = 1 + k/R^2(t_b)H^2(t_B) \,.$$

With $R(t_B) = R(t_A) \exp(H_* \Delta t)$ and $H(t_B) = H(t_A) = $ constant, we find

$$R^2(t_A)H^2(t_A)/R^2(t_B)H^2(t_B) = \exp(-2H_* \Delta t) \leq \exp(-120) = 10^{-52}$$

for $H_* \Delta t \geq 60$. Thus, if $\Omega(t_A)$ is of the order of unity, at the end of the inflationary era we have

$$\Omega(t_B) = 1 \pm 10^{-52} \cong \Omega_i \,,$$

i.e., the value of Ω is driven rapidly toward 1 in the inflationary era no matter what value it had before, and there is no need for fine-tuning. This behavior is most easily understood by recalling that a value of $\Omega = 1$ corresponds to a space that is geometrically flat ($k = 0$). The rapid inflationary expansion causes the space to become flatter just as the surface of a balloon becomes flatter the more it is inflated. The mechanism driving Ω toward 1 is so effective that the inflationary model predicts the value of Ω today to be almost exactly equal to 1 (see, however, Fow85). Thus, a reliable determination of Ω would provide a crucial test of the inflationary model.

Of course, in contemplating the inflationary model there arise a number of questions: what drives the exponential expansion of the universe during the inflationary era? How does the universe gracefully end this inflation and return to the standard big-bang universe? Can the scheme of the inflationary universe be realized from the microphysics, e.g., of GUTs? It turns out that the inflationary scenario is allowed by GUTs and is even quite plausible, but it is not required by GUTs. Whether inflation occurs or not depends sensitively on undetermined parameters in GUTs (Gut81, Lin82, Alb82, Alb82a, Gut84). One of the essential ingredients in applying GUTs to the very early universe is the phenomenon of spontaneous symmetry breaking (And72). In GUTs symmetry breaking is accomplished by including in the formulation of the theory a special set of postulated fields known as the "Higgs fields," which are described by an order parameter ϕ. The symmetry is unbroken when all the

Higgs fields have a value of zero ($\phi = 0$), but it is spontaneously broken whenever at least one of the Higgs fields acquires a nonzero value. Interactions lead to local alignments, and the correlations cause a lowering of the total energy. It is thus possible to formulate the theory in such a way that a Higgs field has a nonzero value in the state of lowest energy density, which in this context is known as the "true vacuum." At high temperatures, thermal fluctuations drive the equilibrium value of the Higgs fields to zero, resulting in a transition to the symmetric phase. While the true vacuum energy density is known to be small today ($\rho_{vac} < 10^{-5}$ GeV cm^{-3}; Dol81), the inflationary scenarios assume that the unbroken vacuum of the universe had an enormous energy density,

$$\rho_{vac} = bT_c^4 \simeq 10^{103} \text{ GeV cm}^{-3} ,$$

at $kT_c \simeq 10^{15}$ GeV with $T_c \simeq 10^{28}$ K.

The inflationary model assumes as initial conditions that the very early universe included at least some regions that were extremely hot and also expanding. In such a hot region the Higgs fields would have a value of zero, and the total energy density is given by

$$\rho_{tot} = \rho_{rad} + \rho_{vac} = aT^4 + bT_c^4 ,$$

where ρ_{rad} includes matter and radiation and ρ_{vac} is connected with the "old" cosmological constant $\Lambda \neq 0$ (Wei72). As the region continued to expand and cool, it would approach a peculiar state of matter known as the "false vacuum." This state of matter has never been observed, but it has properties that are predicted by quantum field theories. The temperature, and hence the thermal component of the energy density, would rapidly decrease (supercooling at $T < T_c$ with $\rho_{tot} \simeq \rho_{vac} =$ constant), and the energy density of the region would be concentrated entirely in the Higgs fields. A zero value for the Higgs fields implies a large energy density for the false vacuum. In the classical form of the theory such a state would be absolutely stable, even though it would not be the state of lowest energy density. States with a lower energy density would be separated from the false vacuum by an intervening energy barrier, and there would be no energy available to take the Higgs fields over the barrier. Thus, the universe would be trapped in the false vacuum for $T \ll T_c$. However, in the quantum version of the model, the false vacuum is not absolutely stable. Under the rules of quantum theory all the fields would be continually fluctuating, and such fluctuations would occasionally cause the Higgs fields in small regions of space to "tunnel" through the energy barrier. These Higgs fields will acquire nonzero values, bringing the system to its broken-symmetry phase. They will form "bubbles," which will then start to grow into the surrounding region of the false vacuum. The growth is favored energetically because the true vacuum has a lower energy density than the false vacuum.

The most peculiar property of the false vacuum is probably its pressure, which is both large and negative and which leads to gravitational effects.

Under ordinary circumstances the expansion of any region of the universe would be slowed by the gravitational attraction of the matter within it. In Newtonian physics this attraction is proportional to the mass energy density. According to general relativity, the pressure also contributes to the attraction. For the false vacuum the contribution made by the pressure would overwhelm the energy density contribution and would have the opposite sign. Hence, the bizarre notion of negative pressure leads to the equally bizarre effect of a gravitational force that is effectively repulsive. As a result, the expansion of the region is accelerated, leading to an exponential growth as discussed earlier. Thus, there is an epoch of inflationary expansion (some 10^{-37} s after the big bang) triggered by the energy of the false vacuum, which is stored in the Higgs fields. It is assumed in the inflationary model that the colossal expansion (inflation) occurs for a time period Δt, while the universe is in the false-vacuum state. After this period the universe falls into the true vacuum, i.e., the transition to the broken-symmetry phase will be finally completed. The energy density of the false vacuum is then released (analogous to the energy released when water freezes), resulting in a tremendous amount of particle production, and the region is reheated to a temperature of almost 10^{27} K. The region becomes a hot gas of elementary particles in thermal equilibrium, just as was assumed in the initial conditions for the standard model. From this point on, the scenario coincides with that of the standard big-bang model, and so all the successes of the standard model are retained. However, the new cosmology severely alters our comprehension of the size of the universe: our observable universe may be only an expanding bubble embedded in an otherwise steady-state universe containing many other expanding bubbles (Gut84), i.e., the observable universe may be only a small fraction of the entire universe.

It should be emphasized that in spite of its successes the inflationary scenario is incomplete at the present time. Several "tuning" problems remain, such as the domain wall and magnetic monopole problems (Lan81; for new developments see Lin82, Alb82, Gut84). For example, it is not yet quite clear whether the period of rapid expansion can be gracefully terminated to give rise to the present, noninflationary universe. There is also the issue of inhomogeneities (the smoothness problem), which are necessary for the formation of the structures observed in the universe today. Thus far, the inflationary scenarios have not shed any light on the origin of the density fluctuations necessary for galaxy formation (see, however, Gut85). Clearly, a crucial requirement is that a model correctly predict observable effects. At the present time there is only one prediction of the inflationary scenarios: the present universe should be flat to a high degree ($\Omega = 1$). So far, observational evidence points to a value of Ω less than 1 ($\Omega < 0.4$). If $\Omega = 1$ now, where, then, is the rest of the "dark matter"? All inflationary scenarios also take it for granted that the vacuum energy density played a dominant role in the very early universe ($\rho_{\text{vac}} = bT_c^4$) but is negligible today. It is, however, possible that a fundamental understanding of this term, which will likely involve a deep connection between quantum field theory and gravity, may have some surprises in store. Perhaps the vacuum energy term has always been negligible. Since there is an epoch about which our present understanding is very limited ($t = 0$–10^{-43} s, the quantum gravity epoch), it is conceivable that the peculiarities of the initial conditions of the universe are presented to us at the Planck time as a result of quantum gravitational effects. However, even if the solution of the cosmological puzzles really occurs at the level of quantum gravity, it is possible that the mechanisms closely resemble the inflationary scenarios.

2.2. Nucleosynthesis in the Early Universe

2.2.1. The Quest for Light-Element Creation

As early as 1937, von Weizsäcker (Wei37, Wei38) tried to show theoretically how the heavy elements might be produced. He spoke of the "cooking" of hydrogen during an early supermassive star stage of the universe, after which the system exploded into the expanding universe. Gamow and others pointed out in the late 1940s that the universe could not have existed in such a static, high-temperature state. Instead, Gamow and associates proposed (Alp48, Alp50, Gam53) that the elements were formed largely during the early and very rapid expansion of the universe. Thus, they introduced the basic idea of big-bang nucleosynthesis. As discussed in section 2.1.3, when the universe expanded and cooled to 7.5×10^9 K, the ratio of neutrons to protons froze out at a value of about $\frac{1}{7}$ (Pee66, Wag73, Yan84). This ratio remained at this value as the temperature dropped to 10^9 K, at which point deuterium and heavier nuclear species could be formed. As the temperature dropped well below 10^9 K, all neutrons were converted to protons or incorporated into ^{4}He. Nucleosynthesis via these charged particles stopped, however, because the thermal energies were not sufficient to overcome the Coulomb barriers. Therefore, nucleosynthesis in the early universe could take place only when the universe had a temperature of around 10^9 K, a time when protons and neutrons were available as fuel.

Successive neutron-capture reactions (chaps. 4 and 9) followed by β-decays were thought by Gamow and coworkers to be the principal mechanisms by which essentially all the elements were manufactured. To Gamow the universe was a giant fusion reactor, a primordial cauldron. Such a scenario for the birthplace of the chemical elements (referred to as the universal synthesis hypothesis) was a natural explanation for the relatively similar abundance distribution of the elements observed throughout the universe (chap. 1). However, subsequent discovery of the nuclear stability gaps at mass 5 and mass 8 implied that in this process nucleosynthesis of elements heavier than ^{4}He is practically impossible and thus that the above scenario cannot account for the formation of the heavier elements. An improved theory in which element formation takes place via nuclear transformations in the interiors of stars (Burbidge, Burbidge, Fowler, and Hoyle in 1957 [B^{2}FH] and Cameron, independently, in 1957 [Cam57]) finally superseded scenarios in which heavy-element formation takes place in the big bang itself.

The idea of nucleosynthesis in a cosmological setting was revived in the 1960s when Hoyle and Taylor (Hoy64) recognized that the large amount of ^{4}He present in the universe (Table 2.1) could not have been manufactured in ordinary stars. They also demonstrated that ^{4}He could have been made in a primeval big-bang environment. These results stimulated a series of improved calculations (Pee66, Wag67, Wag73), and it was shown that indeed sizable amounts of some of the lighter nuclides could have been formed in the early universe. The protons and neutrons in the early universe could fuse to make a nucleus of deuterium, most of which would quickly combine to form ^{4}He nuclei. For the temperatures and densities of big-bang nucleosynthesis virtually all of the neutrons present would end up in ^{4}He. Model calculations indicated that, for the entire possible range of densities in the early universe, about 20%–30% of the mass emerging from the big bang was in the form of ^{4}He nuclides. Indeed, the ^{4}He abundance observed throughout the universe is exactly in this narrow range (Table 2.1 and Kun83, Jam83). The remarkable concordance in helium abundance strongly supported the fundamental assumption that the universe went through a period of extremely high temperature and density as part of the big bang. The helium abundance combined with the observation of the 2.76 K universal background radiation, which could be explained only in terms of a sufficiently hot dense period in the early universe, firmly established the big-bang model.

TABLE 2.1 Primordial Abundances of the Light Elements[a]

Nuclide	Number Density $N_i/{}^1$H	Mass Fraction[b]
^{1}H	1.00	0.75
^{2}H (D)	$(1.6 \pm 1.0) \times 10^{-5}$	$(2.5 \pm 1.5) \times 10^{-5}$
^{3}He	$(1.8 \pm 1.2) \times 10^{-5}$	$(4.2 \pm 2.8) \times 10^{-5}$
^{4}He	$(7.5 \pm 0.9) \times 10^{-2}$	0.23 ± 0.02
^{6}Li	$70^{+70}_{-35} \times 10^{-12}$	$300^{+300}_{-150} \times 10^{-12}$
^{7}Li	$900^{+900}_{-450} \times 10^{-12}$	$4600^{+4600}_{-2300} \times 10^{-12}$

a From Aus81 (see also Ste85 and references therein).
b Normalized to $\sum_i X_i = 1$.

2.2.2. Ashes of the Primeval Big Bang

Although the bulk of the deuterium (d), as produced via the reaction $p + n \rightarrow d + \gamma$, was rapidly burned to ^{4}He through reactions such as $d + p \rightarrow {}^3$He $+ \gamma$ and ^{3}He $+ n \rightarrow {}^4$He $+ \gamma$, some small amount may have been left unburned if the density of the universe was so low that the deuterium-burning reactions could not go to completion. Detailed calculations used the rates of a rather large number of reactions, which synthesize the light nuclei from protons and neutrons. Such calculations show (Wag73, Sch77) that nucleosynthesis takes place when the temperature is near 10^9 K and the baryon density near 10^{-5} g cm^{-3}. Fig. 2.3 shows the resulting abundances of the various light nuclear species, synthesized in the early universe, relative to all the matter in the universe, as a function of the present average universal density of baryons, ρ_B. The present density is directly related to the baryon density at the time of nucleosynthesis, because the universe is adiabatic with

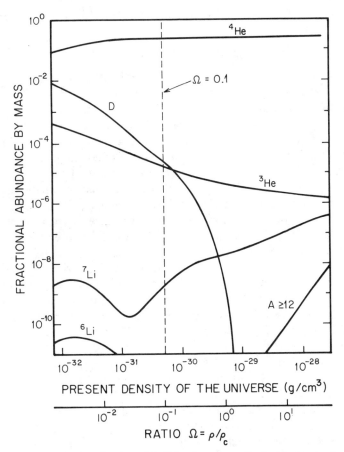

FIGURE 2.3. Shown is the mass fraction of the light elements synthesized in the standard big-bang model (Wag73, Got74, Yan84) as a function of the present baryon density of the universe, or, equivalently, as a function of the present density in terms of the critical density ρ_c ($\Omega = \rho/\rho_c$). The abundance of ^{4}He is almost independent of baryon density, but that of deuterium (D) is quite sensitive to it. Note also that the heavy elements ($A \geq 12$) have mass fractions far below observed values, strongly indicating that their origin be attributed to later nucleosynthesis such as that occurring in stellar interiors.

$\rho \propto T^3$ (sec. 2.1.2), with T now equal to 2.76 K and nucleosynthesis occurring at 10^9 K. The calculated abundances depend then only on the present value of the baryon density and the temperature of the microwave background radiation. From a comparison with observed abundances (Table 2.1), it is clear that only nuclides ^{4}He, ^{2}H ($=$ D), ^{3}He, and ^{7}Li are made primarily by big-bang nucleosynthesis. It is significant to note that these nuclides are the ones most difficult to produce in correct abundances by processes within stars or by spallation reactions in the interstellar medium (for example, ^{2}H/^{1}H $< 10^{-17}$ in stars; chap. 6); thus they are most likely direct ashes of the fireball phase of the universe.

The baryon density ρ_B is not accurately determined by astronomical measurements (Got74, Dol81):

$$1 \times 10^{-31} \text{ g cm}^{-3} \leq \rho_B \leq 1 \times 10^{-29} \text{ g cm}^{-3} \,,$$

where the lower limit is obtained from estimates of the mass associated with clusters of galaxies and the upper limit from modern values of the Hubble constant and the age of the universe. Owing to this uncertainty and owing to the strong dependence of the abundances on ρ_B (Fig. 2.3), one cannot make a priori predictions of nucleosynthesis in the big bang. The exception is the observed abundance of ^{4}He, which is approximately reproduced over the allowed range of densities. This concordance was one of the triumphs of the hot big-bang model. For the other nuclides, one must invert the arguments. Assuming that a nuclide is produced in the big bang, one determines the value of ρ_B required to reproduce the observed abundance of each nuclide. If the picture is valid, the resulting densities should be consistent within their uncertainties. Such an impressive concordant value for ρ_B has indeed been found near $\rho_B \simeq 5 \times 10^{-31}$ g cm^{-3} (Wag73, Sch77, Ste79, Ste85). If this baryon density represents the total density of matter in the universe, we have $\rho/\rho_c = 0.10$ and the universe is open and will continue to expand forever. Thus, nuclear physics, from which one gains knowledge of reaction rates among the light nuclei (chaps. 3–5), is also able to provide important cosmological information.

We have seen that in the first few minutes after the start of the big bang the rapidly expanding early universe cooled just enough to allow thermonuclear reactions to proceed, resulting in the synthesis of ^{4}He and, to a much lesser extent, other light nuclides such as ^{2}H, ^{3}He, and ^{7}Li. Within only a few more minutes, expansion reduced the temperature and density still further, well below the levels needed to sustain further nuclear reactions. Thus, the universe cooled too quickly for carbon and heavier elements to form. The nuclear processing, though incomplete, was sufficient to account for the present abundance of ^{4}He, ^{2}H, ^{3}He, and most of ^{7}Li in the universe. Thus, Gamow's original idea of an inevitable connection between a hot big-bang origin of the universe and nucleosynthesis of heavier elements out of nucleons has been fully established. However, big-bang nucleosynthesis is restricted to light elements and does not produce all elements as originally suggested (sec. 2.2.1). Any description of the evolution of the early universe during the element-producing epoch requires a model within which the primordial nucleosynthesis can take place. The favored standard big-bang model (Wag73) contains a number of ad hoc assumptions (sec. 2.1.2). However, primordial nucleosynthesis provides a unique opportunity to test the assumptions of the standard model, serving as a probe of the physical conditions during epochs in the early universe which would otherwise be completely obscured. The standard model of primordial nucleosynthesis is certainly the simplest model. It explains the facts remarkably well, its predictive power is substantial, and in principle it contains no free parameters. Thus it makes an ideal starting point for more elaborate models.

To sum up, from the early hot state of the universe predominantly hydrogen (^{1}H) and helium (^{4}He) emerged, with the latter comprising about 23% by mass. A generous supply of hydrogen also was provided, which later was to be incorporated in stars to serve as nuclear fuel (sec. 2.5). Within about 10^9 years after the big bang, galaxies and stars would form out of this material, and only then would significant production of elements heavier than ^{4}He begin in the interior of the stars. In fact, stars which from other evidence are believed to be very old contain essentially only the two elements hydrogen and helium, with the abundance of metallic elements being up to an order of magnitude lower than solar abundance.

2.2.3. Implications of Primordial Nucleosynthesis

Having achieved an understanding of primordial nucleosynthesis, one can then turn to implications of this understanding of light-element creation to other fields such as cosmology and particle physics. The number of implications arrived at from the abundances of the big-bang nuclides are considerable and have proved more fruitful than one might have imagined. They include determinations of properties such as the age of the universe and the value of the Hubble constant, the anisotropy in universal expansion and the inhomogeneity of the universe (density fluctuations), the time dependence of the gravitational constant G and tests of alternative theories of gravity, the allowable masses of neutrino species and hence the allowable number of quark flavors, the mean baryon density of the universe, and whether the universe will continue to expand forever. Detailed discussions of these implications are given elsewhere (Ste79, Sch77, Dol81, Ste85). In the following we will discuss only two of these implications.

We have already concluded in section 2.2.2 that the baryon density of the universe ($\rho_B \simeq 5 \times 10^{-31}$ g cm^{-3}) is far below the critical density ($\rho/\rho_c = 0.10$), and thus the present expansion will continue forever. Another possibility is that the mass of the baryons does not at present dominate the mass of the universe. To put this possibility in perspective (Ste79, Tay80), note that the number of neutrinos is approximately the same as the number of photons

($N_\gamma \simeq 100$ cm^{-3}). Thus, if any one of the neutrino species had a finite mass m_ν, the neutrino mass density would be

$$\rho_\nu \simeq 2 \times 10^{-31} m_\nu \text{ g cm}^{-3}$$

(m_ν in units of eV/c^2) and the universe would be closed ($\rho_\nu = \rho_c$) if $m_\nu c^2 \geq$ 25 eV. Astronomical observations provide an upper limit of $\sigma \leq$ 1×10^{-29} g cm^{-3}, leading to the constraint of $m_\nu c^2 \leq 50$ eV. Clearly, experimental determinations of neutrino masses are of the utmost importance to cosmology and to the solution of other astrophysical puzzles (chap. 10).

The observed abundance of ^{4}He allows one to set some upper limit on the number of neutrino species (Sch77, Ste79, Tay80, Ste85). The basic arguments involved are quite simple (Aus81). Early in the big-bang expansion, the density is dominated by radiation, i.e., by the contributions of massless or relativistic particles. Additional massless particles such as new types of neutrinos lead to a greater density, which in turn increases the rate of the universal expansion or equivalently shortens the expansion time ($t_{\text{exp}} \propto 1/\rho^{1/2}$). Among the effects of an increased expansion rate is an increase in the neutron-to-proton ratio (sec. 2.1.3). Since most of the neutrons are eventually incorporated into ^{4}He, a higher n/p ratio leads to a larger ^{4}He abundance. Quantitatively, an additional neutrino species changes the abundance $X(^4\text{He})$ by roughly $+0.014$. Detailed calculations for the standard big-bang model, with e and μ neutrinos only, show (Yan79) that for $X(^4\text{He}) \leq 0.25$ (Table 2.1) and the concordant value for the baryon density, only one additional neutrino species could be tolerated, presumably that associated with the τ lepton. If one accepts the validity of these constraints, then all possible types of lepton species have been discovered, i.e., there are no new neutrinos to be found. Accepting further the relationship of lepton and quark generations (Clo83), then only the "top" quark remains to be found.

The above calculations (Yan79) assumed the neutron half-life to be 10.6 minutes. A recent remeasurement yields (Bon78) a half-life of 10.13 ± 0.09 minutes (see, however, Byr80), implying that the weak interaction is stronger than had been assumed. Accepting this result means that the weak-interaction rates freeze out at a lower temperature T_* (sec. 2.1.3), leading to a smaller $(n/p)_*$ ratio and thus to a decrease in the ^{4}He abundance. In this case another neutrino species would be allowed to be "squeezed in." Clearly, an improved knowledge of the neutron half-life is desirable, and experimental efforts in nuclear laboratories would help in reducing the uncertainties of some of these implications. Another example is the calculated abundance of ^{7}Li, which is uncertain to a factor of 2–3 owing to the uncertainties in the rates of the reactions ^{3}H(α, γ)^{7}Li, ^{7}Li(p, α)α and ^{7}Be(n, p)^{7}Li. Again, nuclear physics could help to improve the precision of predicted big-bang abundances and thus the reliability of possible implications.

The above discussions show clearly that there are strong interconnections among nuclear physics, light-element nucleosynthesis, cosmology, and particle physics and that it is no longer possible to consider these fields in isolation. For example, a new measurement of the neutron half-life may change the estimated limit on the number of possible neutrino species (and number of quarks) in nature, through the intermediary of ^{4}He nucleosynthesis in a standard big-bang cosmology.

2.3. The Formation of Galaxies

The remarkable isotropy of the universe background radiation implies that the universe was totally homogeneous and isotropic at the time radiation decoupled from matter (Fig. 2.1). Clearly, the present universe is quite granular, containing as it does planets, stars, galaxies, and clusters of galaxies. If large-scale clumping began before the moment of decoupling, the background radiation should exhibit bright and dark spots corresponding to these clumps. The apparent absence of such features indicates that large-scale structures had not yet formed at the moment of decoupling. Such clustering (formation of galaxies) must then have happened subsequently, several hundred thousand years after the start of the big bang. Can the path from homogeneity of the early universe to the rich assortment of present-day structures be traced? Why does matter tend to aggregate in bundles of the particular sizes observed in the billions of galaxies? Why do galaxies have a limited hierarchy of shapes? Why do spiral galaxies rotate like giant pinwheels? Why and how did quasars form? These and many other questions remain to be answered. Some astronomers believe that the paucity of isolated galaxies and the presence of large voids may provide direct evidence for establishing the relative times of formation of galaxies, clusters of galaxies, and clusters of clusters.

One scenario for galaxy formation (Ree70, Sil83) is based on gravitational instability. This scenario maintains that galaxies condensed out of the hot cosmological gas that expanded from the big bang, and, if a region of the early universe happened to have a density higher than that of the surrounding regions, it would produce a net gravitational attraction to material outside the region. When condensed, it would then attract more matter from the less dense region and continue to contract under the influence of its own gravity. What began as a fluctuation in a fairly homogeneous universe would eventually snowball into a huge inhomogeneity. At a time equivalent to a redshift of 1000 (moment of decoupling), gas pressure, which tends to oppose collapse, was stronger than gravity over dimensions equivalent to the mass of about 100,000 suns (sec. 2.5.1). Over larger dimensions gravity was much stronger. Slight fluctuations in the density distribution of the gas on these larger scales grew larger in response to gravitational attraction and became increasingly irregular. Eventually the gas became dense enough to collect into vast sheets of material called "droplets," which then fragmented into galaxies. According to this popular model, material equivalent to clusters and superclusters coalesces first as concentrations of gas, and only then do galaxies and stars appear. The model is thus consistent with the lack of isolated galaxies. The scale of the perturbations probably ranges from the mass of a globular cluster of stars ($\simeq 10^6\ M_\odot$) to the mass of a large aggregate of galaxies ($\simeq 10^{15}\ M_\odot$).

The perturbations that are the sizes of galaxies ($\simeq 10^{11}\ M_\odot$) are called protogalaxies. Initially, each protogalaxy would expand with the rest of the universe, but it would do so at a slightly lower rate. After a few hundred million years, it would stop expanding even though the universe continued to expand. At that point the protogalaxy would in effect begin to collapse on itself to form a primeval galaxy, which subsequently would evolve into a regular galaxy.

Within the galaxies smaller clouds collapsed to form clusters, stars, and possibly other objects. All objects in a galaxy are subject to a net gravitational pull toward its center, the nucleus. Although the rotation of a galaxy slows the infall of matter, and momentary outbursts (i.e., supernovae) may eject matter from the center, frictional forces inevitably cause the center to acquire a mass density much higher than that of the outer regions. The ultraluminous objects known as quasars (chap. 1) may represent an extreme of this process. In fact, quasar formation may be a natural phase through which most galaxies pass as they evolve from primeval to normal galaxies.

A competing model assumes that galaxies and stars are formed first out of the homogeneous primordial gas. Small irregularities in their distribution are slowly amplified by the operation of the long-range gravitational forces. The end result of such an amplification would again be the superclusters of galaxies seen today. Present observational evidence does not allow a clear choice to be made between these models. Neither model explains the origin of the density nonuniformities initiating the formation of galaxies (sec. 2.1.7), nor do they explain the origin of the rotation of galaxies. Independent of the models, it is clear that if there is angular momentum, contraction of a galaxy will lead to more rapid rotation, which in turn will tend to form the observable relatively flat disk.

2.4. Physical State of the Stellar Interior

The vast majority of stars reveal no changes in their properties (luminosity, mass, radius, chemical composition of the outer layers) over long time intervals. This constancy of the stars implies that the stellar interior must be in a state of hydrostatic and thermal equilibrium. Thus the physical state of the stellar interior is governed by these equilibrium conditions, which must be fulfilled throughout the volume of the star. To the extent that one can ignore effects due to rotation, pulsation, tidal distortion, and large-scale magnetic fields, and can regard the stars as spherically symmetric objects, the physics of their hot gaseous interiors is relatively simple (Edd26, Scha39, Sch58).

2.4.1. Hydrostatic Equilibrium

The first condition which must be fulfilled throughout the stellar interior is that of hydrostatic equilibrium. At any point in the interior of a star, the internal pressure must be high enough to support the weight of the outer layers. Hence, the difference of pressure between two adjacent points in the stellar interior (the pressure gradient) will be given by the weight of the material in the shell lying between these two points. Expressing this more quantitatively, we have in the nonrelativistic limit (Sch58)

$$dP(r)/dr = -GM(r)\rho(r)/r^2 ,$$ (2.1)

where $P(r)$ is the total gas pressure at the radial distance r, G is the gravitational constant, $\rho(r)$ is the density at the distance r, and $M(r)$ is the mass contained inside the sphere of radius r:

$$M(r) = \int_0^r 4\pi r^2 \rho(r)dr .$$ (2.2)

The sign in equation (2.1) expresses the fact that the gravitational force is directed inward and the internal pressure outward. If the internal pressure is too low, the star will contract; if it is too high, the star will expand. Owing to the gravitational force, the internal pressure must be a monotonically decreasing function of the distance r from the center, if equilibrium is to prevail. If one knew how the density $\rho(r)$ of the stellar material diminishes with distance from the center, the mass and pressure distributions could be calculated throughout the interior. Lacking this knowledge, one can still use these equations to gain some insight into the order of magnitude of these quantities. Applying the hydrostatic condition (eq. [2.1]) to a point in the star midway between the center and surface and assuming, in the roughest approximation, that $M(r)$ is one-half the stellar mass M and r is one-half the stellar radius R, the density there is given by

$$\bar{\rho} = (M/2)/[(4\pi/3)(R/2)^3] \simeq M/R^3 ,$$

which is about 4 times the mean density ρ_s of the star. Further, let us take for the differential dr the stellar radius and for dP the difference between the central pressure $P(0)$ and the surface pressure $P(R) = 0$. We then obtain

$$P(0) \simeq 8\rho_s GM/R ,$$

from which we find for the sun $P(0) \simeq 2.2 \times 10^{16}$ dyn cm^{-2}.

If equilibrium does not occur, then we have in the nonrelativistic limit (Fow64)

$$\rho\ddot{r} = -G(M\rho/r^2) - dP/dr$$

where the right-hand side of the equation determines whether an expansion will take place or whether the star implodes. For the case $dP/dr = 0$, free fall occurs. In this case the equation reduces to

$$\ddot{r} = -GM/r^2 \, ,$$

and integration leads to

$$\dot{r} = -[(2GM/r) - (2GM/r_0)]^{1/2}$$

for the boundary condition $\dot{r}(r_0) = 0$. If a characteristic free-fall time τ is defined by the logarithmic time derivative of the density $[\rho = \rho_0 \exp(t/\tau)]$, using the equations

$$d \ln \rho = -3d \ln r \, ,$$

$$M = (4\pi/3)\rho_0 r_0^3 = (4\pi/3)\rho r^3 \, ,$$

the free-fall time τ is given by

$$\tau = (d \ln \rho/dt)^{-1} = [1/(24\pi G\rho_0)]^{1/2}\{(\rho/\rho_0)^{2/3}[(\rho/\rho_0)^{1/3} - 1]\}^{-1/2} \, .$$

At time $t = \tau$ we have $\rho = e\rho_0$ ($e = 2.72$), and the free-fall time can be written as

$$\tau \simeq 1.03/(24\pi G\rho_0)^{1/2} \simeq 460/\rho_0^{1/2} \text{ s} \, ,$$

where ρ_0 is in units of g cm^{-3}. This time is very short in terms of the usual times for cosmological processes. Thus, once free fall begins, very large densities can be built up very quickly and at an ever-increasing rate (chap. 8).

A star spends most of its life in a state of precise balance between enormous opposing forces and energy flows (see below). The force of gravity works to compress the star to ever higher densities and is exactly balanced by the thermal pressure of the hot plasma in the star's core. Without such supporting pressure the sun, for example, would collapse in less than an hour.

2.4.2 Equation of State of Normal Stars

An equation of state is a relation between the pressure, temperature, and density of the matter under consideration. In most ordinary stars, the hot gaseous material in the interior is described fairly well by the equation of state of an ideal gas, which can be put in the form

$$P(r) = (k/m)\rho(r)T(r) \, , \tag{2.3}$$

where $k = 1.38 \times 10^{-16}$ ergs K^{-1} is the Boltzmann constant and m is the mean molecular weight of the particles in the gas ($m \simeq \frac{1}{2}m_H$; see below). A useful generality that follows from this equation is that if the temperature decreases toward the stellar surface, so does the pressure. Another argument for a decreasing temperature from the center to the surface of the star is provided by the fact that heat (energy) has to flow from the hotter to the cooler regions in the star, i.e., outward (sec. 2.4.6).

Applying equation (2.3) to the center of the star, we take for the density $\rho(0) = 2\rho$ and combine it with the above estimate for $P(0)$. We find then for the temperature

$$T(0) \simeq (mG/k)(M/R) \,,$$

and for the sun we obtain $T(0) \simeq 1.1 \times 10^7$ K. More precise calculations lead to $T(0) = 1.5 \times 10^7$ K and $\rho(0) = 150$ g cm^{-3}, representing more or less typical values for the temperature and density in stellar interiors. In the laboratory a gas at such high pressure and density would cease to conform to the ideal gas law. However, in the very hot interior of a star, where the matter is nearly completely ionized, the stripped atomic nuclei and the free electrons occupy only a small fraction of the available room. This is analogous to a card house, which fills a much larger volume than the cards themselves. Thus, even at densities far exceeding that of solid lead, such a hot medium is a gas and the ideal gas law continues to apply everywhere in the star. Thus, for normal stars we need not consider the complex physics of solids and fluids.

The gas pressure $P_{gas}(r)$ at any point inside the star is produced by the motion of the gas particles (ions and electrons). Additional pressure is generated by the outward flow of radiation. It is exerted by the myriad of photons that are flying about. This radiation pressure $P_{rad}(r)$ is directly proportional to the energy density of radiation at that point, which is given by Stefan's law:

$$P_{rad}(r) = \tfrac{1}{3}aT^4 \,,$$

where the radiation constant a is given by $a = 7.565 \times 10^{-15}$ ergs cm^{-3} K^{-4} and is related to the Stefan constant σ by the equation

$$\sigma = ac/4$$

(with c the speed of light). The total pressure is then determined by

$$P_{tot}(r) = P_{gas}(r) + P_{rad}(r) \,,$$

which represents the equation of state of nondegenerate matter.

For the above estimates, one finds $P_{rad}(r = R/2) \simeq 2 \times 10^{13}$ dyn cm^{-2}. Thus, for low-mass stars such as the sun, the radiation pressure can be neglected compared with the gas pressure. However, in massive stars with $M \geq 10\ M_\odot$ the radiation pressure plays an important role. The stellar mass required for both pressures to be equal at the center can be estimated—within the above approximations—to be (Cha84)

$$M \simeq (12ck^4/\sigma G^3 m^4)^{1/2} \simeq 62\ M_\odot \,,$$

which is in the range of the most massive stars observed. It was suggested by Eddington (Edd26), based on similar considerations, that there is an upper limit on the mass of a stable (nonrotating) star of $M \leq 100\ M_\odot$ (Lar71, Cha84). Supermassive stars with $M \geq 100\ M_\odot$ are general-relativistically unstable (Ibe63, Fow66, Fri73, Fri80). The above considerations show that the magnitude of stellar masses can be estimated using well-known physical laws.

2.4.3. Effects of the Chemical Composition

In using the equation of state of an ideal gas for stars (eq. [2.3]), one must take into account that the mean molecular weight m of the particles in the gas changes as a function of distance r from the center. This change occurs because near the surface of the star most atoms are fully recombined, whereas toward the hotter interior they are progressively more ionized. Furthermore, nucleosynthesis in the stellar interior changes the chemical composition in the burning zones ($H \rightarrow He \rightarrow C, O \rightarrow \cdots$).

As more and more electrons are stripped off the ions, the mean molecular weight in the interior decreases, since the same amount of mass is, on the average, divided among more particles. For completely ionized hydrogen atoms, one finds in the interior $m = \frac{1}{2}m_H$ and for helium atoms $m = \frac{4}{3}m_H$, while the average value m for the metallic elements (all except H and He) are all nearly the same at $m = 2m_H$. If, then, the gas composition is described by the mass fractions X (H), Y (He), and Z (metals) with $X + Y + Z = 1$, the mean molecular weight m is given by the equation

$$m = [1/(2X + \tfrac{3}{4}Y + \tfrac{1}{2}Z)]m_H . \tag{2.4}$$

Consider two stars having the same size, density, and density distribution, the first star composed mainly of iron ($Z = 1$) and the second entirely of hydrogen ($X = 1$). By equation (2.3) ($T \propto m$) the temperature at a particular point in the metallic star would have to be 4 times higher than that in the hydrogen star to produce the same gas pressure. If the sun were composed mainly of metallic gases, its central temperature would be 4 times higher than if it contained only hydrogen, and the hydrogen-only sun would shine—because $L \propto T^4$—only about 1% as brightly as the metallic sun. Thus the chemical composition of a star (usually given in terms of X and Y [or X and Z], since $Z = 1 - X - Y$) is an important factor in determining the interior temperatures and other properties of a star. Since the mean molecular weight $m(r)$ is variable in the stellar interior, this must be taken into account in constructing models.

There appears to be no major convective mixing of the material in the stellar interior with its surface material. Thus, since there is no nuclear burning at the surface in the early stages of stellar evolution, it seems safe to assume that the spectroscopically determined surface composition (sun: $X = 0.73$, $Y = 0.25$, $Z = 0.02$) is practically identical with the initial composition during the main-sequence evolution. Of course, as a result of nuclear burning (sec. 2.6), the chemical composition in the interior will change with time. At present, for the center of the sun, one uses $X = 0.42$, $Y = 0.56$, $Z = 0.02$ (chap. 10).

2.4.4. Stored Energy of a Star

All stars pour out prodigious amounts of energy from their surfaces. We are all aware of the flood of light and heat which the earth receives from the sun. A major historical problem was to identify the source of energy which could supply this radiation over the very long lifetime of the stars. It is well known that neither stars nor any other objects can continue to shine for long, simply because they are extremely hot. In fact, energy must be continuously supplied from some source to maintain the high temperature. It is also well to remember that the chemical energy which plays the major role in everyday human activities counts very little in stars and the universe as a whole.

The thermal energy, E_T, stored in a star is given approximately by the integral over the star of the thermal energy per unit mass of an ideal monatomic gas:

$$E_T = \int_0^R [\tfrac{3}{2}(k/m)T]\rho(r)4\pi r^2 dr \simeq +\overline{[\tfrac{3}{2}(k/m)T]}M .$$

The total gravitational energy, E_G, can be determined by integrating, over the entire star, the energy needed to move 1 gram of stellar matter from its surface to infinity:

$$E_G = \int_0^R [-GM(r)/r]\rho(r)4\pi r^2 dr \simeq -\overline{[GM(r)/r]}M .$$

Using numerical values for the sun, the order-of-magnitude values derived from the above rough figures are $E_T = 2.0 \times 10^{48}$ ergs and $E_G = -3.8 \times 10^{48}$ ergs. That these two energies differ by nearly a factor of 2 follows directly from hydrostatic equilibrium. Multiplying equation (2.1) by $4\pi r^3$ and integrating over the star, partial integration on the left-hand side and the use of equation (2.3) lead to the result

$$2E_T = -E_G . \tag{2.5}$$

Thus, large gravitational energies imply large kinetic energies, and the latter mean high temperatures. This relation is known as the virial theorem of classical mechanics.

In a more general form of the virial theorem, E_G stands for the potential energy of the system and E_T for the internal kinetic energy, which includes the kinetic energy of rotation. The virial theorem states that, when a bound configuration is formed, half the released gravitational potential energy is stored as internal energy and the other half is lost from the system. Thus, for a planet in a circular orbit about a star, the kinetic energy of its motion is half the gravitational potential energy which has been released in binding the star and the planet together. In a nonrotating contracting star (ideal gas law assumed) the internal energy is one-half the released gravitational potential energy, which is stored in the form of heat, E_T. The other half will be lost by radiation from the stellar surface. Thus the net energy available for radiation from the surface is just equal to the thermal energy. The time τ over which this energy will cover the radiative surface losses is given by the ratio of this stored energy to the star's luminosity. For the sun we find

$$\tau = E_T/L \simeq 4.4 \times 10^7 \text{ y} .$$

This limit to the duration of sunshine, often called the Kelvin-Helmholtz contraction time, is far too short, since geological dating places the beginning of life on the earth at least 10^9 y in the past (Bar71b). From the above, one must conclude that the thermal and gravitational energies of a star are insufficient to account for the radiation losses over the whole life of a star, although they play an important role in short, critical phases of stellar evolution (beyond the main sequence).

The third source of stored energy in a star is nuclear energy, E_N. Nuclear reactions release energy by the conversion of mass according to the well-known relationship $E = mc^2$. One might suppose that the total available nuclear energy of a star is $E_N = Mc^2$ (where M is the total stellar mass). This, however, is an overestimate, since this amount of energy would be released only if nuclei were completely annihilated. Actually, the energies released in such reactions are equivalent to the relative mass defects (chap. 3). One finds that the stored nuclear energy in a star is $E_N = 0.008Mc^2$. For the sun this is approximately 1.4×10^{52} ergs, which is over 1000 times as much as the thermal gravitational energies. Using this value and the known luminosity of the sun, we find that this energy source is sufficient to cover the losses by radiation from the sun's surface over a time interval of

$$\tau = E_N/L \simeq 1.2 \times 10^{11} \text{ y} ,$$

which is ample to provide energy over the times arrived at by geological dating. Thus nuclear reactions, specifically those converting hydrogen into helium, are the source of the enormous energy released by stars. Through these reactions atoms of lighter chemical elements are built up into atoms of heavier ones in the stellar interiors.

Gravitational contraction, although it does not provide the long-range energy source, is nevertheless of importance in stars, since it provides the energy youthful stars need to heat up to the point where their supply of nuclear energy can be released. Each burning phase requires a different temperature; a period of gravitational compression and concomitant heating separates each phase from the next. Gravitational contraction ultimately assumes the major role when the nuclear fuel supplies are exhausted (Fig. 2.9). For example, if the iron core of a massive star collapses to the size of a neutron star ($R \simeq 7$ km), about 4×10^{53} ergs of gravitational energy is released in a few seconds (free-fall time); hence $L/L_\odot \simeq 10^{20}$. This is one possible source of the enormous energy involved in a supernova explosion, which can outshine the entire galaxy for a period of days or even weeks and which leaves an expanding shell of nuclear debris. The best-known example in our Galaxy is the Crab Nebula in Taurus (Fig. 1.10).

2.4.5. Thermal Equilibrium

The condition of hydrostatic equilibrium (eq. [2.1]) is not sufficient to ensure a stable star. Thermal equilibrium must also be considered. Thermal equilibrium is obtained within a system when all parts of the system have reached the same temperature and there is no further flow of energy. Such thermal equilibrium cannot hold within a star, since, as we have already seen, the interior temperatures in the main-sequence stage are of the order of 10^7 K, while the surface layers are observed to have temperatures of the order of only several thousand K. Furthermore, energy flowing through the surface in the form of radiation (the star's luminosity) precludes attaining perfect thermal equilibrium.

It follows from energy conservation that the energy lost at the surface must be replaced by energy released in nuclear reactions occurring throughout the stellar interior. This condition may be expressed by the equation

$$L = \int_0^R \epsilon(r)\rho(r)4\pi r^2 dr , \tag{2.6}$$

where $\epsilon(r)$ is the energy released from nuclear processes per gram of stellar matter per second (chap. 3). This nuclear energy production depends on the temperature, density, and chemical composition at the radial distance r.

The above equation ensures balance for the whole star. However, the same type of balance also must hold everywhere within the star. Clearly, energy gains in one section and energy losses in another section would lead to a change in the temperature structure of the interior, and the star would become unstable. The energy balance of a spherical shell of radius r may be written as

$$dL(r)/dr = \epsilon(r)\rho(r)4\pi r^2 , \tag{2.7}$$

where $L(r)$ is the energy flux through the sphere with radius r. The left-hand side of this equation represents the net loss to the shell caused by the excess of flux leaving the shell through the inner surface. The right-hand side represents the energy produced within the shell by nuclear processes (or by gravitational potential-energy release). This equation represents the third of the basic equilibrium conditions (the two others are eqs. [2.1] and [2.3]), which must hold throughout the stellar interior.

To summarize: the energy lost by radiation from a star's surface and by neutrino emission from its core (chaps. 6–8) would gradually rob it of its thermal pressure and would thus lead to gravitational collapse, if such losses were not replenished by energy from thermonuclear reactions.

2.4.6. Energy Transport Mechanisms

Above we considered only the conditions the energy flux must fulfill to balance the energy sources. The actual flux is determined by the mechanisms which transport the energy: conduction, convection, and radiation. For all three the energy flux is governed by the temperature gradient. In general, energy flows from hotter places to cooler places. A rough estimate for the sun gives a temperature gradient of

$$\Delta T / \Delta r \simeq [T(0) - T(R)]/R = 3 \times 10^{-4} \text{ K cm}^{-1}.$$

This relatively small gradient is not necessarily constant throughout the sun. We must, therefore, consider all three mechanisms in determining the temperature gradient needed to provide the energy flux necessary to fulfill the energy-balance condition (eq. [2.7]).

Energy transport by thermal conduction is familiar to anyone who stirs a cup of coffee with a metal spoon. This mechanism of energy transport is not important in most stellar interiors because, at the high gas densities encountered there, the mean free paths for the ions and electrons are extremely short compared with the stellar radius. This mechanism does become important, however, under conditions in which the electrons form a degenerate gas. In this case the electrons have very long mean free paths, and there is very effective electron conduction. This occurs, for example, in white dwarfs.

Another form of energy transport is radiative transfer, which is the dominant mechanism, for example, in the major part of the solar interior (below a depth of some 100,000 km from the solar surface). To understand how this process works, think of a place halfway down from the surface to the center of the sun. The temperature at this level is about 3×10^6 K, the atoms are well stripped of electrons, and the ions and free electrons are darting this way and that at high speeds. The emission of electromagnetic radiation, generated by a number of microscopic processes (Sch58), is governed by Stefan's law, i.e., the rate of emission varies as the fourth power of the temperature. Using Wien's law ($\lambda_{max} T = 0.29$ cm K) at 3×10^6 K, a typical photon has a wavelength of only about 10 Å (1 Å = 10^{-8} cm), in the X-ray part of the spectrum. Such a photon does not move far before it is absorbed, because the number density of ionized atoms is very large. In this layer of the solar interior the mean free path of a photon is only about 1 centimeter, and therefore the matter acts as a very effective barrier to the flow of radiation. The picture, therefore, is one in which photons are emitted in all directions, each traveling only a short distance before being stopped by the matter present. Now, if the temperature were constant at all points, Stefan's law tells us that the rate of emission of radiant energy would also be a constant everywhere. For every photon leaving a point in one direction, another would arrive from that direction to replace it. If energy is to flow outward, there must be a temperature gradient. Because of this radial temperature gradient, if one looks from any interior point toward the center, one sees radiation coming from a slightly hotter region, and, conversely, if one looks toward the surface, one sees radiation coming from a slightly cooler region, a mean free path away. The average energy of photons emitted from the hotter, inner region is slightly greater than the average energy of photons emitted from the outer, cooler region, which make the reverse trip. Thus, there is a net flow of radiant energy outward, down the temperature gradient, transporting energy from the high-temperature region to the low-temperature region according to the proportionality $L(r) \propto [d(T^4)/dr] \propto T^3(dT/dr)$. This is the principal mechanism for energy transport in the interior of normal stars. The analogy with the mechanism of conduction is close, except that here the energy is in the form of photons rather than the kinetic energies of atoms, ions, and electrons.

To complete the picture halfway down into the sun, assume that your eyes are responsive to X-rays and survey the scene around you. The surroundings are extremely foggy, or opaque. Because the radiation you see originates not much more than a centimeter away, you may justly claim that the visibility is only about 1 centimeter. Photons reaching you from an interior point a centimeter away come from a slightly hotter layer than do those coming from a corresponding exterior point. If there were no fogginess, or opacity, you would see an overwhelmingly greater amount of radiation coming to you from the center. Because of the fogginess in the solar interior, however, photons reach you from all directions in nearly equal numbers—but not quite. The greater the fogginess, or opacity, the more photons are obstructed in their journeys. Thus, if the flow of energy outward is to be maintained in spite of the interference of the matter, there must be a sufficiently steep temperature gradient. By analogy, imagine a large table covered with marbles with a line dividing the table in two. Tilting the table slightly will cause the marbles to cross the line at a certain rate; inclining the table at a larger angle will cause the rate of flow to increase. Analogously, the net outward flow of radiation is related to the steepness of the temperature gradient. Next, suppose that a number of nails are driven into the table. Tilting it slightly as before, note that the flow of marbles is slower because of their collisions with the nails. To maintain the original rate of flow, the table must be inclined at a larger angle. Similarly, the temperature gradient in a star is greater, the larger the obstructiveness or opacity κ of stellar matter. Clearly, the magnitude of the energy flux depends on the opacity of the stellar gas. If the opacity is low, energy is easily transported.

It is found (Sch58, Cox68, Har73) that the energy flux $L(r)$ is related to the temperature gradient and opacity of stellar matter by the relation

$$L(r) = -4\pi r^2 (4ac/3)[T(r)^3/\kappa(r)\rho(r)](dT/dr) . \tag{2.8}$$

This equation represents the fourth basic equilibrium condition. Note that in radiation transport the photons generated at the high temperatures in the center of the star (chiefly in the X-ray part of the spectrum) are continually emitted and reabsorbed and gradually degraded to longer and longer wavelengths as they proceed outward, and—in the case of the sun—they finally emerge from the surface as visible sunlight.

Radiative transfer is a dominant form of energy transport only if the temperature gradient (or, equivalently, the opacity) is not too large. If the opacity is so great that radiation is unable to carry the full load of energy transport, then convection comes to the rescue (Sch58). As in a pot of boiling water, energy in these regions moves outward, in part, through a circulating system of rising hot bubbles and sinking cooler bubbles. This is apparently the dominant mechanism by which energy is carried upward in the outermost part of the solar interior. If a bubble is slightly hotter than its surroundings, its density must be less than that of its surroundings because the pressure inside and outside the bubble is the same. The bubble will then rise, like a cork released under water. Another bubble that is cooler than its surroundings will be denser and will sink like a stone released under water. Thus, in convection, masses of relatively hot gas are physically moved from the hotter regions to the cooler regions as cooler gas moves into the hotter regions and energy moves upward, whereas the mass of matter at each level stays the same as time goes by. In convection, the temperature gradient is related to the pressure gradient by the expression (Har73)

$$dT/dr = (1 - 1/\gamma)[T(r)/P(r)](dP/dr)$$

$$= (1 - 1/\gamma)T(r)(d \ln P/dr) ,$$
(2.9)

where γ is the ratio of specific heats ($\gamma = c_p/c_v = 5/3$ for a monatomic gas). The above equation is obtained for adiabatic convection by differentiation of the well-known adiabatic relationship $T \propto P^{(\gamma - 1)/\gamma}$. When one is constructing a model of the stellar interior, it is customary to compute temperature gradients from both equations (2.8) and (2.9) and to use the lower in absolute value of the two values thus obtained (Sch58).

2.4.7. Magnitude of Stellar Luminosities

Calculations show that for stars in the mass range of the sun, energy is transported predominantly via radiation, except for regions near the surface and possibly near the center. We may then use the radiative equilibrium condition (eq. [2.8]) to estimate the stellar luminosity. As before with $\bar{r} = R/2$, $\bar{T} = T(0)/2$, $\bar{\rho} = 4\rho_s$, $dT = T(0)$, $dr = R$, and $\bar{\kappa} = 1$ cm^2 g^{-1}, we obtain

$$\overline{L(r)} \simeq 4\pi(R/2)^2(4ac/3)[T(0)^3/32\kappa\rho_s][T(0)/R] \propto T(0)^4 .$$
(2.10)

For the sun $\bar{L}/L_\odot \simeq 8.9$, and, clearly, this estimate is high. Most likely the value used for the temperature at $\bar{r} = R/2$ is too high. More realistic calculations reveal $T(R/2) = 3 \times 10^6$ K ($T \simeq T(0)/4$), hence $L/L_\odot \simeq 1.1$. These estimates then lead us to luminosities within the general range of the observed values as discussed in chapter 1.

2.4.8. The Mass-Luminosity Relation and Stellar Lifetimes

To the same approximation as applied above, one may use the radiative equilibrium condition to extract a relation between the luminosity and the mass of a star. The density within a star is related to the radius and mass according to the proportionality

$$\rho \propto M/R^3 .$$

Inserting this relation in equation (2.1) leads to

$$P \propto M^2/R^4 .$$

With these proportionalities, the equation of state of an ideal gas transforms to

$$T \propto M/R .$$

Inserting the proportionalities into equation (2.8) and using the opacity relation $\bar{\kappa} \propto 1/M$ (Sch58), one finds for stars with $M \leq 10\ M_\odot$

$$L \propto M^4 . \tag{2.11}$$

We see that the dependence on radius cancels out, and we arrive at an approximation to the theoretical mass-luminosity relation. This result indicates that the luminosity of a low-mass star increases as the fourth power of its mass. In general such a relation depends on the opacity as well as on the equation of state. For example, using the Kramer opacity relation (Sch58)

$$\kappa = \kappa_0 \rho T^{-3.5} ,$$

one finds $L \propto M^{5.5}$ and the condition $\kappa = $ constant leads to $L \propto M^3$. Finally, for massive stars with $M \gg 10\ M_\odot$ the pressure is dominated by radiation ($P \propto T^4$) and one arrives at the relation $L \propto M$. In general, observation confirms these relationships (Fig. 1.25).

Since the nuclear energy stored in a star is proportional to its mass ($E_N \propto M$), the time over which nuclear burning occurs, which is equivalent to the star's lifetime, is given for stars with $M \leq 10\ M_\odot$ by the relation

$$\tau = E_N/L \propto M^{-3} . \tag{2.12}$$

A star with $M/M_\odot = 10$ consumes its nuclear fuel a factor 10^3 times faster than the sun and lives a correspondingly shorter time (i.e., if $\tau_\odot \simeq 10^{10}$ y, then $\tau \simeq 10^7$ y). A lifetime 1000 times longer than that of the sun would be expected for stars with $M/M_\odot = \frac{1}{10}$.

2.4.9. Stellar Stability

Using the estimates just discussed, we have found that the luminosity of a star is determined not only by the rate of energy generation but also by the radiative equilibrium condition. To summarize: the internal pressure must counteract the gravitational force according to the hydrostatic equilibrium condition (eq. [2.1]), and to have a high enough internal pressure, the internal temperature must be correspondingly high. The gradient between the internal and surface temperatures leads to a net radiation flux according to equation (2.7). The flux will be fixed by the radiative equilibrium condition whether or not the energy lost is restored via nuclear energy production in the interior. If the nuclear energy generated is less than the energy lost by radiation, it can only be made up by gravitational energy generated by contraction. In our discussions of stored energy of a star (sec. 2.4.4) we saw that only one-half the gravitational energy released during a contraction is available for radiation from the surface. The other half goes to increase the thermal energy. The contraction will stop when the overall rate of nuclear energy generation is equal to the radiative surface losses, that is, to the luminosity of the star. Thus within a star the nuclear energy production is balanced against the radiation loss. This balance is achieved not by the changes in luminosity but rather by changes in the nuclear energy production governed by appropriate contraction or expansion. Thus a star is a gravitationally controlled thermonuclear reactor in the sky.

One important objective of studies of stellar interiors is to understand the process by which the energy is produced and how it is likely to affect the future of the star. Special circumstances exist in which the nuclear energy production cannot adjust to the radiation loss simply by moderate expansion or contraction. This occurs when the internal densities become so high that the equation of state for an ideal gas no longer applies. This happens in white dwarfs and neutron stars.

2.4.10. Equation of State for Degenerate Matter

Hot gaseous matter in low-mass main-sequence stars can be described by the ideal gas law. However, in the advanced stages of stellar evolution, the electrons are often in a so-called degenerate state, and the equation of state is more complicated.

When a star has consumed its nuclear fuel, it will collapse gravitationally, greatly increasing its density. Consider a volume of such a compressed gas. Given sufficient time for energy to flow out, the temperature will remain constant. Assume that initially the material is fully ionized by pressure-induced ionization of the atoms and that both the ions and the electrons have Maxwell-Boltzmann distributions (chap. 3). When very high densities are reached, the energy distribution of the electrons must be altered for quantum mechanical reasons associated with Fermi-Dirac statistics. Since the electrons are fermions (spin $\frac{1}{2}$ particles), the Pauli exclusion principle requires that no two electrons can be placed in any single quantum state (where spin is included). The number of quantum states accessible to the electrons in our material is truly enormous, but the number of states in a unit energy interval is progressively reduced as the material shrinks and its volume decreases. Eventually a condition is reached where there are fewer quantum states available for electrons at lower energies than demanded by the Maxwell-Boltzmann distribution. Under these circumstances some of the electrons must remain at higher energy. As the process continues, the structure of the material forces more and more electrons into higher energy states, until nearly all of the quantum states are filled up to an energy level called the Fermi level, E_F. Under these circumstances, the electrons form a degenerate gas, and the pressure they exert—because of their rapid motion—is very much higher than an electron gas having a Maxwell-Boltzmann distribution. In order to compress the system further, electrons have to be put into higher, unfilled levels, which requires a lot of energy. The electrons thus offer a stronger resistance to compression, a mechanism that stabilizes white dwarfs. Loosely speaking, when ordinary matter is compressed, higher density is achieved by squeezing out the empty space between atoms. In the core of a white dwarf this process has reached its limit: the atomic electrons are pressed tightly together; the electrons offer powerful resistance and become essentially incompressible.

Using the Pauli exclusion principle, we find that the number of states available for electrons compressed into a space of volume V is given by (Cha39, Har73)

$$dN_e = 2[4\pi p^2 dp(V/h^3)] \ ,$$

where h^3 represents the unit of phase space ($\delta x \delta p_x = h$). The factor of 2 is a result of the fact that each state can be occupied by two electrons of opposite spins. For nonrelativistic conditions and strong degeneracy, the total density of states is obtained by integrating up to the Fermi level, resulting in

$$N_e/V = [16\pi(2^{1/2})/3h^3]m_e^{3/2}E_F^{3/2} \ .$$

This density can be expressed in terms of the chemical composition of the star (sec. 2.4.3):

$$N_e/V = (\rho/m_H)(X + Y/2 + Z/2) = (\rho/m_H)(1 + X)/2 \ ,$$

and hence

$$E_F = \{[3\rho(1 + X)h^3]/[32\pi m_H m_e^{3/2}(2^{1/2})]\}^{2/3} \ . \tag{2.13}$$

The great bulk of the energy possessed by electrons under these circumstances is not thermal energy in the ordinary sense, for it cannot be radiated away. If the temperature were allowed to drop, the electrons would simply fill all of the available quantum states up to the Fermi energy. At absolute zero, no electrons would be above the Fermi energy level; the electron gas would continue to exert a strong pressure and would have a very high thermal conductivity. Under normal circumstances of low densities, the lowest quantum states up to the Fermi level can be filled only by reducing the temperature of the gas to absolute zero. At high densities this situation is also approached but is nearly independent of the actual gas temperature. Such a closely packed gas of fermions is said to be degenerate. It follows from the above that for $E_F \geq kT$ the criterion for degenerate behavior of matter is

$$T \leq (N_e/V)^{2/3}(h^2/8m_e k) .$$

The electron pressure, as for a simple gas, is the momentum transfer per unit area, or

$$dP_e = (\text{momentum}) \times (\text{velocity}) \times (\text{number density})$$

$$= 2pv(dN_e/6V) .$$

Inserting the equations arrived at previously and carrying out the integration leads to the following equation for the nonrelativistic electron pressure (Fow26):

$$P_e^{nr} = \tfrac{1}{40}(3/2\pi)^{2/3}[(1 + X)\rho/m_H]^{5/3}(h^2/m_e) . \tag{2.14}$$

Similarly, for relativistic conditions ($E = pc$, $v = c$) the relativistic electron pressure is (Cha31, Cha84)

$$P_e^r = \tfrac{1}{8}(3/16\pi)^{1/3}[(1 + X)\rho/m_H]^{4/3}ch . \tag{2.15}$$

Note that both equations of state of a degenerate electron gas depend only on density; however, in degenerate matter the pressure is not thermally produced but rather is a consequence of the Pauli principle.

One can define a critical density ρ_c for the condition that $P_e^{nr} = P_e^r$ and obtain

$$\rho_c = [(5^3/2^5)3\pi^2][1/(1 + X)][m_H/(\hbar/m_e c)^3]$$

$$= 3.82 \times 10^6/(1 + X) \text{ g cm}^{-3} . \tag{2.16}$$

The quantity $m_H/(\hbar/m_e c)^3$ corresponds to the density of protons which have been compressed to a distance equal to the electron Compton wavelength $\lambda_e = \hbar/m_e c = 3.9 \times 10^{-11}$ cm. Equation (2.14) can now be rewritten as

$$P_e^{nr}/\rho c^2 = [5(1 + X)/32](m_e/m_H)(\rho/\rho_c)^{2/3} \qquad \text{for } \rho < \rho_c ,$$

and equation (2.15) as

$$P_e^r/\rho c^2 = [5(1 + X)/12](m_e/m_H)(\rho/\rho_c)^{1/3} \qquad \text{for } \rho \geq \rho_c .$$

These equations show that the pressure is produced by the electrons while the rest energy ρc^2 arises from the nuclei.

In ordinary matter, composed of electrons and ions, the total pressure is

$$P_{tot} = P_e + P_i .$$

Because in the stellar interior there are many more electrons than ions and because the electrons—because of their much smaller mass—have the larger velocities, the pressure will be due mostly to electrons, and $P_{tot} \simeq P_e$.

2.4.11. Theory of White Dwarfs

Using the hydrostatic equilibrium condition in its approximate form (sec. 2.4.1)

$$P(0) = P \simeq GM^{2/3}\rho^{4/3}/2^{1/3} \tag{2.17}$$

and the nonrelativistic electron pressure (eq. [2.14]), the following relation between stellar mass and density is obtained for $\rho < \rho_c$:

$$M(\rho) \simeq (5^{3/2}/2^7)[(1 + X)m_e c^2/m_H G]^{3/2}(\rho^{1/2}/\rho_c) . \tag{2.18}$$

Similarly, for $\rho \geq \rho_c$ the relativistic electron pressure (eq. [2.15]) leads to

$$M(\rho) \simeq (5^{3/2}/2^7)[(1 + X)m_e c^2/m_H G]^{3/2}(1/\rho_c^{1/2}) . \tag{2.19}$$

This relativistic mass is independent of density, which means that it is also independent of radius and, thus, it is a limiting mass:

$$M_c \simeq (6^{1/2}\pi/32)(1 + X)^2(1/m_H^2)(c\hbar/G)^{3/2}$$

$$\leq 1.8 \ M_\odot \qquad \text{for } X = 1 . \tag{2.20}$$

More realistic calculations give as an upper limit $M_c \leq 1.4 \ M_\odot$ for $X = 0$ (Cha35, Cha84), which is called the Chandrasekhar limit. More massive stars cannot be supported solely by electron degeneracy pressure, no matter how small they are. In this case, gravitation wins and the star will collapse; no stable configuration exists. This discovery of Chandrasekhar (Cha31, Cha32) and Landau (Lan32) was based on the recognition that the dependence of pressure on density changed in going from nonrelativistic to relativistic conditions, giving rise to a finite limit on the mass of a star. Equations (2.18) and (2.19) prescribe the mass range of white dwarfs. Indeed, no white dwarf has ever been found whose mass is greater than the Chandrasekhar limit. Since their mean distance is about 10 ly, they represent a significant part of the mass of the universe.

It is interesting to note that the Chandrasekhar limit contains the Planck constant $\hbar$. Thus the Planck constant determines not only the structure of the microworld (atoms, nuclei) but also the mass scale and the inner structure of the stars in the universe. The results also show that the features of stars are not accidentally arrived at, but are a consequence of well-known physical principles. In many ways white dwarfs resemble giant atoms. As in an atom, the electrons cannot radiate, since they find their way to lower energy blocked by other electrons. The matter is thus effectively very cold in spite of its high internal energy.

The limit on mass and density also places a limit on the size of white dwarfs:

$$R_c = (3M_c/4\pi\rho_c)^{1/3} \simeq 8000 \text{ km} \qquad (\text{for } X = 1) , \tag{2.21}$$

which is comparable to the size of the earth ($R_\oplus = 6378$ km). It is easily shown that

$$MR^3 = M_c\, R_c^3\ ,$$

and thus that the size of white dwarfs decreases with increasing mass. In summary, white dwarfs have masses comparable to that of the sun but with diameters more like those of planets, implying that they have exceedingly high densities. The visible white dwarfs are presumed to be old stars which have consumed all their nuclear fuel (sec. 2.5).

Equations (2.17) and (2.18) imply that with decreasing pressure the density of stellar matter approaches zero, while the density of planetary matter is finite at zero pressure, $\rho(P = 0) = \rho_0$. While actual density depends on the chemical composition, there is a characteristic density for such matter

$$\rho_0 \simeq (m_H/r_B^3) \simeq 10 \text{ g cm}^{-3}$$

(r_B is the Bohr radius), which is determined by the electromagnetic force (atomic physics) and not by the gravitational force. If the mass and hence the pressure pass a certain threshold value, the atomic structure collapses, and we enter the regime of degenerate matter discussed above. This threshold can be used to define the boundary between planets and white dwarfs. The threshold mass is obtained by inserting the above density ρ_0 in equation (2.18): $M(\rho_0) \simeq 10^{-3}\, M_\odot$. Thus this mass represents the maximum mass for planets ($M_{\text{Jupiter}} = 0.95 \times 10^{-3}\, M_\odot$) and the minimum mass for white dwarfs. Although white dwarfs can exist only in a relatively narrow mass range of

$$10^{-3}\, M_\odot \le M_{\text{WD}} \le 1.4\, M_\odot\ ,$$

cool matter can range from the hydrogen atom with $m_H = 1.6 \times 10^{-24}$ g up to planets with $M_p = 2 \times 10^{30}$ g, that is, over a range of 10^{54}.

2.4.12. Neutron Stars

If the density continues to increase much beyond that of a white dwarf, the Fermi energy of the electrons becomes truly high (E_F can be many MeV or even many tens of MeV). Under these circumstances, the energetic electrons react with protons whether inside or outside nuclei. The net result of these and other interactions is the conversion of ordinary nuclei into essentially a neutron gas. At even higher densities, on the order of 10^{15} g cm^{-3}, these neutrons (also fermions), as with electrons, will form a degenerate gas, since again there will be too few quantum states available to maintain a Maxwell-Boltzmann distribution. Thus at very high densities, the gas pressure will have a large contribution from degenerate neutrons. This is the density regime of the so-called neutron stars.

Despite the fact that the actual temperature in the central region of these stars is in the range of a million to a billion degrees, their physical properties again are those of a gas near zero temperature. The neutrons cannot move freely in space because at these high densities matter behaves like matter near 0 K. Thus, while white dwarfs resemble giant atoms, neutron stars resemble giant nuclei held together by the gravitational force, not by the nuclear force.

Replacing the quantities X, m_H, and m_e in equations (2.16) and (2.20) with 1, m_n, and m_n, respectively, we obtain $\rho_c = 1.2 \times 10^{16}$ g cm^{-3}, $R_c \geq 6.6$ km, and $M_c \leq 1.8 \, M_\odot$, showing again that there is an upper limit on the stellar mass (and a lower limit on stellar size) that can be stabilized by the degenerate neutron gas, which is—in our rough approximation—identical with the Chandrasekhar limit for white dwarfs. Thus neutron stars are objects measuring a few kilometers in size but having a mass on the order of that of the sun. Detailed calculations of the conditions of equilibrium for neutron stars were first done by Gamow and by Oppenheimer and Volkoff (Gam37, Opp39). The value of the critical mass was found to be $M_{NS} = 0.7 \, M_\odot$. This mass limit is very difficult to calculate with great accuracy because the pressure is modified by the nuclear forces between the neutrons, and these forces are not well understood.

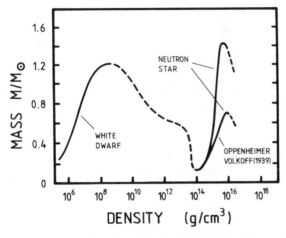

FIGURE 2.4. Equilibrium mass as a function of density (Ree74). The solid portion of the curve is the region where stable stars reside. On the dashed portions, stars are unstable against collapse.

Figure 2.4 shows a plot of the equilibrium mass of stars as a function of density. There is no "classical" equivalent of the relationship shown in this figure. Ordinary matter in quantities of stellar size cannot be supported unless it is very hot (sec. 2.4.2). Thus, as a quantity of matter cools, it will become a planet like Jupiter, if its mass is less than about $10^{-3} \, M_\odot$, while at higher masses it will become a white dwarf, as shown by the solid curve in Figure 2.4 ($M \propto \rho^{1/2}$). Near densities of 10^8–10^9 g cm^{-3}, the matter reaches the Chandrasekhar mass limit. Beyond this point the slope of the curve becomes negative, and there are no stable configurations (*dashed line*) until the slope becomes positive again near 10^{13} g cm^{-3}. This instability can be understood as follows. Suppose that a star with a mass of about 0.6 $M_\odot$ and a density of 10^{12} g cm^{-3} has its radius slightly reduced. The central density then begins to increase, but the star is too massive for its new central density, causing yet an additional increase in density; and the cycle continues. Before the degeneracy pressure of the neutrons becomes strong, most of the electrons will have been absorbed by the protons of the system, thus reducing very greatly the electron degeneracy pressure. No objects stable against gravitational collapse can exist in nature if their central densities are in this intermediate region where the electron degeneracy pressure has been weakened but the neutron degeneracy pressure has not yet become strong. The star cannot stabilize until it reaches the positive slope at a density of about 10^{15} g cm^{-3} (Fig. 2.4), which is in the domain of neutron stars. The star has "collapsed" from its original density of 10^{12} g cm^{-3} to a density of 10^{15} g cm^{-3}, and its radius has shrunk by a factor of 10. The great variety of stars existing in nature implies significant differences in the equation of state for the various density regimes.

Because of the small diameters of neutron stars, they can spin very rapidly, usually about once per second, although some may spin faster. As they spin, any surface emission features will produce a lighthouse effect. All kinds of radiation—radio waves, visible light, and even sometimes X-rays—periodically sweep past an external observer. These stars are known as pulsars. There is a pulsar within the Crab Nebula (Fig. 1.12) that turns very rapidly, over 30 rotations per second; and, as it turns, the radiation from it (thought to be synchrotron radiation) appears to go on and off like a beam from a lighthouse.

2.4.13. Black Holes

Once stellar collapse gives rise to densities greater than the critical density for neutron stars, there is no mechanism known that can stop the further collapse (Opp39a, Ruf71, Cha83, Cha84). What happens during this final collapse? We can get a clue by calculating the potential energy of a single proton on the surface of a collapsing star of one solar mass:

$$\Phi = \frac{GM_\odot m_p}{R} \simeq GM_\odot^{2/3} \rho^{1/3} m_p \ .$$

Using a density of 10^{18} g cm^{-3}, one finds that the surface potential of a proton exceeds its rest-mass energy of 938 MeV. This density corresponds to the condition $R_S/R \geq 2$, where $R_S = 2GM/c^2$ is the well-known Schwarzschild radius (Wei72). As an object approaches this radius, the energy of any light it emits is diminished as a result of the gravitational redshift according to the relation $\Delta v/v = R_S/2R$. When the object crosses the Schwarzschild radius (called also the "event horizon"), it disappears entirely, having lost its ability to transmit light to any external observer. Thus the surface defined by the Schwarzschild radius is said to be closed, since an object passing through it cannot escape. Such objects are called "black holes." All one can detect of a black hole (a singularity in the spacetime continuum) is its gravitational field (Ruf71; see, however, Haw74).

Note that this condition of "blackness" arises when the density greatly exceeds nuclear density. However, the effect is very general and independent of the particular density. Table 2.2 lists the Schwarzschild radius R_S for a variety of bodies of a given mass and average density. To become a black hole, the earth and the sun would have to shrink to sizes of a few mm and km (Fig. 2.5 and Table 2.2) and a galaxy would be confined to a region only a fraction of its normal size. Presumably this is the ultimate fate of our Galaxy (if it has not been disrupted by collisions with other galaxies). Interestingly, when it occurs, the density will be so low that the laws of matter on a local scale will be completely unaffected, and objects will pass into the black hole without notice. In fact, the above could describe our universe as it is now, that is, we may at this moment be living in an enormous black hole. It follows from the above that the condition of "blackness" of a black hole is neither surprising nor spectacular. What is dramatic, but only for objects of about 1 $M_\odot$, is that a black hole can be accompanied by such enormous gravitational forces that the nuclear forces, the strongest known forces at short distances, are overcome.

TABLE 2.2 Characteristics of Black Holes

Object	$M/M_\odot$	R_S	ρ (g cm^{-3})
Earth	3×10^{-6}	9 mm	2×10^{24}
Sun	1	3 km	2×10^{16}
Galaxy	10^{11}	0.03 ly	2×10^{-6}
Universe	10^{22}	3×10^9 ly	2×10^{-28}

What happens beyond this point is pure speculation. Unless some new laws of physics intervene, matter will shrink to a singular point. Unfortunately, even if such a point were close by, we would never see it. According to our present understanding of physics, the gravitational force near such a collapsed object would be so great that, from a region surrounding it fixed by the Schwarzschild radius, essentially nothing, not even radiation produced by the singularity, could escape. Probes could come only as close as several kilometers before being swallowed up within the Schwarzschild radius (Ree74, Ruf75). Material in the strong gravitational field just outside the black hole would be accelerated to high velocities and heated by collisions to very high temperatures. As a result, a black hole should be surrounded by a small but intense disklike source of radio and X-ray radiation, looking rather like the compact radio sources described in chapter 1.

A black hole is thought to be created when a very massive star ($M \gg 10\, M_\odot$) that has exhausted its nuclear fuel can no longer generate the internal pressure needed to counterbalance the inward pull of gravity. Such a star would not have become stabilized in its gravitational collapse at the dimensions of a neutron star but would have continued until all its matter was crushed to near infinite density (chap. 8).

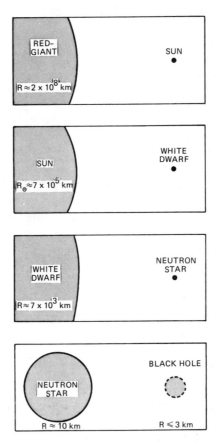

FIGURE 2.5. Comparison of sizes of stellar objects for $M = M_\odot$. Note that neutron stars are not far removed from the conditions of a black hole.

2.5. The Lives of the Stars

The energy emitted by stars is recognized as being nuclear in origin, but with the nuclear-burning processes ignited and moderated by the energy derived from the gravitational contraction that accompanies the process of stellar evolution. Stars are born, live, and die: a mass of gas becomes unstable, contracts, condenses into smaller units, forms stars, and is then irreversibly transformed. Finally, some of the mass is returned to the interstellar gas and the cycle starts again. Each cycle leaves its mark on the galaxy (sec. 2.6).

2.5.1. Birth of Stars in Interstellar Clouds

A gaseous cloud becomes unstable and starts to collapse when the gravitational energy is larger than the thermal energy of the constituent molecules and atoms of the cloud:

$$E_G \simeq G(M^2/R) \geq E_T = \tfrac{3}{2}kT(M/m) ,$$

where M, R, and T are the total mass, radius, and temperature of the cloud and m is the mean molecular weight. This condition—called the Jeans criterion—can be expressed in a more compact form as

$$M \geq 3.7(kT/Gm)^{3/2}\rho^{-1/2} .$$

Hence, clouds of low temperature and high density require smaller total masses in order to become gravitationally unstable. Typically, interstellar clouds (primarily neutral hydrogen) have a density of about 100 atoms cm^{-3} and temperatures of about 100 K. Hence the instability occurs when the mass of the cloud is greater than $2 \times 10^4 \, M_\odot$.

As this massive cloud contracts, the density increases rapidly in the central region. If the temperature can be kept nearly constant by radiating away the excess thermal energy resulting from the gravitational contraction (sec. 2.4), increasingly smaller portions of the collapsing cloud fulfill the Jeans criterion. These regions can then fragment into independent collapsing clouds, if there are some inherent anisotropies (inhomogeneities). Such a process could lead to the birth of individual stars. These stars illuminate the remainder of the gas cloud, leading to the astronomically observed phenomenon of bright young stars (some no more than a few million years old) embedded in interstellar clouds of gas and dust (Fig. 2.6). These considerations imply that stars are not born alone but rather are born in association with other stars, leading to large families of stars observed as clusters. For example, the globular cluster M13 (Fig. 1.7) contains about 1 million stars and is about 10 billion years old. This cluster must have formed very early in the universe when conditions of $\rho \simeq 10^{-21}$ g cm^{-3} and $T \simeq 10^4$ K prevailed. Applying the Jeans criterion to such a primordial cloud, one finds the required mass to be greater than $5 \times 10^6 \, M_\odot$, which is consistent with observation.

Unlike giant clouds of gas and dust, smaller clouds such as the solar nebula cannot form a star spontaneously because the thermal pressure of the gas is sufficient to prevent gravitational collapse. It has been proposed that such smaller clouds of gas require some violent external influence, such as a shock wave, to exert additional pressure on the cloud to provide the needed compression. The density of the cloud then increases to the point where the gravitational pressure does overcome the gas pressure, the cloud contracts, and stars are born. The possibility that such a violent event may have triggered the birth of our sun is discussed in chapter 10. Understanding of the processes of

FIGURE 2.6. Shown is the Great Nebula in Orion (M42 or NGC 1976), a glowing cloud of gas with a diameter of a few hundred light-years. There is enough mass in this gigantic cloud to form a very large number of stars, perhaps as many as 10^5. It is only one of many such clouds found in the Milky Way, in which the density of interstellar matter is from 10^4 to 10^7 times greater than elsewhere in the Galaxy. Apparently, even now stars are being born in these clouds (Wyn81). (Lick Observatory photograph.)

star formation remains one of the most complex pursuits of contemporary astrophysics (Zei78, Str75, Str76, Woo78b, Hab79).

2.5.2. From Nebulae to the Main Sequence

Star formation begins when a cool cloud of gas in the interstellar medium collapses to a high density. At first, the internal heat resulting from the contraction is readily radiated away (and is absorbed by the cold interstellar grains) because of the relatively high transparency of the gas in the infrared. At this stage the gas is essentially in gravitational free fall. As the gas becomes denser, its opacity increases, and the energy released in the collapse can be stored in the interior. The gas settles down to form a star when the stored energy is one-half the released gravitational potential energy (virial theorem; sec. 2.4.4). The other half of the energy is lost from the surface through radiation. Because the center of the star is relatively hot and the surface layers relatively cool, energy flows from the center toward the surface. The evolutionary segment beginning with the start of collapse through establishment of quasi-hydrostatic equilibrium takes only 30–100 years. At this point the star becomes visible, being larger and brighter than it will be when it reaches the main sequence (Fig. 2.7). According to Hayashi (Hay62, Hay66) a star destined to be similar to the sun when it reaches the main sequence will be about 300 times brighter and slightly redder at this protostar stage than it will be when it reaches the main-sequence stage.

As energy is radiated from the surface, the star shrinks just enough to provide not only the energy that has been radiated away but also an equal amount to be added to that already stored. Thus the interior of the star

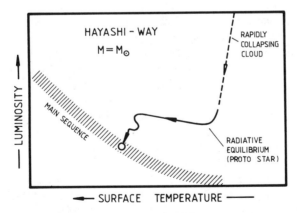

FIGURE 2.7. Schematic diagram of the evolutionary track, called the Hayashi way, for a solar-type star. Radiative equilibrium sets in at the sharp break in the track. The sites of star formation are cold gaseous nebulae in interstellar space. The formation of protostars might have been observed via infrared astronomy, where objects as cool as 650 K or even 150 K seem to have been detected in gaseous nebulae (Faz76).

shrinks at a rate which is governed by the rate of energy loss from the surface. This stage of stellar evolution is referred to as the Kelvin-Helmholtz stage. If the history of a star were plotted on an H-R diagram, one would see that, with increasing temperature and very little change in surface brightness (luminosity), the star evolves nearly horizontally toward its position on the main sequence (Fig. 2.7). The time required to evolve to this stage of stellar life ranges between 10^7 and 10^8 years. The time scales and paths in the H-R diagram for individual stars, which vary depending upon the stellar mass, are referred to as Hayashi evolutionary tracks (Hay62, Hay66, Ibe65, Ibe67, Ibe70). When a group of youthful stars approaches the main sequence, the most massive stars settle in the bluest parts of the sequence. They arrive in advance of the others because they contract more rapidly. The less massive stars arrive later and array themselves in order of decreasing mass along the redder part of the main sequence.

2.5.3. Main-Sequence Stars

In a shrinking star the conversion of gravitational energy leads to a continual heating of the stellar interior. Eventually, the temperature at the center reaches about $(1-2) \times 10^7$ K, and a new source of energy appears in the core of the nascent star as thermonuclear reactions among hydrogen nuclei (protons) begin. The amount of energy released through conversion of hydrogen to helium, called "hydrogen burning," is very large compared with the amount of gravitational potential energy released before hydrogen burning starts. According to the virial theorem, the star will settle into a long-lived state in which the internal thermal energy remains one-half of the released gravitational potential energy; but now nuclear energy generated in the central region flows to the surface and is radiated into space. The large mass (mainly hydrogen) of a star together with the large amount of energy generated in hydrogen burning allows it to shine with prodigious energy output throughout its long lifetime. Gravitational contraction is, temporarily, halted; and for a long period the star changes little in size, temperature, or luminosity. This long period is reflected by the so-called main-sequence region of the H-R diagram. The majority of stars visible in space are at this stage of evolution occupying a restricted region as shown in Figure 1.24. With the hydrogen remaining in its interior, the sun, already approximately 5 billion years old, can continue on the main sequence with its present power output an equal period into the future. Life on earth is therefore possible for a long time.

Since stars on the main sequence differ in mass, they also differ in their luminosity—the rate at which they emit light and heat. A star with a mass ratio of $M/M_\odot = 10$ has a luminosity ratio of $L/L_\odot \simeq 10^4$. To compensate for this prodigious energy output, the more massive stars must convert hydrogen into helium far more rapidly than the sun. Because of this rapid use of hydrogen, the supply of this element is depleted in a comparatively short time. Whereas the hydrogen in the sun will last for about 10 billion years, hydrogen in some stars of larger mass will last for no more than a few tens of millions of years, a time span much shorter than the age of the Milky Way (estimated to be about 12–18 billion years old; chap. 9). Thus hydrogen in such very massive stars is consumed on a time scale that is, cosmically speaking, rather short. Consequently, since the beginning of the universe there has been ample time for several generations of hot, massive stars to have lived and died.

Some protostars do not have sufficient mass ($M \leq 0.1\ M_\odot$) to ignite the self-sustaining hydrogen burning and evolve downward on the H-R diagram completely outside the main sequence. They are highly degenerate and shine only on account of the energy resulting from gravitational contraction. These stars evolve directly into "brown or black dwarfs" (Jupiter is on the margin of being such an object), where the gravitational collapse is stopped indefinitely by the pressure of the degenerate electrons.

It is interesting to note that stellar model calculations indicate that a low-mass star is more dense at its center than a high-mass star. These results can be understood using the approximate relation derived in section 2.4.8:

$$T \propto M/R .$$

With this result the mean density is given by the proportionality

$$\rho \propto M/R^3 \propto M(T/M)^3 .$$

Since a nuclear-burning phase proceeds approximately at a fixed temperature (independent of stellar mass), one arrives at the mass-density relationship

$$\rho \propto M^{-2} .$$

Thus, for a given nuclear-burning phase, the smaller the mass of the star, the higher the mean density (and thus the stronger the electron degeneracy). Note that the above relationship is valid only as long as the mass of the core is smaller than the Chandrasekhar mass limit ($M_{\text{core}} \leq 1.4\ M_\odot$).

2.5.4. Endpoints of Stellar Evolution

A star remains in its original position on the main sequence as long as the nuclear energy generated in its interior is sufficient to sustain its luminosity. When the hydrogen in the core is nearly exhausted, the core will have become predominantly helium (^{4}He). Having run out of hydrogen fuel, the whole star will rapidly resume its gravitational contraction until intense nuclear hydrogen burning is ignited in a shell surrounding the core. As a consequence, more energy is generated in the core than can be radiated away at the surface; and the outer layers around the core (the envelope) gradually expand and cool. Thus, while the star becomes brighter, its surface becomes cooler (redder). The star begins to move upward and to the right in the H-R diagram (Fig. 2.8) and becomes a red giant (Ibe74, Cam76, Ibe84a). At a central temperature of about 10^8 K, the sun will have become a red giant about 30 times its present diameter and about 100 times its present luminosity. Within the inner 50% of its mass its hydrogen will be exhausted, and its central temperature will become high enough for helium to ignite in its degenerate interior. At this stage temperature and density are decoupled. This means that, if temperature and density are sufficiently high at the onset of helium burning, a small increase in temperature will result in a very large increase in the energy production rate, leaving the density and hence the pressure unaffected. Since, under these conditions, the expansion cooling mechanism is not effective and the helium gas becomes highly explosive, the star experiences detonations, often referred to as helium flashes.

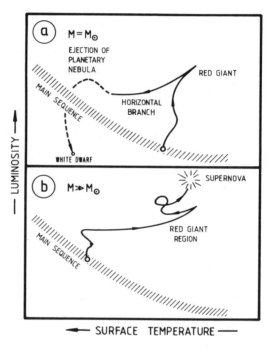

FIGURE 2.8. Schematic and simplified diagram of the evolutionary track of a solar-mass star and a more massive star. For more detailed discussions of single-star evolution see Cam76 and Ibe84a.

With little fuel remaining, the sun will presumably contract as quickly as it expanded. The sun may then move down and to the left on the H-R diagram to the horizontal branch. The stars in this region tend to pulsate, producing variations in light output that characterize them as variable stars as discussed in chapter 1 (Ibe84a). The rhythmic variation can be traced to instability in a zone near the surface of the star, where helium and hydrogen are only partially ionized. The degree of ionization depends sensitively on pressure and temperature. Any instability that increases ionization subtracts energy from the flow of energy reaching the surface, and the star dims. Ionization, however, increases the number of free particles in a given volume, so that the shell simultaneously starts to expand. As the shell expands it cools and electrons recombine with nuclei. The recombination releases energy, and the star again brightens. As the star hunts unsuccessfully for equilibrium, its luminosity fluctuates.

The star may finally eject its envelope, in which case it will create a planetary nebula (chap. 1) and move down in the H-R diagram to become a white dwarf (Fig. 2.8) the size of the earth, a faintly glowing cinder. From this point on, the sun is stabilized indefinitely against further shrinkage. It will gradually cool and fade until it becomes a black dwarf (a dead star). The time from the white dwarf stage to the black dwarf stage is quite long, probably on the order of 500 billion years.

The evolutionary picture given for the sun holds for all stars with mass ratios in the range $M/M_\odot = 0.1$–1.4. White dwarfs are found frequently in the universe and represent the final stage of evolution of normal stars such as the sun. As noted before, protostars with $M/M_\odot \leq 0.1$ never initiate self-sustaining energy production but evolve directly into the black dwarf stage. It was also noted that stars with $M > 1.4\ M_\odot$ cannot die in this way. A violent adjustment must take place, since degenerate objects cannot support collapsing masses greater than the Chandrasekhar limit. Stars of moderate mass ($M \leq 8\ M_\odot$) are thought to shed their excess masses rather gently, probably ending with a fling in which a cloud known as a planetary nebula is thrown off (Fig. 1.14). Other stars shed their excess material into space via the nova phenomenon. Mass loss through high-velocity winds (type O and Wolf-Rayet stars) and through low-velocity winds (M supergiants) is also very important (Cam76, Ibe84a and references therein). In all cases these stars work their way toward the Chandrasekhar mass limit and eventually become white dwarfs.

Massive stars with $M \geq 8\ M_\odot$ have shorter and more spectacular lifetime patterns (Cam76). Their evolution involves a continued release of gravitational potential energy through shrinkage of the interior, with pauses in the shrinkage when new nuclear fuels are ignited which can then supply, for a time, the flow of energy toward the surface. Generally speaking, the ashes of one set of nuclear reactions become the fuel for the next set. Thus helium, the ashes of hydrogen burning, eventually ignites, and carbon-oxygen burning begins at a still higher temperature. Stars in such advanced stages of burning can attain extremely high central temperatures, perhaps billions of degrees, when they become supergiants. In these stars practically all the known chemical elements may be created.

This process of shrinkage followed by nuclear fuel burning (Fig. 2.9) cannot continue indefinitely. There are only a finite number of stages of nuclear burning possible (sec. 2.6) before all of the available thermonuclear energy is released. Each successive stage of the evolutionary process generates less energy than the previous one. In addition, energy losses due to neutrinos become large in the advanced stages. All the stages taken together are like a

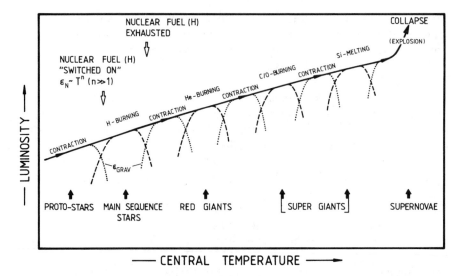

FIGURE 2.9. The life of a star is composed of alternative stages of gravitational (ϵ_{GRAV}) and nuclear (ϵ_N) energy generation. Because of the nuclear burning, the gravitational contraction is interrupted several times. The length of these interruptions depends partly on the nature of the nuclear fuel being consumed. During these nuclear stages, the evolution of the macroscopic parameters of a star (L, T, R) slows down substantially (giving rise to the H-R diagram). These stages are also characterized by a change in the chemical composition of the star.

limited number of energy checks which the star can cover. Once they are cashed, no more energy can be obtained from nuclear processes. This situation prevails when the core reaches the pure iron stage, where the material consists mainly of iron group elements, for which the binding energy per nucleon is maximum (chap. 3). Thus, after the nuclear fuels have been completely exhausted, the only prospect in store for massive stars is continued shrinkage until they become dynamically unstable and collapse.

As the core of a massive star collapses gravitationally, photodisintegration of iron nuclei and electron capture occur, removing electrons from the gas and consuming energy. Both effects speed up the collapse, and vast amounts of energy are lost to the star by escaping neutrinos (sec. 2.4.12 and chap. 8). These energy losses can be compensated for only by an extremely rapid gravitational collapse of the iron core, leading finally to an enormous explosion. This scenario describes the Type II supernova phenomenon (Baa34, Hoy60, Fow64), which represents the brief final event in the evolution of a massive star (chap. 8). In the process the star is destroyed, except possibly for a high-density central core, which is left behind as a remnant in the form of either a neutron star or a black hole.

The gravitational energy released by the contracting matter during the formation of a neutron star is converted partly into rotational kinetic energy as the collapsing star spins faster and faster to conserve angular momentum. A comparable amount of energy escapes in the blast of neutrinos and in the explosive expansion of the star's outer envelope, which forms the typical expanding cloud of a supernova remnant. The energy stored in the rotation of the neutron star escapes gradually in streams of high-energy particles. In some way, not yet understood, the streaming particles generate beams of electromagnetic radiation (synchrotron radiation), which, as the neutron star rotates, are observed as the characteristic periodic signals of pulsars. The fastest one, the Crab Nebula pulsar, rotates 30 times per second. Any object held together exclusively by its own gravity will fly apart at such a rotation rate unless its density exceeds 10^{12} g cm^{-3}. (Terrestrial objects such as tops and flywheels that rotate more rapidly are held together by the much stronger electromagnetic force.) The discovery of a neutron star in the Crab Nebula seems, then, to support the above picture of the evolution of massive stars (chap. 8).

2.6. The Origin of the Chemical Elements

As just described, a main-sequence star fuses hydrogen into helium, releasing the energy needed to compensate for the energy loss from the surface and to stabilize the star for a long period. As the hydrogen is exhausted, further contraction of the central region of the star occurs. This results in the temperature rising, which, in turn, sets off higher temperature reactions (Fig. 2.9). The residual helium undergoes reactions producing the elements C and O and a very small amount of Ne. This is the next stage in the evolutionary process, wherein reactions among simpler atoms result in more complex ones. At higher temperatures, C and O themselves undergo reactions, creating a host of well-known elements: Na, Mg, Al, Si, S, and Ca. The Ne is photodisintegrated between C and O burning. At even higher temperatures, the common metals appear: Fe, Ni, Cr, Mn, and Co. These are the main processes that give rise to the elements with comparatively large abundances. Elements of low abundance are also created inside stars in minor secondary reactions (Tru84).

We see, then, how the evolution of nuclei is associated with the evolution of stars themselves. In nuclear processes taking place inside stars, the basic components of the materials of our everyday world are produced. The stars are intensively hot furnaces ("cosmic cauldrons"), in which all the materials of the universe have been forged. These same processes also serve as the energy source of the stars, enabling them to emit radiation from their surfaces over enormous periods of time. The network of nuclear reactions, transforming matter from hydrogen into heavier atoms like Fe, can provide stars with only a limited amount of energy.

The ejection of material into space at the end of the life of a massive star contributes to a cyclic process that begins with the condensation of a gas cloud like the Orion Nebula (Fig. 2.6) and ends with the return of matter to a similar cloud like the Crab Nebula (Fig. 1.10). The matter so returned differs in its composition from the original material. It now contains more complex atoms. So the composition of the gas clouds themselves is changed by the nuclear processes occurring within stars. Stars that are now forming in the Milky Way must, therefore, contain more complex atoms than stars formed earlier. The logical consequence of this cyclic evolution is that material is transferred back and forth between stars and gas clouds as the ejected material becomes available for the creation of new stars. The process is shown schematically in Figure 2.10. Note that a supernova can blow an amount of material into space equal to several solar masses. A normal nova returns less than 0.001 $M_\odot$ in a single explosion; but normal novae are far more numerous, and explosions may occur repeatedly in the same star.

STATE OF AFFAIRS

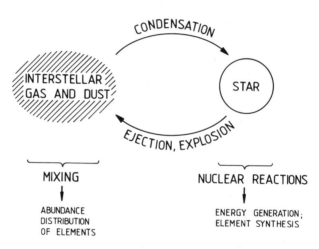

FIGURE 2.10. Shown is the cycling process (Hoy46) in which matter is transferred back and forth between stars and the interstellar gas. Each cycle enriches the interstellar material with the metal elements, leading finally to the observed abundance distribution (chap. 1).

The Milky Way is estimated to be about 15 billion years old. Hence, by the time the solar system was formed, the above cyclic process had already been operating for about 10 billion years. The complex atoms of our local world, in the abundances given in chapter 1, were produced during this long time span. The C and N in our bodies, the oxygen we breathe, the iron in our blood, were all generated inside stellar furnaces at remote epochs in the past. The parent stars of our complex materials are by now dead white dwarfs or neutron stars, which we cannot trace or identify.

The above theory, that the elements have been formed and are still being synthesized in the interiors of evolving stars, is known as the "stellar nucleosynthesis of the elements," as proposed in 1957 by Burbidge, Burbidge, Fowler, and Hoyle (B^2FH) and Cameron (Cam57). Recently, several supernova remnants have been studied by the *Einstein X-Ray Observatory* (*HEAO 2*). The supernova remnant in Cassiopeia revealed bright regions corresponding to fast-moving knots of heated gas. These knots appear to contain material that has been enriched in O and S (as found optically) by nuclear processes in the interior of the star and subsequently thrust outward by the explosion. Detailed spectra also reveal the presence of Mg and Si. Future studies should provide clues to the details of a supernova explosion and to how the interstellar gas is mixed with the stellar debris and heated by it (chap. 10).

Finally, let us point to another general argument in favor of the stellar synthesis theory. The universal abundance distribution of the elements requires that the elements, however they were formed, be so distributed at least on a galactic scale. Stars do this by ejecting material in various phases of their evolution—as red giants, supernovae, or novae. Primordial theories certainly distribute material on a cosmic scale, but a difficulty is that the distribution ought to have been spatially uniform and independent of time, once the initial phases of the universe were past. There is now convincing evidence that this is not the case (except for H and He): all stars do not have the same elemental composition. One of the triumphs of the stellar synthesis theories is that it not only predicts variations of elemental abundances, but also shows great promise of quantitatively explaining these variations.

2.7. Evolution of Binary Systems

Stellar evolution becomes more complicated when the stars involved are the members of close binary systems—pairs of stars that revolve around a common center of gravity. Probably more than one-half of the stars in our Galaxy are members of binary systems, but only a small percentage of the binaries are close. One scenario is that these systems originate in the early stages of the formation and evolution of protostars. If the protostar is rotating as it contracts, instabilities can arise that will split the rotating mass (a pear-shaped figure that would break apart) to form a contact binary. Let us commence our considerations of these close binary systems with their mutual gravitational potential and then continue with their evolution, which will depend on the mass of the objects involved.

2.7.1. Effects of Gravity on Stars in Contact

Consider a close binary system with two stars of different mass. Clearly in this system (Fig. 2.11) there is a point on the line joining the two stars where the gravitational potentials are equal. A bit of matter at this point has the same probability of falling toward one star as it does of falling toward the other. When centrifugal acceleration is taken into account, a figure-eight equipotential surface called the Roche surface is defined (Fig. 2.11). Matter finding itself on that surface can go to either star. If the material of the two stars is well within the critical Roche surface, the binary system is stable. When one of the stars at some stage in its life begins to expand as part of its normal evolution on the main sequence, it may fill its Roche surface. Further expansion ceases as material is transferred through the common point (neck) of the Roche surface and comes under the gravitational influence of the companion star. In this case a gas "stream" may form. The expelled material may swirl around the companion star, eventually falling into it. The Roche surface (lobe) is then the boundary beyond which an expanding star in a binary system begins to transfer matter to its companion star. Some material is also pumped out of the system through the opposite ends of the figure eight. Most of the ejected material will, however, stay near the stars and envelop them in a thin

CLOSE BINARY SYSTEM

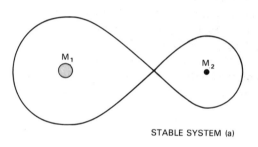

STABLE SYSTEM (a)

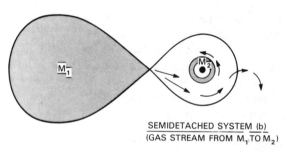

SEMIDETACHED SYSTEM (b)
(GAS STREAM FROM M_1 TO M_2)

FIGURE 2.11. Schematic diagrams of close binary systems (Gur75). The solid line is the critical Roche surface, where the attraction toward the two stars is equal. The relative sizes of the two lobes of the figure eight are governed by the ratio of the masses of the two stars, with the large lobe around the more massive star. If the stellar matter is within the critical surface, the system is stable (*top*). In the lower part of the figure, the more massive star is shown in the expansion phase—filling its Roche surface. With further expansion material flows to the compact companion. This is called a semidetached system.

shell. The resulting object resembles an egg with a double yolk. It is a star with two separate cores burning within a single atmosphere.

There are many close pairs of stars where the separations are small compared with the diameter of the stars themselves. Because of the enormous mutual tides, their gaseous atmospheres bulge toward each other, and their material is mixed via mutual gas streams. Thus the two stars form a close pair, and their appearance and evolution will be qualitatively different from those of single stars. Clearly, a massive binary star cannot evolve in its normal way, since it may lose a significant part of its mass to the companion star while expanding. This mass transfer from one star to the other will alter the structure of both stars as well as their subsequent evolution. In the interaction between the close pairs hydrogen-rich material flows from the atmosphere of one star into that of the other. The matter falling through the gravitational field of a compact companion star can be a substantial energy source much greater even than nuclear burning. For example, a mass m falling to the surface of a neutron star ($M = M_\odot$; $R = 10$ km) would liberate an amount of energy equal to $0.1mc^2$, regardless of what the falling material is. This may explain the enormous energies involved in eruptive stars as discussed in chapter 1. The concept of mass transfer in close binary systems is important in the evolution and energy budget of the system. Theoretical studies indicate (sec. 2.4.11) that before a massive star can reach the white dwarf stage, its evolutionary end point, the star must reduce its mass to a little more than the mass of the sun. Many overmassive stars which shed their excess mass in spectacular nova explosions are thought to be part of such a binary system (see below).

2.7.2. The Nova Phenomenon

The discovery that the nova phenomenon is directly related to the existence of very close binary systems in which one member is a white dwarf provided a plausible explanation for nova outbursts. It has been suggested (War76, Egg76, Fri77, Gal78, Tru82, Tru84) that nova outbursts are caused by the transfer of matter from the normal star of the binary pair onto the degenerate white dwarf companion (Fig. 2.12), which had previously converted virtually all its available fuel (hydrogen and helium) into carbon and oxygen (sec. 2.4.11). Because of its high density, the white dwarf has a very strong gravitational field, and gas accreting onto its surface has very high velocity, resulting in extremely high temperatures upon impact. When the temperature becomes high enough to ignite nuclear reactions, energy is violently released. The energy is released suddenly because the nuclear reactions proceed as in a degenerate gas. In a normal gas explosive energy release does not occur because the gas expands with increasing temperature, causing the temperature of the gas to fall. The nuclear reaction rates, which are a sensitive function of temperature (chap. 4), fall along with it. This thermodynamic effect controls the burning and enables all normal stars to radiate energy at a constant level for billions of years. As discussed in section 2.4.10, reactions in degenerate matter are not automatically controlled by expansion. Rather, the reactions proceed at an ever-increasing rate which releases an enormous amount of energy culminating in an explosive outburst.

Suppose that in a binary system involving a white dwarf, the white dwarf's companion is a star of normal size that has exhausted most of the hydrogen in its core and has begun to expand as the site of thermonuclear reactions moves outward. As it expands on its way to becoming a red giant, its shape is distorted by the gravitational attraction of the white dwarf. Gas (mainly hydrogen) streaming off the expanding star is swept into an accretion disk (in the plane of the star's orbit) that swirls around the white dwarf. Ultimately, most of the gas spirals into the white dwarf at extremely high velocity, raising the temperature of the dwarf's surface and setting the stage for the nova outburst. The accretion of gas onto the white dwarf is continuous, driven by the steady expansion of the outer layers of the companion star. This gas mixes with the material in the outer layers of the white dwarf (mainly helium with an admixture of carbon and oxygen), material that originated in the dwarf's interior. A white dwarf normally lacks hydrogen because its hydrogen was converted into heavier elements via nuclear reactions earlier in its evolution. The buildup of a fresh supply of hydrogen on the surface of the white dwarf is

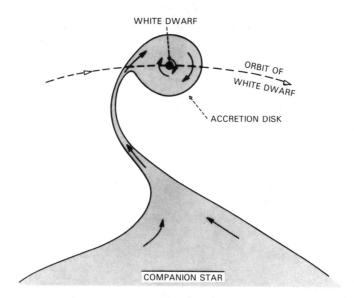

FIGURE 2.12. A nova binary system is represented schematically as seen from above the plane of the orbit (Wil81). The white dwarf member is a star that has contracted to roughly the size of the earth. Its companion is a star of normal size that has started to evolve toward becoming a red giant. As this star expands, the gas streaming off it is swept into an accretion disk that swirls around the white dwarf, ultimately spiraling onto its surface.

significant because nuclear reactions involving hydrogen proceed more rapidly than those involving other light nuclei. It is the addition of fresh hydrogen to the heavier elements already present in the dwarf that makes possible a new series of nuclear reactions.

The initiation of the fusion reactions requires a temperature of about 20×10^6 K. At that temperature individual nuclei are able to collide, fuse, and release energy. The energy necessary to raise the temperature is, as pointed out above, provided by the impact of the gas spiraling at high velocity onto the surface of the white dwarf. The time required for the temperature to reach the level necessary for fusion depends on several factors, including the rate at which the expanding companion star feeds gas into the accretion disk. Typically tens of thousands of years are required. Eventually a temperature is reached at which the degenerate material on the surface of the white dwarf, seeded with fresh hydrogen, begins to support fusion reactions. The chain of reactions is exactly the same as the one that occurs in the core of normal stars more massive than the sun. They are the reactions of the CNO cycles (chap. 6) and helium burning (chap. 7). Experimental tests of this nova model consist of observing expanding shells of novae. Observations have shown, indeed, that the shells are richer than normal stars in heavier elements such as C, N, and O, as expected from the operation of the CNO cycles and helium burning (Wil81, Wil82). Although there are uncertainties in some of the details, there is now general agreement that nova eruptions are caused by thermonuclear reactions (or thermonuclear runaways) on the surface of white dwarf stars in close binary systems (Tru84). Other cycles such as the NeNa cycle and the MgAl cycle (chap. 6) can also occur.

2.7.3. X-Ray Stars

The transfer of gas from a normal star into a nearby degenerate compact companion star in a rotating binary system is not unique to the nova phenomenon. A similar process is involved in the X-ray stars (chap. 1). These stars are also binary systems; however, in this case the compact object is not a white dwarf but rather is thought to be a neutron star or perhaps, in a few cases, a black hole. The matter in a neutron star is far more condensed (and in a black hole immensely more condensed) than that in a white dwarf. The large high-density mass of a neutron star creates a gravitational field about 100 billion times more powerful than that near the surface of the earth. As a result, the gas spiraling onto the compact object via an accretion disk is accelerated to extremely high velocities, which results in the emission of radiation, primarily in the X-ray region of the spectrum. Hence, in order to make a star that can generate X-rays at the prodigious power levels observed, nature must combine a superdense gravitationally collapsed object, such as a neutron star, and a supply of matter that allows the accretion process to proceed at an appropriate rate. In the case of a black hole, only the X-rays emitted by the infalling matter while it is outside the Schwarzschild radius will be observed.

Let us now turn to massive close binary systems where the transfer of matter influences the evolution of the two companions. Early in their life the companions evolve as isolated stars with minor perturbations, if any (Fig. 2.13). Each star is contained within its own gravitational potential well. There is a radical change, however, when star A, the more massive member of the close binary pair, and thus the faster evolving of the pair, ends the hydro-

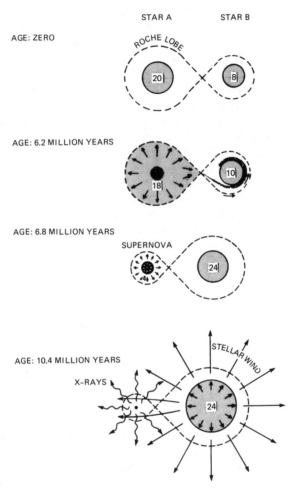

FIGURE 2.13. Schematic diagram of the evolution of a close binary system (Gur75). The dashed lines are the Roche lobes, and the shaded areas are the regions occupied by the stars. The system begins with two stars of masses 20 and 8 $M_{\odot}$, both still on the main sequence (age zero). At 6.2 million years, star A has swelled to fill its Roche lobe and begins to transfer mass to its companion B. This first mass transfer is completed at 6.8 million years, leaving behind a 4 $M_{\odot}$ helium star which explodes as a supernova. The supernova remnant becomes an X-ray source when star B becomes a supergiant.

gen burning in its core and begins burning the hydrogen in its shell. The star then turns off the main sequence and expands as it ascends the red giant branch of the H-R diagram. At some point the outer envelope of star A overflows the potential barrier (the Roche lobe) that separates it from its partner, star B, and mass is transfered to star B on a grand scale (Fig. 2.13). The result is that star A is stripped of its outer hydrogen-rich envelope, leaving only its highly evolved helium core. Such helium stars are very hot and luminous. They resemble the Wolf-Rayet stars which are formed in many close binaries (chap. 1). A helium star of this type moves rapidly to the late stages of evolution and a supernova explosion, just as if it were an isolated helium star.

Meanwhile star B, having inherited the envelope of star A, now has sufficient mass to hold the system together through the eventual supernova explosion. The system is now a binary consisting of the newly massive star B and the remnant of star A, now a neutron star or a black hole. Star B continues to evolve and begins to feed matter back onto the neutron star by emitting high-speed particles (the stellar wind, driven by the enormous radiation pressure of star B) or by expanding until its envelope overflows its gravity barrier. On the reverse trip, matter is drawn onto the gravitational potential well of star A, which is now so deep that a freely falling proton would arrive at the surface of the neutron star with a kinetic energy of more than 100 MeV. This energy is 14 times the energy released per proton when hydrogen fuses to helium. Indeed, the process is so efficient that an accretion rate of only 1 $M_\odot$ per 10^9 y is sufficient to provide all the power radiated by a typical high-luminosity X-ray star. Moreover, at this low rate of accretion the density of the inflowing material is small enough to allow X-rays emitted near the neutron star to escape from the system with little loss by absorption.

Actually the flow of accreting matter toward a neutron star is retarded by shock waves, magnetic fields, centrifugal forces, and radiation pressure. Instead of moving inward radially, the matter forms a rotating accretion disk of hot plasma. The material spirals through the accretion disk and is finally guided to the surface at the magnetic poles of the neutron stars (" hot spots "), with each proton losing its energy not as a single 100 MeV γ-ray but ultimately as many X-rays with energies mostly in the range below 20 keV. Since the magnetic poles usually are not on the axis of rotation, the regions of intense X-ray emission rotate off-axis. As a result, the X-ray emission, like the emissions from pulsars, seems to pulse periodically with periods of a few seconds. Because black holes cannot have a surface or a strong magnetic field, they are ruled out of consideration as sources of pulsating X-rays. Hence, X-ray sources in class 1 objects (chap. 1) are rotating neutron stars with powerful magnetic fields. The magnetic fields that guide the flow of accreting material cause extremely regular and revealing pulsations in their X-ray emission. Class 1 objects themselves result from the evolution of massive close binary systems. These young stars, located in the spiral arms of the Galaxy, are rich in interstellar gas and dust. As discussed above, a fast-pulsating X-ray source is a spinning neutron star which is orbiting closely around a massive normal star. The observed longer term X-ray variations, with periods of a few days, are caused by the periodic eclipsing of the neutron star by the much larger companion star.

Class 2 X-ray stars, showing neither regular pulsation nor eclipsing, are found in the center of the Galaxy in globular clusters, regions associated with old stars. They are characterized by aperiodic energetic X-ray bursts in which as much X-ray energy is emitted in a few seconds as the sun radiates at all wavelengths in two weeks. Analysis of the spectra of the X-ray bursts provides persuasive evidence that the X-ray–emitting object is also a neutron star. The absence of regular pulsations indicates that the neutron stars are old and hence have only weak magnetic fields (slow rotation). The source of matter from the continuous X-ray emission, as well as the X-ray bursts, is again a binary normal star, but of low mass ($M \leq M_\odot$). Such a low-mass star is not luminous enough to be detected directly by optical telescopes, which also explains the absence of detectable eclipsing. This star burns its nuclear fuel very slowly and hence must also be an old star. The nature of the X-ray bursts is thought to be caused by uncontrolled thermonuclear reactions on the surface of the neutron star; i.e., the bursts are the result of a gigantic thermo-nuclear flash similar to that of a nova. They may be initiated by interruptions in the flow of matter, which could pile up and then suddenly be dumped onto the surface of the neutron star (Con78, Mar80, Lew81).

2.7.4. Supernovae of Type I

Supernovae have been divided by astrophysicists into two major classes (Kir76). Supernovae of Type I (SNIs) are high in heavy elements (oxygen to iron) and very low in hydrogen. Their variation of luminosity with time, called the light curve, follows a rather definite pattern (Gur76): the luminosity rises to a maximum of about 10^{43} ergs s^{-1}, and then drops by an order of magni-tude. The light curve has a peak in its luminosity which lasts about 50 days, then declines exponentially with a half-life of about 80 days. Since no point X-ray sources or X-ray nebulae have been detected in the SNI remnants, it is believed (Hel84) that they do not leave any compact remnant such as a neutron star. SNIs are observed in all types of galaxies, including elliptical galaxies, and are not concentrated in spiral arms as are supernovae of Type II (SNIIs) (Tri82, Whe82). SNIIs contain some heavy elements but are primarily hydrogen. They have more variable light curves, often with a "broader" maximum followed by a "plateau." The luminosity remains constant for 50–100 days and then declines very steeply. The Crab Nebula (Figs. 1.10 and 1.12) provided the first direct evidence that at least some SNIIs leave neutron stars as remnants. For some time it was thought that all SNIIs produced a neutron star. However, recent surveys (Hel83, Hel84) indicate that only about 20% of all supernova remnants contain a neutron star in the form of a pulsar. Assuming that supernova explosions are divided roughly equally between

Type I and Type II, a hypothesis consistent with extragalactic supernova studies, it is concluded that only about 40% of SNIIs produce an active pulsar. For the remaining SNIIs one must assume that they become black holes or neutron stars that do not have a magnetic field sufficient to create a pulsar. There is as yet no evidence of what fraction of SNIIs become black holes.

As early as 1960 Hoyle and Fowler (Hoy60) proposed that supernovae of Type I might be triggered by explosive carbon burning in degenerate cores. This idea was subsequently developed into the carbon detonation supernova model (Arn69a) and later into the carbon deflagration supernova model (Nom76). In the currently most popular models accreting white dwarfs in binary systems are assumed to be the progenitors of SNIs (Whe82a, Ibe84, Woo84). Various scenarios have been proposed to explain how matter from a companion star may fall onto the white dwarf, and also to explain how the white dwarf starts nuclear burning of the fresh supply. In one of the simplest scenarios, matter accretes onto the white dwarf and is converted quickly by hydrogen and helium burning into carbon and oxygen, thus increasing the mass of the carbon and oxygen core. Eventually, the carbon ignites at the center and burns in a wave that travels outward at a supersonic velocity leading to a detonation (Cou48). The detonation wave totally disrupts the star, leaving no remnant. Most of the star burns to ^{56}Ni, which will later decay radioactively to ^{56}Co and ^{56}Fe. This decay is important in understanding the light curve of SNIs (Bor50, Col69, Arn79, Che81, Sut84). However, these detonating white dwarf models do not synthesize appreciable amounts of intermediate mass elements (O, Si, Ca) and therefore cannot account for the existence of these elements in the outer layers of SNIs (Bra84). Nomoto and collaborators (Nom76, Nom84) found, however, that under certain conditions (i.e., a relatively rapid mass accretion rate of $\dot{M} \geq 4 \times 10^{-8} \, M_\odot$ per year) the burning is not explosive. The burning wave runs out through the star like a flame but at a subsonic velocity with respect to the unburned material. This is called a deflagration wave (Cou48). Even though the burning is much more peaceful than in a detonation, the white dwarf is completely disrupted, again leaving no compact remnant behind. The explosive nuclear burning (carbon to nickel) produces about 10^{51} ergs, which is sufficient to account for the observed expansion velocities of about 10^4 km s^{-1} in the supernova remnants. In the course of the deflagration, about one solar mass of ^{56}Ni is synthesized in the inner layer of the star, which decays to ^{56}Co and then to ^{56}Fe over a period of months. The energy furnished by this radioactive decay is sufficient to power the SNI light curves. In the outer layers of the star substantial amounts of intermediate mass elements are synthesized in the decaying deflagration wave, and are also ejected into space. Optical spectra of various intermediate mass elements found in the light curves of SNIs near maximum light have been reproduced in considerable detail by this mechanism (Bra84, Nom 84). It has also been suggested that SNIs may produce a significant fraction of the elements from Si to Ca in the Galaxy in addition to the iron-peak elements which may supplement the production of these elements in massive stars (supernovae of Type II; chap. 8). Although the carbon deflagration model in accreting C + O white dwarfs can account for many of the observed features of SNIs, whether or not it is consistent with other observational constraints (i.e., a neutrino burst of about 10^{50} ergs s^{-1} and of about 1 s duration is predicted; Nom84) as well as with some of the assumed physics input (i.e., the propagation speed of the deflagration wave) needs more investigation.

At present it appears plausible that supernovae of Type I are associated with white dwarfs in binary systems and thus are not part of the evolution of a singular star. In contrast, supernovae of Type II are believed to be the explosive death of singular stars with masses in the range $M/M_\odot \simeq 8$–100 (chap. 8).

3

Definitions and General Characteristics of Thermonuclear Reactions

Thermonuclear reactions play a key role in understanding energy production and nucleosynthesis of the elements in stars (chap. 2). A star is born when interstellar gas, mainly hydrogen and helium, condenses and, as a result of the conversion of gravitational to thermal energy, also heats up. When the temperature and density at the center become high enough, nuclear reactions commence in the most easily burned nuclear fuel, hydrogen. The energy released by the nuclear reactions stabilizes the star by establishing a balance between thermal pressure and gravitational force which persists without a large temperature change until the exhaustion of that particular nuclear fuel. The star then contracts, again converting gravitational energy to thermal energy, until the temperature and density become high enough to ignite the next available fuel. An examination of the abundance data (chap. 1) and Coulomb barrier penetration considerations (chap. 4) led Burbidge, Burbidge, Fowler, and Hoyle (B^2FH) and Cameron (Cam57) to postulate that a series of more or less discrete nucleosynthetic processes might have taken place. These postulations have withstood experimental and theoretical investigations over the years with important but not basic changes. These processes are

1. Hydrogen burning (conversion of hydrogen to helium);
2. Helium burning (conversion of helium to carbon, oxygen, and so on);
3. Carbon, oxygen, and neon burning (production of $16 \leq A \leq 28$);
4. Silicon burning (production of $28 \leq A \leq 60$);
5. The s-, r-, and p-processes (production of $A \geq 60$); and
6. The l-process (production of the reactive light elements D, Li, Be, and B).

These processes will be discussed in detail in chapters 6–10. In this chapter we describe definitions and general characteristics of thermonuclear reactions, and in chapter 4 we discuss the actual procedures in determining stellar reaction rates. The laboratory equipment and experimental techniques are described in chapter 5.

3.1. Source of Nuclear Energy

Research in nuclear physics has provided abundant information on the properties of nuclei. The resulting picture is that a nucleus of mass M_n is composed of Z protons with mass ZM_p and N neutrons with mass NM_N, where $M_N \simeq ZM_p + NM_N$. Modern high-energy physics shows that protons and neutrons consist of quarks. However, the quarks are confined, and nuclei are produced and decay into protons and neutrons in combinations of these two nucleons. Nuclei with the same number of protons but a different number of neutrons and hence different masses are called isotopes. The notation describing a particular isotope is usually written in the form $^A_Z X_N$, where the atomic number A is the total number of nucleons ($A = Z + N$). Examples of the use of this notation for atomic isotopes throughout the periodic table are

$$^1_1 H_0 , \qquad ^4_2 He_2 , \qquad ^{40}_{20} Ca_{20} , \qquad ^{238}_{92} U_{146} .$$

Since A and X uniquely define a given isotope, the subscripts are usually dropped, and the shorter notation

$$^1 H , \qquad ^4 He , \qquad ^{40} Ca , \qquad ^{238} U$$

is used. For historical reasons, the nuclei of the hydrogen isotopes ^1H, ^2H, and ^3H are named protons (p), deuterons (d) and tritons (t) and the nucleus of the ^4He isotope is called an α-particle (α). Otherwise, atoms and nuclei are designated by the same symbol. Here, the notation H as well as the notation p will sometimes be used for protons and D as well as d for deuterons.

Since it is difficult to measure the mass of a bare nucleus (i.e., completely stripped of all its surrounding electrons), most mass measurements are carried out in atomic mass measuring experiments (mass spectrographs). In these measurements only a few of the electrons are stripped off and what is measured is essentially the atomic mass M_a. The nuclear mass M_n is then obtained from the atomic mass through the relation

$$M_a = M_n + Zm_e - B_e(Z)/c^2 , \tag{3.1}$$

where m_e is the rest mass of the electron. The quantity $B_e(Z)$ represents the total binding energy of the orbital electrons (Sev79), which is related to a mass equivalent of $\Delta M_e = B_e(Z)/c^2$ via the Einstein mass-energy relationship $E = mc^2$. The mass of a hydrogen atom ^1H in energy units is about 10^9 eV, while $B_e(Z) = 13.6$ eV; hence the correction due to the binding energy of the electron is of the order of 10^{-8}. While the total binding energy $B_e(Z)$ of the electron cloud in heavier atoms is much larger (empirical relation: $B_e(Z) \simeq 15.7 Z^{7/3}$ eV), the correction remains of the same order of magnitude.

Early mass measurements by Aston (Ast27), though less precise than those

TABLE 3.1 Binding Energies for Several Nuclei[a]

Nucleus	Total Binding Energy ΔE (MeV)	Binding Energy per Nucleon $\Delta E/A$ (MeV)
^{2}D	2.22	1.11
^{4}He	28.30	7.07
^{12}C	92.16	7.68
^{16}O	127.62	7.98
^{40}Ca	342.05	8.55
^{56}Fe	492.26	8.79
^{238}U	1801.70	7.57

[a] From Wap85.

obtained with modern techniques (Bie55, Wap85), showed that the total mass of a nucleus is less than the sum of the masses of the constituent nucleons. The difference $\Delta M_n = M_n - ZM_p - NM_N$ is referred to as the nuclear mass defect, which is equivalent to an energy of $\Delta E = \Delta M_n c^2$. The quantity ΔE is the energy released in assembling a nucleus from its constituent nucleons and is gained at the expense of the mass of the nucleus. In turn, it is the energy required to separate the nucleus into its constituent nucleons and is referred to as the binding energy ΔE of the nucleus. Table 3.1 lists the binding energy for a representative sample of nuclei throughout the periodic table as well as their mean binding energy per nucleon, defined as $\Delta E/A$. A graphical representation of $\Delta E/A$ in Figure 3.1 shows that nuclei near iron have the largest nuclear binding, while lighter and heavier nuclei are less tightly bound. As a conse-

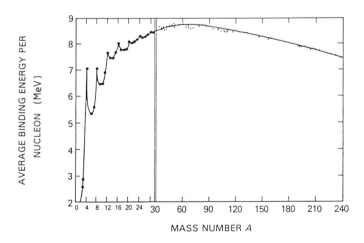

FIGURE 3.1. Binding energy per nucleon shown as a function of atomic mass number A. The highest nuclear binding energy (stability) is found for elements near iron.

quence of these variations in binding energy, nuclear energy can be liberated either by combining lighter nuclei into heavier ones or by splitting heavier nuclei into lighter ones. The first process is known as fusion and the latter one as fission. An example of a fusion process would be the combining of four protons to produce a ^{4}He nucleus (chap. 6). As can be seen from Fig. 3.1, this would release about 7 MeV per nucleon for a total energy release of 28 MeV. An example of a fission reaction is the splitting of a ^{238}U nucleus into two ^{119}Pd nuclei with a total energy release of 240 MeV. Since here there are 238 nucleons involved, the energy release per nucleon is only 1 MeV (Fig. 3.1); hence the fusion process is much more efficient than the fission process in terms of energy per nucleon or energy per unit of nuclear mass. Here on earth, the production of nuclear energy via the fission process in a controlled way is accomplished using reactors. Controlled energy production via the fusion process has not yet been realized terrestrially.

In stellar objects, the light elements hydrogen (^{1}H) and helium (^{4}He) are by far the most abundant (chap. 1). It is fusion involving these elements which predominates in the production of nuclear energy in stars. In this process, the production of nuclear energy and the synthesis of elements proceeds through fusion reactions until all light nuclei are converted to iron (chaps. 6–8). The simplest type of nuclear reaction may be written symbolically as $1 + 2 \rightarrow 3 + 4$ or $x + A \rightarrow y + B$ [the latter is often written as A(x, y)B]. The symbol x (or 1) represents the projectile and A (or 2) the target nucleus which together constitute the entrance channel, while y (or 3) and B (or 4) represent the emerging nuclei which together constitute the exit channel. If the nuclei in the entrance channel have nuclear masses $M_{n,1}$ and $M_{n,2}$ and those of the exit channel have nuclear masses $M_{n,3}$ and $M_{n,4}$, energy conservation leads to the nuclear reaction Q-value Q_n, defined as

$$Q_n = (M_{n,1} + M_{n,2} - M_{n,3} - M_{n,4})c^2 . \tag{3.2}$$

If the Q-value is positive, there is a net production of energy in the reaction and the reaction is said to be exothermic. Conversely, if the Q-value is negative, a minimum energy equal to Q has to be provided to make the process go and the reaction is said to be endothermic. Clearly, for fusion processes in stars, reactions having positive Q-values are most important (sec. 3.6).

If the nuclear masses are known, the Q-value of a given reaction can be calculated using equation (3.2). However, as pointed out above, it is the atomic mass M_a of a given isotope that is usually measured and tabulated (Wap85). In terms of atomic masses, the atomic Q-value

$$Q_a = (M_{a,1} + M_{a,2} - M_{a,3} - M_{a,4})c^2 \tag{3.3}$$

is related to the nuclear Q-value Q_n by

$$Q_a = Q_n + m_e c^2(Z_1 + Z_2 - Z_3 - Z_4) + B_e(Z_1) + B_e(Z_2) - B_e(Z_3) - B_e(Z_4) .$$

Because of charge conservation in a nuclear reaction, the second term vanishes and the Q-values based on atomic and nuclear masses differ only by the differences of electron binding energies involved in the entrance and exit channels:

$$Q_a = Q_n + \Delta B_e .$$ (3.4)

In general one can neglect the ΔB_e correction. Of course, if $Q_n \simeq \Delta B_e$, such a correction will be significant (Yok83, Yok85).

Note that the standard Q-value of a nuclear reaction results from the mass difference between the initial and final atoms. The mass tables list atomic masses, not nuclear masses. As just shown, it makes little difference, except when positrons are involved in the reaction. For example, the reaction $p + p \to d + e^+ + \nu$ (chap. 6) leads to $Q = (M_H + M_H - M_D)c^2 = 1.44$ MeV, while the nuclear masses give $Q' = (M_p + M_p - M_d)c^2 = 0.93$ MeV, the kinetic energy shared by the e^+ and ν. Thus, the use of atomic masses automatically includes the e^+ annihilation energy.

In general one finds in tabular form not the atomic mass but rather the atomic mass excess ΔM, in units of energy, defined as

$$\Delta M = (M - AM_u)c^2 ,$$

where M is the atomic mass, A is the atomic number, and M_u is the atomic mass unit (amu), with $M_u c^2 = 931.50$ MeV (Wap85). The latter quantity is defined as one-twelfth of the mass of a neutral ^{12}C atom ($M_u = 1.66 \times 10^{-24}$ g; see Appendix). Table 3.2 gives the atomic mass excesses ΔM for the lightest isotopes. For example, using Table 3.2, the Q-value of the $^3He(^3He, 2p)^4He$ reaction is found to be $Q = 12.860$ MeV, illustrating the large amount of energy liberated when lighter nuclei are fused to form heavier nuclei. The ratio $\Delta M/M$ is of the order of 0.1%–1% for all nuclei and represents an extremely large mass conversion factor.

3.2. Cross Section

The Q-value of a nuclear reaction is the energy liberated for each event. This value is known for nearly all possible nuclear reactions from the mass tables (Wap85). More difficult to measure is the probability that a given nuclear reaction will take place. This probability is used to determine how many reactions occur per unit time and unit volume and hence, together with the Q-value, provides important information on energy production in stars.

We know that when one is shooting at a target the probability of hitting it increases as its area increases. By analogy, one may associate with each nucleus a geometrical area, which is directly related to the probability of a projectile interacting with that nucleus. This area is referred to as the cross section for the reaction. Classically, this cross section σ is equal to the combined geometrical area of the projectile and the target nucleus. If the projectile and the target nucleus have radii R_p and R_t, respectively, the cross section may be written as $\sigma = \pi(R_p + R_t)^2$. Nuclear physics research has revealed that the average radius R of a nucleus is related to the atomic number A according to the relationship $R = R_0 A^{1/3}$, where R_0 is nearly a constant equal to 1.3×10^{-13} cm (Bla62, Bri82). With this relation the geometrical cross sections for three extreme cases are

$$^1\text{H} + \,^1\text{H}: \qquad \sigma = 0.2 \times 10^{-24} \text{ cm}^2 \,,$$

$$^1\text{H} + \,^{238}\text{U}: \qquad \sigma = 2.8 \times 10^{-24} \text{ cm}^2 \,,$$

$$^{238}\text{U} + \,^{238}\text{U}: \qquad \sigma = 4.8 \times 10^{-24} \text{ cm}^2 \,.$$

Note that all cross sections are of the order of 10^{-24} cm^2. Thus, for conve-

TABLE 3.2 Atomic Mass Excess ΔM of the Lightest Isotopes[a]

Z	Element	A	Mass Excess (keV)[b]
0	n	1	8071.37
1	H	1	7289.03
		2	13135.82
		3	14949.91
		4	25840.00
2	He	3	14931.32
		4	2424.92
		5	11390.00
		6	17592.30
		7	26110.00
3	Li	4	25120.00
		5	11680.00
		6	14085.60
		7	14906.80
		8	20945.40
		9	24953.90
4	Be	6	18374.00
		7	15768.70
		8	4941.73
		9	11347.70
		10	12607.00
		11	20174.00
5	B	7	27870.00
		8	22920.30
		9	12415.80
		10	12050.78
		11	8668.00
		12	13369.50
		13	16562.30
6	C	12	0.00

[a] From Wap85.
[b] Note that the values for the mass excess decrease with increasing mass and eventually become negative.

nience, a new unit of area, the barn (b), equal to 10^{-24} cm^2 has been defined for cross sections.

In the above classical treatment, the cross section depends only on the geometrical area of the projectile and the target nucleus. In reality (as discussed in more detail in chap. 4), since nuclear reactions are governed by the laws of quantum mechanics, the geometrical cross section $\sigma = \pi(R_p + R_t)^2$ must be replaced by the energy-dependent quantity $\sigma = \pi \lambdabar^2$, where $\lambdabar$ represents the de Broglie wavelength reflecting the wave aspect of quantum mechanical processes (Bla62):

$$\lambdabar = \frac{m_p + m_t}{m_t} \frac{\hbar}{(2 m_p E_l)^{1/2}} ,$$

with E_l the laboratory energy of the incident projectile of mass m_p, and m_t the mass of the target nucleus. In some cases the "Coulomb and centrifugal barriers" related to nuclear charge and angular momentum act to inhibit the penetration of the projectile into the nucleus. This situation also leads to a strong energy dependence of the cross section. Other energy-dependent effects may also play a role, and the situation can become quite complex. In addition to the effects just discussed, the cross section for a particular nuclear reaction depends on the nature of the force involved and can vary by several orders of magnitude. That the variation in cross section is predominantly a result of differing interaction forces is demonstrated by the following examples (chap. 6):

Strong (nuclear) force:

^{15}N$(p, \alpha)^{12}$C , $\sigma \simeq 0.5$ b at $E_p = 2.0$ MeV ,

Electromagnetic force:

^{3}He $(\alpha, \gamma)^7$Be , $\sigma \simeq 10^{-6}$ b at $E_\alpha = 2.0$ MeV ,

Weak force:

$p(p, e^+ \nu)d$, $\sigma \simeq 10^{-20}$ b at $E_p = 2.0$ MeV .

As the cross section decreases, the challenge to the experimentalist to make precise cross-section measurements increases enormously (chap. 5). Indeed, of the examples shown above, only the first two have been experimentally determined. The third cross section is a theoretical but quite reliable estimate (chap. 6).

3.3. Stellar Reaction Rate

As just discussed, nuclear cross sections are in general energy-dependent or, equivalently, velocity-dependent. Thus $\sigma = \sigma(v)$, where v represents the relative velocity between the projectile and the target nucleus. In nuclear reactions, as in other physical interactions, it is the relative velocity, not the individual velocities, which enters into the cross section.

Consider a stellar gas with N_x particles per cubic centimeter of type X and N_y particles per cubic centimeter of type Y, with relative velocities v. Since the cross section for a nuclear reaction between nuclei depends only on the relative velocity v, one can consider particles of either type X or type Y as the projectiles. If nuclei X are arbitrarily chosen as the projectiles moving with the velocity v, then the nuclei Y must be considered at rest. Consequently, the projectiles see an effective reaction area F equal to the cross section for a single target nucleus $\sigma(v)$ multiplied by the number of target nuclei N_y per cubic centimeter: $F = \sigma(v)N_y$. Since each projectile sees this area F, the total number of nuclear reactions occurring depends on the flux J of incident particles defined as the product of the number of particles per cubic centimeter (unit volume) times their velocity v: $J = N_x v$. The rate of nuclear reactions r is therefore given by the product of both quantities

$$r = N_x N_y v \sigma(v) \,, \tag{3.5}$$

where r is in units of reactions per cubic centimeter per second.

In a stellar gas, as in other gases, the velocity of the particles varies over a wide range of values, given by the probability function $\phi(v)$, where

$$\int_0^\infty \phi(v)dv = 1 \,. \tag{3.6}$$

Since $\phi(v)dv$ represents the probability that the relative velocity v between the partners of a nuclear reaction has a value between v and $v + dv$, the product $v\sigma(v)$ in equation (3.5) has to be folded with this velocity distribution $\phi(v)$ to arrive at a value for $v\phi(v)$ averaged over the velocity distribution, $\langle \sigma v \rangle$:

$$\langle \sigma v \rangle = \int_0^\infty \phi(v)v\sigma(v)dv \,. \tag{3.7}$$

The bracketed quantity $\langle \sigma v \rangle$ is referred to as the reaction rate per particle pair. For exothermic reactions $(Q > 0)$, the integral extends from $v = 0$ to $v = \infty$, while for endothermic reactions $(Q < 0)$ the integral starts at the threshold velocity $v_T \propto Q^{1/2}$. The total reaction rate r is then

$$r = N_x N_y \langle \sigma v \rangle \,. \tag{3.8}$$

The product $N_x N_y$ represents the total number of pairs of nonidentical nuclei X and Y. It can be shown that this product, and hence the total reaction rate for nonidentical particles, is a maximum if $N_x = N_y$. For identical particles, the number product must be divided by 2, otherwise each pair would be counted twice. This situation is accounted for by introducing the Kronecker symbol δ_{xy} in formula (3.8), giving

$$r = N_x N_y \langle \sigma v \rangle (1 + \delta_{xy})^{-1} \,. \tag{3.9}$$

It is the task of the experimentalist and theorist to determine the quantity $\langle \sigma v \rangle$ at the relevant stellar velocities. Most frequently direct measurements at these velocities are not possible, and theoretical extrapolations are needed (chaps. 4 and 5).

Note that frequently in stellar model calculations one finds not the number densities N_i but rather the matter density ρ (g cm^{-3}) and the mass fraction X_i (or mole fraction Y_i) of nuclei i in the stellar medium (i.e., their atomic abundance in terms of mass). These quantities are related through the expression

$$N_i = \rho N_A X_i / A_i = \rho N_A Y_i \,,$$

where $N_A = 6.02 \times 10^{23}$ mole^{-1} is Avogadro's number and A_i is the atomic mass of the nuclear species i in amu units (numerically equal to the mass in grams of one gram atom: $A(n) = 1.008665$, $A(^1\text{H}) = 1.007825$, $A(^4\text{He}) = 4.002603$ (Wap85). For example, the mass fractions X_i of ^{1}H and ^{4}He in the sun's interior are 0.73 and 0.25, respectively. Using the above expression, one finds that the number densities are by contrast in the ratio of $N(^1\text{H}):N(^4\text{He}) \simeq 12:1$.

3.4. Mean Lifetime

In the consideration of stellar evolution in terms of time, an important quantity is the mean lifetime of the nuclei in the stellar environment. The mean lifetime of nuclei X against destruction by nuclei Y, $\tau_y(\text{X})$, is defined by the usual statistical equation

$$\left(\frac{dN_x}{dt} \right)_y = -\lambda_y(\text{X})N_x = -\frac{1}{\tau_y(\text{X})} N_x \,. \tag{3.10}$$

This equation can be interpreted as the change in the abundance N_x of nuclei of type X as the result of bombardment by particles of type Y. As usual, the decay constant λ is inversely proportional to the mean lifetime τ. Alternatively, the change in the abundance of nuclei X is related to the total reaction rate by the relationship

$$\left(\frac{dN_x}{dt} \right)_y = -(1 + \delta_{xy})r \,. \tag{3.11}$$

The Kronecker symbol δ_{xy} is required, since for identical particles two particles are destroyed per reaction. If equations (3.9), (3.10), and (3.11) are combined, the mean lifetime due to this reaction is given by

$$\tau_y(\text{X}) = \frac{1}{N_y \langle \sigma v \rangle} \tag{3.12}$$

Analogously, one obtains $\tau_x(\text{Y}) = 1/(N_x \langle \sigma v \rangle)$. Note that the effect of identical particles has canceled out (for the case of 3-body reactions see Fow67, Fow75).

The lifetime of nuclei X depends only on the number of destructive nuclei N_y and the reaction rate per particle pair, $\langle \sigma v \rangle$. For high densities of destructive nuclei and large reaction rates, the lifetime of nuclei X can be quite short. If the environment contains a number of different destructive nuclei i, the mean lifetime of species X is given by the expression $1/\tau(X) = \sum_i 1/\tau_i(X)$.

3.5. Maxwell-Boltzmann Velocity Distribution

In normal stellar matter the stellar gas is nondegenerate and the nuclei move nonrelativistically (chap. 2). The gas is in thermodynamic equilibrium, and the velocities of the nuclei can be described by a Maxwell-Boltzmann velocity distribution,

$$\phi(v) = 4\pi v^2 \left(\frac{m}{2\pi kT} \right)^{3/2} \exp\left(-\frac{mv^2}{2kT} \right), \tag{3.13}$$

which as usual is normalized to unity (eq. [3.6]). Here T refers to the temperature of the gas and m to the mass of the nucleus of interest. The numerator in the exponential term represents the kinetic energy of the nucleus, $E = \frac{1}{2}mv^2$. The function $\phi(v)$ can be rewritten in terms of this energy as

$$\phi(E) \propto E \exp\left(-E/kT \right).$$

At low energies, $E \ll kT$, the function $\phi(E)$ increases linearly with E and reaches a maximum value at $E = kT$. At higher energies, $E \gg kT$, the function $\phi(E)$ decreases exponentially, and asymptotically approaches zero at very high energies (Fig. 3.2). If the temperature T is in units of 10^6 K ($\equiv T_6$), the numerical value for kT, in keV, is $kT = 0.0862T_6$. At room temperature the value of kT (2.6×10^{-5} keV) is small, while at temperatures in the solar interior ($T_6 = 15$) $kT = 1.3$ keV and in supernova events ($T_6 = 5000$) $kT = 431$ keV.

For nuclear reactions in stellar media, the velocities of both of the interacting nuclei X and Y are described by the Maxwell-Boltzmann velocity distribution, that is,

$$\phi(v_x) = 4\pi v_x^2 \left(\frac{m_x}{2\pi kT} \right)^{3/2} \exp\left(\frac{-m_x v_x^2}{2kT} \right),$$

$$\phi(v_y) = 4\pi v_y^2 \left(\frac{m_y}{2\pi kT} \right)^{3/2} \exp\left(\frac{-m_y v_y^2}{2kT} \right).$$

Therefore, the reaction rate per particle pair, $\langle \sigma v \rangle$ (eq. [3.7]), involves a double integral over both velocity distributions:

$$\langle \sigma v \rangle = \int_0^\infty \int_0^\infty \phi(v_x)\phi(v_y)\sigma(v)v\,dv_x\,dv_y ,$$

where v again represents the relative velocity between interacting nuclei. The variables v_x and v_y are related to the variables v and V (the center-of-mass velocity) by the usual kinematic relations. Using the reduced mass μ of the interacting particles, $\mu = m_x m_y/(m_x + m_y)$, and $M = m_x + m_y$ as the total mass, the rate $\langle \sigma v \rangle$ can be written in terms of the variables v and V as

$$\langle \sigma v \rangle = \int_0^\infty \int_0^\infty \phi(V)\phi(v)\sigma(v)vdVdv , \qquad (3.14)$$

where the transformed velocity distributions $\phi(V)$ and $\phi(v)$ are

$$\phi(V) = 4\pi V^2 \left(\frac{M}{2\pi kT}\right)^{3/2} \exp\left(\frac{-MV^2}{2kT}\right),$$

$$\phi(v) = 4\pi v^2 \left(\frac{\mu}{2\pi kT}\right)^{3/2} \exp\left(\frac{-\mu v^2}{2kT}\right). \qquad (3.15)$$

Note that when written in terms of the new variables v and V, the rate equation $\langle \sigma v \rangle$ is still factorable into two independent Maxwell-Boltzmann distributions with the usual normalization (eq. [3.6]). Since the nuclear cross section $\sigma(v)$ depends only on the relative velocity v between the interacting nuclei, one can immediately integrate equation (3.14) over the velocity V. This yields

$$\langle \sigma v \rangle = \int_0^\infty \phi(v)\sigma(v)vdv .$$

Inserting formula (3.15), one obtains

$$\langle \sigma v \rangle = 4\pi \left(\frac{\mu}{2\pi kT}\right)^{3/2} \int_0^\infty v^3 \sigma(v) \exp\left(-\frac{\mu v^2}{2kT}\right)dv .$$

Using the center-of-mass energy $E = \frac{1}{2}\mu v^2$, this equation can finally be written in the form

$$\langle \sigma v \rangle = \left(\frac{8}{\pi\mu}\right)^{1/2} \frac{1}{(kT)^{3/2}} \int_0^\infty \sigma(E)E \exp\left(-\frac{E}{kT}\right)dE . \qquad (3.16)$$

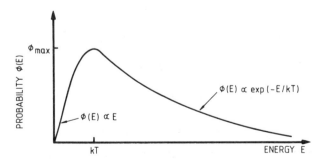

FIGURE 3.2. Shown schematically is the Maxwell-Boltzmann energy distribution of a gas characterized by the temperature T. The distribution exhibits a maximum at $E = kT$.

3.6. Inverse Reactions

At low stellar temperatures, nuclear reactions occur predominantly for those reactions having positive Q-values:

$$1 + 2 \to 3 + 4, \qquad Q > 0.$$

As discussed above, in this notation, 1 and 2 represent nuclei of types 1 and 2 in the entrance channel, and 3 and 4 the nuclei in the exit channel. As the stellar temperature rises, the number of particles with energies $E \geq Q$ increases and the inverse process, where the Q-value is negative,

$$3 + 4 \to 1 + 2, \qquad Q < 0,$$

becomes progressively more significant. In order to understand stellar evolution, one must know the rates at which both processes proceed.

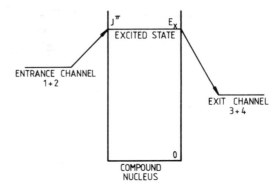

REACTION : 1+2 ⟶ C ⟶ 3+4 + Q (Q>0)

FIGURE 3.3. Schematic diagram of a nuclear reaction initiated by particles 1 and 2 in the entrance channel. This reaction proceeds through an excited state in the compound nucleus C before decaying into particles 3 and 4 of the exit channel. The excited state in the compound nucleus is characterized by its excitation energy E_x, angular momentum J, and parity π (J^π).

A nuclear reaction might be visualized as shown in Figure 3.3. In this example, the nuclear reaction $1 + 2 \to 3 + 4$ proceeds through an excited state in the compound nucleus, and its cross section is given by

$$\sigma_{12} = \pi \lambdabar_{12}^2 \frac{2J + 1}{(2J_1 + 1)(2J_2 + 1)}$$

$$\times (1 + \delta_{12}) |\langle 3 + 4 | H_{II} | C \rangle \langle C | H_I | 1 + 2 \rangle|^2 , \qquad (3.17)$$

where the double indices refer to the particles involved in the entrance channel. The expression for the cross section can be interpreted as consisting of the product of the following terms.

1. The term $\pi\lambda_{12}^2$ reflects the quantum mechanical character of a cross section.

2. The term $(2J + 1)/[(2J_1 + 1)(2J_2 + 1)]$ is a statistical factor, where J represents the angular momentum of the excited state in the compound nucleus and J_1 and J_2 represent the spins of nuclei 1 and 2 of the entrance channel. It is well known from quantum mechanics that a state with angular momentum J has $2J + 1$ magnetic substates. The statistical factor is arrived at by "summing over the final states and averaging over the initial states" (Bla62). The summing reflects the fact that the probability for the process increases as the number of final states available goes up. In the present example, the excited state in the compound nucleus represents the compound state for the entrance channel, where there are $2J + 1$ possible substates. The entrance channel involves a total of $(2J_1 + 1)(2J_2 + 1)$ initial substates. The probability of finding particles 1 and 2 in a particular substate is inversely proportional to $(2J_1 + 1)(2J_2 + 1)$. The product of the two probabilities leads to the statistical factor given above. It is frequently designated by the letter ω.

3. The term $(1 + \delta_{12})$ reflects the fact that the cross section is doubled for identical particles in the entrance channel. It cancels the term $(1 + \delta_{12})$ in the denominator for the reaction rate. Identical particles must not be counted twice, but their cross section doubles because they cannot be distinguished as projectile and target.

4. The final term of equation (3.17) depends on the force involved in the process. Its magnitude varies enormously as one goes from weak to electromagnetic to nuclear interactions. In quantum mechanical processes, each step of the process is described in terms of matrix elements (Bla62). Because of the interim state involved in the example (Fig. 3.3), the reaction is referred to as a two-step (or resonant) process; therefore, its description involves the product of two matrix elements. The first matrix element, $\langle C | H_I | 1 + 2 \rangle$, describes the transition from the entrance channel with particles 1 and 2 to the excited state in the compound nucleus C. The interaction operator responsible for this transition is designated by H_I. Similarly, the second matrix element, $\langle 3 + 4 | H_{II} | C \rangle$, describes the decay of the compound nucleus C into the exit channel involving particles 3 and 4. The interactions involved in the entrance and exit channels H_I and H_{II} need not be the same.

By analogy, the cross section for the inverse reaction can be written as

$$\sigma_{34} = \pi\lambda_{34}^2 \frac{2J + 1}{(2J_3 + 1)(2J_4 + 1)}$$

$$\times (1 + \delta_{34}) |\langle 1 + 2 | H_I | C \rangle \langle C | H_{II} | 3 + 4 \rangle|^2 . \tag{3.18}$$

The matrix elements in equations (3.17) and (3.18) are identical except for the reversed order, which reflects the fact that the processes go in opposite directions. In general, laws governing reactions do not change when the direction of the reaction is reversed. This is known as the principle of time-reversal invariance (Bla62, Fra74). For processes involving the strong and electromagnetic interactions, there is to date no experimental evidence that this invariance is violated. As long as the cross section depends on these two interactions, the matrix elements are identical, and the ratio of the two cross sections is

$$\frac{\sigma_{12}}{\sigma_{34}} = \frac{m_3 m_4 E_{34}(2J_3 + 1)(2J_4 + 1)(1 + \delta_{12})}{m_1 m_2 E_{12}(2J_1 + 1)(2J_2 + 1)(1 + \delta_{34})}. \tag{3.19}$$

In arriving at this ratio, the quantity λ_{ik}^2 has been replaced by the equivalent quantity $\hbar^2/(2\mu_{ik} E_{ik})$, where μ_{ik} is the reduced mass and E_{ik} is the center-of-mass energy. These relations are in nonrelativistic form. Note that while the above ratio was arrived at using a two-step process as the model (Fig. 3.3), the properties of the intermediate state, the compound nucleus, do not appear. Indeed, this result is completely general. In many systems it is easier to measure a nuclear reaction cross section in the reverse direction. By invoking the time-reversal invariance theorem implied in equation (3.19), one can, by measuring the cross section in one direction, obtain the reaction cross section for the reverse direction between the same two specific nuclear states as well.

In order to arrive at the analogous ratio for the reaction rates, equation (3.16) is first written for the two processes:

$$\langle \sigma v \rangle_{12} = \left(\frac{8}{\pi\mu_{12}}\right)^{1/2} \frac{1}{(kT)^{3/2}} \int_0^\infty \sigma_{12} E_{12} \exp\left(-\frac{E_{12}}{kT}\right) dE_{12},$$

$$\langle \sigma v \rangle_{34} = \left(\frac{8}{\pi\mu_{34}}\right)^{1/2} \frac{1}{(kT)^{3/2}} \int_0^\infty \sigma_{34} E_{34} \exp\left(-\frac{E_{34}}{kT}\right) dE_{34}.$$

Then, using equation (3.19) and the relation $E_{34} = E_{12} + Q$ $(Q > 0)$, the ratio of reaction rates per particle pair is found to be

$$\frac{\langle \sigma v \rangle_{34}}{\langle \sigma v \rangle_{12}} = \frac{(2J_1 + 1)(2J_2 + 1)(1 + \delta_{34})}{(2J_3 + 1)(2J_4 + 1)(1 + \delta_{12})} \left(\frac{\mu_{12}}{\mu_{34}}\right)^{3/2} \exp\left(-\frac{Q}{kT}\right). \tag{3.20}$$

TABLE 3.3 Ratio of Reaction Rates per Particle Pair

Temperature T_9	Ratio $\langle \sigma v \rangle_{34}/\langle \sigma v \rangle_{12}$		Factor[a] $(\mu c^2/kT)^{3/2}$
	$Q = 2.0$ MeV	$Q = 8.0$ MeV	
0.2	4.0×10^{-51}	2.5×10^{-202}	1.26×10^7
0.5	6.9×10^{-21}	2.3×10^{-81}	3.18×10^6
1.0	8.3×10^{-11}	4.8×10^{-41}	1.12×10^6
2.0	9.1×10^{-6}	6.9×10^{-21}	3.97×10^5
5.0	9.6×10^{-3}	8.6×10^{-9}	1.00×10^5
10.0	9.8×10^{-2}	9.3×10^{-5}	3.55×10^4

[a] In the case of photodisintegration, the ratio $\langle \sigma v \rangle_{34}/\langle \sigma v \rangle_{12}$ has to be multiplied by the factor $(\mu c^2/kT)^{3/2}$. The values are quoted for $\mu c^2 = 931.50$ MeV $= 1$ amu (see Appendix).

Since the quantities preceding the exponential term are of order unity, the ratio is dominated by the exponential term, which can be written as $\langle \sigma v \rangle_{34}/\langle \sigma v \rangle_{12} \simeq \exp(-11.605 Q_6/T_9)$, where Q_6 is in MeV and T_9 is in units of 10^9 K. Table 3.3 gives examples of the magnitude of this ratio for two Q-values and a number of stellar temperatures. Clearly, the inverse reaction plays a role only at high stellar temperatures and is very sensitive to the Q-value.

The total reaction rate in a stellar environment is the difference between the two rates:

$$r = r_{12} - r_{34} = \frac{N_1 N_2}{1 + \delta_{12}} \langle \sigma v \rangle_{12} - \frac{N_3 N_4}{1 + \delta_{34}} \langle \sigma v \rangle_{34} \,,$$

which, with equation (3.20), leads to

$$r = \frac{\langle \sigma v \rangle_{12}}{1 + \delta_{12}} \left[N_1 N_2 - N_3 N_4 \frac{(2J_1 + 1)(2J_2 + 1)}{(2J_3 + 1)(2J_4 + 1)} \left(\frac{\mu_{12}}{\mu_{34}} \right)^{3/2} \exp\left(-\frac{Q}{kT} \right) \right].$$

(3.21)

For equilibrium conditions ($r = 0$), we find the ratio of the products of elemental abundances to be

$$\frac{N_3 N_4}{N_1 N_2} = \frac{(2J_3 + 1)(2J_4 + 1)}{(2J_1 + 1)(2J_2 + 1)} \left(\frac{\mu_{34}}{\mu_{12}} \right)^{3/2} \exp\left(\frac{Q}{kT} \right). \tag{3.22}$$

In an important class of nuclear reactions in stars, one of the particles in the exit channel is a photon. This type of process is called a radiative capture reaction, an example of which is

$$^{12}C + p \rightarrow {}^{13}N + \gamma \quad (Q = +1.95 \text{ MeV}) \,.$$

Capture reactions are important in stellar processes because they all have positive Q-values (except for nuclei far off the valley of stability) and therefore contribute to the stellar energy production. Furthermore, these reactions, though slow, are very important in the synthesis of the elements (chaps. 6–9). At high stellar temperatures, the intensity of thermal photons with high energies increases, and photons can initiate the inverse reaction. For the above example it is described as

$$^{13}N + \gamma \rightarrow {}^{12}C + p \quad (Q = -1.95 \text{ MeV}) \,.$$

Here a heavier nucleus is split into lighter nuclei, in a process referred to as photodisintegration. This process is analogous to the photoionization of atoms, except that, because of the higher nuclear binding energies, higher energy photons are required. These photons have energies of the order of MeV, compared with eV for atomic ionization. The reaction rate for the photodisintegration is given by

$$r_\gamma = N_3 N_\gamma \langle \sigma v \rangle_{3\gamma} \,,$$

where the energy distribution of the photons is described by the Planck radiation law. With the total number of photons N_γ per unit volume given by

$$N_\gamma = \frac{8\pi^4}{13c^3h^3}(kT)^3,$$ (3.23)

the ratio of reaction rates per particle pair become (Fow67, Cla68)

$$\frac{\langle\sigma v\rangle_{3\gamma}}{\langle\sigma v\rangle_{12}} = \left(\frac{169}{8\pi^5}\right)^{1/2}\frac{(2J_1+1)(2J_2+1)}{2J_3+1}\left(\frac{\mu c^2}{kT}\right)^{3/2}\exp\left(-\frac{Q}{kT}\right).$$ (3.24)

The first two terms are of order unity, and again the ratio is dominated by the exponential term, $\exp(-Q/kT)$, but multiplied by the ratio of rest-mass energy to thermal energy, $(\mu c^2/kT)^{3/2}$. Since $\mu c^2 \geq 1$ GeV, the factor $(\mu c^2/kT)^{3/2}$ is very large at all stellar temperatures of interest (Table 3.3). Consequently, the photodisintegration process, the inverse of a capture reaction, plays a major role in stars, particularly during the late stages of evolution (chap. 8), when it destroys elements previously synthesized. Combining equations (3.12), (3.23), and (3.24), the mean lifetime of nucleus 3 against photodisintegration is found to be

$$\tau_\gamma(3) = \frac{1}{N_\gamma\langle\sigma v\rangle_{3\gamma}} = \left(\frac{2\pi\hbar^2}{\mu kT}\right)^{3/2}\frac{2J_3+1}{(2J_1+1)(2J_2+1)}$$

$$\times \exp\left(\frac{Q}{kT}\right)\frac{1}{\langle\sigma v\rangle_{12}},$$ (3.25)

where $\langle\sigma v\rangle_{12}$ is the reaction rate per particle pair of the capture reaction. Photodisintegration rates are harder to measure, and frequently the rate of the inverse reaction, the capture reaction, is determined first. The capture reaction rate is then used to find the photodisintegration rate using equation (3.24).

At high stellar temperatures, a significant fraction of reacting nuclei may be in excited states, and hence reactions can take place involving excited nuclei (chap. 8). If thermodynamic equilibrium is established between the excited states and the ground state of nuclei, which is the usual case in stars where excited states are relevant, the statistical factors $(2J_1+1)$ and $(2J_2+1)$ in equation (3.19) must be replaced by a "partition function" (Fow67 and chap. 4) reflecting the Boltzmann factor:

$$G_i = \sum_e (2J_{ie}+1)\exp(-E_{ie}/kT),$$ (3.26)

where the index i refers to nuclei involved in the entrance channel and the index e to the various excited states of those nuclei. The quantities J_{ie} and E_{ie} are the angular momentum and excitation energy of the excited state, respectively. The sum in equation (3.26) includes the ground state. The rate $\langle\sigma v\rangle_{12}$ must be averaged over all combinations of excited states of nuclei 1 and 2 and

also must be summed over all excited states of the final nuclei 3 and 4. Similar
equations hold for $\langle \sigma v \rangle_{34}$ (Fow67, Fow75, Har83).

3.7. Energy Production

The energy liberated per nuclear reaction is given by the Q-value, which—
when coupled with the reaction rate r—will provide the total stellar energy
production $\epsilon_{12} = Q r_{12}$ (in ergs s^{-1} cm^{-3}). Usually this equation is written in a
form that involves the density of stellar matter $\rho: \epsilon_{12} = Q r_{12}/\rho$ in ergs s^{-1} g^{-1}.
At high stellar temperatures the inverse reaction becomes important: $\epsilon_{34} = -Q r_{34}/\rho$, and the net energy production in a star is the difference between the
quantities ϵ_{12} and ϵ_{34}:

$$\epsilon_{net} = \epsilon_{12} + \epsilon_{34} = (r_{12} - r_{34})Q/\rho . \tag{3.27}$$

In processes involving neutrino production, the neutrinos usually leave the
scene in the stellar interior at the speed of light, emerge from the stellar surface
a few seconds later, and move outward into space. Thus, the neutrino energy
must be subtracted from the Q-value when determining stellar energy production rates because the neutrino energy is not deposited within the star. If
positrons are created in the processes, they soon find opposite partners, electrons, and the two particles annihilate each other, creating γ-rays. This annihilation energy is retained in the stellar medium. The standard Q-value based on
atomic masses automatically includes this energy.

4

Determination of
Stellar Reaction Rates

An understanding of most of the critical stellar features, such as time scales, energy production, and nucleosynthesis of the elements, hinges directly on the magnitude of the reaction rate per particle pair, $\langle \sigma v \rangle$, derived in chapter 3:

$$\langle \sigma v \rangle = \left(\frac{8}{\pi \mu}\right)^{1/2} \frac{1}{(kT)^{3/2}} \int_0^\infty \sigma(E) E \, \exp\left(-\frac{E}{kT}\right) dE \, . \tag{4.1}$$

This equation characterizes the reaction rate at a given stellar temperature T. As a star evolves, its temperature changes, and hence the reaction rate $\langle \sigma v \rangle$ must be evaluated for each temperature of interest. Since there are usually many different nuclear reactions involved at each stellar temperature, reevaluation of these reaction rates is tedious and time-consuming, taxing the capability of even the largest computers. It is therefore desirable and important to obtain analytic expressions for $\langle \sigma v \rangle$ in terms of temperature T (Fow67, Fow75, Har83). The mathematical approach used in arriving at such an analytic expression is determined by the energy dependence of the cross section, $\sigma(E)$. This energy dependence reflects the reaction mechanism involved in the process. Nuclear reaction mechanisms can be nonresonant or resonant; both types are defined and discussed in what follows.

4.1. Neutron-Induced Nonresonant Reactions

Neutron-induced nuclear reactions play an important role in studies of the early phase of the universe as well as in stars, where they are involved in stellar evolution as well as in nucleosynthesis of the elements (chap. 9). Because of the short lifetime of free neutrons (10 minutes), there are none present in the protostellar cloud. They must therefore be produced by nuclear reactions within the star. Some well-known neutron-producing reactions are $^{13}C(\alpha, n)^{16}O$, $^{18}O(\alpha, n)^{21}Ne$, and $^{22}Ne(\alpha, n)^{25}Mg$. Reactions of this kind take place, for example, in the helium-burning phase of stellar evolution (chap. 7). In a stellar environment the neutrons produced in these reactions are quickly thermalized through elastic scattering (in about 10^{-11} s; All71). Their velocities are then described by the Maxwell-Boltzmann distribution. Consequently, the formalism developed in chapter 3 is applicable.

Most neutron-induced nuclear reactions involve only two particles in the exit channel and therefore can be written in the notation

$$A(n, x)B ,$$

where x may be any one of a number of species, for example, γ, p, or α. Assuming a two-step process (Fig. 3.3) the expression for the cross section can be written as

$$\sigma_n \simeq \lambda_n^2 |\langle B + x | H_{II} | C \rangle \langle C | H_I | A + n \rangle|^2 . \qquad (4.2)$$

Each matrix element represents a probability for this step in the reaction to occur and is usually in the form of a partial width. The partial width of the entrance channel, $\Gamma_n(E_n)$, is determined by the neutron energy E_n alone. In the exit channel, the partial width $\Gamma_x(Q + E_n)$ is related to the sum of the Q-value and neutron energy E_n. Equation (4.2) may now be written in the form

$$\sigma_n(E_n) \simeq \lambda_n^2 \, \Gamma_n(E_n) \Gamma_x(Q + E_n) . \qquad (4.3)$$

For thermal neutron energies $Q \gg E_n$, and since Γ_x is generally not extremely sensitive to energy, one can approximate the term $\Gamma_x(Q + E_n)$ by $\Gamma_x(Q)$, which is constant. The remaining energy dependence of $\sigma_n(E_n)$ is thus found in the first two terms. The term λ_n^2 is the de Broglie wavelength squared, which is inversely proportional to the neutron energy E_n or, equivalently, to v_n^2. The energy dependence of the formation width $\Gamma_n(E_n)$ is described by the relation $\Gamma_n(E_n) \propto v_n P_{l_n}(E_n)$, which is discussed in section 4.4 (eq. [4.65]). If the entrance channel involves the orbital angular momentum l_n of the neutron, the neutron must penetrate a centrifugal barrier. The term $P_{l_n}(E_n)$ in the above relation represents the probability of this occurring for a neutron energy E_n. For neutrons with $l_n = 0$ no centrifugal barrier inhibits the nuclear reaction. At thermal energies, neutrons with $l_n > 0$ do not penetrate the centrifugal barrier in any appreciable number, and consequently the nuclear reaction is dominated by neutrons with $l_n = 0$ (s-wave neutrons), for which the probability for penetration is unity $[P_0(E_n) = 1]$, and the energy dependence of $\Gamma_n(E_n)$ reduces to $\Gamma_n(E_n) \propto v_n$. For s-wave neutrons, the cross-section formula (4.3) is thus given by the relation

$$\sigma_n(E_n) \propto \frac{1}{v_n^2} v_n = \frac{1}{v_n} , \qquad (4.4)$$

which is the familiar $1/v$ law (Fig. 4.1). Note again that this result is completely general, i.e., independent of the assumption of a two-step process. A reaction in which the cross section varies slowly as a function of energy, for example,

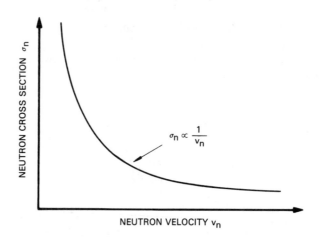

FIGURE 4.1. Cross section for s-wave neutrons at thermal energies follows the $1/v$ law.

according to the $1/v$ law (Fig. 4.1), is called "nonresonant." Resonant reactions, which vary often very rapidly over a small energy range, will be discussed in section 4.3.

If the reaction follows the $1/v$ law, the product $\sigma_n v_n$ is a constant. It follows immediately that the reaction rate per particle pair, $\langle \sigma v \rangle$, is also a constant:

$$\langle \sigma v \rangle = \text{constant} . \tag{4.5}$$

In principle, the reaction rate is determined by measuring the cross section at a single energy in the appropriate thermal neutron energy range, which is at $E_n = 30$ keV (chap. 9). However, in order to verify the $1/v$ law, measurements over a wider energy range are often desirable.

With increasing neutron energy, neutron partial waves with $l_n > 0$ contribute more significantly to the cross section. Thus, the product σv becomes a slowly varying function of neutron velocity, which one can expand into a Maclaurin expansion in terms of v or, equivalently, $E^{1/2}$ around zero energy:

$$\sigma v = S(0) + \dot{S}(0)E^{1/2} + \tfrac{1}{2}\ddot{S}(0)E + \cdots . \tag{4.6}$$

The dotted quantities represent derivatives with respect to $E^{1/2}$. The parameters $S(0)$, $\dot{S}(0)$, $\ddot{S}(0)$, ... are determined experimentally and can be found in tabular form (Fow67, Fow75, Mac65, and chap. 9 below). The energy dependence of the cross section $\sigma(E)$ from equation (4.6) is

$$\sigma(E) = \left(\frac{\mu}{2E}\right)^{1/2}\left[S(0) + \dot{S}(0)E^{1/2} + \frac{1}{2}\ddot{S}(0)E + \cdots \right]. \tag{4.7}$$

Inserting this result in equation (4.1), and carrying through the integration, one obtains for the reaction rate per particle pair

$$\langle \sigma v \rangle = S(0) + \left(\frac{4}{\pi}\right)^{1/2} \dot{S}(0)(kT)^{1/2} + \frac{3}{4}\ddot{S}(0)kT + \cdots \tag{4.8}$$

4.2. Charged-Particle–Induced Nonresonant Reactions

In a protostellar cloud, the light elements such as hydrogen (^{1}H) and helium (^{4}He), produced in the big bang, are by far the most abundant. As the cloud contracts under the force of gravity, the temperature at the center rises until it becomes sufficiently high to cause fusion reactions between hydrogen nuclei (chaps. 2 and 6). Since fusion reactions have positive Q-values, one might imagine that the nuclear burning will occur at relatively low temperature. In reality, however, a temperature of about 10^7 K is required (chap. 2). This high temperature is needed because nuclei are positively charged and repel each other with a force proportional to the nuclear charge. The potential energy for this repulsive force has the form

$$V_C(r) = \frac{Z_1 Z_2 e^2}{r}, \tag{4.9}$$

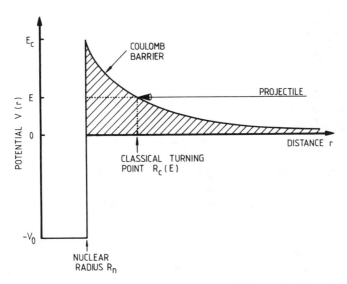

FIGURE 4.2. Schematic representation of the combined nuclear and Coulomb potentials. A projectile incident with energy $E < E_C$ has to penetrate the Coulomb barrier in order to reach the nuclear domain. Classically, the projectile would reach the closest distance to the nucleus at the turning point R_c.

where the symbols Z_1 and Z_2 represent the integral charges of the interacting nuclei and r is the distance between them. This potential, when combined with the potential for the very strong and attractive nuclear force, which comes into play at distances equal to the nuclear radius $R_n = R_1 + R_2$ (chap. 3), leads to an effective potential shown in Figure 4.2. The shaded area in Figure 4.2 represents a barrier, the Coulomb barrier, which inhibits nuclear reactions. For the $p + p$ reaction, the effective height E_C of this Coulomb barrier ($e^2 = 1.44 \times 10^{-10}$ keV cm) is $E_C = 550$ keV.

Classically, the $p + p$ reaction can occur only when the energy of the protons exceeds 550 keV. This energy corresponds to a stellar temperature ($E = kT$) of $T = 6.4 \times 10^9$ K ($T_9 = 6.4$). If no nuclear reactions occur until this temperature is reached, which in principle is possible through gravitational contraction, then when it is reached all pairs of nuclear particles should react instantaneously and stars—rather than burning their nuclear fuel for billions of years—should experience a catastrophic explosion. Consequently, if this is the case, there would be no stars today and the evolution of life as we know it would not have occurred. In the above treatment it was assumed that all particles had energies of $E = kT$, which occurs at the maximum of the Maxwell-Boltzmann distribution (Fig. 3.2). One might argue that nuclear reactions can occur between particles whose energies are represented by the high-energy wing of the Maxwell-Boltzmann distribution, and therefore such a high stellar temperature might not be necessary. For a lower stellar temperature of $T_9 = 0.01$, a calculation shows (eq. [3.13]) that the number of particles with an energy $E = 550$ keV compared with those with $E = kT = 0.086 \times 10$ keV $= 0.86$ keV is in the ratio

$$\frac{\phi(550)}{\phi(0.86)} = 3 \times 10^{-275} . \tag{4.10}$$

Clearly, the number of high-energy particles is not adequate to produce the energy radiated by stars.

Although Eddington, on the basis of nuclear physics experiments performed in Rutherford's laboratory in the early 1920s, speculated that the enormous amount of energy liberated in nuclear reactions could be the major source of energy in the sun and other stars, the classical condition required for nuclear reactions was an insurmountable obstacle. This obstacle was removed when Gamow (Gam28) and independently Condon and Gurney (Con29) showed that quantum mechanically there is a small but finite probability for the particles with energies $E < E_C$ (Fig. 4.2) to penetrate the Coulomb barrier. This phenomenon of barrier penetration, sometimes referred to as the tunnel effect, is of such considerable importance in stellar burning that a sketch of its quantum mechanical origin is in order.

In quantum mechanics the square of the wave function, $|\psi(r)|^2$, gives the probability of finding the particle at the position r. At the classical turning point R_c of the Coulomb barrier (Fig. 4.2) it is $|\psi(R_c)|^2$. While classically an incident particle cannot penetrate the barrier beyond this point, quantum mechanically one finds that the squared wave function has a finite value at the nuclear radius R_n of $|\psi(R_n)|^2$. The ratio of the two quantities represents the probability that incoming particles penetrate the barrier, which is the basis of the tunnel effect:

$$P = \frac{|\psi(R_n)|^2}{|\psi(R_c)|^2} . \tag{4.11}$$

Solving the Schrödinger equation for the Coulomb potential leads to (Bet37)

$$P = \exp\left\{ -2KR_c\left[\frac{\arctan (R_c/R_n - 1)^{1/2}}{(R_c/R_n - 1)^{1/2}} - \frac{R_n}{R_c} \right] \right\}, \tag{4.12}$$

with

$$K = \left[\frac{2\mu}{\hbar^2} (E_C - E) \right]^{1/2} . \tag{4.13}$$

The tunneling probability P for the $p + p$ reaction using equation (4.12) is given in Table 4.1. As one expects, the probability approaches unity at energies near the Coulomb barrier and falls off very rapidly at energies below the Coulomb barrier. At an energy of 1 keV, which is near the maximum of the Maxwell-Boltzmann distribution for a stellar temperature $T_9 = 0.01$, the tunneling probability is very small, $P = 9 \times 10^{-10}$, but it is still sufficiently large compared with the requirement imposed in the classical limit (eq. [4.10]) to account for nuclear energy production in stars.

TABLE 4.1 Tunneling Probability for the Reaction $p + p$ as a Function of Center-of-Mass Energy E

E (keV)	Tunneling Probability	
	Exact Expression	Approximation
1	8.9×10^{-10}	2.5×10^{-10}
2	5.6×10^{-7}	1.6×10^{-7}
5	1.7×10^{-4}	5.0×10^{-5}
10	3.1×10^{-3}	9.2×10^{-4}
20	2.4×10^{-2}	7.1×10^{-3}
50	0.14	4.4×10^{-2}
100	0.35	0.11
200	0.64	0.21
500	0.99	0'37
550[a]	1.00	0.39

[a] Height of the Coulomb barrier for this reaction. Here it was assumed that there is no centrifugal barrier ($l = 0$).

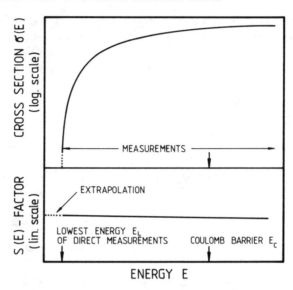

FIGURE 4.3. Cross section $\sigma(E)$ of a charged-particle–induced nuclear reaction drops sharply with decreasing energy E (by many orders of magnitude) for beam energies below the Coulomb barrier E_C, thus effectively providing a lower limit E_L to the beam energy at which experimental measurements can be made. Extrapolation to lower energies is more reliable if one uses the $S(E)$ factor.

At low energies where $E \ll E_C$ or, equivalently, where the classical turning point R_c is much larger than the nuclear radius R_n, equation (4.12) can be approximated by the simpler expression

$$P = \exp\left(-2\pi\eta\right) . \qquad (4.14)$$

The quantity η is called the Sommerfeld parameter and is equal to

$$\eta = \frac{Z_1 Z_2 e^2}{\hbar v} , \qquad (4.15)$$

In numerical units the exponent is

$$2\pi\eta = 31.29 Z_1 Z_2 \left(\frac{\mu}{E}\right)^{1/2} ,$$

where the center-of-mass energy E is given in units of keV and μ is in amu. This approximate expression for the tunneling probability (Table 4.1) is commonly referred to as the Gamow factor.

Primarily because of the exponential behavior of the probability for tunneling, the cross section for charged-particle–induced nuclear reactions drops rapidly for energies below the Coulomb barrier:

$$\sigma(E) \propto \exp\left(-2\pi\eta\right) .$$

Another non–nuclear energy–dependent term involves the de Broglie wave-length (chap. 3):

$$\sigma(E) \propto \pi \lambdabar^2 \propto \frac{1}{E} \, .$$

Using both relations, one can express the cross section as

$$\sigma(E) = \frac{1}{E} \exp\left(-2\pi\eta\right)S(E) \, , \tag{4.16}$$

where the function $S(E)$, defined by this equation, contains all the strictly nuclear effects. It is referred to as the nuclear or astrophysical S-factor. For nonresonant reactions this factor is a smoothly varying function of energy which varies much less rapidly with beam energy than the cross section (Fig. 4.3). Because of these characteristics, the factor $S(E)$ is much more useful

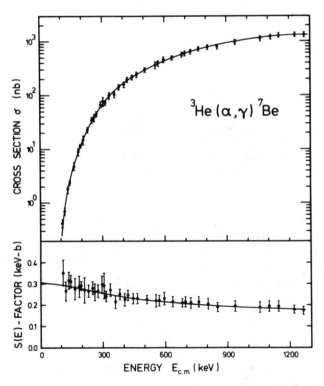

FIGURE 4.4. Energy dependence of the cross section $\sigma(E)$ and the factor $S(E)$ for the ^{3}He$(\alpha, \gamma)^7$Be reaction (Krä82). The line through the data points represents a theoretical description of the cross section in terms of the direct-capture model. This theory is used to extrapolate the data to zero energy. Data from other sources (chap. 6) give a higher absolute scale (40% difference).

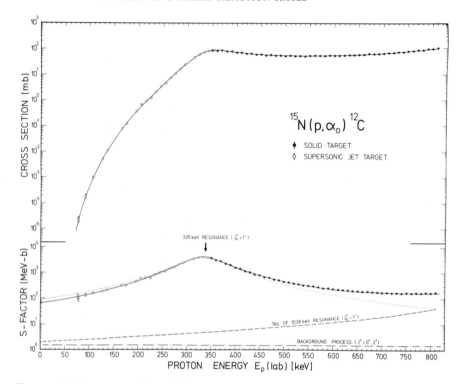

FIGURE 4.5. Energy dependence of the cross section $\sigma(E)$ and the astrophysical $S(E)$ factor for the $^{15}\text{N}(p, \alpha)^{12}\text{C}$ reaction (Red82). The line through the data points is a theoretical fit to the data, which is used in the extrapolation of the data to zero energy. In this case the $S(E)$ factor clearly reflects the presence of a broad resonance at $E_{\text{lab}} = 335$ keV (sec. 4.4). Note that an extrapolation of $\sigma(E)$ to stellar energies—without a theoretical understanding of the reaction—is treacherous.

in extrapolating measured cross sections to astrophysical energies (Fig. 4.3). Figures 4.4 and 4.5 illustrate the smoothly varying character of the $S(E)$ factor for the reactions $^3\text{He}(\alpha, \gamma)^7\text{Be}$ and $^{15}\text{N}(p, \alpha)^{12}\text{C}$ and show the usefulness of $S(E)$ in extrapolating to zero energy.

If equation (4.16) is inserted in equation (4.1) for reaction rate per particle pair $\langle \sigma v \rangle$, one obtains

$$\langle \sigma v \rangle = \left(\frac{8}{\pi \mu} \right)^{1/2} \frac{1}{(kT)^{3/2}} \int_0^\infty S(E) \exp \left[-\frac{E}{kT} - \frac{b}{E^{1/2}} \right] dE , \qquad (4.17)$$

where the quantity b, which arises from the barrier penetrability, is given by

$$b = (2\mu)^{1/2} \pi e^2 Z_1 Z_2 / \hbar = 0.989 Z_1 Z_2 \mu^{1/2} \ (\text{MeV})^{1/2} . \qquad (4.18)$$

The quantity b^2 is also called the Gamow energy, E_G. Since for nonresonant reactions $S(E)$ varies smoothly with energy, the energy dependence of the integrand in equation (4.17) is governed primarily by the exponential term. The penetration through the Coulomb barrier gives rise to the term $\exp(-b/E^{1/2}) = \exp[-(E_G/E)^{1/2}]$, which becomes very small at low energies. The other exponential term, $\exp(-E/kT)$, which vanishes at high energy, is a measure of the number of particles available in the high-energy tail of the Maxwell-Boltzmann distribution. The product of the two terms leads to a peak of the integrand near the energy E_0, which is usually much larger than kT. Although the Maxwell-Boltzmann distribution has a maximum at an energy $E = kT$, the Gamow factor shifts the effective peak to the energy E_0. The peak is referred to as the Gamow peak (Fig. 4.6).

For a given stellar temperature T, nuclear reactions take place in a relatively narrow energy window around the effective burning energy of E_0 (Fig. 4.6). Frequently, the $S(E)$ factor is nearly a constant over the window

$$S(E) = S(E_0) = \text{constant} , \tag{4.19}$$

and for this condition equation (4.17) reduces to

$$\langle \sigma v \rangle = \left(\frac{8}{\pi\mu}\right)^{1/2} \frac{1}{(kT)^{3/2}} S(E_0) \int_0^\infty \exp\left(-\frac{E}{kT} - \frac{b}{E^{1/2}}\right) dE . \tag{4.20}$$

By taking the first derivative of the integrand in equation (4.20), one can find the energy E_0 for which the integrand has its maximum value:

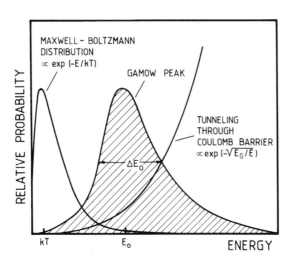

FIGURE 4.6. The dominant energy-dependent functions are shown for nuclear reactions between charged particles. While both the energy distribution function (Maxwell-Boltzmann) and the quantum mechanical tunneling function through the Coulomb barrier are small for the overlap region, the convolution of the two functions results in a peak (the Gamow peak) near the energy E_0, giving a sufficiently high probability to allow a significant number of reactions to occur. The energy of the Gamow peak is generally much larger than kT.

$$E_0 = \left(\frac{bkT}{2}\right)^{2/3} = 1.22(Z_1^2 Z_2^2 \mu T_6^2)^{1/3} \text{ keV} .$$ (4.21)

This quantity E_0 is the effective mean energy for thermonuclear fusion reactions at a given temperature T. For example, at a stellar temperature $T_6 = 15$ (sun), the effective burning energy E_0 for several representative reactions is

$p + p$: $E_0 = 5.9$ keV ,

$p + {}^{14}N$: $E_0 = 26.5$ keV ,

$\alpha + {}^{12}C$: $E_0 = 56$ keV ,

${}^{16}O + {}^{16}O$: $E_0 = 237$ keV .

From these examples it is clear that the Gamow peak is shifted toward higher energy as the nuclear charge (i.e., Coulomb barrier) increases. It should be noted that the rate of the $p + p$ reaction is determined not only by the Coulomb barrier but also by the strength of the weak force (chap. 6).

The maximum value of the integrand in equation (4.20) is found by inserting the expression for E_0:

$$I_{max} = \exp\left(-\frac{3E_0}{kT}\right) = \exp(-\tau) ,$$ (4.22)

where the dimensionless parameter τ is given numerically by

$$\tau = \frac{3E_0}{kT} = 42.46(Z_1^2 Z_2^2 \mu/T_6)^{1/3} .$$ (4.23)

For the examples used above and $T_6 = 15$, the value of I_{max} is

$p + p$: $I_{max} = 1.1 \times 10^{-6}$,

$p + {}^{14}N$: $I_{max} = 1.8 \times 10^{-27}$,

$\alpha + {}^{12}C$: $I_{max} = 3.0 \times 10^{-57}$,

${}^{16}O + {}^{16}O$: $I_{max} = 6.2 \times 10^{-239}$.

Since the reaction rate $\langle \sigma v \rangle$ is proportional to I_{max}, the rate of nuclear reactions depends strongly on the Coulomb barrier, as these numbers show. Even if there are many species of nuclei present in a star at a particular time, those with the smallest Coulomb barrier (and hence the largest value of I_{max}) will be consumed most rapidly and will account for most of the nuclear energy generation. In this phase, the species with higher Coulomb barriers do not contribute in any significant way to energy production. When the nuclei with the smallest Coulomb barrier have been consumed, a star will contract gravitationally until the temperature rises to a point where nuclei with the next lowest Coulomb barrier can burn. The burning of this new fuel and the energy produced stabilize the star against further contraction. These well-defined epochs in stellar evolution are referred to as the stages of nuclear burning: hydrogen burning, helium burning, heavy-ion burning (chap. 3). The existence of these epochs depends critically on the fact that the nuclear reaction rates are very sensitive to the Coulomb barrier.

The exponential term in the integrand of equation (4.20) can be fairly well approximated by a Gaussian function:

$$\exp\left(-\frac{E}{kT} - \frac{b}{E^{1/2}}\right) = I_{\max} \exp\left[-\left(\frac{E - E_0}{\Delta/2}\right)^2\right]. \tag{4.24}$$

In this approximation, the $1/e$ width of the peak is the effective width Δ of the energy window, wherein most of the reactions take place (Fig. 4.6). This width is determined by matching the second derivatives of equation (4.24), with the result

$$\Delta = \frac{4}{3^{1/2}} (E_0 kT)^{1/2} = 0.749(Z_1^2 Z_2^2 \mu T_6^5)^{1/6} \text{ keV}. \tag{4.25}$$

The result shows that the half-width $\Delta/2$ is roughly given by the geometrical mean of E_0 and kT and, since $kT \ll E_0$, the half-width $\Delta/2$ is always smaller than E_0. An approximate value of the integral in equation (4.20) is given by the width Δ times the height $I_{\max}$ of the Gamow peak:

$$(\Delta)I_{\max} = \frac{4}{3^{1/2}} (E_0 kT)^{1/2} \exp\left(-\frac{3E_0}{kT}\right). \tag{4.26}$$

For the examples used above, with $T_6 = 15$, the quantities $\Delta/2$ and $(\Delta)I_{\max}$ are

$p + p$:	$\Delta/2 = 3.2$ keV ,	$(\Delta)I_{\max} = 7.0 \times 10^{-6}$ keV ,
$p + {}^{14}\text{N}$:	$\Delta/2 = 6.8$ keV ,	$(\Delta)I_{\max} = 2.5 \times 10^{-26}$ keV ,
$\alpha + {}^{12}\text{C}$:	$\Delta/2 = 9.8$ keV ,	$(\Delta)I_{\max} = 5.9 \times 10^{-56}$ keV ,
${}^{16}\text{O} + {}^{16}\text{O}$:	$\Delta/2 = 20.2$ keV ,	$(\Delta)I_{\max} = 2.5 \times 10^{-237}$ keV .

The numerical values for the product $(\Delta)I_{\max}$ again demonstrate how dramatically the reaction rate $\langle \sigma v \rangle$ changes as the Coulomb barrier increases.

Nuclear-burning reactions take place predominantly over the energy window $E = E_0 \pm \Delta/2$. This energy window clearly increases with the Coulomb barrier of the nuclei involved. It is over this energy range where information regarding the nuclear processes must be obtained. The major problem in nuclear astrophysics results from the fact that this favored energy band near E_0 is generally at energies too low for direct measurement of the cross section $\sigma(E)$ or, equivalently, of $S(E)$. The standard solution to this problem is to measure $S(E)$ over a wide range of energies and to the lowest energies possible (chap. 5) and to extrapolate the data downward to E_0 with the help of theoretical and other arguments (chaps. 6–8). Thus, an extrapolation formula or procedure becomes a crucial necessity.

With the approximation in equation (4.24), the reaction rate becomes

$$\langle \sigma v \rangle = \left(\frac{2}{\mu} \right)^{1/2} \frac{\Delta}{(kT)^{3/2}} S(E_0) \exp \left(- \frac{3E_0}{kT} \right). \tag{4.27}$$

Inserting the parameter τ (eq. [4.23]) and evaluating the numerical constants, one obtains

$$\langle \sigma v \rangle = 7.20 \times 10^{-19} \frac{1}{\mu Z_1 Z_2} \tau^2 \exp (-\tau) S(E_0) \text{ cm}^3 \text{ s}^{-1},$$

where the factor $S(E_0)$ is in units of keV b and μ is in amu. This result is the nonresonant reaction rate per particle pair.

It is frequently of interest to know the dependence of the reaction rate $\langle \sigma v \rangle$ on a power of the temperature T. Using the above expression, $\langle \sigma v \rangle \propto \tau^2 \exp (-\tau)$, one finds

$$\langle \sigma v \rangle \propto T^{\tau/3 - 2/3}. \tag{4.28}$$

For the examples used above, the temperature dependence near $T_6 = 15$ is

$p + p$:	$\langle \sigma v \rangle \propto T^{3.9}$,	$E_C = 0.55$ MeV,
$p + {}^{14}\text{N}$:	$\langle \sigma v \rangle \propto T^{20}$,	$E_C = 2.27$ MeV,
$\alpha + {}^{12}\text{C}$:	$\langle \sigma v \rangle \propto T^{42}$,	$E_C = 3.43$ MeV,
${}^{16}\text{O} + {}^{16}\text{O}$:	$\langle \sigma v \rangle \propto T^{182}$,	$E_C = 14.07$ MeV.

The results demonstrate a dramatic sensitivity of the reaction rates to temperature, which is again directly related to the heights E_C of the Coulomb barriers.

Shown in Figure 4.7 is the Gamow peak for the $p + p$ reaction at a stellar temperature $T_6 = 15$. Also shown is the approximation derived by using a Gaussian distribution. Compared with the Gaussian distribution, the exact form is slightly asymmetric around the energy E_0. This is a general feature of the Gamow peak, which is accounted for when calculating the integral in equation (4.20) by the factor $F(\tau)$:

$$\int_0^\infty \exp \left(- \frac{E}{kT} - \frac{b}{E^{1/2}} \right) dE = F(\tau) \pi^{1/2} \frac{\Delta}{2} \exp (-\tau). \tag{4.29}$$

Equation (4.29) can be written in terms of τ only. Recalling that τ is much larger than unity,

$$\tau = \frac{3E_0}{kT} \gg 1, \tag{4.30}$$

one can expand the transformed equation (4.29) in a Maclaurin series with respect to $1/\tau$. The result is

$$F(\tau) = 1 + \frac{5}{12}\frac{1}{\tau} - \frac{35}{288}\frac{1}{\tau^2} + \cdots , \tag{4.31}$$

where in general the correction provided by the $1/\tau$ term is sufficient. For the examples used previously ($T_6 = 15$), the correction term $F(\tau)$ in first order is

$p + p$: $F(\tau) = 1.030$,

$p + {}^{14}N$: $F(\tau) = 1.0068$,

$\alpha + {}^{12}C$: $F(\tau) = 1.0032$,

${}^{16}O + {}^{16}O$: $F(\tau) = 1.00076$,

showing that the correction is small. Using only the first-order correction, the reaction rate $\langle \sigma v \rangle$ (eq. [4.27]) becomes

$$\langle \sigma v \rangle = \left(\frac{2}{\mu}\right)^{1/2} \frac{\Delta}{(kT)^{3/2}} S(E_0)\left(1 + \frac{5}{12\tau}\right) \exp\left(-\tau\right) . \tag{4.32}$$

If the $S(E)$ factor is described by a slowly varying function of energy E (Figs. 4.4 and 4.5) rather than by a constant as assumed above (eq. 4.19), it can be expanded in a Taylor series. Since the energy E_0 of the Gamow peak is temperature-dependent, for ease of computation the expansion is usually done around zero energy:

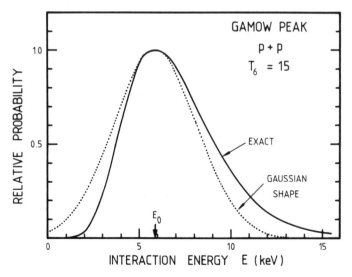

FIGURE 4.7. Curves for the Gamow peak for the $p + p$ reaction at $T_6 = 15$, as obtained from the exact expression and from the approximation using the Gaussian function.

$$S(E) = S(0) + \dot{S}(0)E + \tfrac{1}{2}\ddot{S}(0)E^2 + \cdots . \tag{4.33}$$

Again the dot represents a derivative with respect to energy, and the terms $S(0)$, $\dot{S}(0)$, and $\ddot{S}(0)$ are obtained from fits to experimental data. If the above expansion for $S(E)$ is inserted in equation (4.17) for the reaction rate $\langle \sigma v \rangle$, one arrives at a sum of integrals related to each term in equation (4.33). For each of these integrals, one uses the procedure described above, that is, the exact exponent in each integrand is replaced by the Gaussian function and the correction factor $F(\tau)$ is applied. As a result of this procedure, one can replace the constant $S(E_0)$ factor in equation (4.27) by an effective S-factor (Bah62, Fow67):

$$S_{\text{eff}}(E_0) = S(0)\left[1 + \frac{5}{12\tau} + \frac{\dot{S}(0)}{S(0)}\left(E_0 + \frac{35}{36}kT \right) + \frac{1}{2}\frac{\ddot{S}(0)}{S(0)}(E_0^2 + \frac{89}{36}E_0 kT) \right].$$

$$\tag{4.34}$$

The first two terms inside the brackets, $1 + 5/12\tau$, reflect the correction factor $F(\tau)$ for the asymmetry of the Gamow peak. The remaining terms result from the assumption that the $S(E)$ factor is a slowly varying function of energy, and are generally much more important compared with the second term. Using the experimental data obtained for the $^{12}C(p, \gamma)^{13}N$ reaction (chap. 6), the coefficients in equation (4.32) are (Fow67, Fow75) $S(0) = 1.40$ keV b, $\dot{S}(0)/S(0) = 3.04 \times 10^{-3}$ keV^{-1}, and $\ddot{S}(0)/2S(0) = 1.33 \times 10^{-5}$ keV^{-2}. At a stellar temperature $T_6 = 30$, the Gamow peak occurs at $E_0 = 37.9$ keV, and the effective S-factor (eq. [4.34]) is $S_{\text{eff}}(E_0) = 1.62$ keV b. The reaction rate per particle pair (eq. [4.27]) for these conditions is $\langle \sigma v \rangle = 3.26 \times 10^{-35}$ cm^3 s^{-1}. For a stellar density $\rho = 15$ g cm^{-3} and an 80% hydrogen abundance by weight, the mean lifetime (eq. [3.12]) of ^{12}C nuclei against destruction by protons is $\tau_p(^{12}C) = 138$ y. The lifetimes of other nuclei involved in the CNO cycle are discussed in chapter 6.

Since the quantities τ and E_0 are related to temperature T (eqs. [4.21] and [4.22]), one can express the effective S-factor (eq. [4.34]) in terms of temperature alone as follows:

$$S_{\text{eff}}(E_0) = S(0)(1 + \alpha_1 T^{1/3} + \alpha_2 T^{2/3} + \alpha_3 T + \alpha_4 T^{4/3} + \alpha_5 T^{5/3}) .$$

The reaction rate,

$$\langle \sigma v \rangle = \left(\frac{2}{\mu}\right)^{1/2} \frac{\Delta}{(kT)^{3/2}} S_{\text{eff}}(E_0) \exp\left(-\frac{3E_0}{kT} \right),$$

is an analytic expression, which also can be written in terms of temperature alone:

$$\langle \sigma v \rangle = AT^{-2/3} \exp\left(-BT^{-1/3}\right) \sum_{n=0}^{5} \alpha_n T^{n/3} . \tag{4.35}$$

Equation (4.35) is in the form given often in compilations (Fow67, Fow75, Har83, Cau85).

The $S(E)$ factor for nonresonant nuclear reactions may sometimes exhibit a rapidly varying change with energy. In this instance, the Maclaurin expansion is not applicable, and one must develop other techniques for arriving at an analytic expression for the reaction rate (Fow67, Fow75). For example, if $S(E) = S(0) \exp(-\alpha E)$, equation (4.17) leads to

$$\langle \sigma v \rangle = \left(\frac{8}{\pi\mu}\right)^{1/2} \frac{1}{(kT)^{3/2}} S(0) \int_0^\infty \exp\left(-\alpha E - \frac{E}{kT} - \frac{b}{E^{1/2}}\right) dE \ .$$

The first two terms in the exponent can be combined through the relation $\alpha E + E/kT = E/kT_{\text{eff}}$, whereby the effective temperature of the integrand is defined as $T_{\text{eff}} = T/(1 + \alpha kT)$. The reaction rate $\langle \sigma v \rangle$ (eq. [4.27]) then becomes

$$\langle \sigma v \rangle = \left(\frac{8}{\pi\mu}\right)^{1/2} \frac{\Delta_{\text{eff}}}{(kT)^{3/2}} S(0) \exp(-\tau_{\text{eff}})$$

where the quantities τ_{eff} and Δ_{eff} are given by equations (4.23) and (4.25) with T replaced by T_{eff}. Examples can be found in the literature (Fow67, Fow75), where other techniques have been used to find analytic expressions for the reaction rate $\langle \sigma v \rangle$. In those cases where the energy dependence of the $S(E)$ factor is extremely complicated, one must carry out a numerical integration of the reaction rate as a function of temperature and attempt to fit the form of these rates to an appropriate function of temperature.

In the above treatments, it was assumed that the Coulomb potential of the target nucleus as seen by the projectile is that resulting from a bare nucleus and thus would extend to infinity (Fig. 4.8). For nuclear reactions induced in

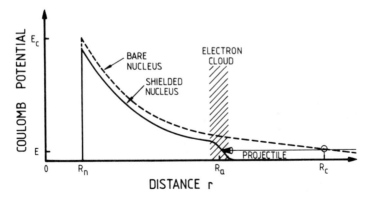

FIGURE 4.8. Shown in an exaggerated and idealized way is the effect of the atomic electron cloud on the Coulomb potential of a bare nucleus. This potential is reduced at all distances and goes essentially to zero beyond the atomic radius R_a. The effect of this electron shielding on an incident projectile is to increase the penetrability through the barrier, and thus also the cross section.

the laboratory, the target nuclei are in the form of atoms. The atomic electron cloud surrounding the nucleus acts as a screening potential. As a result, the total potential goes to zero outside the atomic radius (Fig. 4.8). An incoming projectile sees no repulsive Coulomb force until it penetrates beyond the atomic radius; thus it effectively sees a reduced Coulomb barrier. The electrostatic potential of the electron cloud at distances less than the atomic radius R_a is constant, with the approximate value $\phi_a = Z_1 e/R_a$, and consequently the total Coulomb potential within the atomic radius is $\phi_{tot} = (Z_1 e/r) - (Z_1 e/R_a)$. The effective height E_{eff} of the Coulomb barrier seen by the incoming projectile (Fig. 4.8) is then

$$E_{eff} = \frac{Z_1 Z_2 e^2}{R_n} - \frac{Z_1 Z_2 e^2}{R_a}. \tag{4.36}$$

This equation shows that the effect of the electron shielding on the height of the Coulomb barrier is in the ratio of nuclear to atomic radii, that is, $R_n/R_a \simeq 10^{-5}$. In general, this shielding correction is negligible. However, when the classical turning point R_c of an incoming projectile for the bare nucleus is near or outside the atomic radius R_a (Fig. 4.8), the magnitude of the shielding effect becomes significant. Since the classical turning point is related to the projectile energy E by the equation $E = Z_1 Z_2 e^2/R_c$, the condition $R_c \geq R_a$ can be written as

$$E \leq U_e = \frac{Z_1 Z_2 e^2}{R_a}. \tag{4.37}$$

For the worst case, setting the atomic radius R_a equal to the radius of the innermost electrons of target or projectile ($R_a \simeq R_H/Z$), one finds energies E from equation (4.37) for the previously discussed examples to be

$p + p$: $E \leq U_e = 0.029$ keV ,

$p + {}^{14}N$: $E \leq U_e = 1.41$ keV ,

$\alpha + {}^{12}C$: $E \leq U_e = 2.07$ keV ,

${}^{16}O + {}^{16}O$: $E \leq U_e = 14.75$ keV .

The penetration through a shielded Coulomb potential at projectile energy E_s is equivalent (Fig. 4.8) to that of bare nuclei at energy $E = E_s + U_e$. With equation (4.16) the enhancement ratio in cross sections $\epsilon = \sigma(E)/\sigma(E_s)$ for the $p + p$ reaction is found to be [assuming $S(E) =$ constant and $U_e = 0.029$ keV]

$E_s = 100$ keV , $\epsilon = 2.7 \times 10^{-3}\%$,

$E_s = 20$ keV , $\epsilon = 2.1 \times 10^{-1}\%$,

$E_s = 5$ keV , $\epsilon = 2.3\%$,

$E_s = 1$ keV , $\epsilon = 33.0\%$.

Since the energy window around the Gamow peak for these examples, and indeed in general, occurs at much higher projectile energies, the shielding correction can be disregarded, and laboratory experiments can be regarded as measuring essentially the cross section $\sigma(E)$ for penetration through a Coulomb barrier for the bare nuclei.

For the high temperatures occurring in stars, the atoms are in most cases completely stripped of their atomic electrons, and one might imagine that electron screening has no effect on nuclear reactions in stars. However, the nuclei are immersed in a sea of free electrons which tend to cluster in the area of a nucleus, resulting in an effect similar to the one resulting from the atomic orbital electrons as discussed above. For the condition that kT is much larger than the Coulomb energy between the particles, the stellar gas is called "nearly perfect." For such a stellar gas, the electrons tend to cluster into spherical shells around a nucleus at the Debye-Hückel radius R_D (Sal54, Ich82, Ich84):

$$R_D = \left(\frac{kT}{4\pi e^2 \rho N_A \xi}\right)^{1/2}, \tag{4.38}$$

where N_A is Avogadro's number. The quantity ξ is given by the equation

$$\xi = \Sigma_i (Z_i^2 + Z_i)\frac{X_i}{A_i}, \tag{4.39}$$

where the sum is over all positive ions and X_i is the mass fraction of nuclei of type i. For a stellar gas consisting of ^{1}H and ^{12}C in equal mass fractions at a temperature of $T_6 = 100$ and $\rho = 100$ g cm^{-3}, one finds $R_D = 5.4 \times 10^{-9}$ cm $= 5.4 \times 10^4$ fm, where 1 fm (fermi) $= 10^{-13}$ cm. For this example the Debye-Hückel radius R_D is the same order as the atomic radius R_a, and consequently electron screening effects are negligible. However, as the stellar density increases, the Debye-Hückel radius decreases and the shielding effect becomes more important. The shielding effect reduces the Coulomb potential and eases the penetration of the Coulomb barrier. Thus, it increases the cross sections and thus the thermonuclear reaction rates within the star:

$$\langle\sigma v\rangle_{\text{shielded nuclei}} = f\langle\sigma v\rangle_{\text{bare nuclei}} . \tag{4.40}$$

The quantity f is referred to as the electron shielding factor, which according to calculations (Sal54) generally varies between 1 and 2 for typical densities and compositions. However, at high densities the quantity f can be enormous (Ich82, Ich84). Note that in this book all expressions derived for $\langle\sigma v\rangle$ are for bare nuclei and must be multiplied by the factor f for stellar conditions.

When calculating reaction rates $\langle\sigma v\rangle$ from equation (4.17) for shielded nuclei with energy E_s, the energy E for bare nuclei must be replaced by the equation $E_s = E - U_e$, where the quantity U_e is the energy of the electron screening potential. For energies of the Gamow peak E_0 much greater than the energy U_e, one obtains for the electron screening factor f (eq. [4.40]) the analytic expression

$$f = \exp\left(\frac{U_e}{kT}\right).$$

(4.41)

For relatively low stellar densities, the number of screening electrons at the Debye-Hückel radius R_D is nearly equal to the number of protons in the nucleus (Z_1). In this case, the quantity U_e is given by $U_e = Z_1 Z_2 e^2/R_D$. For the $p + {}^{12}C$ interaction (see above), U_e is equal to 0.16 keV, which is much smaller than the energy $E_0 = 37.9$ keV of the associated Gamow peak (for $T_6 = 30$). The screening factor for this example is $f = 1.064$, supporting the conclusion reached above that under these conditions screening effects are negligible.

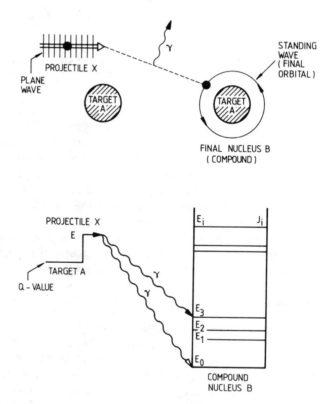

FIGURE 4.9. Illustrated is a capture reaction $A(x, \gamma)B$, where the entrance channel $A + x$ goes directly to states in the final compound nucleus B with the emission of γ-radiation. This process is called a direct-capture reaction and can occur for all energies E of the projectile x.

4.3. Reactions through Isolated and Narrow Resonances

The previous two sections involved nonresonant reactions only. One such reaction, the capture reaction, may be visualized as shown in Figure 4.9. Here, the incoming projectile x is characterized by a plane wave, which goes directly to a standing wave with orbital angular momentum l in the final compound nucleus B. In the transition a photon with energy $E_\gamma = E + Q - E_i$ is emitted. The process is entirely electromagnetic and is similar to the well-known bremsstrahlung process (Bla62). The cross section for γ-ray emission is described by a single matrix element,

$$\sigma_\gamma \propto |\langle B|H_\gamma|A + x\rangle|^2 \, ,$$

where the transition from the entrance channel A + x to the final compound nucleus B is mediated by the electromagnetic operator H_γ. Since this process involves a single matrix element, it is called a single-step process. Alternatively, it is referred to as a direct process (direct capture), because it represents a direct transition from the initial state to the final state. This direct-capture process is nonresonant because it can occur at all projectile energies with a cross section varying smoothly with energy (Fig. 4.4). There are other types of nonresonant direct reactions, such as stripping, pickup, charge exchange, and Coulomb excitation (Bla62, Fra74).

In contrast to the nonresonant reactions discussed so far, there is another type of reaction mechanism in which an excited state E_r of the compound nucleus is first formed in the entrance channel, which subsequently decays to lower-lying states (Fig. 4.10). This process occurs only if the energy of the entrance channel, $Q + E_R$, matches the energy E_r of the excited state in the compound nucleus (Fig. 4.10):

$$E_R = E_r - Q \, . \tag{4.42}$$

In this equation, the Q-value is constant and the process can occur for all excited states E_i above the energy of the Q-value (the threshold energy), i.e., whenever the energy of the projectile E_R fulfills the condition of equation (4.42). Since the reaction occurs only at these fixed values E_R of the projectile energies, the process is referred to as a resonant reaction. If the excited state E_r decays by the emission of γ-rays to a state E_f at lower energy, the cross section σ_γ is described by the product of two matrix elements:

$$\sigma_\gamma \propto |\langle E_f|H_\gamma|E_r\rangle|^2 |\langle E_r|H_f|A + x\rangle|^2 \, , \tag{4.43}$$

where the matrix element involving the operator H_f describes the formation of the compound state E_r and the other matrix element the subsequent γ-decay of the state E_r. Since this process involves two matrix elements, it is referred to as a two-step process. Near the resonant beam energies E_R, the cross sections may be very high (Fig. 4.10). Each matrix element represents the probability for this step in the reaction to occur and is usually represented by a partial width Γ_i. Equation (4.43) can now be written as

$$\sigma_\gamma \propto \Gamma_a \Gamma_b ,\tag{4.44}$$

where Γ_a (Γ_b) is the partial width for the formation (decay) of the compound state. The decay width Γ_b can refer to particle as well as γ-ray emission. The magnitude of the resulting cross section can vary widely depending on the force laws and properties of the nuclear states involved in each step of the reaction.

In previous discussions, the quantity $\pi \lambdabar^2$ has been referred to as the essentially geometrical cross section and, except for statistical effects, is the maximum value the cross section can attain. To arrive at an expression for these quantities, note that a collision between a projectile and a target nucleus can be described in terms of the momentum of the projectile $p = \hbar k = \hbar/\lambdabar$ and the impact parameter b (Fig. 4.11). The orbital angular momentum involved in

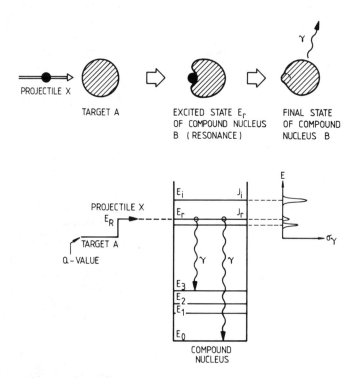

FIGURE 4.10. Illustrated is a capture reaction A(x, γ)B, where the entrance channel A + x forms an excited state E_r in the compound nucleus B at an incident energy of E_R. The excited state E_r decays into lower-lying states with the emission of γ-radiation. This process is called a resonant capture reaction and can occur only at selected energies where $Q + E_R$ matches E_r. At these resonant energies E_R, the capture cross section σ_γ may exhibit large maxima.

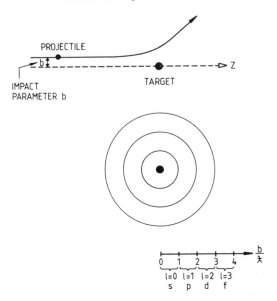

FIGURE 4.11. Schematic representation of the collision of a projectile with a target nucleus characterized by the impact parameter b. The lower part of the figure is a representation of the various areas associated with the orbital angular momenta l in the entrance channel.

this collision is $L = bp = b\hbar/\lambdabar$. In the semiclassical limit, the orbital angular momentum is quantized in the form $L = l\hbar$, where the quantum number l is restricted to integer values. Thus, one can think of the shadow of the nucleus on a plane normal to the incident beam as being divided into zones of differing impact parameters b according to $b = l\lambdabar$ (Fig. 4.11). Each zone will be characterized by an orbital angular momentum l. The area of each zone represents the maximum possible reaction cross section, if all particles inside this zone are absorbed and no outgoing waves exist. The maximum possible reaction cross section for a given value of l is given by the area between successive concentric rings about the nucleus:

$$\sigma_{l\,max} = \pi b_{l+1}^2 - \pi b_l^2 = (2l + 1)\pi\lambdabar^2 . \tag{4.45}$$

The value of $\pi\lambdabar^2$ in terms of center-of-mass energy E (in units of keV) is

$$\pi\lambdabar^2 = \frac{656.6}{\mu E}\ b . \tag{4.46}$$

Note that the $(2l + 1)$ term in equation (4.45) should not be interpreted as meaning that the physics of the process favors high orbital angular momenta. It is simply a statistical factor reflecting the fact that there are more particles with high values of l in an incident plane wave. The above treatment can be generalized to include the spins of the projectile (J_1) and the target nucleus

(J_2). In this case, the statistical factor $(2l + 1)$ in equation (4.45) is replaced by the more general statistical factor ω (chap. 3) of

$$\omega = \frac{2J + 1}{(2J_1 + 1)(2J_2 + 1)}, \tag{4.47}$$

where J is the angular momentum of the excited state in the compound nucleus (Fig. 4.10). Clearly, for $J_1 = J_2 = 0$, one has the condition $J = l$, and ω reduces to $(2l + 1)$. One must also include a term $(1 + \delta_{12})$ in equation (4.45), because the cross section is increased by a factor of 2 in the case of identical particles in the entrance channel. Combining these effects, the maximum cross section is

$$\sigma_{max} = \pi \lambda^2 \frac{2J + 1}{(2J_1 + 1)(2J_2 + 1)} (1 + \delta_{12}). \tag{4.48}$$

Resonant phenomena occur quite frequently in physical systems. Such systems are described by a discrete set of values of a parameter (called eigenvalues) for which the system has a maximum response when driven by an external force. An example of such a system is the damped oscillator. If the oscillator is characterized by the eigenfrequency ω_0 and the damping factor δ, the response of the oscillator to an external force with frequency ω is given by the equation

$$\text{Response} \propto \frac{f}{(\omega - \omega_0)^2 + (\delta/2)^2}, \tag{4.49}$$

where the quantity f, called the oscillator strength, is determined by the restoring force of the oscillator. When plotted as a function of frequency ω, the response function has a Lorentzian form with a maximum at ω_0 and a frequency range usually given in terms of the full width at half-maximum, δ. In all real physical systems there is a dissipative force, which causes the oscillations to die out when the driving force is removed. The magnitude of this force determines how long the system continues to oscillate. In most systems, the slowing down of the oscillator is described by an exponential function, $\exp(-t/t_0)$, where the parameter t_0 is the mean lifetime of the oscillator. This mean lifetime t_0 is related to a frequency band $\Delta\omega$ through the relationship $\Delta\omega = 1/t_0$. The frequency band $\Delta\omega$ is identical with the width δ introduced above. The response function (eq. [4.49]) for resonant phenomena in nuclear reactions is given in terms of a cross section. Nuclear cross sections are measured as a function of energy, and equation (4.49) is written in the analogous form

$$\sigma(E) \propto \frac{\Gamma_a \Gamma_b}{(E - E_R)^2 + (\Gamma/2)^2}. \tag{4.50}$$

An excited state of a nucleus is characterized, as in the case of the damped oscillator, by a mean lifetime t_0 or, equivalently, by an energy width Γ. These conjugate quantities are related through the uncertainty principle by the equation $\Gamma t_0 = \hbar$. The energy analogue of the quantity δ in equation (4.49) is therefore the width Γ (eq. [4.50]). Furthermore, the analogue of the oscillator strength f in equation (4.49) is the product $\Gamma_a \Gamma_b$ (eq. [4.44]), which is determined by the force laws governing the nuclear reaction. The total width Γ is given by the sum of the partial widths of all open, i.e., energetically allowed, decay channels:

$$\Gamma = \Gamma_a + \Gamma_b + \cdots . \tag{4.51}$$

Combining equations (4.48) and (4.50), one arrives at the relation

$$\sigma(E) = \pi \lambda^2 \frac{2J + 1}{(2J_1 + 1)(2J_2 + 1)} (1 + \delta_{12}) \frac{\Gamma_a \Gamma_b}{(E - E_R)^2 + (\Gamma/2)^2} . \tag{4.52}$$

This is the Breit-Wigner formula for a single-level resonance. All energies and widths are in the center-of-mass system. As pointed out previously, the partial widths Γ_a and Γ_b, and thus the total width Γ (eq. 4.52), are all energy-dependent (eqs. [4.65] and [4.67]). It should be emphasized that equation (4.52) is valid only for isolated resonances, defined as resonances for which the separation of the nuclear levels is large compared with their total widths. A resonance is defined as "narrow" if the width Γ of the resonance is much smaller than the resonance energy E_R ($\Gamma \ll E_R$).

A more rigorous derivation of the Breit-Wigner formula leads Blatt and Weisskopf (Bla62) to a result which is identical with equation (4.52), except that in those cases where the partial width Γ_a of the entrance channel has nearly its maximum possible value (eq. [4.65]), the term $E - E_R$ must be replaced by $E - E_R - \Delta(E)$. The quantity $\Delta(E)$ is the Thomas-Lane correction term (Bla62), which represents the energy shift between the energy at which the resonance is observed and that of the corresponding state in the compound nucleus system.

It follows from the Breit-Wigner formula that if the nuclear reaction A(a, b)B takes place ($\sigma_r \propto \Gamma_a \Gamma_b$), then the elastic scattering reaction A(a, a)A also occurs ($\sigma_e \propto \Gamma_a \Gamma_a$). The ratio of the two cross sections is given by $\sigma_r/\sigma_e = \Gamma_b/\Gamma_a$. If $\Gamma = \Gamma_a + \Gamma_b$ the reaction cross section σ_r is a maximum for the condition $\Gamma_a = \Gamma_b = \Gamma/2$ and is equal to the elastic scattering cross section $\sigma_e = \sigma_r$. Usually one finds $\Gamma_a \gg \Gamma_b$ and hence $\sigma_e \gg \sigma_r$. Note that the total scattering cross section is the sum of the elastic cross section and the reaction cross section.

Whether or not a resonant state can be formed via a given reaction channel depends also on angular momentum and parity conservation laws (selection rules). Angular momentum conservation requires the sum of the spins of the particles in the channel, j_1 and j_2, plus their relative orbital angular momentum l, to add up to the angular momentum of the resonant state J:

$$j_1 + j_2 + l = J .$$

For spinless particles ($j_1 = j_2 = 0$) this relation reduces to $l = J$. Parity conservation (Bla62, Fra74) requires the fulfillment of the relation

$$(-1)^l \pi(j_1)\pi(j_2) = \pi(J) ,$$

where $\pi(j_1)$ and $\pi(j_2)$ are the parities of the reacting particles and $\pi(J)$ is the parity of the resonant state. For spinless particles with $\pi(j_1) = \pi(j_2) = +1$ it follows that $(-1)^l = \pi(J)$. In this case the parity of the resonant state is determined by the orbital angular momentum of the channel, and such states are said to have natural parity. Conversely, for $(-1)^l \neq \pi(J)$ the state cannot be formed in this reaction channel, since the resonant state has unnatural parity. Such examples are discussed in chapter 7.

With the knowledge of the energy dependence of the cross section for a narrow resonance ($\Gamma \ll E_R$), the stellar reaction rate per particle pair (eq. [4.1]) is given by

$$\langle \sigma v \rangle = \left(\frac{8}{\pi \mu} \right)^{1/2} \frac{1}{(kT)^{3/2}} \int_0^\infty \sigma_{BW}(E)E \, \exp\left(-\frac{E}{kT} \right) dE ,$$

where the quantity $\sigma_{BW}(E)$ is the Breit-Wigner cross section (eq. [4.52]). For a narrow resonance (Fig. 4.12) the Maxwell-Boltzmann function, $E \exp(-E/kT)$, changes very little over the resonance region, and its value at $E = E_R$ can be taken outside the integral:

$$\langle \sigma v \rangle = \left(\frac{8}{\pi \mu} \right)^{1/2} \frac{1}{(kT)^{3/2}} E_R \exp\left(-\frac{E_R}{kT} \right) \int_0^\infty \sigma_{BW}(E)dE . \qquad (4.53)$$

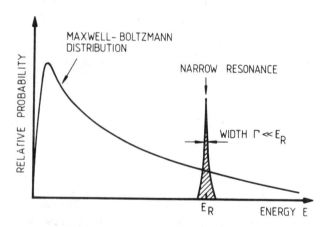

FIGURE 4.12. Shown schematically is the Maxwell-Boltzmann energy distribution for a given stellar temperature T and the cross section for a narrow resonance ($\Gamma \ll E_R$). In this case the Gamow peak coincides with the resonance energy region.

The integration of the Breit-Wigner cross section yields, for a narrow resonance with $\lambdabar^2 \simeq \lambdabar_R^2$ and with negligible energy dependence of the partial and total widths,

$$\int_0^\infty \sigma_{BW}(E)dE = \pi \lambdabar_R^2 \, \omega \Gamma_a \Gamma_b \int_0^\infty \frac{1}{(E - E_R)^2 + (\Gamma/2)^2} \, dE$$

$$= 2\pi^2 \lambdabar_R^2 \, \omega \, \frac{\Gamma_a \Gamma_b}{\Gamma} \, . \tag{4.54}$$

The product of the statistical factor ω, where

$$\omega = \frac{2J + 1}{(2J_1 + 1)(2J_2 + 1)} (1 + \delta_{12}) \, ,$$

and the width ratio $\gamma = \Gamma_a \Gamma_b / \Gamma$ is referred to as the strength of a resonance:

$$\omega\gamma = \omega \, \frac{\Gamma_a \Gamma_b}{\Gamma} \, . \tag{4.55}$$

Clearly, the resonance strength $\omega\gamma$ refers to the integrated cross section. Since the maximum resonance cross section σ_R is obtained at $E = E_R$ with the value (eq. [4.52])

$$\sigma_R = \sigma(E = E_R) = 4\pi \lambdabar_R^2 \, \omega \, \frac{\Gamma_a \Gamma_b}{\Gamma^2} \, ,$$

the integrated cross section (eq. [4.54]) can also be written as

$$\int_0^\infty \sigma_{BW}(E)dE = \frac{\pi}{2} \Gamma \sigma_R \, ,$$

that is, the area under a resonance curve is nearly equal to the product of the width Γ times the height σ_R. Combining equations (4.53), (4.54), and (4.55) leads to

$$\langle \sigma v \rangle = \left(\frac{2\pi}{\mu kT} \right)^{3/2} \hbar^2 (\omega\gamma)_R \exp\left(-\frac{E_R}{kT} \right) f \, , \tag{4.56}$$

which is the stellar reaction rate per particle pair for a narrow resonance. As in the case of nonresonant reactions, we have taken into account electron screening effects in stars by the screening factor f (sec. 4.2). Thus the reaction rate is calculated, for the narrow resonance case, from the experimentally measurable resonance parameters (chap. 5): the strength, the width, and the resonance energy.

Note that for a narrow resonance nuclear burning in stars takes place at the resonance energy E_R (Fig. 4.12), that is, the Gamow peak (Fig. 4.6) is now identical with the resonance peak. When a nuclear reaction has several narrow resonances, their contributions to $\langle \sigma v \rangle$ are simply summed:

$$\langle \sigma v \rangle = \left(\frac{2\pi}{\mu k T} \right)^{3/2} \hbar^2 f \Sigma_i (\omega \gamma)_i \, \exp \left(-\frac{E_i}{kT} \right). \tag{4.57}$$

Clearly, the concept of a Gamow peak is not applicable in this situation. However, the concept of the most effective stellar energy range $E_0 \pm \Delta/2$ is useful in these cases as well, because it shows which resonances are a priori most important. The result of the sum in equation (4.57) can often be represented by an analytic function of the form

$$\langle \sigma v \rangle = \alpha_1 \, \exp \left(-\alpha_2 / kT \right) + \alpha_3 \, \exp \left(-\alpha_4 / kT \right),$$

where the quantities α_i are parameters (Fow67, Fow75) which characterize the contributions of the various resonances.

The strength $\omega \gamma$ of a resonance and thus the stellar rate $\langle \sigma v \rangle$ also depend critically on the effects of the Coulomb barrier. To illustrate this, consider a charged-particle–induced capture reaction A(x, γ)B with only two open reaction channels represented by the partial widths $\Gamma_a = \Gamma_x$ and $\Gamma_b = \Gamma_y$ and thus $\Gamma = \Gamma_x + \Gamma_y$. The γ-width is at most of the order of $\Gamma_y \leq 1$ eV (sec. 4.4). It can vary by a few orders of magnitude depending on the nuclear structure of the resonance state and the role of selection rules. The charged particle width Γ_x of the entrance channel can reach maximum values of the order of MeV at resonance energies E_R near the Coulomb barrier (Fig. 4.14). At these energies, one finds the condition $\Gamma_x \simeq \Gamma$ and thus $\omega_y \simeq \omega \Gamma_y$, that is, the strengths ω_y of resonances near the Coulomb barrier depend only on the Γ_y width and are therefore of the order of 1 eV or somewhat smaller. At energies E_R far below the Coulomb barrier, penetration through the barrier reduces the particle width Γ_x very rapidly to values $\Gamma_x \ll \Gamma_y$ (Fig. 4.14) and thus $\Gamma_y \simeq \Gamma$. The resonance strengths of such low-energy resonances depend therefore only on the particle width Γ_x, that is, $\omega_y \simeq \omega \Gamma_x$. These strengths decrease very rapidly with decreasing resonance energy E_R due to the barrier penetration factor contained in Γ_x. The exponential term in equation (4.57) shows that for a given stellar temperature T resonances with energies E_R near kT will dominate the stellar reaction rate $\langle \sigma v \rangle$. Therefore, for low stellar temperatures, it is extremely important to know the locations and strengths of low-energy resonances. Because of the dependence $\omega \gamma \simeq \omega \Gamma_x$, a laboratory study of these low-energy resonances is very difficult and represents a challenge to the experimentor (chaps. 6 and 7).

To illustrate the importance of low-energy resonances in stellar environments, consider a resonance at $E_R = 2$ MeV in the reaction ^{14}N(p, γ)^{15}O ($Q = 7.3$ MeV) with the following properties: an angular momentum of $J = \frac{1}{2}$ of the resonance state E_r, formation via s-wave protons ($l = 0$), a proton width near maximum $\Gamma_p = 0.1$ MeV (i.e., for a nuclear structure parameter or reduced width $\theta_l^2 = 0.1$; sec. 4.4), and a dipole radiation $E_y = 9.3$ MeV near maximum width $\Gamma_y = 1$ eV. For these conditions and with

$J_1(\text{proton}) = \frac{1}{2}$ and $J_2(^{14}\text{N}) = 1$, the resonance strength is $\omega\gamma = 0.33$ eV. If this same resonance was located at $E_R = 10$ keV, the γ-width Γ_γ would be reduced in proportion to E_γ^3 (sec. 4.4),

$$\Gamma_\gamma(10 \text{ keV}) \simeq \left(\frac{7.3}{9.3}\right)^3 \Gamma_\gamma(2 \text{ MeV}) = (0.48) \times 1 \text{ eV} = 0.48 \text{ eV} ,$$

and the proton width in proportion to the Gamow factor (eq. [4.14]),

$$\Gamma_p(10 \text{ keV}) \simeq \frac{\exp(-66.91)}{\exp(-4.73)} \Gamma_p(2 \text{ MeV}) = (9.88 \times 10^{-28})(0.1 \text{ MeV})$$

$$= 9.88 \times 10^{-23} \text{ eV} ,$$

hence $\omega\gamma(E_R = 10 \text{ keV}) = 3.29 \times 10^{-23}$ eV. The ratio of reaction rates (eq. [4.56]) for $T_6 = 30$ is

$$\frac{\langle\sigma v\rangle(E_R = 10 \text{ keV})}{\langle\sigma v\rangle(E_R = 2 \text{ MeV})} = 1.6 \times 10^{312} .$$

The mean lifetime of nuclei as well as other properties of stars would change dramatically for such a difference in resonance energies E_R. The above example emphasizes the importance of resonances near the particle threshold for stellar reaction rates and stellar properties. Clearly the existence of possible states near the particle threshold of the compound nucleus has to be studied experimentally using several other reactions, which form this compound nucleus at the relevant excitation energies. Measurement of the appropriate quantities (i.e., resonance energy E_R, angular momentum l and J, and partial width Γ_p or reduced particle width θ_l^2) needed to calculate $\langle\sigma v\rangle$ frequently demands nuclear studies that have only subtle connections to the reaction of interest to astrophysics (chap. 6 and 7). For neutron-induced resonance reactions, there is no Coulomb barrier, and $\omega\gamma \simeq \omega\Gamma_\gamma$ holds for nearly all resonance energies E_R. This strength can more easily be determined experimentally (chap. 9).

For nuclei in states with long enough lifetimes so that they may participate in stellar nuclear reactions, it is important to know the number N_{12} of such nuclei available at any stellar temperature T. For a narrow resonance, the production rate of such nuclei is proportional to the number of nuclei in the resonance state produced by the formation channel with N_1 and N_2 nuclei at the resonance energy E_R. At high resonance energies E_R, the formation width Γ_a is frequently much larger than the decay width Γ_b, and hence $\Gamma \simeq \Gamma_a \gg \Gamma_b$. In this case, one has $\omega\gamma \simeq \omega\Gamma_b$, and the total reaction rate r_{12} (eqs. [3.9] and [4.56]) is given by

$$r_{12} = \frac{N_1 N_2}{1 + \delta_{12}} \langle\sigma v\rangle = \frac{N_1 N_2}{1 + \delta_{12}} \left(\frac{2\pi}{\mu kT}\right)^{1/2} \omega\Gamma_b \hbar^2 \exp\left(-\frac{E_R}{kT}\right) .$$

For the above condition $\omega\gamma \simeq \omega\Gamma_b$, the decay width Γ_b determines the lifetime τ_b of the N_{12} nuclei. Since in statistical equilibrium the rate of decay will just balance the rate of production (chap. 3),

$$r_{12} = \frac{N_{12}}{\tau_b} = \frac{\Gamma_b}{\hbar} N_{12} \, .$$

The combination of the two equations leads to the number of nuclei N_{12} in the resonance state,

$$N_{12} = \frac{N_1 N_2}{1 + \delta_{12}} \left(\frac{2\pi}{\mu k T} \right)^{3/2} \hbar^3 \omega \exp\left(-\frac{E_R}{kT} \right). \tag{4.58}$$

This equation is referred to as the Saha equation and is valid for all types of reactions for which an equilibrium can be established. The exponential term is the usual Boltzmann factor (for other cases see Fow67, Fow75, and chap. 7).

For high stellar temperatures equilibrium conditions exist for many excited states E_i of nuclei. These states participate in an important way in stellar nuclear reactions. The number of nuclei N_{12} in these excited states is given by the Saha equation and is related to the partition function discussed in chapter 3. An example is the very important reaction $3\alpha \to {}^{12}C$ (triple-alpha process; chap. 7), where two α-particles first form an unstable ${}^8Be$ nucleus which subsequently captures a third α-particle to form the stable nucleus ${}^{12}C$. In this case it is important to know the number of ${}^8Be$ nuclei (N_{12}) available for the second reaction to occur. The ${}^8Be$ is formed predominantly in its ground state, which represents a $J = 0$ resonance at $E_R = 94$ keV in the $\alpha + \alpha$ formation channel. For stellar conditions $\rho = 10^5$ g cm^{-3}, $X({}^4He) = 1.0$ and $T_6 = 100$ (helium burning in red giants; chap. 7), the Saha equation (4.58) leads to $N({}^8Be)/N({}^4He) \simeq 5.2 \times 10^{-10}$. The number of ${}^8Be$ nuclei relative to ${}^4He$ nuclei, though small, is sufficiently large to bridge the mass 5 and mass 8 instability gaps and to produce ${}^{12}C$ nuclei via the subsequent ${}^8Be(\alpha, \gamma){}^{12}C$ reaction.

4.4. Reactions through Broad Resonances

A resonance for which $\Gamma/E_R \geq 10\%$ is called a broad resonance. For such resonances, the cross section $\sigma(E)$ extends over a wider energy range than in the case of narrow resonances (Fig. 4.12), and the evaluation of the stellar reaction rate $\langle \sigma v \rangle$ (eq. [4.1]) must explicitly take into account the energy dependence of the cross section. If the cross section at the resonance energy E_R, that is, $\sigma_R = \sigma(E = E_R)$, as well as the total width $\Gamma_R = \Gamma(E = E_R)$, are known from experiments (the usual case), the resonance cross section $\sigma(E)$ at any other energy E can be calculated using equation (4.52):

$$\sigma(E) = \sigma_R \frac{E_R}{E} \frac{\Gamma_a(E)}{\Gamma_a(E_R)} \frac{\Gamma_b(E)}{\Gamma_b(E_R)} \frac{(\Gamma_R/2)^2}{(E - E_R)^2 + [\Gamma(E)/2]^2}, \tag{4.59}$$

where the total width $\Gamma(E)$ at energy E is given by

$$\Gamma(E) = \Gamma_a(E) + \Gamma_b(E) + \cdots .$$

Clearly the calculation of the cross section via equation (4.59) requires a knowledge of the energy dependence of the partial and thus the total widths.

Because time-reversal invariance holds for electromagnetic and nuclear forces (chap. 3), the partial width for any step in the nuclear reaction involving these forces is independent of the direction of the process. Thus, the partial width for the formation of a compound nucleus C through an entrance channel A + x is identical with the partial width for the decay of the nucleus C into the channel A + x. First consider the decay of a compound nucleus C by particle emission. The process is shown schematically in Figure 4.13, where the upper part represents the decay of an excited state E_r of the compound nucleus C into a residual nucleus A with the emission of particle x, and the lower part is a schematic plot of the radial wave function for the particle x. In general the particle must penetrate through both the Coulomb barrier (eq. [4.9]) and the centrifugal barrier as given by

$$V_{\mathrm{cf}}(r) = \frac{L^2}{2\mu r^2} = \frac{l(l+1)\hbar^2}{2\mu r^2}, \tag{4.60}$$

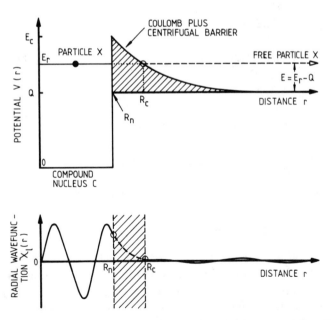

FIGURE 4.13. An excited state at excitation energy E_r of the compound nucleus C, which can decay into the residual nucleus A by the emission of particle x. The lower part is a schematic plot of the radial wave function for the particle x relating its standing-wave character inside the nucleus to its exponential decay inside the barrier and its small amplitude outside the barrier.

where l is the quantum number of the orbital angular momentum. The tunnel effect through both barriers is (Fig. 4.13)

$$P_l(E, R_n) = \left| \frac{\chi_l(R_c)}{\chi_l(R_n)} \right|^2 ,$$

where the index l reflects the angular momentum dependence. For a free particle, the distance R_c (Fig. 4.13) must be replaced by infinity:

$$P_l(E, R_n) = \left| \frac{\chi_l(\infty)}{\chi_l(R_n)} \right|^2 . \tag{4.61}$$

The decay rate $\lambda = 1/\tau$ of the excited state E_r in the compound nucleus is the probability per unit time that the particle x passes through a large spherical surface at a distance r, where r approaches infinity. This rate λ is determined by the product of the probability of finding particles x at this surface and their flux at this surface, which is the velocity of the free particle, or, equivalently, $(2E/\mu)^{1/2}$:

$$\lambda = \lim_{r \to \infty} \left(\frac{2E}{\mu} \right)^{1/2} \int_\Omega |\psi(r, \Theta, \phi)|^2 r^2 d\Omega .$$

If the wave function is written in the form

$$\psi(r, \Theta, \phi) = \frac{\chi_l(r)}{r} Y_{lm}(\Theta, \phi) ,$$

where the $Y_{lm}(\Theta, \phi)$ are spherical harmonics with the normalization

$$\int_\Omega |Y_{lm}(\Theta, \phi)|^2 d\Omega = 1 ,$$

the decay rate λ reduces to

$$\lambda = \lim_{r \to \infty} \left(\frac{2E}{\mu} \right)^{1/2} |\chi_l(r)|^2 = \left(\frac{2E}{\mu} \right)^{1/2} |\chi_l(\infty)|^2 .$$

Inserting equation (4.61) in the above expression, one obtains

$$\lambda = \left(\frac{2E}{\mu} \right)^{1/2} P_l(E, R_n) |\chi_l(R_n)|^2 . \tag{4.62}$$

The Pauli principle and other effects combine to produce an effectively uniform probability density of particles within the nucleus. One can therefore determine the probability for a particle to be near the nuclear surface by simply taking a ratio of the volume of a thin layer near the nuclear surface to the total volume of the nucleus. This leads to

$$|\chi_l(R_n)|^2 \simeq \frac{3}{R_n} . \tag{4.63}$$

This quantity represents the maximum probability of finding the particle x in the excited state E_r (Fig. 4.13) at the nuclear surface. It assumes that the excited state has a single configuration, namely,

$$\psi = |A \otimes x\rangle ,$$

which one might think of as being particle x orbiting around the residual nucleus A. In general this might not be the case, since other configurations are also possible:

$$\psi = \alpha_1 |A \otimes x\rangle + \alpha_2 |B \otimes y\rangle + \cdots ,$$

where α_i^2 is the probability of finding the excited state in the configuration i and $\Sigma_i \alpha_i^2 = 1$. Equation (4.63) must be modified to take this into account:

$$|\chi_l(R_n)|^2 = \frac{3}{R_n} \alpha_i^2 = \frac{3}{R_n} \theta_l^2 , \qquad (4.64)$$

where α_i^2 has been replaced by the more commonly used notation θ_l^2. The dimensionless number θ_l^2 is generally determined experimentally and contains the nuclear structure information. The quantity θ_l^2 is called the reduced width of a nuclear state, and clearly $\theta_l^2 \leq 1$. This limit is frequently referred to as the Wigner limit. Using equation (4.64), the decay rate (eq. [4.62]) becomes

$$\lambda = \left(\frac{2E}{\mu} \right)^{1/2} P_l(E, R_n) \frac{3}{R_n} \theta_l^2 ,$$

and since $\lambda = 1/\tau = \Gamma/\hbar$, one obtains for the partial width

$$\Gamma_l(E) = \frac{3\hbar}{R_n} \left(\frac{2E}{\mu} \right)^{1/2} P_l(E, R_n)\theta_l^2 .$$

This result holds for Wigner's "uniform" nucleus (Bla62). A more refined derivation (nuclear oscillator model) leads to almost the same result (Bla62):

$$\Gamma_l(E) = \frac{2\hbar}{R_n} \left(\frac{2E}{\mu} \right)^{1/2} P_l(E, R_n)\theta_l^2 . \qquad (4.65)$$

Calculations of the energy dependence of the partial width $\Gamma_l(E)$ requires knowledge of the penetration factor $P_l(E, R_n)$ defined by equation (4.61). Clearly, $P_l(E, R_n)$ involves radial wave functions only at distances $r \geq R_n$, and since only the ratio of wave functions is required, the penetration factor is independent of the nuclear potential. However, calculations for realistic nuclear wells such as the Woods-Saxon potential give larger penetration factors (Mic70 and references therein). The enhanced reflection factor can be included in θ_l^2. The solution of the Schrödinger equation at these distances is a linear combination of the regular and irregular Coulomb functions, $F_l(r)$ and $G_l(r)$ (Abr65):

$$\chi_l(r) = AF_l(r) + BG_l(r) ,$$

where A and B are constants. Since $\chi_l(r)$ must be an outgoing wave, one finds the condition $A = iB$, and the penetration factor $P_l(E, R_n)$ is

$$P_l(E, R_n) = \left| \frac{\chi_l(\infty)}{\chi_l(R_n)} \right|^2 = \frac{1}{F_l^2(E, R_n) + G_l^2(E, R_n)} . \qquad (4.66)$$

With equation (4.66), the energy dependence of the partial width $\Gamma_l(E)$ (eq. [4.65]) is completely determined. One can also calculate the absolute value of the partial width for any combination of energy E, orbital angular momentum l, nuclear radius R_n, and reduced width θ_l^2. Figure 4.14 shows the maximum partial widths ($\theta_l^2 = 1$) as a function of energy for the reaction channel $^{16}O + p \rightleftarrows {}^{17}F$, where the double arrows reflect the fact that the partial width is identical for both directions of the process. The partial width $\Gamma_l(E)$ is calculated for a number of l-values. In the case of s-waves ($l = 0$), the partial width $\Gamma_0(E)$ rises very rapidly with energy, attaining nearly a constant value of the order of 1 MeV above the Coulomb barrier. The energy dependence is very similar to that of the Gamow factor (Table 4.1). With increasing orbital angular momentum l, the centrifugal barrier becomes large compared with the Coulomb barrier ($V_C = 2.5$ MeV). For example, for $l = 8$ one finds (eq. [4.60]) $V_{cf} = 58$ MeV. It is therefore understandable that the partial width $\Gamma_l(E)$ at a given energy E drops very rapidly with increasing values of l and

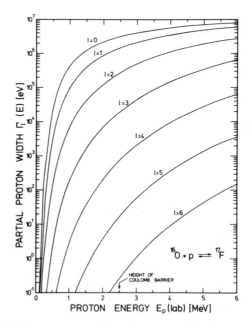

FIGURE 4.14. The maximum partial width $\Gamma_l(E)$ of the reaction channel $^{16}O + p \rightleftarrows {}^{17}F$ for increasing values of the orbital angular momentum l ($R_n = 4.6$ fm). The energetic location of the Coulomb barrier is also indicated. Note the logarithmic scale of the ordinate.

that for a given value of l the partial width $\Gamma_l(E)$ will again approach values of the order of MeV at energies E near the height of the centrifugal barrier (e.g., $E \simeq 58$ MeV for $l = 8$).

At sub–Coulomb energies $(E \ll E_C)$, the penetration factor $P_l(E, R_n)$ is related to the Gamow factor $P_{l=0}(E, R_n) = \exp(-2\pi\eta)$ (eq. [4.14]) by the approximate expression

$$\frac{P_l(E, R_n)}{P_0(E, R_n)} = \exp\left[-2l(l + 1)\left(\frac{\hbar^2}{2\mu Z_1 Z_2 e^2 R_n}\right)^{1/2}\right]$$

$$= \exp\left[-7.61l(l + 1)/(\mu Z_1 Z_2 R_n)^{1/2}\right],$$

where μ is in amu and R_n is in fermis. In the example $^{16}O + p \rightleftarrows {}^{17}F$ ($R_n = 4.57$ fm, $\mu = 16/17$, $Z_1 = 1$, $Z_2 = 8$), one finds values of $P_l/P_0 = 4.2 \times 10^{-4}$, 5.4×10^{-12}, and 2.7×10^{-41} for $l = 2$, 4, and 8, respectively. These results again show the fast drop in $P_l(E, R_n)$ or $\Gamma_l(E, R_n)$ with increasing values of l. Note that the above approximation overestimates the decrease in $P_l(E, R_n)$ with increasing l by substantial factors (Fow85a).

If the excited state E_r in the compound nucleus (Fig. 4.10) decays to a lower-lying state at energy E_f by γ-ray emission ($E_\gamma = E_r - E_f$), the energy dependence of the partial width has the form (Bla62)

$$\Gamma_L(E_\gamma) = \alpha_L E_\gamma^{2L+1}, \tag{4.67}$$

where the letter L refers to the multipolarity of the emitted γ-radiation and α_L is a constant for each value of L. When $L = 1$ (dipole radiation), $\Gamma_L(E_\gamma)$ takes the form

$$\Gamma_{\text{dipole}}(E_\gamma) = \alpha_1 E_\gamma^3,$$

which gives the energy (frequency) dependence for dipole radiation. If the decay from state E_r to state E_f involves only a single particle, it is referred to as a single-particle transition. For the lowest multipolarities the constants α_L are known (Bla62), and thus the partial γ-widths (frequently referred to as Weisskopf widths) are

Electric dipole:　　　　　$\Gamma(E_\gamma) = 6.8 \times 10^{-2} A^{2/3} E_\gamma^3$,

Magnetic dipole:　　　　$\Gamma(E_\gamma) = 2.1 \times 10^{-2} E_\gamma^3$,

Electric quadrupole:　　$\Gamma(E_\gamma) = 4.9 \times 10^{-6} A^{4/3} E_\gamma^5$,

Magnetic quadrupole:　$\Gamma(E_\gamma) = 1.5 \times 10^{-8} A^{2/3} E_\gamma^5$,

where the partial widths are in eV for E_γ in MeV and A is the atomic number. Since these single-particle γ-widths represent upper limits on γ-ray widths (analogous to the Wigner limit; see above), experimental γ-widths are often expressed in units of the Weisskopf widths leading to γ-ray strengths in "Weisskopf units" (W.u.). If the wave functions of the states involved are complex, coherence effects can result in much larger strengths than the Weisskopf limit. The energy dependence of these γ-widths is not nearly as dramatic as it is for particle emission (Fig. 4.10). For a given γ-ray energy E_γ, the dependence on the multipolarity L, which is the analogue to the l-dependence in particle emission, is also very strong. One should note that the absolute value for these γ-widths is of the order of eV or smaller, as compared with MeV for the partial width of particle emission.

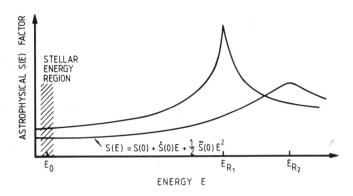

FIGURE 4.15. Energy dependence of the $S(E)$ factor for two broad resonances with maxima at energies E_{R_1} and E_{R_2}. The stellar energy region is also indicated, where E_0 refers to the Gamow energy.

Knowledge of the energy dependence of the partial width (eqs. [4.65], [4.66], and [4.67]) allows one to calculate the cross section of a resonance at any energy E (and in particular at stellar energies; eq. [4.59]). If the resonance energy E_R is near or above the Coulomb barrier, the partial widths for the entrance and exit channels, $\Gamma_a(E)$ and $\Gamma_b(E)$, vary very little over the resonance region ($E = E_R \pm \Gamma_R/2$), and the product $\sigma(E)E$ has a Lorentzian shape. For charged-particle–induced resonances well below the Coulomb barrier, the partial width for the entrance channel, $\Gamma_a(E)$, varies very rapidly over the resonance region (Fig. 4.14). In general, the partial width for the exit channel varies much more slowly, since the energy of the emitted particle is increased by an amount equal to the Q-value of the reaction. Since the term $\Gamma_a(E)/\Gamma_a(E_R)$ in equation (4.59) varies very rapidly and nonlinearly over the resonance region (Fig. 4.15), the cross section is no longer a Lorentzian shape but is asymmetric with respect to E_R, that is, the cross section varies much more rapidly for energies below E_R than for energies above E_R. An experimental example of this is shown in Figure 4.6. The asymmetric resonance shape in $\sigma(E)$ is removed to a large extent, if the cross section is expressed in terms of the $S(E)$ factor (sec. 4.2):

$$S(E) = \sigma(E)E \exp(2\pi\eta),$$

as shown in Figure 4.5 and sketched in Figure 4.15. The low-energy tail of a broad resonance is a smoothly varying function of energy, and the formalism developed for nonresonant reactions (i.e., Gamow peak at energy E_0) can be applied at the low-energy wing,

$$\langle \sigma v \rangle = \langle \sigma v \rangle_{\text{NR}} \quad \text{for } E_0 \ll E_R ,$$

where the index NR refers to the nonresonant formalism (sec. 4.2).

At stellar energies near the resonance energy E_R (Fig. 4.15), the calculated values for the reaction rate $\langle \sigma v \rangle$ (eq. [4.1]) at a number of stellar temperatures, using experimentally determined cross-section values $\sigma(E)$, are fitted to the analytic function

$$\langle \sigma v \rangle_R = \alpha_1 T^m \exp\left(-\frac{\alpha_2}{kT}\right) \quad \text{for } E_0 \simeq E_R ,$$

where α_1, α_2, and m are adjustable parameters. In most cases, one finds $m \simeq -3/2$ and $\alpha_2 \simeq E_R$, and the analytic function is very similar to that for a narrow resonance (eq. [4.56]). If interference effects between resonant and nonresonant reaction mechanisms can be neglected, the reaction rate $\langle \sigma v \rangle$ at any stellar temperature (or equivalently at any energy E_0) may be written as

$$\langle \sigma v \rangle = \langle \sigma v \rangle_{\text{NR}} + \langle \sigma v \rangle_R .$$

For several broad and noninterfering resonances, the reaction rate $\langle \sigma v \rangle$ is given simply by the sum of the reaction rates for each resonance:

$$\langle \sigma v \rangle = f_{\text{cut}} \Sigma_i \langle \sigma v \rangle_{\text{NR}_i} + \Sigma_i \langle \sigma v \rangle_{R_i} . \tag{4.68}$$

The term f_{cut} is introduced to truncate the total nonresonant reaction rate so that it approaches zero at energies near the first resonance (E_{R_1} in Fig. 4.15):

$$f_{\text{cut}} = \exp\left(-E_{R_1}/kT\right) . \tag{4.69}$$

This term is required because the reaction rate $\langle \sigma v \rangle$ for energies near and above the first resonance is usually determined entirely by the resonant terms $\Sigma_i \langle \sigma v \rangle_{R_i}$ in equation (4.68). Of course, for the nonresonant reactions described in secs. 4.1 and 4.2, there is no cutoff term.

In cases where the angular momentum J and the parity π of broad overlapping resonances are the same, interference effects between these resonances must be taken into account. Such cases will be discussed in chapters 6 and 7.

4.5. Subthreshold Resonances

Consider an excited state at energy E_r in a nucleus, where the energy E_r is smaller than the reaction Q-value for decay into a reaction channel with nuclei A and x (Fig. 4.16). Clearly, the excited state at energy E_r cannot decay into this reaction channel, or, equivalently, it cannot be formed through this reaction channel, since its resonance energy $E_R = E_r - Q$ is negative, that is, E_R is

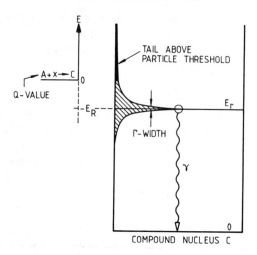

FIGURE 4.16. An excited state of a compound nucleus at an energy E_r, which lies energetically below the particle threshold Q for decay into nuclei A and x. The excited state is shown to decay by γ-ray emission and is characterized by a width Γ. Because of this width Γ, the state E_r extends energetically to both sides of E_r on a rapidly decreasing scale.

below the particle threshold Q with a value $-E_R$ (Fig. 4.16). Such a state is referred to as a subthreshold state or, with regard to a resonant formation via the reaction channel A + x, as a subthreshold resonance.

Every excited state E_r has at least one allowed (i.e., open) reaction channel through which it can decay, namely, the decay via γ-ray emission (Fig. 4.16) or two γ-rays and e^+–e^- pair emission for 0^+ to 0^+ transitions. This unavoidable decay implies that the excited state is characterized by a mean lifetime or, equivalently, a width Γ. In the case of γ-ray emission only, one finds $\Gamma = \Gamma_\gamma$, where Γ_γ is at most of the order of 1 eV (eq. [4.67]). If there are other particle channels (B + y) open, into which the state can decay (or, equivalently, from which it can be formed), the total width Γ can attain values of the order of 1 MeV (eq. [4.65]), where

$$\Gamma = \Gamma_\gamma + \Gamma_y + \cdots .$$

The concept of a width Γ, associated with an excited state, means that the state, though predominantly located near the energy E_r (Fig. 4.16), can nevertheless be "seen" at other energies above as well as below the value of E_r. Thus, the high-energy wing of a subthreshold state extends above the particle threshold (Fig. 4.17). The cross section in this energy region is described by the Breit-Wigner expression (eq. 4.52)

$$\sigma(E) = \pi \lambdabar^2 \omega \, \frac{\Gamma_1(E)\Gamma_2(E + Q)}{(E - E_R)^2 + [\Gamma(E)/2]^2} ,$$

or, equivalently, in terms of the $S(E)$ factor,

$$S(E) = E \exp (2\pi\eta)\pi\lambdabar^2 \omega \frac{\Gamma_1(E)\Gamma_2(E + Q)}{(E - E_R)^2 + [\Gamma(E)/2]^2} \cdot$$

The formation width $\Gamma_1(E)$ is given by equation (4.65):

$$\Gamma_{1,l}(E) = \frac{2\hbar}{R_n} \left(\frac{2E}{\mu}\right)^{1/2} P_l(E, R_n)\theta_l^2 \, ,$$

and the penetration function $P_l(E, R_n)$ by equation (4.66):

$$P_l(E, R_n) = \frac{1}{F_l^2(E, R_n) + G_l^2(E, R_n)} \cdot$$

Consider, as an example, the s-wave formation ($l = 0$) at energies E near zero. The asymptotic behavior of the Coulomb functions $F_0(E, R_n)$ and $G_0(E, R_n)$ has the following form (Abr65):

$$F_0(E, R_n) \simeq 0$$

$$G_0(E, R_n) \simeq 2 \exp (\pi\eta)\left(\frac{\rho}{\pi}\right)^{1/2} K_1[2(2\eta\rho)^{1/2}] \, ,$$

where $K_1(x)$ is the modified Bessel function of order unity with argument $x = (8Z_1Z_2 e^2 R_n \mu/\hbar^2)^{1/2} = 0.513(\mu Z_1 Z_2 R_n)^{1/2}$, and $\rho = (2\mu E/\hbar^2)^{1/2}R_n$. The penetration function $P_0(E, R_n)$ then becomes

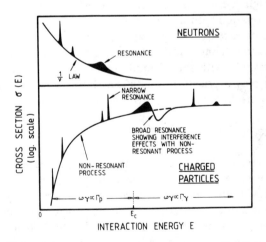

FIGURE 4.17. Schematic diagram showing dependence of the total cross section $\sigma(E)$ on the interaction energy E for neutrons and for charged particles. In both cases, resonances are superposed on a slowly varying nonresonant reaction yield, where broad resonances often exhibit interference effects with the nonresonant process or other broad resonances of the same J^π values.

$$P_0(E, R_n) \simeq \frac{1}{G_0^2(E, R_n)} = \frac{\pi}{4\rho K_1^2(x)} \exp(-2\pi\eta),$$

and the formation width,

$$\Gamma_{1,0}(E) = \frac{\pi\hbar^2}{2\mu R_n^2} \frac{1}{K_1^2(x)} \exp(-2\pi\eta)\theta_0^2.$$

The $S(E)$ factor near zero energy can then be written as

$$S(E) = \frac{\pi^2\hbar^4}{4\mu^2 R_n^2} \frac{1}{K_1^2(x)} \omega\theta_0^2 \frac{\Gamma_2(E + Q)}{(E - E_R)^2 + [\Gamma(E)/2]^2},$$

showing that the $S(E)$ factor has a finite value at $E = 0$ that drops off rapidly with increasing energy E. However, since $\Gamma_{1,0}(E) \simeq \exp(-2\pi\eta)$, the cross section $\sigma(E)$ approaches the value zero as the energy E goes to zero, but $S(E)$ remains finite.

In the ^{20}Ne$(p, \gamma)^{21}$Na reaction ($Q = 2432$ keV) there is a subthreshold state at $E_R = -7$ keV with the properties $l = 0$, $J = \frac{1}{2}$, $\Gamma_\gamma = \Gamma = 0.31$ eV and $\theta_0^2 = 0.9$ (chap. 6). In this case, one finds $\omega = 1$, $\mu = 0.95$, $Z_1 = 1$, $Z_2 = 10$ and $R_n = 4.8$ fm, which leads to $x = 3.47$ and $K_1^2(x) = 5.30 \times 10^{-4}$. At energy $E = 0$, using the E_γ^3 dependence for dipole radiation (eq. [4.67]), that is, $\Gamma_\gamma = \Gamma = (2432/2425)^3 \times 0.31$ eV $= 0.313$ eV, leads to an $S(0)$ factor of $S(0) = 21,800$ keV b. Since θ_0^2 is near maximum, the term in the denominator $(E - E_R)^2$ has to be corrected by the Thomas-Lane factor (sec. 4.3), which reduces $S(0)$ by a factor 4.78; hence $S(0) = 4560$ keV b. With increasing energy E, the $S(E)$ factor drops rapidly [for example, $S(E = 100$ keV$) = 19.5$ keV b] reflecting the high-energy wing of the subthreshold resonance. Since proton-induced nonresonant capture reactions have values of $S(0) \leq 100$ keV b, a subthreshold resonance can completely dominate the total stellar reaction rate at low temperatures. Thus, subthreshold resonances, in particular those near the particle threshold with low orbital angular momenta l and large reduced widths θ_l^2, may play a crucial role in stellar burning. Experimental investigation of such subthreshold resonances will be discussed in chapters 6 and 7.

4.6. Summary

In the previous sections nonresonant and resonant mechanisms involved in nuclear reactions have been discussed separately. However, in practice, a given reaction involves both types of reaction mechanisms (Fig. 4.17), and can exhibit interference effects, in particular near broad resonances. In charged-particle–induced reactions, the cross section for both reaction mechanisms drops rapidly (on an exponential scale) at low energies owing to the effect of the Coulomb barrier, and it becomes more difficult to measure the relevant cross sections. In contrast, for neutron-induced reactions the cross section is very large and increases with decreasing energy (resonances are superposed on a smooth nonresonant yield following the $1/v$ law), and experimental measurements of $\sigma(E)$ can be carried out directly at the relevant stellar energies.

With improved experimental techniques (chap. 5) direct measurements of $\sigma(E)$ for charged-particle–induced reactions can be extended toward lower energies, but in practice one hardly reaches the relevant stellar energy regions for quiescent stellar burning. The observed energy dependence of $\sigma(E)$, or equivalently of $S(E)$, must therefore be extrapolated (Fig. 4.18) into the stellar energy region (essentially to zero energy). Of course, the basis for extrapolation will be improved if extremely low energy data are available. However, these extrapolated data represent only lower limits of the stellar reaction rate. If there are resonances near the particle threshold at $-E_R$ or $+E_R$, they can completely dominate the reaction rate for low stellar temperatures. The experimental investigation of such resonances near the particle threshold is described in chapters 6 and 7.

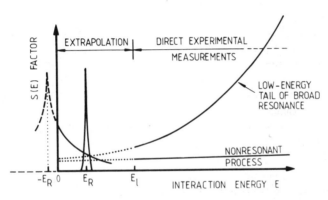

FIGURE 4.18. In charged-particle–induced reactions, the steep drop of the cross section $\sigma(E)$ owing to the Coulomb barrier results in a lower limit in energy, E_l, below which direct measurements cannot be carried out. The data are usually presented in the form of the $S(E)$ factor and extrapolated to zero energy with the guidance of theory and other arguments. If there is a subthreshold resonance at $-E_R$ or a resonance only slightly above the particle threshold, the $S(E)$ factors can completely dominate the stellar burning. The extrapolated $S(E)$ factor in this case is only a lower limit.

5

Laboratory Equipment and Techniques in Nuclear Astrophysics

As discussed in chapters 3 and 4, the reaction rate per particle pair, $\langle \sigma v \rangle$, of nuclear processes in stars is essential to the understanding of many stellar phenomena. This rate depends strongly on stellar temperature, and its analytic expression is governed by the energy dependence of the cross section $\sigma(E)$. Clearly, the absolute value of $\sigma(E)$ must also be known. The objective of this chapter is to discuss the problems encountered in the experimental study of charged-particle–induced nuclear reactions occurring within stars and their various solutions. The study of neutron-induced reactions and their unique features will be discussed in chapter 9.

A schematic diagram of a typical setup for studying charged-particle–induced nuclear reactions is shown in Figure 5.1. Ions of interest are produced in an ion source and are brought to the desired energy using an accelerator. An analyzing system sorts out the ions according to energy and ion species to form an ion beam which is deflected into the beam transport system. This system usually consists of slits, quadrupole lenses, steerers, beam profile monitors, and Faraday cups. These elements are used to diagnose and appropriately fix properties of the beam that is directed onto the target and finally stopped in a Faraday cup or a beam calorimeter. The incident projectiles interact with the target nuclei (essentially at rest), causing nuclear reactions. The resulting reaction products, particles or photons, are observed using detectors. The pulses from the detector are amplified and appropriately shaped. The data may then be accumulated, stored, and displayed.

The observed counting rate in the detector is proportional to the cross section for the nuclear reaction. To extract quantitative information, knowledge of characteristics of the ion beam such as the absolute energy, the energy spread, and the number of incident projectiles per time unit is needed, as is knowledge of the characteristics of the target and detection systems, such as the number of active target nuclei, the angular distribution of the resulting radiation, and the detector efficiency.

This chapter is intended as an introduction to the field of experimental nuclear astrophysics. It attempts to point out many of the important experimental problems as well as some practical considerations for doing experiments. It is not intended to be a compendium on the hardware and instrumentation required for experimental nuclear astrophysics. Much of the information is straightforward nuclear physics, since nuclear astrophysics is to a large extent simply nuclear physics applied to the interesting problems of astrophysics. However, because of the nature of the astrophysical problems, there are many special requirements and considerations (e.g., measurements at energies far below the Coulomb barrier), which are not encountered in ordinary nuclear physics. We have therefore tried to emphasize how one addresses these special problems and considerations.

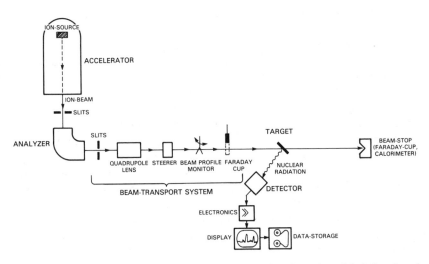

FIGURE 5.1. Schematic diagram of a typical setup for studying charged-particle–induced nuclear reactions.

5.1. Ion Beams

Charged-particle–induced nuclear reactions occur in stars over a wide range of temperatures. While the initial stage of stellar burning, hydrogen burning (chap. 6), involves temperatures of about $T_9 = 0.01$–0.4, the later phases of stellar evolution are characterized by explosive environments with temperatures as high as about $T_9 = 1$–5 (chap. 8). The corresponding stellar energies E_0 (chap. 4) of a particular nuclear reaction compared with the height of the associated Coulomb barrier E_C are in the range of $E_0/E_C = 0.01$–1 for the above range of stellar temperatures. This, then, is the range of ion-beam energies, E, of interest to nuclear astrophysics, the energies below the Coulomb barrier or the sub-Coulomb energies. It is in this energy range that information on the absolute cross section $\sigma(E)$ of a given nuclear reaction must be obtained. Because of the strong energy dependence of the probability of penetrating the Coulomb barrier, the cross section $\sigma(E)$ drops by many orders of magnitude for energies between the Coulomb barrier and the point at which the ratio E/E_C is about 0.01. It follows that experimental investigation of the cross section $\sigma(E)$, at energies far below the Coulomb barrier, requires high ion-beam currents. The need for high currents establishes the requirements for ion sources, accelerators, beam transport systems, and beam integration. These characteristics are described in detail in what follows.

5.1.1. Ion Sources and Beam Formation

The characteristics of the ion source determine to a great extent the limitation on performance of an ion-beam facility. For this reason a considerable amount of development effort has been devoted to improving ion sources. An important characteristic of an ion source is the variety of ion species it can provide. Experiments in nuclear astrophysics involve a wide range of ion species; those involving light ions such as hydrogen and helium are particularly important. It is important that the ion source produces high currents, since this allows one to extend measurements to lower beam energies. The size of the ion-emitting area of the source is also important, since it determines the minimum spot diameter to which the resulting ion beam can be focused. The beam diameter, the beam intensity, and the angular divergence of the beam emerging from the source taken together fix the "emittance" and "brightness" of an ion source (Lej80). It is desirable that the energy distribution of the ions emitted by the source (the energy spread) be a minimum. Also of importance are the mass and charge spectrum of the ion beam, the lifetime of the source, and its power consumption. Well-engineered ion sources always represent a compromise among all these requirements, and in some cases most of these requirements can be met (Wil73, Alt81). We will discuss below some of the most widely used sources of positive ions. A sputter source used for the production of negative ions also will be described.

5.1.1.1. Electron-Impact Ion Source

The first step in making an ion beam is to produce ions. While there are a number of different ways to produce ions (Wil73), the most common one uses the electron-impact ionization process. In this process electrons passing through a gas collide with the atoms, transferring sufficient energy to ionize some of them, i.e., the amount of energy transferred to some atoms exceeds their ionization energy W_i; the maximum in the ionization process is realized when the electron energy is about 3 times W_i (Gry65). The way in which the electrons are created and accelerated to energies high enough to ionize the atoms is different for different ion sources. The essentials of electron impact ion sources are illustrated in Figure 5.2a. As shown, in the thermionic emitter–

(a) ELECTRON IMPACT ION SOURCE

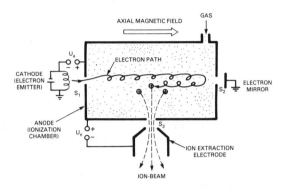

(b) ARC-DISCHARGE ION SOURCE

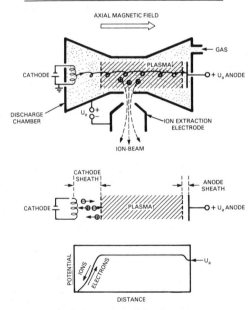

FIGURE 5.2. (a) Schematic drawing of an oscillating electron-impact ionization source. The gas to be ionized is allowed into the chamber through a pressure-regulating valve. In the case of solid materials, an oven is used to produce an atomic vapor. (b) Schematic drawing of an arc-discharge ion source, which is an extension of the electron-impact source. Shown also is a plot of the potential versus the distance between the cathode and the anode. The shape of the curve is caused by plasma sheaths.

type ion source, the cathode (the filament) is heated by an external power supply to a temperature high enough to cause electron emission as described by the Richardson equation (Wil73). The electrons are accelerated by the anode voltage U_a ($\simeq 200$ V) and guided into the ionization chamber containing gas at a low pressure ($\simeq 10^{-4}$ torr). An axial magnetic field guides and concentrates the electrons to a region somewhat smaller than the ionization chamber. The electrons spiral around the magnetic flux lines, thereby significantly increasing their effective path length and enhancing their probability of colliding with a gas atom before striking the anode. An electrode at the same potential as the cathode is placed near the exit slit (S_2) to repel the electrons back into the ionization chamber. It is called an electron mirror. In this way, the electrons may execute several transits through the chamber before being collected, further enhancing the ionization efficiency. Such ion sources produce high currents of positive ions. The ions are extracted normal to the electron flow using an extraction electrode (U_e, of a few kV). The electric field extends through the hole S_3, which is a few millimeters in diameter, into the ionization chamber and has a shape which focuses and accelerates the ions, thus forming a beam of ions. In other designs, the ions are extracted along the axis of electron flow. About one-half of the ions created in the chamber will be extracted as useful current; the remaining ions will bombard the cathode.

5.1.1.2. Arc-Discharge Ion Source

If the gas pressure in an electron-impact–type ion source is increased to a few 10^{-2} torr, an arc discharge is struck between the cathode and the anode (Fig. 5.2b). The discharge current increases by 3–4 orders of magnitude, to about 1 ampere, causing a high density of charged particles (electrons and positive ions). Such a highly ionized state of matter with quasi-neutrality is called a plasma (Sch25, Ses73). The most important distinction between a plasma and a normal gas is that a plasma is a fairly good conductor, i.e., it can sustain a current. The creation of an arc plasma can be achieved in two ways, depending on the mechanism of emission of the electrons. The electron source can be either a hot (thermionic) cathode (Fig. 5.2b) or a cold cathode type (Fig. 5.3b). Usually an axial magnetic field is used to enhance and concentrate the discharge currents. Note that the hot-filament cathode often has a short lifetime because of chemical reactions and sputtering of the cathode surface (Alt81).

Because of their low mass, electrons have high mobility, and they move rapidly through the gas toward the anode. As they pass through, they ionize the gas, forming positive ions and secondary electrons. The positive ions move toward the cathode, but because of their larger mass and lower mobility, they take a much longer time than do the electrons. Equilibrium is established in this double-streaming current with approximately equal numbers of positive ions and electrons but with a slight excess of ions, resulting in a positive space charge. The voltage applied between cathode and anode results in a region around the cathode having a space-charge limited potential, and a boundary layer is formed which is called a plasma sheath. Under equilibrium conditions, the plasma is almost a uniform mixture of gas, drifting ions, and electrons, with a very low potential gradient. In such a plasma most of the current is carried by the electrons because of their much higher mobility.

Because of their high mobility, the electrons also move rapidly out of the center of the plasma toward the chamber wall (and anode), giving the wall a negative charge and thus a negative potential with respect to the plasma. It repels the electrons and attracts the positive ions diffusing through the plasma to the wall, so that very near the wall (and anode) there is another sheath. Thus ions and electrons are continually lost from the discharge and must be replaced by new ones produced by ionizing collisions between electrons and neutral gas atoms. To sustain the discharge in the arc plasma, the primary electron current from the cathode must be sufficient to maintain the balance between the ionization rate and the ion loss rate.

Since the potential of the plasma becomes 5–10 V positive relative to the wall, the positive ions in the plasma are forced outward, drifting toward the wall, where they can pass through an exit hole (Fig. 5.2b). An extraction electrode (U_e, a few kV) helps to pull the ions out of the chamber. Other plasma-type ion sources extract the ions through a canal in the cathode or the anode (Fig. 5.3). Practically all sources of positive ions rely on some form of gaseous discharge for ion production (Wil73).

5.1.1.3. Duo-Plasmatron Ion Source

An elegant version of the hot cathode plasma source is the duo-plasmatron ion source (Ard56). In this source an arc discharge is struck between a thermionic cathode and the anode at a voltage (U_a) of a few 100 V (Fig. 5.3a) in a gas with a pressure of about 10^{-2} torr. Near the anode, a conical intermediate electrode with an aperture of a few millimeters and a negative voltage U_i of about 100 V relative to the cathode constricts the plasma on the anodic side to a small volume, forming a type of plasma bubble. The concave shape of the plasma bubble focuses the electrons coming from the cathode region, resulting in a volume with a high density of electrons and positive ions located between the intermediate electrode and the anode. The extraction potential (U_e = 10–30 kV) pulls the ions out of the plasma through a small hole (1 mm diameter or less) in the anode. The most important feature of this source is the concentration of an intense axial magnetic field in the discharge region (exit hole) by an ingenious combination of an electromagnet and electrodes. In this structure the plasma-forming electrodes are also the poles of a small electromagnet (Fig. 5.3a). This inhomogeneous magnetic field further compresses the plasma volume, and thus the plasma is formed only in the region around the anode aperture. In subsequent extraction a well-focused ion beam is formed. The intense magnetic field also has a focusing effect on the ions. This ion source provides beams with high current (e.g., 0.5 mA H^+), excellent optics, and small energy spread. The problems are cooling the source and the cathode lifetime.

5.1.1.4. Penning Ion Source

The problem of the short lifetime of thermionic cathodes in plasma sources is avoided in the cold-cathode Penning ion source (Fig. 5.3b; Pen37, Fin40). If the pressure ($\simeq 10^{-2}$ torr) and the potential between the anode and the cold cathode ($\simeq 500$ V) are high enough, a self-sustaining discharge occurs. Electrons released from the cathode, by secondary emission, are accelerated toward the cylindrical anode. Constrained by the axial magnetic field, they pass through the (electric) field-free volume inside the anode and out toward the other cathode, where they are reflected. This process leads to multiple passes of the electrons through the potential well, resulting in the formation of a plasma column through the center of the anode, which terminates in a thin

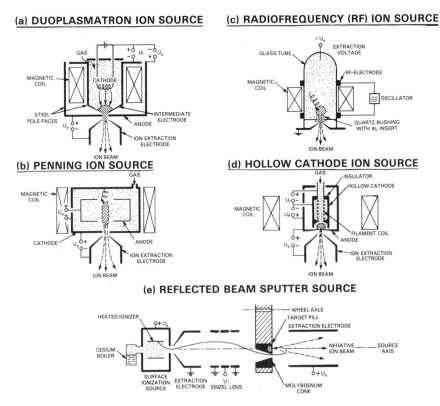

FIGURE 5.3. Schematic diagrams of plasma sources producing positive-ion beams (a–d). Practical designs of four arc-discharge sources are shown. These sources produce high beam currents with good ion-optical quality and low energy spread. Shown also (e) is the reflected beam sputter source producing intense negative ions for a large variety of elements.

sheath at both cathode surfaces. Positive ions formed in the discharge are attracted to the anode and are pulled from the plasma through a hole in one of the cathodes by the electric field of the extraction electrode ($U_e \simeq$ 10–30 kV). The cold-cathode Penning source (Wil73, Bau81, Sch84) has the advantages of long lifetime, low power consumption, high singly charged ion currents (e.g., 1 mA He^+ or Ar^+), relatively high currents of multiply charged heavy ions (e.g., 100 μA Ar^{++}), good beam optics, and an energy spread of less than 80 eV per charge state.

5.1.1.5. RF-Ion Source

Plasma sources have been built using an electrodeless discharge produced by a radio-frequency (rf) field (Tho46), called rf-ion sources (Fig. 5.3c). The gas is enclosed in a glass tube, and an rf field is produced inside the tube by exterior coils. The rf field causes naturally present electrons to oscillate in helical orbits through the gas, thus ionizing the gas atoms. An axial magnetic field increases the efficiency of the source by concentrating the plasma in the neighborhood of the exit canal. The ions (mainly light ions, e.g., 1 mA H^+) are extracted by a superposed electric field between the extraction canal and the metallic electrode at the top of the source (U_e = 3–10 kV).

5.1.1.6. Hollow-Cathode Ion Source

Another type of plasma ion source uses a hollow-cathode structure (Fig. 5.3d) to create electrons, which are directed into the discharge region by means of an appropriate electrode structure and an axial magnetic field. In this ion source large electron currents are concentrated in a small volume; thus smaller sources can be designed. The first ion source of this type was built by Sidenius (Sid65). Present sources include an oven to vaporize solids and other improvements (Cle84). These improvements made available a large variety of high-current ion beams from both solid and gaseous materials (Table 5.1). The ion source has a small emittance and an energy spread of less than 20 eV.

TABLE 5.1 Ion-Beam Currents from a Hollow-Cathode Ion Source[a]

Ion Species	Ion Current (μA)	Ion Species	Ion Current (μA)
H^+	250	$^{52}Cr^+$	70
H_2^+	250	$^{74}Ge^+$	70
$^4He^+$	250	$^{75}As^+$	80
$^{11}B^+$	50	$^{107}Ag^+$	100
$^{14}N^+$	80	$^{107}Ag^{++}$	10
$^{31}P^+$	100	$^{120}Sn^+$	50
$^{40}Ar^+$	250	$^{197}Au^+$	120

[a] As measured on the target of the 350 kV Münster accelerator facility (Cle84).

5.1.1.7. Sputter Ion Source

Ion sources have been developed for the production of negative ions required for tandem accelerators (sec. 5.1.2). There are essentially four processes for generating negative ions (Wil73, Mid74): charge exchange, high-voltage dissociation, direct extraction, and sputtering. The most versatile type in terms of the ion species which can be generated is the sputter source with the reflected-beam modification (Hor68, Mid74, Bra77, Bra81). The first step in a sputter source (Fig. 5.3e) is to create a beam of positive cesium ions in a surface ionization source using the Langmuir effect. When an atom of low ionization energy W_i has contact with a hot solid surface (called the ionizer) of a material with a high work function W_a, the least-bound electron of the absorbed atom is lost to the surface and the atom departs as a positive ion. The ratio of positive ions N_i to the number of neutral atoms N_0 in the evaporating material is given by the relation

$$N_i/N_0 = \exp\left[-(W_i - W_a)/kT\right] .$$

If cesium atoms ($W_i = 3.89$ eV) are placed on a 1000 °C surface of tungsten ($W_a = 4.53$ eV), the ratio is $N_i/N_0 = 355$, and hence 99.7% of the evaporating cesium atoms are positively charged. The cesium atoms, provided as a vapor from a boiler, move through an aperture to the ionizer. In other designs (Bra77) a porous tungsten ionizer is used, in which the cesium atoms diffuse through the ionizer (heated to 1000 °C); when an atom leaves the surface of the ionizer, an electron remains on the surface, and the emitted cesium atoms are positively charged. By the use of an einzel lens (sec. 5.1.3), the extracted cesium ions ($\simeq 30$ keV) are focused to a small spot and, in the original sputter source, are steered onto the edge of a target cone containing the element of interest. Bombardment of the target material with energetic cesium ions results in a large fraction of the sputtered ions being negatively charged. These negative ions are extracted through a hole in the target cone by an electrode. In the modified design of Brand (Bra77), the cesium beam is steered through an off-axis molybdenum cone (Fig. 5.3e). The beam is then reflected by a positive ion extraction electrode and sharply focused on a pill of target material located on the back side of the molybdenum cone. This "reflected beam sputter source" improved the emittance of the negative ion beam and eliminated some disadvantages of the standard type of sputter source. It has provided high-current beams of many heavy-ion species with good ion-optical qualities (Bra81). New developments (Bra81, Mid83) have increased further the beam currents for heavy ions by nearly an order of magnitude (e.g., 100–150 μA for ion species with high electron affinity).

As we have just seen, there are a variety of ways to create an ion beam. The resulting ion beam must be manipulated by elements of an injection system (Fig. 5.4) to make it suitable for acceleration. To begin with, the plasma, from which the ions are extracted, determines the initial trajectories of the ions and their initial energy spread. Both of these quantities affect greatly the sub-sequent focusing of the ion beam. In a plasma source the ions are extracted mainly from the plasma sheath. The potential drop across the plasma sheath is only a few volts, so the energy spread of the ions is also a few electron volts. The shape of the sheath makes it a strong immersion lens and as such is a critical part of the optics of the total ion-beam transport system because it is the first and strongest optical element in the system. The area of the extraction hole is also important, since this ultimately determines the minimum beam diameter possible at the target. The ions usually emerge from the source through a small (0.5–2.0 mm diameter) hole. They are then accelerated, colli-mated, and steered further to form an ion beam for injection into the acceler-ator. This beam manipulation is carried out (Fig. 5.4) by optical elements such as an extraction electrode, an einzel lens, a steerer, and an analyzing device (sec. 5.1.3). The manipulation can produce either a nearly parallel beam of large diameter or a beam of small spot size with a large angular divergence. The requirements of a parallel and narrow ion beam cannot be fulfilled con-currently (Lej80), and compromises specific to the experimental situation (i.e., collimating slit systems) must be made.

The space outside the ion source must be kept at a high vacuum ($\leq 10^{-5}$ torr) in order to reduce discharging and beam loading of the accelerator tube and to minimize beam neutralization and scattering, particularly when high ion-beam currents are used. For a single-ended accelerator this requirement is not easily met, owing to space and power restrictions in the terminal, but recently specially designed ion getter pumps and turbopumps have been used. The purity of the ion beam is of importance for low cross section measure-

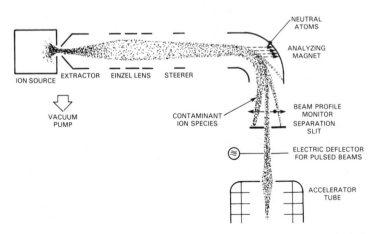

FIGURE 5.4. Schematic diagram of the ion injection system forming a suitable ion beam for sub-sequent acceleration. In single-ended electrostatic accelerators, the control of the components of the ion source injection system is carried out using either insulated rods or light guides. A linear accelerator is assumed, but the ion-source features are much the same for circular accelerators.

ments as well as for optimizing the useful beam for the machine's maximum beam current. Atomic and molecular ions of various species and in different charge states (including neutral species) emerge from the ion source or are created along the beam outside the chamber by charge exchange. To reduce substantially the impurities in the ion beam, either an analyzing magnet or a Wien filter is used (Fig. 5.4). Postacceleration analysis of the beam (sec. 5.1.2) further reduces the impurities arriving at the target. For applications such as the time-of-flight technique (sec. 5.4.4), it is required that ions come from the source in very short bursts, sometimes as short as 1 ns. Such pulses can be produced by sweeping the ion beam rapidly across a fine slit by electric deflection (Fig. 5.4). For the production of high-quality pulsed beams the bunching technique is needed (Fau68).

5.1.2. Accelerators

All accelerators of charged atomic particles, often called "atom smashers" in the newspapers and magazines, consist of a source of ions and some means of producing electric fields to accelerate them. Investigations in experimental nuclear astrophysics have been carried out predominantly using beams of energetic ions produced by electrostatic accelerators (Her59, All74, Bro74, Ble67, Cou68, Per68, Eng74). The Van de Graaff accelerator (Van31, Van33) has been used most frequently, but the Cockcroft-Walton (Coc32, Bal59), the Dynamitron (Cle60, Cle68), and the Pelletron (Her72, Her74) also have been used extensively. For some special astrophysical investigations, cyclotrons (Law30, Liv62) and linear accelerators (Smi59, Liv62) have been employed.

To study nuclear reactions relevant to astrophysics, the required properties of an accelerator are the following: (1) The capability of delivering ion beams with currents ranging from several hundreds of microamperes to several milliamperes. (2) The production of ion beams with an energy spread of less than 1 keV at several MeV ion energy. This beam quality is usually necessary because the complexity of the cross section $\sigma(E)$ as a function of energy and the steep drop in cross section at sub-Coulomb energies require high-resolution measurements. An energy spread much less than 100 eV is generally not needed because the resolution is limited by broadening due to the thermal motion of the target atoms (Ben68). (3) The ability to vary the energy of the ion beam over a wide range (maintaining high-current capability) and in steps as small as 100 eV. Again, steps smaller than 100 eV are not effective unless very special target techniques are used (Ben68). In practice the voltage of electrostatic accelerators (i.e., ion-beam energy E) can be varied over a dynamical range of 1 order of magnitude ($E/E_{max} \simeq 0.1$–1). Since astrophysical problems require a wider range of energies, such studies have to be carried out using several accelerators. (4) The ability to produce a collimated beam with a well-defined beam envelope. Ion beams with diameters as small as 1 mm should be possible. This feature is of great value when differentially pumped gas targets are used and small apertures are needed (sec. 5.2.2). (5) The ability to produce pulsed ion beams for use in the study of reactions in which capture γ-rays are in competition with fast neutrons (Alt49, Sho49, Dye74).

In general, a well-engineered electrostatic accelerator will meet most of these requirements. A detailed discussion of such devices can be found in the literature (Liv62, Arn67, All74). In what follows we will describe the basic features of only two such devices, the Van de Graaff and cascade-type electrostatic accelerators, which have been and still are the accelerators used most extensively in nuclear astrophysics. Other accelerators will be discussed only briefly. The two types of accelerators to be discussed are essentially large capacitors with the ion source located on the positively charged plate (called the terminal). The ions are accelerated in a single step by the electric field between the plate and the ground to an energy $E = qV$, where q is the charge state of the ions and V is the potential of the plate relative to ground.

5.1.2.1. Van de Graaff Accelerator

In the Van de Graaff accelerator the high voltage is produced by mechanically transporting positive charge from ground to the insulated high-voltage terminal (Fig. 5.5a). This is done using a rotating belt made of insulating material which is driven by a motor. Usually there are two pulleys, one at the base of the machine and one in the terminal. The surface of the ascending belt is charged near the lower pulley, using a set of sharply pointed needles (resembling a comb) powered by a standard transformer-rectifier system (5–40 kV), called the sprayer. The crucial innovation in the design of Van de Graaff's accelerator was the recognition that the charge had to be carried into the interior of the terminal (usually a metallic cylinder with a semispherical cap) before being transferred to the terminal via another comb near the top pulley. The terminal dome acts as a Faraday cage; thus the interior is field-free. The high voltage V of the terminal is given by the relation $V = Q/C$, where Q is the charge transported to and stored on the terminal dome and C is the capacity of the terminal. The voltage increases in proportion to the transported charge, which is controlled by the sprayer voltage. If the belt delivers a current I_{belt} to the terminal, the rate of rise of terminal voltage is $dV/dt = I_{belt}/C$, which may be as much as 1 MV s^{-1}. To maintain a desired voltage at a constant level, the transported charge must be in equilibrium with charge losses (current drains). Charge is lost via the accelerated ion beam (I_{beam}), leakage through the supporting insulators (I_{ins}), corona discharge (I_{cor}), and current flow through the chain of high resistances (I_{res}) connecting the terminal with the ground:

$$I_{belt} = I_{beam} + I_{ins} + I_{cor} + I_{res} .$$

An increase in belt current raises the terminal voltage, and an increase in the current losses lowers the voltage. The desired voltage is established predominantly by applying the correct sprayer voltage. For a given voltage the last three currents are fixed, and the extractable beam current is limited by the amount of charge that can be transported by the belt. It is because of their limited charging capability that belt-driven accelerators have maximum beam currents of 1 mA. The belt pulley also drives a generator (located inside the terminal), which is needed to power the various components of the ion source and the injection system. In the Pelletron accelerator (Her74) the belt is replaced by a charging system consisting of several chains of cylindrical metal pellets joined by links of insulating nylon. The metallic cylinders are charged by induction. These chains provide a more stable transportation of charge, thus improving the ion-beam energy spread. They also are dust-free and have longer lifetimes than the belts. Using several chains, the charging capability is similar to that of a belt.

After processing in the injection system, the ions enter the accelerating tube. This tube must be kept at high vacuum ($\leq 10^{-6}$ torr) to reduce discharging in the residual gas. The tube also must be able to hold the entire high voltage between the terminal and ground. It is thus one of the most critical elements in the accelerator. The tubes in use differ in detail, but basically they are made up of insulating sections separated by metallic electrodes. These electrodes are connected to a resistor chain (Fig. 5.5a), producing a nearly uniform voltage drop along the column. Thus the acceleration of the ions is also uniform along the tube. Such a tube has good breakdown properties, provided that the electrode structure is properly designed. Inclined electrodes are frequently used to prevent internal multiplication of small electron currents (Bro74). The elec-

FIGURE 5.5 Basic features of the Van de Graaff, Cockcroft-Walton, and Dynamitron accelerators.

(a) VAN DE GRAAFF ACCELERATOR

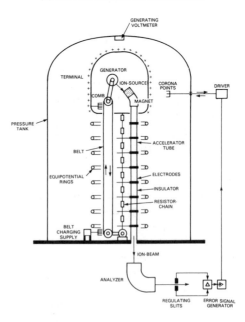

trodes act as a succession of electrostatic lenses of long focal length providing some focusing properties (sec. 5.1.3). As a result, the ion beam at the exit of the accelerator usually has a diameter of less than a few millimeters. It should be pointed out that most of the focusing is accomplished by the first few electrodes. For this reason these sometimes are not inclined. In order to improve the stability of the high voltage, the electric field along the belt and the tube should be as uniform as possible. For this reason, the whole structure is enclosed in a type of cage consisting of metallic circular rings (called the corona rings) enclosing the belt and the tube (Fig. 5.5a). The rings are placed at equal distances along the axis of the accelerator and are connected to the resistor chain. The radius of curvature of all high-potential surfaces, including the terminal dome, is kept as large as possible to reduce field gradients and thus arc discharges.

It was found, from early investigations of electrostatic generators, that in dry air sparking occurs at an electric field of about 30 kV cm^{-1}. The electric field at the surface of a charged sphere of radius R at a potential V is given by the ratio V/R. Thus, for a radius of 1 meter, the maximum voltage is $V_{max} = 3$ MV. This limited the voltage available with the early electrostatic accelerators (Bro74). In fact, in practice the voltage was only about one-third of the theoretical maximum. To overcome this limitation the terminals were very large spheres requiring very large rooms to house them. The situation improved significantly when Herb (Her35, Her37) suggested that the entire accelerator

(b) COCKCROFT-WALTON ACCELERATOR   **(c) DYNAMITRON ACCELERATOR**

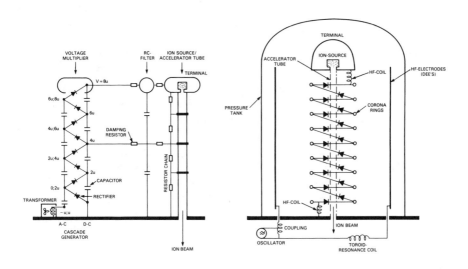

structure be placed in a high-pressure tank filled with electrically insulating gas (today SF_6 gas at 5–8 bars pressure). This immediately led to an improvement of nearly an order of magnitude in the attainable high voltages. It was also found that greatest stability was obtained if the ratio of the diameter of the cylindrical pressure tank to the diameter of the terminal equals the base of natural logarithms, e.

After leaving the accelerator, the beam is usually bent by a highly stabilized magnet which serves as a beam analyzer. It is then passed through a slit system (Fig. 5.5a), which monitors the actual beam energy. If, for example, the voltage has risen above the desired value, the beam is bent less and hits the high-energy slit with greater intensity. The difference in the currents at the two slits is measured and amplified. The error signal produced activates a driver motor, which moves the corona points closer to the terminal. As a consequence, a higher corona discharge current is drawn from the terminal, thus reducing the terminal voltage. With the slit system and improved versions of corona controls (Boe70, Ham79), the high voltage can be kept constant to about one part in a thousand (or better). In some installations using proton beams for target bombardment, the singly charged molecular hydrogen ions from the source are used for voltage control. In a magnetic field these ions are deflected through an angle which is $1/2^{1/2}$ times that for the protons in the analyzer, and this beam is allowed to fall on the slit system used for control. In this way the proton beam can be kept free from confining slits.

5.1.2.2. Cockcroft-Walton Accelerator

Another method of generating high voltages was invented by Schenkel (Sch19) and Greinacher (Gre21). This method uses voltage-multiplier circuits and was used first by Cockcroft and Walton in 1932 (Coc32) in the construction of their accelerator. Thus, machines based on this principle are usually called Cockcroft-Walton accelerators (Fig. 5.5b). In these capacitor-rectifier circuits a transformer feeds alternating current (a-c) of a certain voltage into a rectifying and multiplying apparatus, called the cascade generator. The alternating current travels up the line of capacitors to the left in the cascade generator and is distributed to all the rectifiers (Si-diodes), returning to ground through the capacitors to the right. The direct current (d-c) flows through the rectifiers in series. This generator is thus a circuit for charging capacitors in parallel and discharging them in series. Some rectifiers are used to charge capacitors during one half-cycle of the alternating current, while other rectifiers transfer the charge during the other half-cycle. When no current is drawn, the potential on the capacitors to the right is constant and has the values indicated, where u is the peak secondary voltage of the transformer. These voltages are applied in a series connection across the acceleration tube. The voltage V on top of the cascade generator is—in the example shown—8 times as large as the peak input voltage u. The potentials on the capacitors on the left oscillate between the limits indicated in the figure. By adding units to a total of N capacitors and N rectifiers, one can obtain a voltage multiplication of a factor of N. The whole assembly is also called a cascade accelerator. With a transformer peak voltage of $u = 100$ kV and $N = 4$ capacitors and rectifiers, Cockcroft and Walton (Coc32) were able to produce a proton beam with an energy of 400 keV. Again, high voltages are obtained by transporting charge. In the Van de Graaff accelerator the charges travel smoothly on the moving belt, while in the cascade accelerator they move less smoothly on a type of stairway.

The circuit operates with $V_{max} = Nu$ only if no current is drawn from the cascade generator. With a finite current drain i, the voltage V is reduced below this optimum. Also, a residual voltage ripple ΔV is superposed. The resulting voltage and voltage ripple are given by (Bou37, Lor57)

$$V = Nu - \frac{i}{12fC} \left(N^3 + \frac{9}{4} N^2 + \frac{1}{2} N \right), \tag{5.1}$$

$$\Delta V = \frac{i}{8fC} N(N + 2), \tag{5.2}$$

with f the transformer frequency and C the capacitance of the capacitors. Note that a residual ripple is always present when an alternating voltage is transformed into a constant voltage. Note also that the reduction of the terminal voltage increases linearly with the current drain and with about the third power of the number of cascade units. For a given peak transformer voltage u and current i there is a limit to the number of units N that can be used effectively. Differentiation of equation (5.1) leads to this optimum number of stages (for $N \gg 1$):

$$N_{opt} = (4f\,Cu/i)^{1/2}.$$

It follows that high frequencies f and large capacitances C reduce the ripple ΔV and increase the maximum voltage attainable for a given drain current. In practice one uses $f = 0.5$–10 kHz, $C = 10^{-3}$ to 10^{-8} F, and $u = 100$–300 kV, so that a 1.5 MV accelerator with a beam current of several milliamperes requires 5–10 stages. It should be noted that stray capacitance in the cascade generator as well as the finite resistance of the rectifiers introduces an additional ripple on the high voltage (Eng74). To a large extent, these effects can be damped by installing an RC filter circuit between the cascade generator and the accelerating tube (Fig. 5.5b) or by using symmetrical cascade-rectifier circuits in combination with damping coils (Hei55, Bal59, Har67). The power (several kW) for the ion source is commonly provided by a generator driven from ground via an insulated rotating shaft. The high voltage is stabilized using methods similar to those used for the Van de Graaff accelerator. The effective limit on the high voltage attainable with open accelerators of this type (as well as with a Van de Graaff accelerator) is about 2 MV. Because of their ready accessibility and high-current capabilities (several mA ion-beam current), Cockcroft-Walton type accelerators remain important tools for studies in nuclear astrophysics. As with the Van de Graaff, if the accelerator is placed in a pressure tank, higher voltages can be achieved.

5.1.2.3. Dynamitron Accelerator

Another version of a cascade machine is the "Dynamitron" accelerator (Fig. 5.5c), so named by the manufacturer Radiation Dynamics (Cle60, Cle68). Two electrically insulated electrodes of semicircular cross section (called therefore the "dees") extend inside the pressure tank along the full length of the tank enclosing the entire accelerator assembly. These electrodes act as a huge capacitor, which—together with the exterior toroidal coil— forms an LC resonant circuit oscillating at a frequency around 100 kHz. This oscillatory circuit obtains energy from a powerful Hartley oscillator via a coupling coil. In this way an intense high-frequency field is created between the two electrodes. Insulated parts of the corona ring system act as antennas, in which a high-frequency current is induced. The rings are connected in series by rectifiers, thus forming a chain of rectifiers capacitively coupled to the oscillating tank circuit. Thus the individual partial d-c voltages add up to a high voltage. This chain of rectifiers is connected via high-frequency chokes to the terminal and ground. Because of the parallel application of the high-frequency power, equations (5.1) and (5.2) are modified to

$$V = Nu/k - i(N-1)/f\,Ck\,, \qquad \Delta V = i/f\,Ck\,,$$

where k is the coupling factor. Note that the ripple ΔV is independent of the number of stages ($N_{opt} = \infty$). Because of these features and the high frequency of the oscillating circuit, the ripple of the high voltage is drastically reduced ($\Delta V/V \le 10^{-3}$). Dynamitron accelerators with up to 4 MV terminal voltage have been built. They are characterized by high reliability and convenience, excellent beam quality, and high beam currents (several milliamperes; Ham75, Ham79).

In summary, accelerators of the cascade type (Cockcroft-Walton and Dynamitron) and of the Van de Graaff type produce beams of comparable quality. The cascade accelerators have the advantage of higher beam current, not being limited by the amount of charge that can be transported by a belt or a Pelletron chain. The energy spread of the beams from cascade accelerators arise to a large extent from the high- and low-frequency ripples. Since these ripples are harmonic (Ham79), they can be practically eliminated, and ultrahigh beam energy resolution, desirable in some applications, is possible (Mor69, Goe85).

5.1.2.4. Tandem Accelerator

A development of great importance in the evolution of electrostatic accelerators was the introduction of the "tandem" principle (Ger31, Ben36, Alv51, Ben53). In this innovation two accelerating tubes are contained within one pressure tank (Fig. 5.6a) with a high-voltage terminal at their junction. The terminal is charged positively by a normal belt system within one of the columns or an electronic power supply. A negative ion source outside the pressure vessel supplies a beam of singly charged negative ions. These ions are accelerated by the terminal voltage to an energy equal to the terminal voltage times the negative charge. In the terminal the energetic negative ions are stripped of electrons and converted to positive ions. The stripping is achieved by passage through a canal in which there is an increased gas pressure (gas stripper) or through a thin carbon foil (foil stripper). The positive ions of charge state q are then further accelerated in the second tube, receiving a second energy increment equal to their charge state times the terminal voltage. When they emerge at ground potential, they have an energy equal to $eV(1 + q)$. Thus, using a charge-changing process in the terminal, the ions are accelerated through the same high voltage twice. It is clear that only particles for which negative ions can be created can be accelerated in these double-ended tandem accelerators. The arrangement has most of the advantages of

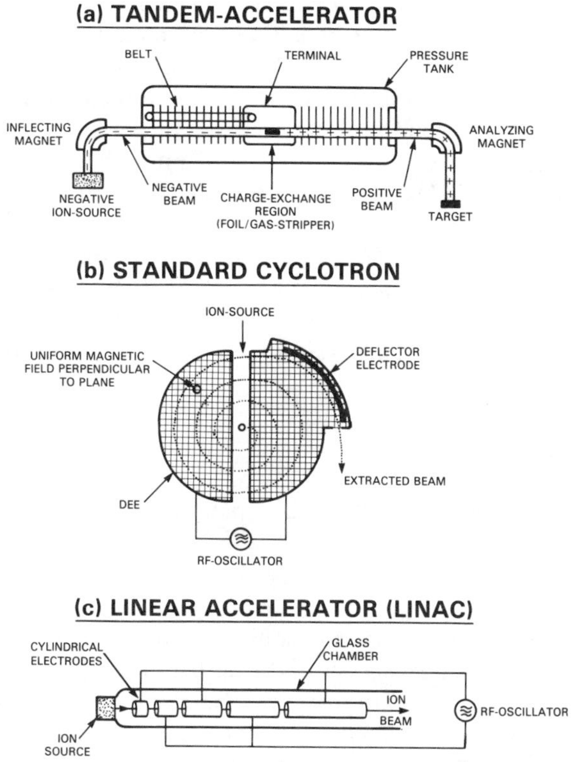

FIGURE 5.6. Shown schematically are the principal elements of a tandem Van de Graaff, a cyclotron, and a linac.

the normal single-ended electrostatic accelerators (Fig. 5.5), together with the manifest improvements of higher beam energies and having the ion source readily accessible outside the tank at ground. Tandem accelerators have been built using both the Van de Graaff (or Pelletron) and Dynamitron high-voltage generation principle (Bro74, Bra81).

5.1.2.5. Cyclotron

In 1930 a new concept for the acceleration of positive ions was introduced by Lawrence (Law30, Law32). It involves the resonant acceleration of ions in a constraining magnetic field. The ion makes many traverses through the accelerating structure, and the accelerator is technically a "magnetic resonance accelerator," or, more concisely, a "cyclotron." The cyclotron accelerates ions to high energy without using a single very high voltage, thereby avoiding the problems of electrical breakdown. In the standard cyclotron (Fig. 5.6b) an electromagnet is used to provide a nearly uniform magnetic field between the flat faces of cylindrical poles of large radius. A vacuum chamber fits between the pole faces, and two hollow copper electrodes are mounted inside the vacuum chamber. These electrodes were originally semicircular in shape, as though a flat, hollow pillbox had been cut in half through a diameter. Because of their shape the electrodes are referred to as D's or "dees." A radio-frequency power supply provides an alternating electric field between the opposing faces of the dees. Ions are accelerated in this region between the dees. To begin with, positive ions are produced near the center of the chamber between the dees by an ion source. They are accelerated toward and into the electrode, which is negatively charged at that instant and accelerated into the magnetic field between the pole pieces. In the electric field–free region between the pole pieces the ions are acted upon only by the uniform magnetic field B and thus travel in a circle in a plane normal to the field. The value of the radius r of any circular path for an ion of mass m, charge qe, and velocity v is obtained by balancing the centrifugal force and the magnetic force:

$$\frac{mv^2}{r} = qevB .$$

The frequency of revolution in the circular path is thus

$$f = \frac{v}{2\pi r} = \frac{qeB}{2\pi m} \tag{5.3}$$

and is referred to as the cyclotron condition or the magnetic resonance principle. This frequency is independent of the particle velocity v of the ion and of the radius r of the circular path and—for nonrelativistic velocities—is a constant for given values of q, B, and m. For the condition of resonance, the magnetic field is adjusted so that the time for an ion (with given q and m) to complete a half-cycle is equal to the time for reversal of the oscillatory electric field in the dees. Thus, after completing the first half-cycle the ion experiences another acceleration, acquires higher velocity, and traverses a path of larger radius through the magnetic field. As long as this resonance condition is fulfilled, the ions are accelerated each time they cross the gap between the dees, traveling in ever-widening semicircles until they reach the maximum radius R permitted by the dimensions of the magnet. If there are N accelerations through a potential difference between the dees of V, the final energy is

$$E = NqeV = (qeRB)^2/2m .$$

If the ion bunches arrive at the cyclotron accelerating gaps slightly after the accelerating field passes its maximum value, a condition called phase stability, i.e., compression of the time spread of the particle bunch, occurs. At the periphery the ions are extracted with the use of electrostatic deflectors. In some cases both electrostatic and magnetic deflectors are used. The maximum energy increases quadratically with the magnetic field B and the radius R. For example, a cyclotron with $B = 10$ kgauss, $R = 1$ m, and $f = 15.3$ MHz provides a proton beam of $E = 48$ MeV.

We see that in a cyclotron the magnetic field forces the particles into ever-increasing circular orbits. The ions are constrained to pass and repass many times through a relatively low accelerating voltage in resonance with the oscillating electric field. Thus their final energy is many times greater than the accelerating voltage. A simple analogy (Liv62) is the garden swing, which can be urged to large amplitude by successive small pushes, each push timed with the natural period of the swing. The standard cyclotron (fixed magnetic field, fixed frequency) was the first of the resonance accelerators. This accelerator tested the principles and basic concepts which with further development led to the modern high-energy accelerators (Liv62). The standard cyclotron provides a variety of ion species with high beam currents over a limited range of energies. They are used in applications where energy resolution is not essential, although recent developments have allowed energy resolution which approaches that of electrostatic accelerators.

5.1.2.6. Linac

Another type of resonance accelerator is the "linear accelerator," often abbreviated to "linac." As the name implies, in the linac the charged particles are accelerated in a straight line (Fig. 5.6c) by an oscillating electric field. Many different electrode structures have been devised and tested (Liv62, Lap70). The early linear accelerators (Wid28, Slo31) consisted of an array of coaxial cylindrical electrodes (drift tubes) of increasing length L separated by small accelerating gaps. These electrodes were aligned along the axis of a long glass vacuum chamber. The electrodes were connected alternately to two bus bars extending along the length of the chamber with power supplied by a radio-frequency power source. The separation L between accelerating gaps is the distance traversed by the particles during one half-cycle of the applied alternating electric field, whence $L = v/2f$, where v is the particle velocity and f the radio frequency. Each time a particle of charge qe crosses a gap, it sees an electric field which gives it an increment of energy $qeV \sin \phi$, where V is the maximum gap voltage and ϕ is the phase at which the particle crosses the gap. During the other half-cycle, the particle is shielded from the decelerating electric field by being inside a drift tube. Early machines of this type were designed to accelerate only particles which arrived at a phase close to $90°$, so the energy gain at each gap was about qeV. Phase stability for a linac requires (opposite to that for a cyclotron) that the ion bunches arrive slightly before the accelerating field reaches its maximum value. After passing through a series of N electrode gaps, the particles attain a final energy $E = NqeV$. The particles emerge in bunches corresponding to peak field at the gaps. We see then that the linac also uses the principle of multiple acceleration, i.e., of resonance acceleration (particles in phase with an oscillating electric field). Historically, the report of Wideroe's linac design (Wid28) provided the initial idea from which the concept of the cyclotron was developed by Lawrence (Law30). The main advantage of the linac lies in its high beam current, low divergence, and relatively small energy spread. In both resonance-type accelerators (linac and cyclotron) energy variation is more complicated and sometimes limited in range.

5.1.3. Beam Transport System

It is in general necessary for the ion beam to travel over a relatively long distance, starting at the ion source and passing through the accelerator and onto the target (Fig. 5.1). The basic purpose of a beam transport system is to convey the ion beam over this path from one point in space to some other point in space with the maximum efficiency. A secondary purpose may be to reduce the emittance of the beam by collimating it with defining apertures along the way. One may also reduce the momentum (or equivalently the energy) spread of the beam by passing it through a momentum- (or energy-) dispersive element in the transport system followed by momentum (or energy) selecting slits. Finally, the transport system may be used to minimize the contamination of the ion beam (sec. 5.1.1) by selecting which species to allow through (reduction in mass spread). Without some selection, a single-ended electrostatic accelerator would dump a whole spectrum of particles onto the target. The individual elements in the transport system may be either electrostatic or magnetic or a combination of the two. The guiding elements will usually be separated by field-free drift spaces. Guiding and controlling ion beams over long distances without significant loss in intensity is rather similar to guiding a light beam through an optical system. The function of the optical components such as lenses and prisms are replaced in an ion-beam transport system by electric or magnetic components which do the same thing. In what follows we will describe qualitatively the most common beam transport devices and compare them with their optical analogues. Detailed discussions of such devices, including their theoretical treatments (Maxwell equations, matrix techniques), can be found in the literature (Liv62, Ste64, Ban66, Sep67, Wol67, Eng74).

The early work on the formation of electron beams in X-ray tubes and electron microscopes showed that a sequence of accelerating electrodes with cylindrical symmetry has lenslike properties. This is a consequence of the radial components of the fields inside and between the accelerating electrodes. These field components will act either to focus or to defocus nonaxial particles in the accelerated beam. A common type of electrostatic lens for ion beams is that produced by the field in the small gap between two coaxial cylinders of the same diameter (Fig. 5.7a). The principle of their focusing properties can best be understood in terms of an optical analogue. Figure 5.7b shows two optical lenses, one convex and one concave, both with the same numerical focal length. In any lens, whether optical or ion-optical, the angle of deflection experienced by a given ray is proportional to the distance from the axis. Therefore, the rays experience a larger angle of deflection in the focusing lens than in the defocusing lens for both cases shown in Fig. 5.7b. A net focusing effect always results, no matter which lens comes first. The action of the electric field in the gap between the cylindrical electrodes on the ion particle path is the same as the action on the light ray by the lens combination. In the accelerating lens (top of Fig. 5.7a) the fields around the gap force the particle inward as it approaches the gap. At the gap it is accelerated, and, since the radial field component increases as the radius increases, it proceeds through slightly weaker radial fields that are pushing it outward. The increased velocity and the decreased radial field component both make the defocusing weaker than was the focusing effect, and the particle is pushed outward less than it was originally deflected inward. Therefore, it leaves the lens with a net inward component of radial velocity. In the decelerating lens (bottom of Fig. 5.7a) the ions are pulled away from the axis as they approach the gap. Leaving the gap with decreased velocity, they enter a focusing field stronger than they felt before, and they again leave the lens with a net inward velocity component. Thus, cylindrical lenses are always converging lenses regardless of which cylinder has the higher potential. The focal lengths for accelerating and decelerating lenses of the cylindrical type differ slightly because of the somewhat different paths traversed. By making the second cylinder larger than the first one in an accelerating lens, called an "asymmetrical lens" in Figure 5.7c, the electric lines of force spread out more in the second cylinder. Such spreading weakens the electric field in the larger cylinder and reduces the defocusing action of the lens to bring the ions to a shorter focus. The same effect can be obtained if a conducting grid is placed across the second cylinder (called a gridded lens; Fig. 5.7d). With a dense grid the field approaches that between a cylinder and a plane, effectively removing the defocusing half of the two-cylinder lens field and thus making the lens strongly focusing. Of course, a grid obstructs a portion of the aperture and somewhat reduces the beam transmission. Often a combination of three cylinders is used with a single potential

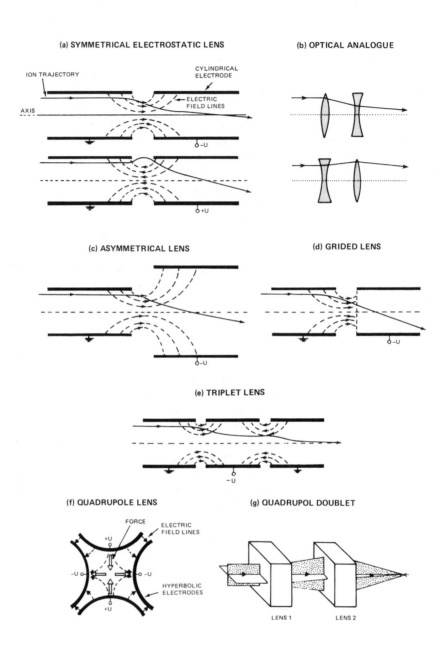

FIGURE 5.7. Shown schematically are the focusing effects of electrostatic cylinder lenses on the trajectories of a positive-ion beam (a, c, d, e). The optical analogue is also shown in (b). An electrostatic quadrupole lens (f) whose four electrodes form rectangular hyperbolae focuses a beam in one plane and equally strongly defocuses it in the perpendicular plane. The combination of two such lenses, the quadrupole doublet (g), results in a focusing in both planes.

applied to the middle one (Fig. 5.7e), the system being referred to as an electro-static triplet lens or an einzel lens. Note that this einzel lens is purely a focus-ing device which does not change the energy of the ions, while the lenses a, c, and d have accelerating and focusing properties.

Of interest to the accelerator designer is the ion-optical behavior of a uniform accelerating field with no radial components. In this system the parti-cle trajectory is modified because the paraxial velocity is continually increas-ing, although the radial velocity component is unaffected. Consequently, a diverging beam tends to become less divergent during acceleration, and a con-verging beam has its convergence slightly decreased. The general tendency during acceleration is to bring the beam closer to parallelism. In this respect, the electrode structure of the accelerating tube (sec. 5.1.2) helps to give the ion beam the desired properties at the exit port of the accelerator. Usually the emergent beam will be approximately parallel and reduced in diameter from more than 1 cm near the ion source to 1–2 mm at full energy. Important to this achievement is precise construction and alignment, large lens apertures, and carefully regulated potentials applied to the lenses. With increasing ion beam current (mA range) space-charge effects play a more and more important role in the focusing properties. Since like charges repel each other, their mutual field is always defocusing. This space-charge field can be taken into account in designing a lens, and many variations of electrode geometries (called Pierce lenses) have been designed for maintaining desired convergence or divergence in dense charged-particle beams (Pie54, Hut67).

A most elegant and efficient technique for restraining the radial motions in beams of charged particles is that of alternating-gradient focusing. Since it gives much stronger restraining forces than most of the other available methods, it has often been described as "strong focusing." The basic principle was first proposed in 1950 by Christofilos and was independently discovered in 1952 by Courant, Livingston, and Snyder (Liv62). The technique is applic-able to both magnetic and electrostatic gradient fields. The specialized applica-tion of this technique to the focusing of linear beams passing through nonuniform electrostatic fields led to the electrostatic quadrupole lens as shown in Figure 5.7f. A particle beam enters the lens perpendicular to the plane of the figure. The four electrodes of the lens are hyperbolic in shape, with the axis of neighbors at 90° extending along the symmetry axis. Voltages of alternating sign are applied to the electrodes. The electric field of this quadru-pole lens is zero on the symmetry axis, increasing linearly with increasing radial distance from the axis. This field has the effect of focusing the beam in one plane and equally defocusing it in the perpendicular plane. If the beam passes a second quadrupole lens of reversed gradients, the net effect of the combination (called a pair of quadrupoles or a quadrupole doublet) is to focus the beam in both planes (two-dimensional focusing action; Eng59, Cha60, Liv62). This principle of alternating-gradient focusing also has an analogue in optics (Fig. 5.7b). An electrostatic quadrupole lens focuses all ions of the same energy to the same point, independent of their masses.

Focusing by axial magnetic fields is also used in beam transport systems. When ions cross a short magnetic field and their paths make an angle with the magnetic field lines, they are deflected in spiral-like paths which, if properly controlled, can bring them to a focus (Fig. 5.8a). The magnetic field is often confined to a short axial gap between the end plates of an iron case. Such short magnetic lenses are used where there is a need to minimize the physical size of the lens system. Also, in the presence of dense space charges (high ion currents), such axial magnetic lenses can keep the beam from expanding. Note also that the circular frequency in the spiral path is independent of the velocity of the ions and that all ions are focused at the same place on the axis. Such a system is said to be free of dispersion. Another magnetic instrument, called the magnetic quadrupole lens (Fig. 5.8c), also focuses particles (Cha60). As with the electrostatic quadrupole, the magnetic poles are shaped in such a way that the magnetic field is zero on the symmetry axis and increases linearly with radial distance from the axis. The Lorentz force in such a field is such that the particles experience a focusing force in one plane and an equal defocusing force in the perpendicular plane. A quadrupole doublet, consisting of two such

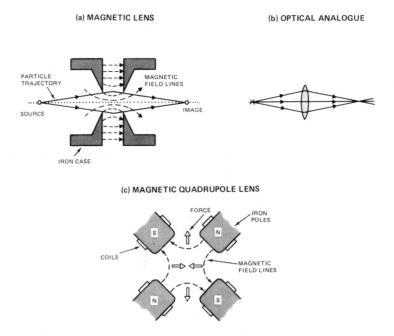

FIGURE 5.8. Cross-sectional view of a magnetic lens, its optical analogue and a magnetic quadrupole lens. The quadrupole lens focuses the beam in one plane and defocuses it in the perpendicular plane. If the ion beam passes a second quadrupole lens with gradients reversed (north and south poles rotated by 90°), one has a quadrupole doublet which focuses in both planes.

lenses with alternating field gradients, again has the net effect of focusing in both planes. Generally such foci can coincide but usually they have different amplifications. Such systems are called astigmatic. Nonastigmatic systems are built using three or more lenses (called multipoles), which not only produce more symmetric results but also reduce other aberrations (Reg67). The combination of magnetic lenses represents one of the most widely used lenses in beam transport systems. Because the field at the center of the quadrupole lens is zero, there is no momentum dispersion in first order. Thus these lenses focus all equally charged particles at the same point for paraxial rays. Nonparaxial particles of the same energy but different momenta (masses) are of course focused at different points.

It frequently happens that the beam axis in a transport system is not coincident with the geometrical axis of the beam pipe either because of small misalignments of the transport elements or because of steering effects caused by fringe fields. For optimum transmission of the beam to the target, the beam must be steered at several places along the transport system to take account of this. This steering is usually done using two sets of uniform-field magnetic deflection coils (steering in both planes).

Several devices discussed so far have been nondispersive, i.e., their effects on a particle's trajectory have been independent of the particle's momentum. Dispersing elements are important in a practical beam transport system. One such device (Fig. 5.9a) is a pair of flat electrostatic plates separated by a distance d and operated at a voltage U producing a uniform electric field between the plates. A particle of energy E, mass m, and charge q entering this field at an angle θ will behave just as a ball thrown in the air behaves under the influence of gravity. The ions moving toward the positive electrode have to move against the electric field. Thus they are slowed down, and the radius of their orbit decreases as they approach the positive electrode. For nonrelativistic particles the distance x from the entrance point to the exit point is traveled via a parabolic path. The distance is given in terms of the parameters of the deflector by the relation (Pie54)

$$x = \frac{2Ed}{qU} \sin 2\theta .$$

It follows, then, that the particles travel farthest horizontally if directed upward at an angle of $\theta = 45°$. Since at this angle the derivative $dx/d\theta$ is zero, it also follows that a particle stream consisting of particles injected at the entrance into the field at various angles around $45°$ is brought to a focus by the field on returning to their starting height at the exit. Logarithmic differentiation of the above equation for a given system (U, d, θ = constant) leads to

$$\frac{\Delta x}{x} = \frac{\Delta E}{E} + \frac{\Delta q}{q} ,$$

and, thus, the uniform electric field has a dispersive action with regard to

(a) UNIFORM FIELD ELECTROSTATIC ANALYZER

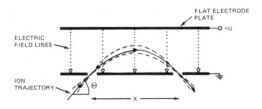

FLAT ELECTRODE PLATE

+U

ELECTRIC FIELD LINES

ION TRAJECTORY

Θ

X

(b) CYLINDRICAL ELECTROSTATIC ANALYZER

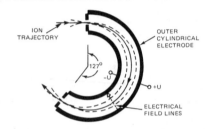

ION TRAJECTORY

OUTER CYLINDRICAL ELECTRODE

127°

-U

+U

ELECTRICAL FIELD LINES

(c) SPHERICAL ELECTROSTATIC ANALYZER

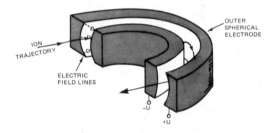

OUTER SPHERICAL ELECTRODE

ION TRAJECTORY

ELECTRIC FIELD LINES

-U

+U

(d) CROSSED-FIELD ANALYZER (WIEN FILTER)

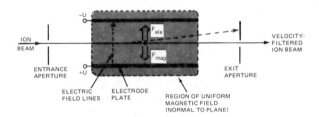

-U

F_{ele}

F_{mag}

ION BEAM

VELOCITY-FILTERED ION BEAM

ENTRANCE APERTURE

+U

EXIT APERTURE

ELECTRIC FIELD LINES

ELECTRODE PLATE

REGION OF UNIFORM MAGNETIC FIELD (NORMAL TO PLANE)

FIGURE 5.9. Schematic diagram of electrostatic analyzers with a uniform (*a*), a cylindrical (*b*), and a spherical (*c*) electric field. The first two analyzers are single focusing, while the third type is double focusing. The crossed electric and magnetic homogeneous fields (*d*) filter the velocity distribution of the incident ion beam to a single value.

energy and charge state but not with regard to particle mass. This is a general characteristic of electrostatic analyzers, which deflect, disperse, and focus particle beams. Because of the energy dispersion of the analyzer, it is called an energy filter. Note also that there is no focusing action of the uniform field analyzer (Fig. 5.9a) in the direction perpendicular to the plane of the figure. An ion that initially has a velocity component in this direction will continue to have this velocity during the motion through the analyzer. Therefore, a point object will have a line image rather than a point image, and the ion-optical element is said to be "single" focusing (focusing in a single plane) or to be astigmatic. A more widely used electrostatic analyzer consists of two concentric cylinders (with radii of curvature r_1 and r_2), which have a constant separation $d = r_2 - r_1$ (Fig. 5.9b). The outer electrode is operated at a positive voltage, $+U$, and the inner electrode at an equal negative voltage, $-U$. An ion of energy E passing through the center of such an electrode system, as defined by entrance and exit slits, is governed by the equation

$$E = \frac{qUr}{d}, \tag{5.4}$$

where $r = (r_1 r_2)^{1/2}$ is the mean radius of curvature of the electrodes. Again, this electrostatic analyzer (War47, Fow47) produces radial focusing if the cylindrical electrodes extend over an angular range of $\pi/2^{1/2} = 127°$. Since the force acting on the ion by the electric field is radial, there is no focusing action in the axial direction. Thus, this element is also single focusing. Note that double focusing, i.e., focusing in radial and azimuthal directions, is achieved for a circular orbit of 180° (Eng74) with an analyzer consisting of concentric spheres (Fig. 5.9c). Because of the engineering problems in the construction of such a device, it has not been widely used. Note also that, if relativistic particles are involved, the above equation is replaced by

$$E = \frac{qUr}{d} \left(1 + \frac{d^2}{24r^2}\right)^{-1} \left(1 + \frac{E}{mc^2}\right)\left(1 + \frac{E}{2mc^2}\right)^{-1}. \tag{5.5}$$

The simplest type of magnetic analyzer (magnetic prism) consists of a uniform magnetic field (Fig. 5.10a) produced between the two pole pieces of a dipole magnet. In this field an ion beam is deflected along a circular orbit whose radius of curvature is determined by the equation

$$r = \frac{mv}{qB} = \frac{p}{qB} = \frac{(2mE)^{1/2}}{qB}, \tag{5.6}$$

where p is the momentum of the particles. For a given magnetic field strength B, logarithmic differentiation leads to

$$\frac{\Delta r}{r} = \frac{\Delta p}{p} + \frac{\Delta q}{q} = \frac{1}{2}\frac{\Delta m}{m} + \frac{1}{2}\frac{\Delta E}{E} + \frac{\Delta q}{q}.$$

(a) SEMICIRCLE MAGNETIC ANALYZER   **(b) SECTOR MAGNETIC ANALYZER**

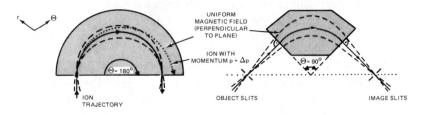

(c) FRINGE FIELD FOCUSING MAGNETIC ANALYZER

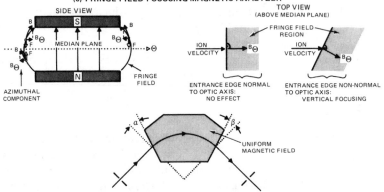

(d) NONUNIFORM FIELD MAGNETIC ANALYZER

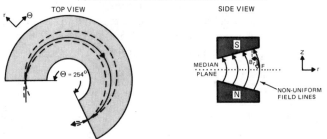

FIGURE 5.10. Conceptual drawings of magnetic analyzers. The fringe fields at the edges of the sector magnet (b) have azimuthal components B_θ, above and below the median plane. When the particle impinges on the magnet edge (the field boundary), which is inclined by more than 90° with respect to the optic axis (c), the velocity and the field B_θ produce the vertical focusing force F. Such homogeneous-field magnets are then double focusing. (d) A nonuniform magnetic field, whose vertical component B_z varies with radius as $B_z \propto r^{-1/2}$ is also double focusing at a deflection angle $\theta = 2^{1/2}\pi = 254°.5$.

One sees that a magnetic field has a dispersive action with respect to charge and momentum (or, equivalently, to mass and energy) and serves as a momentum filter. From Figure 5.10a we see that a particle entering the field with an additional momentum Δp traverses an orbit with a larger radius than one of momentum p and leaves displaced in position and in angle. It follows then that the bending properties and dispersing qualities of a magnet are intimately related. In the relativistic case the above equation becomes

$$r = \frac{(2mE)^{1/2}}{qB} \left(1 + \frac{E}{2mc^2}\right)^{1/2}. \tag{5.7}$$

Like the uniform electric field, the uniform magnetic field has focusing properties (Fig. 5.10a). Particle paths starting from the same point (with identical momentum) but at slightly different angles describe circles of the same radius r and hence are brought to a focus diametrically opposite the point of departure. Thus the flat-field semicircular magnetic analyzer shown in Figure 5.10a is single focusing. Historically, this was the earliest type of magnetic analyzer (Bae10, Rut13), deflecting, dispersing, and focusing the beam. The most common type of flat-field analyzer is the sector (or wedge) magnet with straight entrance and exit edges oriented perpendicular (normal) to the optic axis, defined by the passage of a particle with momentum p making an orbit in the magnet with radius r (Fig. 5.10b). We see from the figure that particles with the same momentum but incident with different angles at the entrance edge are brought to a focus. This focusing effect is due only to the different path lengths within the magnet. The upper (lower) dashed path travels a longer (shorter) distance in the magnetic field than the central ray and hence is deflected more (less), whence they again intersect the central path. In this apparatus the source, the image, and the center of the radius of curvature of the orbit (the bending edge) lie along a straight line, which is known as Barber's rule of magneto-optics (Bar33). This 90° sector magnet is widely used in ion-beam transport systems as a bending magnet for vertical electrostatic accelerators, where it bends a beam of particles with the same mass, velocity, and charge by 90° into a horizontal direction (Fig. 5.1).

Vertical focusing can be provided by quadrupole elements. However, this is often not convenient, especially at the exit of the analyzer where the size of the dispersed beam could require a very large quadrupole gap. More convenient is the use of magnet edges (Fig. 5.10c), which are not perpendicular to the optic axis and which produce edge-focusing effects (Cot38, Cam51, Cro51). To understand this, let us consider the fringe fields at the edges of the sector magnet. Above and below the median plane these fringe fields have azimuthal components B_θ, which are perpendicular to the edge of the magnet. If the optic axis of the ion trajectory orbit is also perpendicular (normal) to this edge, the velocity vector v and the azimuthal magnetic field component B_θ are parallel, producing no focusing force. However, if the edge is not normal to the optical axis, the vector quantities v and B_θ are at an angle, giving rise to a vertical force F and thus to vertical focusing. As seen in Figure 5.10c, vertical focusing is obtained on both sides of the median plane as well as at the entrance and exit edges. Because the orbits of the particles away from the optic axis are altered in such a nonnormal sector magnet, the radial focusing point is modified. The angles α and β at the entrance and exit edges represent the deviations from normal entrance and exit. These angles are free parameters in the design of the magnet. By properly choosing these angles, double focusing at the same point can be achieved (Pen61, Liv69). Thus, a point object forms a point image. The complete optical analogue of such a focusing, deflecting, and dispersing magnet is a set of focusing lenses combined with a prism that deflects and disperses the light beam. An almost completely equivalent system can be attained by placing a vertically focusing quadrupole singlet or doublet (Fig. 5.8c) near the entrance of the dipole (Eng58, Yag64) or by splitting the magnet into multipole magnets of suitably curved edges (e.g., quadrupole dipole dipole dipole [QDDD] magnet; Spe67, Loe73). An alternative method of providing vertical focusing is to make a dipole field whose strength in the vertical direction varies as $B_z \propto r^{-1/2}$ (Fig. 5.10d). It is seen from the figure that the field component B_r is nonzero and is oriented so that the vertical force is always directed toward the median plane. Thus, such an analyzer provides double focusing at a turn angle of $\theta = 2^{1/2}\pi = 254°5$ (Boc33, Sva47, Jud50, Ros50). Interest in magnets of this type for beam transport systems (and nuclear spectroscopy) has diminished because of the mechanical difficulties in making large, suitably shaped pole faces, and because of the problems of nonuniform magnetic saturation of the iron across the pole faces. It is much easier to machine complex shapes on pole edges than on pole faces.

Another device found in beam transport systems is the crossed-field analyzer ($E \times B$ filter) or Wien filter (Fig. 5.9d). In this instrument the charged particles enter the uniform electrostatic field between the plates of a parallel-plate condenser. This condenser is situated between the poles of a magnet (Cha60) in such an orientation that the uniform magnetic field is perpendicular to the electrostatic field and to the direction of motion of the particles. The sense of the magnetic field B is such that the charged particles are deflected in the direction opposite to the deflection produced by the electric field E. By adjusting both fields, a cancellation of deflection can be obtained, that is,

$$q|E| = qv|B| \qquad \text{or} \qquad v = |E|/|B| . \tag{5.8}$$

Thus, for a given strength of both fields, only particles with a velocity v in accord with the above field ratio will pass through the device without deviation—hence the name "velocity filter" (Bai60). The Wien filter is of particular importance for use with ion sources (Fig. 5.4), where one wishes to select the desired ion species out of the many leaving the source. Another application is in eliminating ions created by ionization processes along the accelerating tube. These ions leave the accelerator with a continuous energy distribution ("white" spectrum) and are not completely eliminated by a single subsequent analyzer (Fig. 5.1). A Wien filter can be helpful in further reducing such beam contaminations (sec. 5.1.4) and, since the ion species of interest pass a Wien filter without deflection, such filters can be installed conveniently in the straight sections of the beam transport system. Note that all ions of a given velocity v pass the Wien filter independent of their charge state. Of course, neutral particles will be unaffected, and their elimination requires additional deflecting devices.

It is of the utmost importance to match the acceptance of the various beam transport elements to the emittance of the beam to be transported (Fro67, Lee69, Sil70, Eng74). To adjust the parameters of a beam transport system and thus to optimize its efficiency, it is necessary to measure the position and profile of the beam at various points in the system. For high beam currents such measurements are carried out most efficiently with beam scanners (Hor64, Ros64, Eng74), which are thin rigid wires movable across the beam. All beam-intercepting components of the system, such as slits, apertures, Faraday cups, and beam scanners, have to be specially designed and cooled, if high beam currents (several mA) are to be handled (Ham75, Ham79). It is also necessary that the beam be transported in a vacuum as good as possible, which is maintained in a metal beam tube (Woo71).

5.1.4. Analysis of Beam Properties

In experiments in nuclear astrophysics a precise knowledge of the absolute energy, energy spread, and energy stability (in time) of the ion beam is necessary, particularly at sub-Coulomb energies, because location and width of narrow resonances enter sensitively in determining the stellar reaction rate $\langle \sigma v \rangle$ (chap. 4). Also, the cross sections for charged-particle–induced nonresonant reactions are a steep function of beam energy (chap. 4), and any uncertainty in the absolute beam energy may lead to a large error in the cross section. For example, in the $^{12}C + ^{12}C$ fusion reaction (chap. 8) at $E_{c.m.} = 2.5$ MeV an error of $\Delta E_{c.m.} = \pm 100$ keV translates into an uncertainty in the cross section of a factor of 3. The purity of the ion beam is also very important in nuclear astrophysical experiments. For example, the $^3He(^3He, 2p)^4He$ and $^3He(d, p)^4He$ reactions (chap. 6) at $E_{c.m.} = 30$ keV have relative probabilities for tunneling through the respective Coulomb barriers (chap. 4) of

$$\frac{\exp(-2\pi\eta)(^3He + {}^3He)}{\exp(-2\pi\eta)(^3He + d)} = 1.9 \times 10^{-7} .$$

From these considerations one sees that equal count rates from these reactions would be observed if the ^3He ion beam has a deuterium contamination (e.g., in the form of the mass 3 molecular ion DH^+) of the order of "only" 1.9×10^{-7}. Finally, the charge state of the ion beam must be known in order to determine the number of incident ions from charge measurements (sec. 5.1.5).

For potential-drop electrostatic accelerators the generating voltmeter (Bjo47, Liv62, Boe70) has been for many years the basic instrument for determining the terminal potential and thus the ion-beam energy. It is usually mounted near the high-voltage terminal (Fig. 5.5a). In the rotating voltmeter, an insulated vane (the rotor) rotates at constant speed behind a grounded shield (the stator) shaped so that the vane is covered during half its rotation and exposed to the electric field during the other half. The alternating electrostatic voltage induced on the vane is then a measure of the terminal voltage. The meter is usually calibrated against other standards of potential. In most installations it is used as a qualitative instrument for monitoring operations such as tuning up the generator or in observing spark discharges (conditioning of high voltage). It is also used as an alternative method of stabilizing the accelerator voltage in the absence of an ion beam.

Another method of measuring terminal voltage is to observe the current in the chain of calibrated high resistors running from the terminal to ground (Fig. 5.5). A typical value is 10,000 megaohms per million volts, or a current of 100 μA. An ammeter is inserted at the grounded end of the resistor column. The high voltage can be determined with a precision equal to that of the resistances and the ammeter for the assumption of negligible leakage currents. A problem with this technique is that the resistances change as a function of time, temperature, and overloads and thus frequent recalibration is necessary.

The absolute energy of an ion beam can be determined from a measurement of its velocity using either time-of-flight techniques (sec. 5.4.4) or a Wien filter. If the mass of the ions is known, their kinetic energy is determined. Carefully constructed, large electrostatic or magnetic analyzers also have been used to determine the absolute energy by measurement of the electrostatic or magnetic deflection. These techniques were used (Mar66, Mar68, Rou70) to measure the absolute energies of a few proton-induced nuclear resonances and (p, n) thresholds to a high precision (Table 5.2). These energies are used as standards to calibrate other instruments. Each of the above techniques has its own advantages and disadvantages. For example, while it is much easier to make absolute energy measurements using an electrostatic analyzer, there are associated problems such as the need to condition the plates because of high field-

TABLE 5.2 Absolute Energies of Some Proton-Induced Nuclear Reactions[a]

Reaction	Proton Resonance (keV)	Proton Threshold (keV)
$^{11}B(p, \gamma)^{12}C$	163.1 ± 0.2	...
$^{14}N(p, \gamma)^{15}O$	278.1 ± 0.4	...
$^{19}F(p, \alpha\gamma)^{16}O$	340.46 ± 0.04	...
$^{19}F(p, \alpha\gamma)^{16}O$	872.1 ± 0.2	...
$^{27}Al(p, \gamma)^{28}Si$	991.90 ± 0.04	...
$^{7}Li(p, n)^{7}Be$	...	1880.60 ± 0.07
$^{13}C(p, n)^{13}N$	...	3235.7 ± 0.7
$^{19}F(p, n)^{19}Ne$	...	4234.3 ± 0.8

[a] From Mar68 (see also Fre77).

induced voltage breakdown and the emission of secondary electrons caused by particles striking the plates. The plates are usually installed inside the vacuum envelope. Thus the electrostatic analyzer compared with the magnetic analyzer (located outside the vacuum envelope) is slightly less convenient for general use and considerably less desirable for handling the large beam currents that are often necessary for nuclear astrophysics experiments.

Most commonly a 90° magnetic analyzer (bending magnet) is used, and the absolute energy E is in principle determined from equation (5.7):

$$r = \frac{1}{qBc} (2mc^2E + E^2)^{1/2}$$

if the rest mass m and charge q of the projectiles, the magnetic field B, and the radius of curvature r are known. Note that the magnetic field is usually measured only near the median plane in the center of the analyzer, using a Hall probe or a nuclear magnetic resonance (NMR) probe. However, the practical operating performance of a magnet is more complicated than implied by the above formula, since the extended fringing fields at the magnet entrance and exit affect the actual particle trajectories, making their paths more complex than a simple circular orbit. Other effects, such as constructional defects causing field nonuniformities along the orbit, field saturation, and aberrations of higher order, make a precise absolute calibration of such bending magnets very difficult. For a given geometry of the entrance and exit slits (Fig. 5.10b), which define the ion trajectory in the bending magnet and thus the ion energy, all the " difficult-to-measure " effects are incorporated into a so-called magnetic "constant" K defined by the expression

$$K = qB/(2mc^2E + E^2)^{1/2} . \tag{5.9}$$

This constant is determined using the energies of the recommended standards (Table 5.2). In spite of its name, for a particular device the parameer K might not be independent of the strength of the magnetic field and thus would need to be measured over its full range. Most commonly, the energy calibration is carried out using the precisely known proton-induced reactions at $E_p \leq 4.2$ MeV (Table 5.2). For heavy-ion beams, however, the magnetic fields required are usually far removed from the calibrated field range. Consequently, an absolute energy determination for these beams depends heavily on how the extrapolation of the calibration parameter K toward higher magnetic field strengths is done. However, the precisely known proton-induced reactions can be used to calibrate accurately the magnetic analyzer at high field strengths by reversing the roles of projectile and target nuclei, i.e., a hydrogen gas target is bombarded by the corresponding energetic heavy ions. Using several different charge states of the heavy-ion beam, the same resonance can lead to several calibration points over a wide range of magnetic field strengths. Such an energy calibration has been carried out (Tra79) for the analyzing magnet at the 4 MV Dynamitron tandem accelerator in Bochum over a wide range of

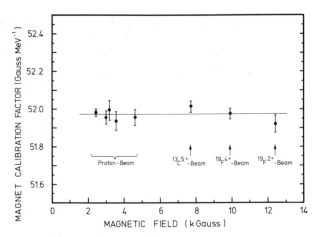

FIGURE 5.11. The magnet calibration constant K is plotted as a function of magnetic field strength (Tra79). The line through the data points assumes a field-independent constant K. The types of projectile ions used in the calibration of the magnetic analyzer are also indicated.

magnetic fields ($B = 2.4$–12.4 kgauss). For this field range the magnetic constant K was observed to be a constant within experimental uncertainties (Fig. 5.11). The results imply a precision in the determination of heavy-ion energies of 4 parts in 10^4, which is sufficient for all purposes of interest. The calibration parameters of a given device might change significantly with time and are critically dependent on the settings of the energy-defining input and output slits. At any given time, a precise beam energy measurement requires one to check the calibration parameters, a lengthy and inconvenient procedure.

Another method was reported (Rol75, Swi77, Fre77, Krä82) for measuring beam energies of light projectiles (protons and helium ions) absolutely at any given value of interest to a precision of the order of 0.4 keV. The method utilizes nonresonant capture reactions with low reaction Q-values, such as $^{12}C(p, \gamma)^{13}N$, $^{16}O(p, \gamma)^{17}F$, and $^{3}He(^{4}He, \gamma)^{7}Be$ in combination with high-resolution Ge(Li) γ-ray detectors (Fig. 5.30). Because of the low Q-value the energies of the capture transitions fall in a region where comparison with precisely known energies of γ-rays from radioactive sources (Mar68) is possible. A measurement of the γ-ray energy E_γ allows one to determine the beam energy $E_{c.m.}$ to high accuracy if the Q-value of the reaction is precisely known:

$$E_\gamma = Q + E_{c.m.} - E_x , \qquad (5.10)$$

where E_x is the excitation energy of the state in the residual nucleus, to which the transition occurs. Because of the smooth yield curve of nonresonant reactions, the capture transition is observable at all beam energies of interest, and thus the beam energy measurement is not restricted to resonant energies and interpolation procedures. In this method, the 90° magnetic analyzer serves only as a beam-energy stabilizing device, and no precise knowledge of its calibration parameter is required. The method can of course be used for a quick check of the calibration constant K over a limited range of field strengths.

It is also important to know the energy spread of the ion beam. It is usually determined using the known features of the thick-target yield curves for narrow resonances (sec. 5.5). An illustrative example is shown in Figure 5.12. Shown are data for γ-ray yield from the 327 keV resonance in the ^{27}Al$(p, \gamma)^{20}$Si reaction obtained (Fre77) using the 350 kV accelerator at Münster. Since the resonance width ($\Gamma \leq 20$ eV) and the estimated Doppler broadening (Don67) of the target atoms ($\xi = 60$ eV for 300 K) are small compared with the observed width (FWHM) of 1.1 ± 0.2 keV, this value represents the energy spread of the proton beam. Another important parameter of an ion beam is its long-term stability. In the example of Figure 5.12 this stability was investigated near the midpoint of the thick-target yield curve and was found to be better than ± 15 eV over a period of several hours (see also Ham79). Finally, as pointed out above, the purity of the ion beam can be of utmost importance for nuclear astrophysical experiments. This requirement can be attained effectively using a Wien filter, as demonstrated in Figure 5.13.

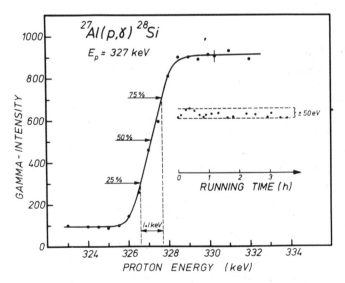

FIGURE 5.12. Thick-target yield curve of the 327 keV resonance of ^{27}Al$(p, \gamma)^{28}$Si (Fre77). The line through the data points is to guide the eye. An experimental energy spread of 1.1 keV is deduced from the 25% and 75% points. For the assumption of a Gaussian energy distribution, the FWHM spread is about 1.6 times the 25%–75% range, i.e., about 1.8 keV. The inset shows the results of the energy stability test.

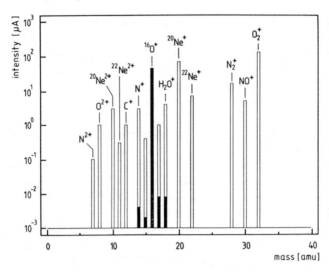

FIGURE 5.13. Mass spectrum of an ion beam with total current of 300 μA, obtained from a gas mixture in an ion source (Cle85). The singly charged ions are preaccelerated to 30 keV by an extraction cone and guided via an einzel lens into a Wien filter, which is followed by a 90° double-focusing magnetic analyzer. The open bars show the mass spectrum with the Wien filter switched off, while the full bars show the mass selectivity of the Wien filter set to $^{16}O^+$. The ion beam is purified to about 1 part in 10^3.

5.1.5. Beam Integration

Since many of the problems in nuclear astrophysics require information in the form of absolute cross sections, it is necessary to have an accurate absolute measurement of the total number of charged particles striking the target during a run. The ion beam must eventually be stopped (Fig. 5.1), and it is usual to make an accurate measurement of the beam current at the same time. The simplest method is to intercept the beam with an insulated metal plate of sufficient thickness to stop the beam. The ion current striking this collection plate is then measured, and by integrating the incident beam current, one obtains a measurement of the total charge accumulation. A number of problems arise when this is attempted (Ynt52, Eng72, Eng74), and precautions must be taken in the design of this beam catcher. First, when a charged particle strikes any material, it liberates secondary electrons from the surface of the material. Second, the ion beam may itself be accompanied by an electron beam, owing to ionizing interactions with residual gas in the vacuum system and with the material of any target foil through which the beam passes. Third, if the particle beam has passed through some target, then it will have been multiply scattered through some small angle. Thus the beam collection plate must have provision for effective suppression of secondary electrons and must subtend a sufficiently large angle at the target to enable it to collect enough of the beam to allow for a measurement of the required accuracy to be made. These considerations lead to the usual designs for beam catchers. The most common design is a cylindrical cup-shaped electrode usually known as the Faraday cup (Fig. 5.14a). In the example shown, the thick metal cup is preceded by an annular electrode held at negative voltage of up to several hundred volts with respect to ground. The sides of the cup are surrounded by a grid (mesh screen) held at the same potential as the annular electrode. The cup itself is operated at ground potential and is supported on high-resistance insulators within a metal vacuum enclosure. The design of a properly shielded Faraday cup, including electrostatic guard rings and suppression grids (or suppressor magnets), can thus ensure that secondary electrons produced in the collimating slits, in the walls of the vacuum vessel, and in the target foil do not enter the Faraday cup and that the secondary electrons produced in the Faraday cup itself do not escape. The effects of secondary electrons on the current measurement are shown in Fig. 5.14a, where a 330 keV proton beam strikes a metal plate surrounded by a cylindrical grid. It is seen that the secondary electrons are suppressed effectively at negative grid voltages above 200 V and that their effects can be very large, comparable to the true ion-beam current. With well-designed Faraday cups the ion-beam current can be recorded with an accuracy of better than 1% (Eng74).

The beam stop at the bottom of the cup (Fig. 5.14a) should be of material in which a minimum amount of radioactive daughters are produced. This problem of the activation of the beam stop and of other beam line components intersecting the beam is very important when high-intensity beams are used and may sometimes determine the upper limit on these beams. An additional problem—from the experimental point of view—is the production of neutrons in the Faraday cup. These neutrons can seriously affect the measurements being made of the reaction from the main target. In general, a high atomic number element which is also chemically inactive, such as gold, is found to be best for reducing the neutron production problem as well as the activation problem. The neutron production problem can be reduced further if the beam stop is well-shielded and placed far away from the main target.

Another problem, which arises, especially at low energies, involves the ionization state of the incident ion. The ionization state is initially well defined by the magnetic (or electrostatic) deflecting analyzer, but after the ions leave the analyzer and traverse field-free drift spaces in the beam tube, this charge state can be altered by a collision in the target or a collision with residual gas molecules in the vacuum system, in particular if the beam path is long. Such charge-state changes can produce erroneous indications by the beam catcher of the number of incident particles (Fig. 5.14a). Thus, one should seek as good a vacuum as possible. For cases where the charge changes occur in the target foil, the most direct solution is to include the target as part of the Faraday cup system, though this requires some additional shielding and secondary electron suppression.

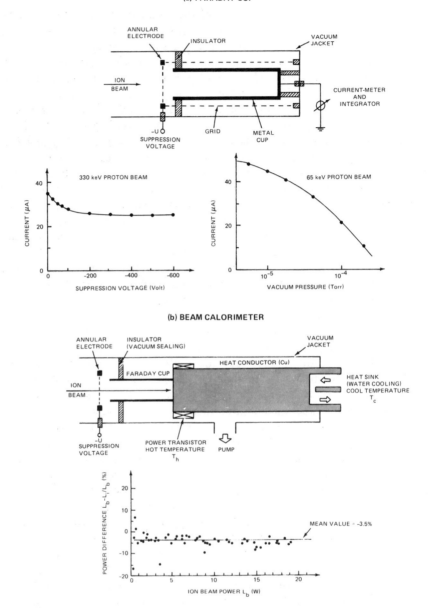

FIGURE 5.14. (*a*) Schematic representation of an electrically suppressed Faraday cup for the collection of a beam of energetic charged particles. Also shown is the beam current measured at a metal plate (surrounded by a cylindrical grid) as a function of suppression voltage. The current measurement of a low-energy proton beam impinging on a Faraday cup can be very sensitive—as shown—to the residual pressure in the beam line. (*b*) Conceptual drawing of the mechanical layout of a beam calorimeter (Vli83). Shown also is a comparison of beam power measurements with the calorimeter (L_b) and the Faraday cup (L_i) for a 20 W calorimeter.

The Faraday cup must also be able to dissipate the power input from the beam. If the singly charged beam is 2 MeV helium ions with an intensity of 500 μA, then the power input is 1 kW. This appears as heat and has to be removed from the collector plate. If the beam is focused to a fine spot of, say, 1 mm^2 area, serious heating of the beam spot will occur, probably sufficient to melt the metal of the collector plate locally and to cause vacuum and electrical insulation problems. Efficient cooling of the collector is thus essential. There is not very much quantitative information in the literature about this subject, but measurements in particular circumstances have been made (Sei67, Han67, Boo67), from which it is clear that the maximum power density which can be safely tolerated is about 0.1 kW mm^{-2}. Hence, for the above example, the beam spot size should be not much smaller than 10 mm^2. Effective cooling can be achieved only if the thickness of the metal of the collector over the area of the beam spot is only slightly greater than that thickness which is necessary to stop the beam. The metal in this region must also be in good thermal contact with the coolant, and the coolant must be in a state of laminar flow without the presence of gas or stream bubbles (Sei67). This requirement is well fulfilled in the design of Hammer (Ham75a), where the collector plate consists of a 1 mm thick copper plate with a very dense set of canals (with diameters slightly less than 1 mm). Water is forced through these canals under a pressure of 20–50 bars.

In the case of gas targets (sec. 5.2.2) charge-exchange effects of the ion beam in the gas before it strikes the Faraday cup make a charge integration difficult if not impossible. Instead, the elastic scattering of the projectiles on the gas target nuclei can be observed with particle detectors to determine both the beam intensity and the target density. This method requires a priori knowledge of the elastic scattering cross section. Furthermore, at low projectile energies the analysis of the elastic scattering peak is hampered by the noise of the detector. These disadvantages are removed in the calorimetric method (Win67, Tho76, Vli83): the beam is stopped in the calorimeter, where the kinetic energy E of the projectiles is converted into heat which is measured by the calorimeter. The total integrated beam power $L_b(t)$ at the calorimeter over a time period t then yields the total number of incident projectiles $N(t)$ over this time period:

$$N(t) = L_b(t)/E .$$

This method then circumvents the problems associated with the charge state of the incident beam, and, when combined with an accurate energy determination, it gives an accurate determination of the number of incident particles in a given run.

The basic principle of the calorimeter shown in Figure 5.14b (Vli83) consists of constant heat transfer between the hot and cool part of the calorimeter, where the hot part is heated either by the incident beam and/or by electrical power. The calorimeter consists of a solid copper cylinder, which is installed in a stainless steel pipe. The front section of the calorimeter serves as a Faraday cup for the incident ion beam, which is used to test the accuracy of the calorimetric method in the absence of gas. In the case of gas targets, only the inner part of the Faraday cup is exposed to the pressure of the gas target, while the outer surface of the entire calorimeter is maintained in a high vacuum. The calorimeter is operated on the principle of a constant temperature gradient $\Delta T = T_h - T_c$ along the axis of the copper cylinder between the hot and cool parts of the calorimeter. The cylinder is heated by power transistors to a temperature T_h in the vicinity of the Faraday cup and cooled to a temperature T_c at the other end of the calorimeter using a water-cooled heat sink. This temperature difference is kept at a constant value by electronic circuits controlling the power of the transistors as well as the water flow. In the absence of an incident ion beam, the transistors supply all the power L_t required to maintain the difference in temperatures ΔT at the selected value. The induced transistor power is removed predominantly by the heat sink and to a negligible extent by other power losses L_l (e.g., radiation loss):

$$L_t = k(T_h - T_c) + L_l .$$

The constant k reflects the thermal properties of the calorimeter and is referred to as the thermal admittance. If an ion beam with power L_b enters the Faraday cup of the calorimeter, the transistors must provide only a reduced power L_t^* to the calorimeter in order to keep the temperature difference $\Delta T = T_h - T_c$ at a constant value:

$$L_t^* + L_b = k(T_h - T_c) + L_l ,$$

hence $L_b = L_t - L_t^*$. Thus, the beam power is determined simply from the difference in transistor powers with and without the ion beam, which can be determined from the electronics of the transistors. Note that because of the constant temperatures in this instrument, the power losses L_l drop out in the subtraction, i.e., this technique is to a large extent independent of surrounding conditions such as gas pressure. Note also that—for a given device—the maximum beam power that can be tolerated safely is determined by the maximum transistor power $L_t = L_b$. The minimum beam power is reached when the difference $L_t - L_t^*$ becomes small compared with L_t, which is usually at $L_b \leq L_t(\text{max})/20$. The differences in beam powers as measured with a 20 W calorimeter (L_b) and with the beam integration on the Faraday cup (L_i) are shown in Figure 5.14b as a function of beam power L_b (Hil84; see also Vli83). As can be seen, the two methods deviate on the average by only 3.5%. This small systematic difference could arise from the problem of absolute temperature measurements (Vli83). It can be seen that the 20 W calorimeter can be used as a reliable and accurate "beam integrator" for beam powers as low as about 1 W.

5.2. Target Features and Target Chambers

The measurement of low cross sections in nuclear astrophysics experiments requires special consideration in the fabrication of targets, particularly if high beam currents are to be used. Of course, the targets should be as stable as possible under heavy beam loads and should be as clean as possible in order to reduce the troublesome effects of contaminant reactions. If a chemical compound is to be used as the target, the latter requirement often demands that in choosing the chemical compound a compromise be made between background yields and maximum reaction yields. If the background level can be tolerated and if the magnitude of its contribution can be determined and taken into account, such a compromise might not seriously affect the final accuracy of the cross-section measurements. It is often necessary to have thin targets available, particularly at sub-Coulomb energies. It is also required that the target thickness be easily variable. The reaction site, i.e., the target volume where reactions take place, should be well defined for angular distribution measurements. For the measurement of excitation functions at extremely low beam energies it is advantageous to have the detectors placed as close as possible to the target volume (called close geometry), in order to increase the detection efficiency and to observe a nearly angle-integrated reaction yield. Most commonly solid and gaseous targets are used in specially designed target chambers. In what follows we will discuss some general features of both types of targets as well as the chambers in which they are installed. For further information the reader is referred to the literature (Hol56, Ric60, Don67, Eng74).

5.2.1. Solid Targets

The making of targets is an extensive subject which cannot be described in detail here. The conversion of the target material into usable targets is accomplished by a variety of techniques. Solid targets are most commonly prepared by rolling from a pellet or a thick sheet, evaporating or sputtering in vacuum onto a backing or a substrate from which the film may be stripped, electroplating onto a backing which is then etched away, depositing from a powder or a powder-liquid suspension, or evaporating a solution or a colloid. These techniques and their many variations are discussed by several authors (Hol56, Ric60, Mug61, Yaf62, Low63, Eng68, Max67, Arn67, Kat69, Fre70). There are many pitfalls; for example, when chemical compounds are evaporated, one should not assume that the composition of the target will be the same as that of the original compound. As noted above, many materials can be made into self-supporting targets by evaporation onto a substrate such as NaCl or teepol and subsequently dissolving the substrate in water. For targets where this is not feasible, it is often possible to evaporate the material directly onto a very thin (10 μg cm^{-2}) previously mounted carbon foil. When very high beam currents are to be used, the targets have to be evaporated onto thick backings of material with high Z and with high melting points. These targets have to be effectively cooled to dissipate the incident beam power.

Having obtained a target, one must determine its thickness, the uniformity of thickness, and the target stoichiometry. Information on the target thickness can be obtained in a number of ways (Yan68, Kat69): the crystal oscillator gauge monitoring the evaporated layers on targets in vacuum, weighing of the target foils, measurement of the energy loss of charged particles in the target, and comparison of the measured yield with the expected yield from a reaction with a known cross section (e.g., elastic scattering at sub-Coulomb energies). Note that evaporating or sputtering onto backing materials gives in most cases a pinhole-free target but produces nonuniformities of the R^{-2} form, where R is the distance from the boat containing the material to be evaporated (or arc or electron-gun anode) to the substrate. Electroplating produces the most uniform targets with well-known thickness. However, this technique is applicable only to a limited range of target materials (Eng74). If the target material is radioactive, the radiation produced can be used, with a well-collimated apparatus scanning the target surface, to measure its uniformity (Buc84a). If the measurements of the elastic scattering yield are made at large angles, the target stoichiometry can be determined using the backscattering method (Chu75, Dec78). The chemical composition of the target for atoms heavier than Mg can also be measured via proton-induced X-ray emission (PIXE) (Dec78, Buc84a). The investigation of lighter atoms usually requires the use of nuclear reactions, which at the same time give a quantitative analysis of the isotopic composition (Buc84a). Isotopically enriched target material can be obtained for most species from various commercial establishments. Of course, the price of some such materials can be "astronomical," as in the case of ^{21}Ne enriched from 0.27% to 90%, which costs about $20,000 per liter (in 1985). Thus, efficient use (e.g., in recirculating gas targets; sec. 5.2.2.) is required. Note that quantitative analyses of the data obtained in several of the above techniques require information on the energy loss and straggling of charged particles in matter. In applications of nuclear reactions the energy loss and straggling of the projectile (and of the reaction products) are also parameters of prime importance, since it is through them that the nuclear information is related to the composition of the target material. A discussion of these features is thus in order.

Heavy charged particles moving through matter lose their energy through collisions with nuclei and with atomic electrons. The greatest part of the energy loss of medium-energy ions occurs by collisions with electrons (Fan63). The accuracy of stopping power data available in updated compilations ranges between 2% and 10% (And77, Zie77). The term "stopping power" usually means the quantity dE/dx, which is the ratio of the energy change to the quantity of matter traversed by the particle. Since dE is negative and dx is, by definition, positive, this quantity is negative. For this reason the stopping power $S(E)$ is defined by

$$S(E) = -dE/dx ,$$

where E is the particle energy (in eV) and dx is the thickness (usually in g cm^{-2}) producing the energy change dE. Another parameter used for calculating energy losses is the stopping cross section ϵ, which is defined by

$$\epsilon = dE/d\rho .$$

It is given in units of energy loss per unit areal density (atoms cm^{-2}), so that ϵ corresponds approximately to the energy loss for a monolayer (eV atoms^{-1} cm^{-2}). The quantity ϵ is thus independent to a large extent of the specific properties of the material. For a pure element the conversion between the two parameters is obtained through the relation

$$\epsilon = 1.66 \times 10^{-24} AS(E)   \text{eV atoms}^{-1} \text{cm}^2 ,$$

where A is the atomic weight of the sample material. Figure 5.15 shows stopping-power cross sections for protons and ^{4}He ions in Al. As seen in the figure, at low energies the stopping power increases with energy to some maximum value, after which it drops, with increasing projectile energy. This energy dependence is a general feature of all projectile-target combinations and differs mainly in the magnitude of energy loss and the location of the maximum. Above the maximum energy loss it is found that, to a good approx-

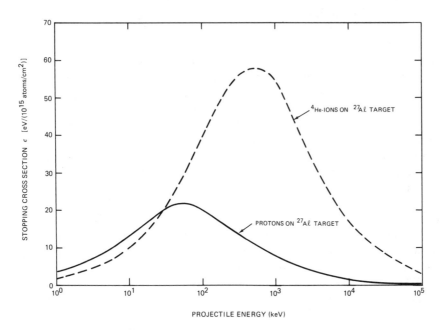

FIGURE 5.15. Stopping-power cross section ϵ for protons and ^{4}He ions in samples of pure aluminum (Zie77, And77).

imation, the stopping power scales (Bethe-Bloch formula; Fan63) quadratically with the atomic number Z_p of the projectiles ($dE/dx \propto Z_p^2$) and linearly with that of the target nuclei Z_t.

For stopping-power calculations a material containing various chemical elements can be treated in first approximation as a combination of atoms, which contribute separately to its stopping power (Fan63). Bragg's rule (Bra05) postulates the linear additivity of energy-loss cross sections for constituents of a compound. In the simple case of a diatomic compound $X_a Y_b$ the stopping cross section is given by

$$\epsilon = a\epsilon(X) + b\epsilon(Y) ,$$

where $\epsilon(X)$ and $\epsilon(Y)$ are the stopping cross sections corresponding to elements X and Y, respectively, and a and b are the stoichiometric coefficients. Similarly, the stopping power $S(E)$ is given by

$$S(E) = w_a S_x(E) + w_b S_y(E) ,$$

where w_a and w_b are the proportions by weight of both elements in the compound:

$$w_a = \frac{aA_x}{aA_x + bA_y} , \qquad w_b = \frac{bA_y}{aA_x + bA_y} ,$$

and A_x and A_y are the atomic weights of the constituents. These formulae are easily generalized to homogeneous mixtures containing more than two atoms. Bragg's rule is an approximation which neglects binding effects. Experimental investigations have shown that Bragg's rule holds for most compounds (Par63a, Pow73, Fen73, Bag74) to within the experimental uncertainty. Note also that the stopping-power formula assumes that for a given path length dx there corresponds exactly an energy loss dE. However, this is only an average value, since the slowing-down process is based on a large number of collisions in which the particle loses energy by discrete amounts which vary in magnitude. Statistical fluctuations of the number of collisions along the particle trajectory are then responsible for the effect known as energy straggling, i.e., unequal energy loss of identical particles traversing the same target under identical conditions. A monoenergetic particle beam of incident energy E_0 will—after passage through a target—be described (Fig. 5.16) by an energy $E = E_0 - \Delta E$ (ΔE = mean energy loss) with an energy distribution around E due to straggling effects (Bes80). For small energy losses the resulting distributions are asymmetric with a high-energy tail, and with increasing energy losses the shape of the distributions tends progressively toward a Gaussian curve (Dec78). In addition to energy loss and energy straggling, the incident particle beam is also characterized by an angular straggling (Fig. 5.16), i.e., a distribution of angles around the incident direction (Eng74). Finally, the path length $R(E_0)$ for which a particle loses all its energy E_0 in the target also displays energy and angle straggling. Thus, the range is distributed around a mean

(a) INTERACTION OF AN ION BEAM WITH MATTER

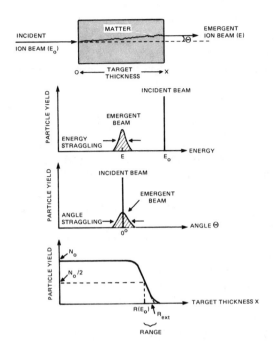

(b) INTERACTION OF PHOTONS WITH MATTER

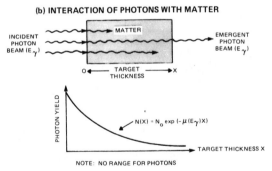

FIGURE 5.16. (a) When a well-defined and monoenergetic ion beam is traversing a target of thickness x, the beam experiences an energy loss $\Delta E = E_0 - E$. Statistical fluctuations in the number of collisions along the particle trajectory lead to energy straggling, angle straggling, and range straggling. (b) A photon beam interacts with matter quite differently, in that the photon either traverses the matter without any interaction or is completely absorbed in a single interaction. Thus the intensity of the photon beam decreases exponentially with the target thickness x, where the attenuation coefficient μ depends on γ-ray energy E_γ. Note that photons, unlike ion beams, have no finite range.

value $R(E_0)$, and extrapolated range values (Fig. 5.16) have to be used to effectively stop all the particles in the target (Dec78).

The yield of a resonant reaction is influenced by the rate at which the incident charged particle loses energy in traversing the target. If the target thickness is much greater than the width of the resonance (both given in energy units), the yield is called a thick-target yield (Y_∞) and is proportional to the number of effective target nuclei per square centimeter (chap. 4 and sec. 5.5.1):

$$Y_\infty = \frac{\lambda^2}{2} \frac{1}{\epsilon} \frac{M+m}{M} \omega\gamma \; . \tag{5.11}$$

For example, a 2 MeV beam of α-particles loses about 4 times more energy than a 2 MeV beam of protons in passing a given target thickness. Consequently, the number of effective target nuclei which can contribute to the resonant yield is 4 times smaller for α-induced resonant reactions than for proton-induced resonant reactions. In addition, the square of the de Broglie wavelength for a 2 MeV α-particle is one quarter of that for a 2 MeV proton, so that the resonant yield is reduced by an additional factor of 4. The lower yield in α-induced resonant reactions due to these effects requires the use of beam currents 16 times higher for the same count rates as in proton-induced resonant reactions. When the target consists of a chemical compound containing active (resonant) and inert atoms, then the resonant thick-target yield is reduced by the amount by which the inert atoms participate in the slowing-down process of the projectiles (Gov60 and sec. 5.5.1):

$$\epsilon_{\text{eff}} = \epsilon_a + \sum_i \frac{N_i}{N_a} \epsilon_i \; , \tag{5.12}$$

where N_a and N_i are the number of active and inactive atoms in the compound, respectively. In this situation it is desirable to choose a chemical compound in which the number of inert atoms is as small as possible compared with the number of active atoms. The inert atom should also be of low Z, because the rate of energy loss increases with Z. Unfortunately, very light atoms can also cause difficulties, because they can interact with the beam and contribute to the background radiation. In order to avoid this difficulty, it is necessary to use a target compound containing a heavy inert atom at the expense of lower yield for the resonant reaction of interest.

The choice of target thickness for a particular experiment is often difficult because proton- and α-induced reactions are usually very strongly resonant, with total widths ranging from μeV up to MeV. In addition, resonances are often close together or overlapping. Thus, an optimum choice of target thickness requires some prior knowledge of the details of the resonance structure of a particular reaction. Consequently, exploratory experiments with targets of different thicknesses are often necessary (Sch83). It is possible to specify a minimum target thickness by determining the energy spread of the ion beam (e.g., Fig. 5.12). There is no advantage in using targets much thinner (in energy units) than the energy spread (sec. 5.5.2). One might also decide upon a maximum target thickness. If the natural width of the resonance is known to be greater than the beam energy spread, then about 90% of the maximum thick-target yield can be obtained by using a target 6 times thicker than the natural width of the resonance (Fow48). These simple criteria can be modified in obvious ways according to the requirements of the particular experiment (sec. 5.5.2).

Studies of the radiation from nuclear reactions are frequently complicated by the presence of contaminants in the target material or in the backing (Don67). In the case of proton-induced capture reactions, the most troublesome contaminant is ^{19}F, which gives rise to 6 and 7 MeV γ-rays from the $^{19}F(p, \alpha\gamma)^{16}O$ reaction (Gov60). The cross section from this reaction is generally several orders of magnitude higher than for (p, γ) reactions, and thus even traces of ^{19}F are troublesome. Another troublesome contaminant is nitrogen. This is because of the high yield of the exothermic $^{15}N(p, \alpha\gamma)^{12}C$ reaction, which produces a 4.43 MeV γ-ray. All reactions of the type $(p, \alpha\gamma)$, $(p, n\gamma)$, $(p, p'\gamma)$, and so on, are sources of background γ-rays in the appropriate beam energy range. The study of (α, γ) reactions is hindered by the neutrons from the reactions $^{13}C(\alpha, n)^{16}O$, $^{17}O(\alpha, n)^{20}Ne$, and $^{18}O(\alpha, n)^{21}Ne$ and the γ-rays of the residual nuclei at the appropriate α-particle energies. In order to minimize contaminants in the target backing, it is necessary to clean and handle the backings carefully and then to test the backings under conditions similar to the conditions of the proposed experiment. The elimination of contaminants from target materials is usually more difficult, since each target material has to be treated as a special case. It is a general rule in studies of low cross sections that one spends nearly as much time studying, identifying, and reducing contaminants as investigating the nuclear reaction of interest. For example, the study of the $^{12}C + ^{12}C$ fusion reaction (chap. 8) was seriously hampered at sub-Coulomb energies by the ubiquitous hydrogen contamination in the carbon targets (Ket77, Ket80). In the γ-ray spectroscopy of this reaction, the carbon targets were evaporated on Ta backings, which were directly water-cooled to dissipate the power of the incident high-intensity ^{12}C beam. The spectra obtained were dominated by background radiation created by the interaction of the ^{12}C beam with the ^{1}H and ^{2}H contaminants (a few percent) in the target. Therefore, the production of carbon targets with low hydrogen content was necessary. As a first step, the hydrogen content and profile in the target and backing was investigated with a ^{19}F beam at the $E(^{19}F) = 6.41$ MeV resonance of the $^{1}H(^{19}F, \alpha\gamma)^{16}O$ reaction. After selection of a proper graphite material with low intrinsic hydrogen content and improved target evaporation techniques, it was found that all targets still had pronounced hydrogen content at the surface of the carbon layer. Subsequent investigations with other target materials revealed similar hydrogen content at the surface. These observations indicated that the water-cooled targets acted as a hydrogen trap in the vacuum system. Because of their low masses, hydrogen molecules represent the major component of the residual gas in a vacuum system. Subsequently, the targets were kept at a temperature of 100 °C by oil flowing behind the backing, after which the targets contained as little as 0.02% hydrogen.

Targets also have been produced by implanting ions into metal backings (Alm61, Wil73, Kei79, Seu85). For example, the study of the ^{14}N$(p, \gamma)^{15}$O reaction in the CNO cycle (chap. 6) is hampered by the ^{15}N$(p, \alpha\gamma)^{12}$C background reaction requiring ^{15}N-depleted ^{14}N targets. Nitrogen targets are usually produced by evaporating a thin layer of Ti onto a thick Ta backing and nitriding the Ti layer in a nitrogen atmosphere, thus forming a TiN target. Unfortunately, isotopically enriched ^{14}N gas (i.e., depleted in ^{15}N from the normal 0.36% abundance) is not commercially available (a rare case), and thus the above technique cannot be applied. As an alternative, ^{14}N ions of 50–100 μA with energies in the range of $E_{lab} = 55$–200 keV were implanted in 0.1 mm thick Ta sheets (Seu85). The target profile was investigated using the $E_R = 278$ keV resonant reaction ^{14}N$(p, \gamma)^{15}$O (natural width $\Gamma = 1.1$ keV). In Figure 5.17a is shown, as an example, the profile of a target obtained after a ^{14}N irradiation dose of 194 μg cm^{-2} with an energy $E_{lab}(^{14}$N$) = 200$ keV. As seen in the figure, the ^{14}N atoms are nearly homogeneously distributed from the surface of the Ta to a depth of 51.5 keV at half-maximum. This is consistent with the range of the ^{14}N ions in Ta. The shallower high-energy tail (as compared with the sharp front edge) is due to the effects of energy and range straggling in both the ^{14}N and the proton beams. The target depth Δ and the γ-ray yield Y_{max} at the plateau of the yield curve have been determined as a function of the amount of implanted ^{14}N. The results (Fig. 5.17b) show that the implantation reaches a saturation above a concentration of ^{14}N of about 50 μg cm^{-2}, which corresponds to a target stoichiometry equal to that of the compound Ta$_2$N$_3$ (see also Kei79). ^{14}N ions were also implanted into a thick Au layer evaporated onto a Ta backing. Results similar to those quoted above were observed, except that the implanted ^{14}N concentration at saturation was a factor of 4–5 smaller than for implantation into Ta. Thus, Au is much less suitable as an implantation material for ^{14}N ions than Ta. The ^{15}N in these targets was found to be reduced by 2 orders of magnitude from the natural abundance level. The targets are extremely stable against high-intensity beams.

Another example is the astrophysically important ^{12}C$(\alpha, \gamma)^{16}$O reaction measurement (chap. 7), which is hampered by the ^{13}C$(\alpha, n)^{16}$O background reaction. The problem of the enormous neutron-induced γ-ray background from this reaction was partially solved by using isotopically enriched ^{12}C targets and a pulsed beam. Time-of-flight techniques were used to discriminate between the prompt capture γ-rays from the target and the neutron-induced γ-ray background (Dye74). In another technique (Ket82) the roles of projectiles and target nuclei were exchanged, with an intense ^{12}C ion beam directed into a windowless ^{4}He gas target. For the measurement of γ-ray angular distributions at very low beam energies (chap. 7), these techniques have their limitations, and thus the use of ^{12}C targets highly depleted in ^{13}C represents the more viable solution (Red85). Such targets were produced (Seu85) by implanting a ^{12}C beam of $E_{lab} = 80$–130 keV into a thick Au layer evaporated onto a 1 mm thick Cu backing. The thickness of the Au layer was chosen such that

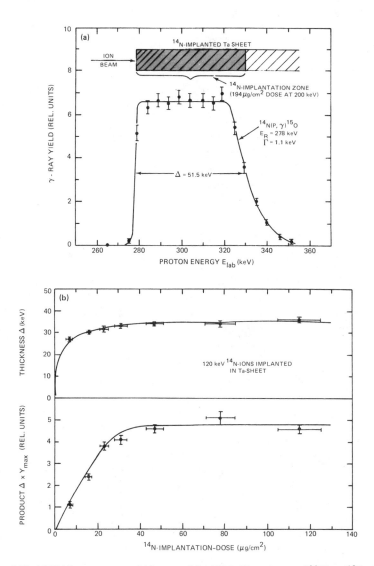

FIGURE 5.17. (a) Thick-target γ-ray yield curve of the 278 keV resonance of $^{14}N(p, \gamma)^{15}O$ obtained with a ^{14}N-implanted Ta sheet. (b) The implantation depth Δ and the product of Δ with Y_{max} (height of the plateau) are shown as a function of implantation dose for a 120 keV ^{14}N ion beam.

an α-particle beam of $E_{lab} \leq 4$ MeV did not reach the "dirty" Cu backing. The ^{12}C implantation was carried out with a series of LN_2-cooled shields, thus minimizing the deposition of carbon onto the target. The profile of the implanted carbon was investigated in situ, using the shape and intensity of the capture γ-ray transition of the $^{12}C(p, \gamma)^{13}N$ nonresonant reaction (i.e., Fig. 5.30). Results similar to those quoted above were found, and targets with concentration ratios of ^{12}C:Au = 5:1 in the implantation zone were produced.

Study of the γ-ray spectroscopy of a nuclear reaction is facilitated by the small attenuation of γ-rays in passing through matter. For example, a sheet of Ta 0.5 mm thick, which is commonly used as a target backing, has approximately 90% transmission for 500 keV γ-rays and 96% transmission for 5 MeV γ-rays. The target materials themselves rarely play a significant role in the absorption of the γ-rays. This is because target thicknesses in the range from 10 to 100 μg cm^{-2} are commonly used. The low attenuation of γ-rays with energy over 500 keV by most materials makes it possible to place the γ-ray detectors outside the (thin stainless steel or brass) vacuum chamber containing the target. This is fortunate and convenient because of the complexity of the Ge(Li) and NaI(Tl) counters used to detect the γ-rays. For studies of γ-rays below 500 keV, special low-Z materials must be used in constructing target chambers. A typical target chamber assembly for γ-ray spectroscopy investigations (Tra75) is shown in Figure 5.18a. In designing a target chamber, several important features must be incorporated. There should be a minimum amount of material between the target and the γ-ray detector to minimize absorption of γ-rays. Cylindrical symmetry for the vacuum envelope is in order when measuring γ-ray angular distributions. This reduces the variation of γ-ray absorption with angle of observation. The low cross section at sub-Coulomb energies requires the use of ion-beam currents over 10 μA and sometimes as high as several hundred μA in order to improve the signal-to-noise ratio. Thus, the power dissipation in the target backing is frequently several hundred watts, so cooling is essential for most target materials. A successful form of cooling at moderate beam loads is the direct cooling of the Ta backing (high melting point) shown in Figure 5.18a. Care must be taken, however, to ensure that there is an adequate flow of water behind the target during an experiment; otherwise the consequences are as spectacular as they are inconvenient. For high beam loads, of the order of a kilowatt, special designs are necessary, including materials of high heat conductivity. As with the beam catcher (sec. 5.1.5), the coolant must be in a state of laminar flow without the presence of gas or stream bubbles. Such a design (Ham75a), which has been used successfully for beam loads up to 8 kW, is shown in Figure 5.18b. The ^{12}C-implanted targets (see above) using such a cooling design are found to be stable except for normal sputtering by the incident beam. The buildup of carbon deposits on targets bombarded with charged particles is a well-known problem. Such a buildup not only degrades the energy of the incident beam before it can strike the true target surface but also introduces a spread in the

(a) SET-UP FOR γ-RAY STUDIES WITH HIGH BEAM POWER

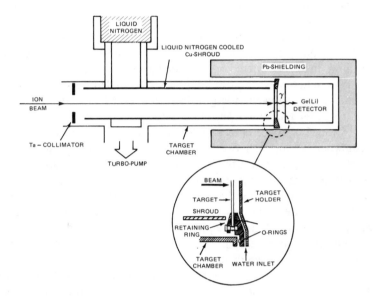

(b) TARGET DESIGN FOR KILOWATT BEAM POWER

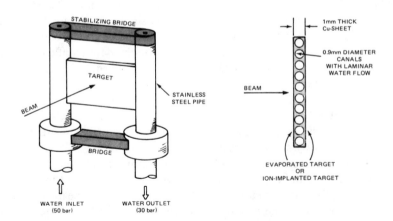

FIGURE 5.18. (*a*) Conceptual drawing of the experimental arrangement for γ-ray spectroscopy of a nuclear reaction. The ion beam is stopped in the target backing, which is directly water-cooled at a pressure of a few bars. A LN_2-cooled Cu shroud extends close to the target. The γ-rays are observed with a Ge(Li) detector in close geometry to the target and surrounded by a heavy lead shielding to suppress the contribution of room background. (*b*) Target design for beam powers in the range of kilowatts.

beam energy due to straggling. Both of these effects are important because of the previously noted requirement for an accurate determination of the energy distribution of the beam. The prevention of the buildup of carbon on the target is particularly important in studies involving (α, γ) reactions, since the ^{13}C component in natural carbon will give rise to very serious neutron background resulting from the $^{13}C(\alpha, n)^{16}O$ reaction. These neutrons interact with the surrounding material, including the materials of the γ-ray detectors, producing γ-rays (Cha65). Carbon buildup can be avoided very effectively by working with a very good vacuum and using LN_2-cooled shrouds in the beam line, particularly immediately in front of the target and in the region surrounding the target (Spe63). Also, care should be taken to minimize the use of materials containing volatile hydrocarbons in the vacuum system near the target chamber. The shroud and the Ta collimator shown in Figure 5.18a are insulated electrically, facilitating the procedure of beam alignment by minimizing the beam current striking them. An accurate measurement of the beam current on the target, and hence of the accumulated charge, requires the suppression of the secondary electrons emitted from the target under beam bombardment conditions. This is accomplished by operating the insulated shroud at a sufficiently high negative voltage.

At high beam energies (i.e., proton energies above 3–4 MeV) the target chamber shown in Figure 5.18a becomes less useful because the interaction of the beam with the heavy-element backing leads to increased background. The solution to this problem is, of course, elimination of the target backing altogether.

Thin self-supporting targets require a target chamber (Dah60, Wal62, Cos64, Don67, Eng74) different from the one shown in Figure 5.18a. In this case, the beam must be allowed to pass through the target and be guided to a shielded beam catcher. Thin solid targets are unfortunately difficult to handle, and high beam currents readily destroy them. An alternative to a fragile, thin solid target is the indestructible gas target.

5.2.2. Gas Targets

In addition to solid targets, it is often desirable and possible to use gas targets, which, compared with solid targets, have many advantages. One advantage is chemical purity, especially for noble gases but also for other gaseous elements and compounds. One way of making gas targets is to contain the gas within a cell with ultrathin foil windows (Fig. 5.19a), allowing the beam to enter and exit and also allowing the reaction products to exit and be detected. As might be expected, the most critical parts of this system are the foil windows. The entrance foil causes not only a significant energy loss of the beam but also a deterioration in the beam quality due to straggling. Another problem is the possible background resulting from reactions taking place in the entrance (and exit) foil. The energy lost by the beam in the foils causes heating, weakening, and finally rupturing of the foils, thus limiting the gas pressure (up to several

(a) GAS TARGET WITH FOIL WINDOWS

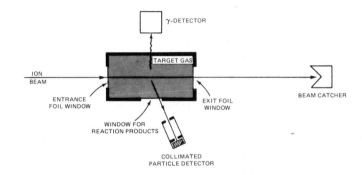

(b) WINDOWLESS DIFFERENTIALLY PUMPED GAS TARGET

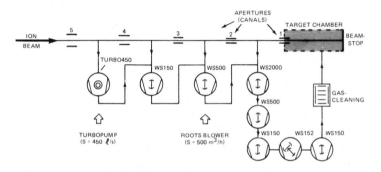

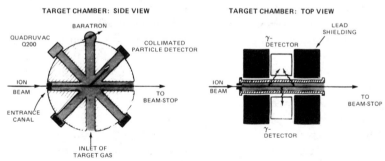

FIGURE 5.19. (*a*) Experimental arrangement of a gas target with foil windows for the passage of the ion beam and of reaction products. (*b*) Conceptual drawing of a windowless differentially pumped gas target (Rol78, Bec82) is shown. The ion beam enters the target chamber through several apertures (canals) of high gas flow impedance. The system has four pumping stages consisting of several Roots blowers (the WS pumps in the figure, e.g., WS2000 with a maximum pumping speed of 2000 m^3 h^{-1} $\simeq$ 560 l s^{-1}) and one turbomolecular pump, which are interconnected to allow recirculation of gas back into the target chamber. The design of a target chamber for γ-ray spectroscopy measurements in close geometry is also shown schematically.

bars) and the usable beam intensities. Typically this latter limitation is about 1 μA for 1 MeV protons. For other particles and energies this limit scales roughly as the inverse of the stopping cross section involved. An ever-present source of trouble is multiple scattering and energy degradation of scattered particles or reaction products in the exit window or windows. This can be reduced by suitable design but can be eliminated only by using a windowless gas target (Fig. 5.19b). Another problem is the protection of the vacuum system in the event of failure of the target window or windows. This usually can be achieved using fast slammer valves (Hah70).

Since it is important to know the energy of the beam in the gas cell, it is necessary to obtain an accurate measurement of the foil thickness and the energy lost by the projectile in the foil. The most direct way to accomplish this is to measure the energetic position of a well-known resonance with and without the foil inserted in front of the target. This measures the energy loss in the foil directly. Using stopping cross sections, this loss can be scaled to any other incident energy for this same foil. An even simpler way is to measure the energy loss of transmitted α-particles from calibrated radioactive sources. Furthermore, if the resonance is narrow (or the energy resolution of the α-particle detector is high), such a measurement can provide information regarding the straggling of the beam in the foil.

To avoid the disadvantages associated with the presence of the thin foil windows, windowless gas targets have been developed in which the window is replaced by one or more apertures through which the beam passes (Sil61, Par64, Gob66, Blo67, Lit67, Bus71, Dia71). These windowless gas targets are known as differentially pumped gas targets and, because of the many stages of high pumping speed required to lower the gas pressure from about $p_0 = 1$ torr in the target chamber to 10^{-6} torr outside, are quite complex. It is of the utmost importance to ensure that this reduction in gas pressure takes place over as short a distance as possible. If the background radiation is to be kept as low as possible, it is of course necessary to ensure that the ion beam does not hit any of the apertures (canals). For this reason, the apertures (in particular near the target chamber) must be larger than the ion-beam diameter. This means large gas flows requiring high pumping speeds. To achieve the pumping speeds necessary for typical target pressures of $p_0 = 0.02–50$ torr, powerful Roots blowers have been used (Dwa71, Rol78, Bec82). In other designs cryo-pumping at liquid helium temperatures has been used as the central pumping stage (All76, Eva82). In many experiments it is necessary to use separated isotopes as targets, which can be expensive (e.g., ^{17}O or ^{21}Ne). Consequently, the high-pressure region of a gas target system should be as small as possible and the target gas charge has to be recirculated continuously. Also, care has to be taken to be able to recover the expensive gas at the end of an experiment as well as to store it temporarily if repairs or exchanges of defective components are required during an experiment. The engineering details of such systems are described in the references quoted.

In the gas target system shown in Figure 5.19*b* the ion beam enters the target chamber through canals of high gas flow impedance and is stopped in a Faraday cup. The main pressure drop occurs across the entrance canal of the target chamber. For hydrogen and helium target gases with $p_0 \leq 2$ torr in the target chamber and for heavier target gases with $p_0 \leq 50$ torr the observed pressure at the beam entrance side of canal 5 (Fig. 5.19*b*) is not influenced by the target gas pressure. This four-stage recirculating pumping system should prevent any measurable gas leak back through canal 5. Canal 2 (Fig. 5.19*b*) has the smallest diameter and thus defines the beam profile. It is directly water-cooled and electrically insulated for beam transmission measurements.

Inevitably, gas recirculation introduces volatile contaminants into the target gas from traces of pump-oil vapor and air leaking into the system. With specially selected and prepared Roots blowers the external leakage rate of the total gas target can be smaller than 5×10^{-9} torr l s^{-1}. The internal leakage can be reduced to 5×10^{-8} torr l s^{-1}, which leads to a contamination of the target gas by about 0.01 torr h^{-1}. A liquid nitrogen–cooled trap in the backing line (for gases such as H_2, He, CO, Ne, and Ar) can further reduce contamination of the target gas. For the noble gases He and Ne, the recirculating gas can be purified extremely well by passing it through a zeolite adsorption trap at LN_2 temperature. Very efficient gas-cleaning elements for removing contaminants such as H_2, N_2, and O_2 have also been developed (Wol83). The cleaning of each target gas has to be treated as a special case. Contamination of the target gas can be monitored via elastic scattering of the projectiles from the target gas or by a mass spectrometer (Fig. 5.19*b*) using a special pressure-reducing valve (Hul82).

The disk-shaped target chamber shown in Figure 5.19*b* has several ports radiating from the center of the chamber. These ports allow one to install well-collimated particle counters for the combined measurement of the beam intensity and target density via elastic scattering of the projectiles from the target gas nuclei. These measurements also allow one to test the purity of the target gas. Other ports are reserved for the inlet of the recirculating gas, the Baratron pressure manometer and the mass spectrometer Q200. The chamber allows for γ-ray spectroscopy measurements in close geometry (Ket82). In order to reduce the contribution of beam-induced background, there is no exit canal of the target chamber, and the beam-stop pipe, surrounded by heavy shielding, is part of the target chamber. The effective target length seen by the γ-ray detectors is then defined by the additional lead shielding around them (Figs. 5.19*b* and 5.45). Clearly, this target chamber cannot be used for γ-ray angular distribution studies, since the target is not well localized in space.

For a target chamber of 50 cm^3 volume and for a gas pressure of 2 torr in the chamber, a total gas charge of approximately 25 cm^3 STP (see Appendix) is required for the system shown in Figure 5.19*b*. This small gas charge allows for the use of expensive isotopically enriched gases. The gas pressure in the target chamber can be measured absolutely to a high precision, and independently of the type of gas used, with a capacitance manometer of the Baratron type. With a gas target one can easily vary the target thickness and determine its numerical value directly from a measurement of the gas pressure and gas temperature. In the extended gas target chamber of Figure 5.19b the gas pressure in the observation region is essentially unmodified by the gas flow through the entrance canal of the target chamber, and the geometrically extended target zone is characterized by a nearly static gas pressure. However, in determining the gas temperature one must take into account beam-heating effects in the gas, which may raise the local temperature along the beam path (Rob61). The influence of intense ion beams on the densities of quasi-static gas targets has been studied (Goe80) in the experimental arrangement shown in Figure 5.19b. Gas target densities along the beam axis depend on the beam diameter, the energy loss per unit length, and the ion-beam current, i.e., the dissipated power in the gas, which raises its temperature. Effects greater than 30% have been seen for dissipated powers greater than 60 mW mm^{-1} (Fig. 5.20). As a consequence, quasi-static gas targets may become less useful under conditions of high dissipated power, depending on specific details.

As noted above, reliable and precise measurements of angular distributions cannot be performed easily with extended gas targets. Therefore, when such

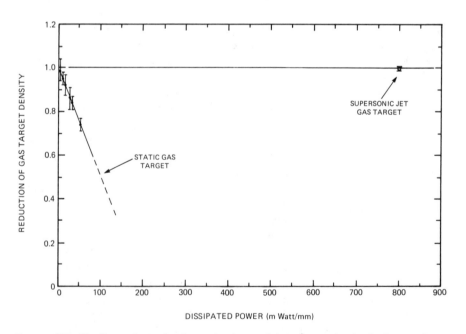

FIGURE 5.20. The figure shows that for an ion beam of 1 mm^2 area the density in a static gas target along the ion-beam path is reduced with increasing dissipated power in the gas (Goe80). In comparison, no effects have been observed in a supersonic jet gas target for dissipated powers up to 800 mW mm^{-1} (Goe85a).

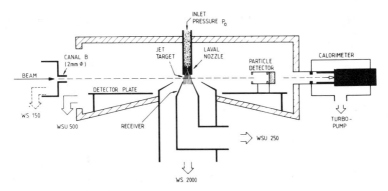

FIGURE 5.21. Schematic diagram of the jet target chamber for particle spectroscopy measurements (Bec82, Red82). The beam axis is defined by canal B in front of the jet target. In this arrangement the jet target zone can be seen from every scattering direction.

information is required, a nearly pointlike gas target of sufficiently high density is required. Such conditions can be fulfilled by a supersonic gas jet. A review of the theoretical and experimental aspects of supersonic jet gas targets as well as the engineering development of such a system (including gas recirculation) and experimental tests of its properties are described by Becker and coworkers (Bec82). In the supersonic jet system shown in Figure 5.21, target gas at a high inlet pressure p_0 (several bars) flows through a Laval nozzle (1 mm diameter at the neck) and expands adiabatically into a vacuum chamber in which the static pressure is several orders of magnitude lower. For such conditions the thermal energy of the gas is transferred to a large extent into kinetic energy of the collective gas flow, forming a geometrically confined supersonic jet stream. After a free expansion of 6 mm, the jet stream is mostly captured by an appropriately formed receiver pumped by a WSU 250 Roots blower. In this arrangement the highest-density region in the supersonic jet near the outlet of the nozzle is chosen as the target zone. The gas outside the target region is pumped differentially using Roots blowers and turbo pumps (Fig. 5.19b). The gas compression up to the inlet pressure p_0 is achieved with the combination of Roots blowers and metal bellows compressors. The adjustable Laval nozzle is located at the center of the scattering chamber.

Information on the radial density distribution in the supersonic jet can be obtained, for example, from proton elastic scattering data. The results of such measurements for N_2 gas at an inlet pressure of $p_0 = 2.3$ bars are illustrated in Figure 5.22. The jet is seen to have a FWHM of $l_j = 3.0$ mm. Similar jet widths were observed for other inlet pressures, demonstrating that the geometrical shape of the supersonic jet flow does not depend on the pressure p_0 (Bec82). At the standard inlet pressure $p_0 = 2.0$ bars (requiring a gas charge of about 0.7 l STP), a target thickness of $N_T = 3.7 \times 10^{17}$ atoms cm^{-2} was measured and found to scale linearly with the inlet pressure p_0 (Red82). Similar

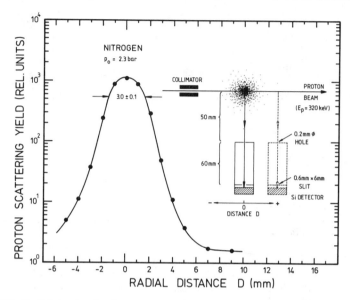

FIGURE 5.22. Radial density distribution of the supersonic jet stream at the selected target area (near the outlet of the Laval nozzle) is shown for nitrogen gas at an inlet pressure of $p_0 = 2.3$ bars (Red82). The data have been obtained via proton elastic scattering as observed in a scanning setup (*inset*). The density distribution drops steeply at the edge of the supersonic jet stream to the ambient density in the target chamber. The line through the data points is to guide the eye.

results were obtained for other light and heavy target gases, using a variety of techniques. It was also verified experimentally that the jet targets developed for particle and γ-ray spectroscopy measurements had acceptable quasi-point qualities, i.e., the yields from regions outside the jet-stream zone are negligible. The compression of the target gas in the recirculating system can be carried out up to several bars, and thus target thicknesses comparable to those of solid targets can be achieved. As one might have expected from the principles of a supersonic gas flow (Bec82), effects of dissipated power on the jet gas density have not yet been observed (Goe85a), at least up to values of 800 mW mm^{-1} (Fig. 5.20). Thus, for high beam intensities and/or gas densities, a supersonic gas jet target is probably superior to a quasi-static gas target.

The complexity in the design of gas targets is rewarded by several advantages over solid targets. (1) Since the target is effectively industructible, high beam currents can be used (at extreme conditions of dissipated power, a jet target may be preferable). (2) The thickness of the target can be readily controlled by varying the gas pressure. This is important at energies far below the Coulomb barrier, where thin and uniform targets are required (Red82). There is also an advantage if the resonance structure of a given reaction is complex

(Sch83). (3) The problem of carbon deposition on the target and the accompanying increased background radiation does not exist. (4) In many cases chemically pure gases can be used (e.g., neon), so that the maximum yield per reacting target nucleus is obtained since no inactive target elements are involved. An example of the high sensitivity of a gas target is demonstrated in studies of the ^{21}Ne$(p, \gamma)^{22}$Na reaction using neon gas of natural isotopic abundance. In these studies resonances were detected even though the active target nuclei comprised only an extremely small fraction of the gas (0.27% ^{21}Ne) (Rol78, Geo82). (5) By mixing various gases (Hul82), one can carry out relative measurements with high precision. For example, by using a mixture of ^{4}He and ^{20}Ne, the energy dependence of the elastic scattering yield for the ^{12}C + ^{4}He system was measured (Ket82) at $E_{lab}(^{12}$C$) \leq 14$ MeV relative to that of the ^{12}C + ^{20}Ne system, where the latter system is known to follow the Rutherford scattering law at the ^{12}C energies involved. (6) The use of an extended transmission-type gas target permits the measurement of angle-integrated yields in close geometry (Ket82) with a high signal-to-noise ratio, since the detectors can be well shielded from beam-intercepting apertures and the beam can be stopped far away from the detection region. Thus, the extended gas target is especially suitable for low cross section measurements. (7) Supersonic jet gas targets have somewhat higher beam-induced background, owing to the closeness of the nozzle and the receiver to the beam axis. They are, however, advantageous for angular distribution measurements (Red82, Krä82). They are also desirable, for instance, where high energy resolution is desired (energetic location and width of resonances [Sch83]; reduction of the Doppler broadening due to the thermal motion of the target atoms; sec. 5.5.2). (8) Another useful feature of a gas target is that background effects can be identified by using different target gases, and they may readily be subtracted using a "gas-in–gas-out" difference method (Ket82). (9) A useful feature of a gas target is that—because of the absence of a backing—the products of a given reaction and the elastically scattered beam particles often can be observed concurrently in a particle detector. This is of particular interest at sub-Coulomb energies, where the elastic scattering yield follows the Rutherford law. Thus, the absolute value of the differential cross section for the reaction is obtained to high accuracy relative to the Rutherford cross section simply from the ratio of count rates of the relevant peaks in the spectrum. (10) Often the background problem of a given reaction can be solved by exchanging the roles of projectiles and target nuclei. An example is the ^{12}C$(\alpha, \gamma)^{16}$O reaction (chap. 7), whose study is hampered by the neutron-induced γ-ray background resulting from the ^{13}C$(\alpha, n)^{16}$O reaction. This background reaction was avoided by bombarding a ^{4}He gas target with a ^{12}C beam (Ket82). This technique of interchanging the roles of projectiles and target nuclei in combination with gas targets has also been used for absolute cross section measurements (Bec82a, and sec. 5.5.6).

5.3. Detectors

An important milestone in the history of nuclear physics was the observation by Becquerel in 1896 (Bec96) of the blackening of a photographic plate in the presence of the salts of uranium. This event provided both the first observation of penetrating radiations and the first method for detecting them. The importance of detectors and methods of detection for nuclear physics research was immediately realized, and an enormous variety of detectors with ever-improving characteristics have since been developed.

As previoulsy noted, at sub-Coulomb energies nuclear reactions have very low cross sections, thus requiring high beam intensities in combination with stable and pure targets. Of equal importance in such studies is the availability of detectors to observe the γ-rays, neutrons, and charged particles resulting from the projectile-target interactions. All detectors operate on the effects caused by the passage of the radiation through matter. Although some such interactions are nuclear, the effects coming from the interaction with the atomic electrons predominate, and the results are governed by atomic physics. As noted in section 5.2.1, heavy charged particles passing through matter experience many interactions. In each interaction they lose a small amount of energy and are scattered through a small angle, i.e., they move almost in a straight path. The energy loss occurs mainly through long-range Coulomb interactions with bound atomic electrons in the material, in which the electrons are raised to higher energy levels (excitation) or completely ejected from the atom (ionization). The latter process dominates if the incident particle has an energy large compared with atomic binding energies; thus, as the particle traverses the matter, its ionization track contains essentially equal numbers of electrons and positive ions. Because of the energy loss the particle will stop after traveling a well-defined distance, the so-called range R (Fig. 5.16a). The higher the particle's energy, the longer the range. Such range-energy relations for several particles in silicon (Zie77) are shown in Figure 5.23a. From the figure one can see that the range depends on the mass, the charge, and the energy of the particle. For example, a 10 MeV α-particle will be stopped by about 0.07 mm of silicon, while 10 MeV protons require a thickness of about 0.7 mm. The relationship between the range and energy is given fairly well by the equation (Gou74)

$$R = aE^b ,\qquad(5.13)$$

where the coefficient a varies with the particle and to a lesser extent with the absorbing material. The exponent b has a value of about 1.7 but also varies slowly with the type of particle and with energy. Electrons lose energy by the same mechanism as do heavy charged particles. However, since these interactions are between particles of equal mass, large energy exchanges result in scattering through large angles. As a consequence, the resulting ionization track is not straight but rather is a complicated irregular trajectory. Thus, only

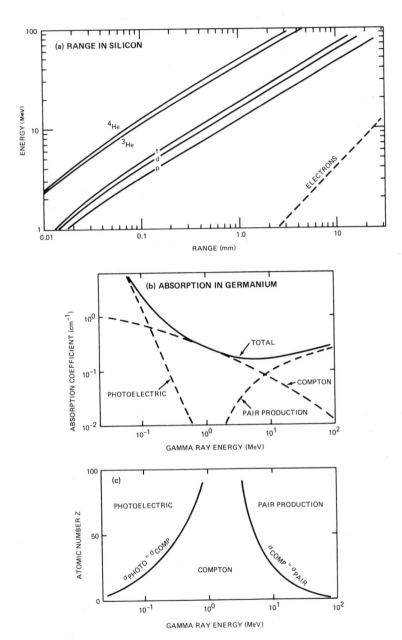

FIGURE 5.23. (a) Range-energy curves for several light particles as well as for electrons in silicon (Zie77, Ber64). (b) Photon absorption coefficients versus energy for germanium (Mar68). (c) Photon interactions with matter via the photoelectric effect, Compton effect, and pair production depend on γ-ray energy E_γ and atomic number Z of the absorbing material, leading to regions in the (Z, E_γ)-plane where each of these effects predominates.

extrapolated values for the electron range are possible (Ber64). Because of its low mass, a 10 MeV electron moves rather rapidly through matter, interacting less effectively than do 10 MeV protons. Thus, to stop 10 MeV electrons in silicon, a thickness of about 25 mm is required (Fig. 5.23a). At very high electron energies ($E \geq 600$ MeV/Z) and for high-Z stopping atoms, bremsstrahlung (also called synchrotron radiation) becomes an important energy-loss mechanism (Bet34).

Neutral particles such as neutrons and high-energy photons (γ-rays) are detected through intermediate interactions in which a charged particle is released which is then detected as discussed above. In a way, these neutral particles either pass through a slab of matter unscattered or are eliminated in a single encounter. It follows, then, that the intensity of a collimated beam of such neutral particles will decrease exponentially with the thickness x of the absorbing material (Fig. 5.16b). The number of particles (γ-rays and high-energy neutrons, but not thermal neutrons) remaining after the beam travels a distance x is given by

$$N(x) = N_0 \exp(-\mu x),$$

where μ is called the absorption coefficient and has the dimension of inverse distance, and $1/\mu$ corresponds to the mean free path length of the radiation in the material. The absorption coefficient is given by the relation

$$\mu = \frac{N_A \rho}{M} \sigma, \tag{5.14}$$

where N_A is Avogadro's number, ρ is the density, M the atomic weight of the absorber element, and σ the cross section for the stopping interaction (for compilations of μ for γ-rays see Sto70). Slow neutrons, for example, are usually detected through the ionizing properties of the α-particles and tritons released in the nuclear reaction ^{6}Li$(n, \alpha)t$. Photons interact with matter via three processes: the photoelectric effect, the Compton effect, and pair production. In the photoelectric effect, a photon of energy E_γ is entirely absorbed by an atom, and an electron is ejected (predominantly from the K shell) with a kinetic energy $E_e = E_\gamma - E_B$, where E_B is the binding energy of the ejected electron. In the Compton effect, the photon scatters from an atomic electron, giving the electron a significant fraction of its energy with a maximum energy transfer of $E_e(\text{max}) = E_\gamma/(1 + m_e c^2/2E_\gamma) \simeq E_\gamma - m_e c^2/2$ for $E_\gamma \gg m_e c^2$. Finally, if $E_\gamma \geq 2m_e c^2 = 1.022$ MeV, pair production may occur, in which case the photon is converted into an electron and a positron. The electron and positron share the γ-ray energy in excess of the 1.022 MeV required to create them. The energy dependence of these processes is quite different, as shown in Figure 5.23b for absorption in germanium. For example, at low photon energies, below 100 keV, the photoelectric effect dominates, the Compton effect is small, and pair production is energetically impossible. In addition to the dependence on photon energy, the cross section for these processes also depends on the atomic number Z of the absorbing material ($\sigma_{photo} \propto Z^5$, $\sigma_{Com} \propto Z$, $\sigma_{pair} \propto Z^2$). As can be seen from Figure 5.23c, at high Z-values the photoelectric effect and pair production are the dominant absorbing mechanisms over a wide range of photon energies, while for small Z-values the Compton effect predominates. Due to the Z-dependence and the density dependence of the absorption coefficient, photons are absorbed most effectively in solid materials with high densities and high atomic numbers Z.

A perfect detector might have the following characteristics (Seg77): (1) 100% detection efficiency at all energies, i.e., a probability of unity for detecting the radiation entering the detector whatever its energy; (2) reasonably large acceptance angles without loss of directional information; (3) sensitivity to all types of nuclear radiations and unique identification of their characteristics such as energy, mass, and charge; (4) insensitivity to background radiation; (5) high count rate r. After detecting an event, the instrument loses its sensitivity for a period known as the dead time t_d. The length of this period sets a limit on the number of events per unit time that an instrument can count. For a perfect detector the fraction of the total time the counter is inactive must be small compared with the interval between events ($rt_d \ll 1$); (6) accurate determination of the time at which an event occurs (say within 1 ns), which is referred to as the time resolution of the detector; (7) good energy resolution, allowing a precise measurement of the energy of the incident radiation and discrimination among particles of almost the same energy; and (8) reasonable cost. Obviously perfection is not of this world, and for a specific experiment requirements often conflict and compromises are necessary in the choice and design of detectors. For example, if high energy resolution is required in a γ-ray detector, a price is paid in low detection efficiency.

An important ingredient in the development of detectors was the availability of auxiliary electronic equipment. These devices provided for objective and precise handling of data as well as fast readout of the detector events. They are today an integral part of any detection system (sec. 5.4.1). In recent years the handling, analyzing and storing of the data have increasingly been carried out using computers. Computers have made possible the recording and analyzing of data from very complex experiments, such as data from coincidence studies employing many detectors. As noted, earlier, the availability of improved experimental techniques frequently determines the direction research follows as well as the rate at which it proceeds.

In what follows we will discuss briefly the detectors most commonly used in experimental nuclear astrophysics. For details of these detectors as well as for the description of other detectors, the reader is referred to the extensive literature (Sie68, Eng74, Pau74, Gou74, Seg77, Bro79, Kno79).

5.3.1. Detectors for Heavy Charged Particles

Ionization chambers (Fig. 5.24a) are among the oldest of nuclear detectors and are extremely versatile. An ionization chamber is a vessel containing some dielectric substance, usually a gas, atoms of which are ionized by charged particles passing through. Only part of the energy of the incident particle goes into ionizing the gas and producing free electrons. A sizable fraction is spent in promoting positive ions or neutral atoms to higher states below their ioniza-

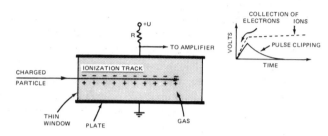

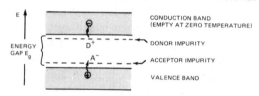

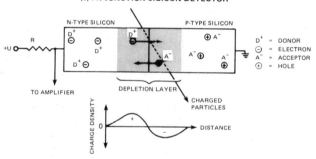

FIGURE 5.24. (a) Schematic representation of an ionization chamber with parallel plates. The comparatively high velocity of the electrons allows them to be collected in a short time compared with the slowly moving positive ions. To provide for high count rates, the contribution of the positive ions to the pulse height is cut off by an RC network. (b) Energy-level diagram for a semiconductor with impurities that can, by thermal excitation, donate electrons to the conduction band or capture electrons from the filled valence band, leaving mobile holes there. (c) In the region of the junction of an N-type and a P-type silicon semiconductor, a layer depleted in charge carriers is created, which is characterized by a space-charge distribution creating an electric field across the layer. An applied reversed-bias voltage increases the depth of the depletion layer. The detection of charged particles in this layer is similar to that of the ionization chamber.

tion energy, and some of it is transformed into detectable scintillation light. The average amount of energy w required to form an ion is remarkably independent of the charge, mass, and velocity of the ionizing particle, but it does depend on the gas which is ionized. For example, it takes an average energy of 26 eV to form an ion pair (positive ion and electron) in argon gas, which—because of the effects mentioned above—is higher than its 16 eV ionization energy. If a charged particle of, say, $E = 1$ MeV is stopped in the chamber filled with argon gas, it will form approximately $n = E/w = 3.8 \times 10^4$ ion pairs along the ionization track, or an electric charge ne equal to 6.1×10^{-15} C. An electric field separates the ion pairs, and the total charge is collected and measured. An ionization current flows through a high resistance R (say, 10^8 ohms), producing a voltage pulse which is amplified by a high-gain linear amplifier to give

$$V = \alpha \frac{ne}{C} = \alpha \frac{e}{C} \frac{E}{w},$$

where α is the voltage gain of the amplifier and C is the capacitance of the chamber plus the input capacitance of the amplifier (typically a few tens of picofarads). Thus, if a charged particle dissipates all its energy in the gas of the ionization chamber, its energy can be determined from the magnitude of the voltage pulse V. In practice a relatively short charge-collection time is achieved by collecting only the electrons, which have velocities about 1000 times greater than those of the positive ions. A short time constant circuit of $CR = T = $ a few milliseconds, usually incorporated in the early stages of the low-noise amplifier, clips the pulse so that the slower pulse due to the positive ions is removed (Fig. 5.24a). The price paid is that the resulting output voltage V depends upon the site of the ionization track. This problem can be obviated by introducing an auxiliary grid close to the anode, referred to as the Frisch grid (Eng74, Seg77).

There are many possible geometrical arrangements for ionization chambers, which can be matched to the requirements of a specific experiment. For example, a gas-filled cylindrical chamber with a central thin wire as the anode is used. If the applied electric field is sufficiently high, the electrons gain enough energy during the collection process to cause further ionization. In this case the charge collected in the external circuit is still proportional to the energy of the ionizing particle, but is much greater than the original total ionization charge produced by the particle. Such instruments are referred to as proportional counters. If the electric field is high enough, a uniform charge output can be produced which is independent of the initial ionization. Such detectors are known as Geiger-Müller counters and are of course not useful for energy measurements.

One advantage of ionization chambers is that they can be built in large sizes, thus providing large acceptance angles. Directional information on the incident particles can also be provided if the anode of the chamber consists of a multitude of wires (Eng74). These instruments have a 100% detection efficiency, and the energy of the particles can be determined provided they are stopped in the gas. For example, in an air-filled chamber at standard temperature and pressure (STP) a 10 MeV proton has a range of about 1 m, while a 10 MeV α-particle is stopped in only about 10 cm. Thus, the spectroscopy of high-energy protons requires either rather large ionization chambers or gas at high pressure. The particles have to enter the chamber through thin foils, thus setting limits on the gas pressure used, particularly for foils of large area (i.e., large solid angles). As a consequence, ionization chambers are used most commonly for spectroscopic studies of short-range particles, such as heavy charged particles with relatively low energies. They are also used as energy-loss detectors (ΔE detectors) in telescopes designed for particle identification (sec. 5.4.1), especially if large solid angles are required. By suitable choice of the gas, spectroscopic studies of charged particles are facilitated by the rather low sensitivity of the ionization chamber to background radiation such as γ-rays and neutrons. The thickness of the window foil sets a lower energy limit on detectable charged particles, and particles passing through the foil are reduced in energy by an amount equal to the energy lost in the foil. The beam is also degraded as a result of energy straggling. In addition to energy degradation due to straggling, the energy resolution (FWHM) is further degraded by broadening arising from other sources. One major source is the spread δ_i due to the variation (statistical fluctuations) in the number of ion pairs produced by a particular monoenergetic particle.

For example, for an α-particle with $E_\alpha = 5$ MeV stopped in an argon-filled gridded ionization chamber ($w = 26$ eV, with a Fano factor $F = 0.4$, see below) the FWHM spread δ_i is expected to be about 27 keV. Note that in proportional counters the statistical fluctuations in the charge amplification further increase the energy spread δ_i; thus these detectors have worse energy resolution than nonamplifying ionization chambers. Another major source of beam degradation is the spreading δ_r due to rise-time variations in the electron component of the pulses from the detector. This is a result of differing charge collection times, which depend on the site of the ionization track. This spread is estimated to be of the same order as δ_i. Thus gridded ionization chambers have relatively poor energy resolution. Due to the rise-time variations, they also have poor time resolutions, of the order of 1 ms, and thus are not well suited for coincidence experiments. The above characteristics also lead to increased dead time ($t_d \simeq 1$ ms) for ionization chambers, which limits them to experiments with low count rates. For example, for a count rate $r = 10^3$ counts s^{-1} the observed count rate r_0, as determined by the relation $r_0 = r/(1 + rt_d)$, is reduced by about 50%. Ionization chambers are often used as detectors with magnetic spectrometers (sec. 5.4.2).

Many of the disadvantages of gas-filled ionization chambers can be removed if the gas is replaced by a solid (most commonly Si or Ge). One immediate advantage is that, because the density of the solid is very much higher than that of the gas, the stopping power and the range (Fig. 5.23a) are significantly improved. As with the ionization chamber, positive and negative charges are produced by the ionizing radiation passing through the insulator, and the resulting charges are accelerated toward the electrodes. An insulator normally does not contain charge carriers; however, once charge carriers are produced, they can move through the insulator. To understand this and the principle of operation of the solid-state ionization chamber (Si and Ge counter), one must be familiar with the rudiments of the band theory of solids, which will be discussed briefly (You68, Eng74, Gou74).

In a solid the effects of interaction between individual atoms cause the atomic energy levels of the outer electrons to split into a large number of levels, forming a band as shown in the energy-level diagram of Figure 5.24b. The electrons in the band do not belong to an individual atom but rather are shared by all atoms of the solid. The solid is then an insulator or a conductor, depending upon whether or not the highest band is filled. In an intrinsic semi-conductor the energy gap E_g between the last filled band, called the valence band, and the next band, the conduction band, is sufficiently small ($E_g = 1.1$ eV for Si, $E_g = 0.67$ eV for Ge) that by thermal excitation some electrons are lifted to the conduction band, giving the material a small intrinsic conductivity. The electrical conductivity of the tetravalent Si (and Ge) can be increased considerably by adding a minute trace (typically about 10^{13} atoms cm^{-3}) of a pentavalent element (P, As, Sb). These pentavalent elements give rise to additional localized energy levels within the forbidden gap. Of particular interest are those levels which lie very close to the conduction band, making it easier to promote electrons into the conduction band. A so-called N-type (for negative carrier) semiconductor is thus produced, in which the increased conductivity is due to electrons donated by pentavalent atoms to the conduction band. The impurities are therefore referred to as donors. The current results from the electrons, while the positive ions remain fixed. Another possibility is the P-type (for positive carrier) semiconductor, which is produced by adding a trace of a trivalent element (B, Ga). These atoms have a strong affinity for electrons in the valence band, since their energy levels localized close to the valence band can easily remove electrons from the band. These impurities are referred to as acceptors, where the resulting negative ion is locally fixed and the missing electron in the valence band, or hole, as it is called, moves like a positive electron through the crystal under the influence of an applied electric field.

At a junction between a N-type and a P-type semiconductor (Fig. 5.24c), the mobile charge carriers of each type of semiconductor diffuse across the junction into the other type of semiconductor. The carriers recombine partially in the transit region and produce a layer depleted of free-charge carriers. This region is characterized by a space-charge distribution caused by the fixed donors and acceptors producing an electric field and an accompanying potential drop across the junction. The junction acts like a diode, so that the depletion layer and the space-charge distribution can be increased by putting a reversed-bias voltage across it, since positive (negative) charge carriers are pushed away from the positive (negative) electrode until the potential drop across this depletion layer equals the bias voltage. The higher the bias voltage, the larger the depletion layer will be, the limit being set by electrical breakdown of the semiconductor. The depletion layer is thus characterized by a high resistivity and a high electric field. Carriers produced by thermal excitation are immediately swept out of the depletion layer. An ionizing particle passing through this charge-depleted region will produce along its path negative and positive charge carriers (electron-hole pairs) by lifting electrons from the lower band into the conduction band. These charge carriers are then separated by the electric field and move readily through the material, producing a current pulse at the ends of the detector, which is registered as a voltage pulse across the load resistor R as in an ionization chamber. Since the collection times for electrons and holes are similar, both electrons and holes are collected and the height of the voltage pulse is a direct measure of the energy loss of the particle in the depletion layer. Methods have been developed (Eng74, Seg77) for locating the depleted region near the surface of the detector. These detectors are called surface barrier detectors and are produced by providing a thin, heavily doped surface layer of carrier type opposite to that of the bulk material. Totally depleted transmission detectors are produced by having heavily doped surface layers of opposite types on their surfaces. In addition, position-sensitive semiconductor detectors have been developed (Gou74).

In solid-state detectors for charged-particle spectroscopy, Si crystals have been used most commonly because they have low intrinsic conductivity as a result of their relatively large energy-band gap $E_g = 1.1$ eV. This means that the detector can be operated conveniently at room temperature without excessive leakage current. It is important to note that the leakage current flowing through the material must be very small, so that the tiny current pulse created by the ionizing radiation can be detected. For example, the operation of Ge detectors at room temperature is impossible because of the large rate of production of hole-electron pairs by thermal excitation of electrons across the $E_g = 0.67$ eV gap. Thus, when a Ge detector is used, it must be cooled to LN_2 temperature. If very low leakage current is necessary, even Si detectors must be cooled to low temperatures. Silicon detectors of the surface-barrier or transmission type are commercially available in thicknesses from a few microns up to several millimeters. The active surface areas for all but the thinnest of these detectors range from about 0.1 to 10 cm^2. The gold or aluminum surface contacts at one or both faces of these detectors typically are made less than 0.2 μm thick, and for precise energy measurements of charged particles correction must be made for energy loss in the dead layers ($=60$ keV for 1 MeV α-particles in 0.2 μm thick Al).

One of the most important properties of any energy-measuring device is its resolution ΔE, which is usually defined as the full width at half-maximum

(FWHM) of the peak produced by a number of particles of identical energy E. If the resolution ΔE is poor, individual peaks within ΔE will not be resolved, and information will be lost. When energies are measured through pulse heights, the quantity actually measured is the number of ion pairs $(n = E/w)$ produced in the ionization process (w = average energy spent to form an ion pair). This number n is—as pointed out earlier—subject to statistical fluctuations in the random energy-loss process. A 1 MeV particle stopped in argon gas ($w = 26$ eV) creates on average $n = 3.8 \times 10^4$ ion pairs. If stopped in Si ($w = 3.5$ eV) or Ge ($w = 2.9$ eV), the number of electron-hole pairs is $n = 2.9 \times 10^5$ or 3.4×10^5, respectively. Note that the energies w for Si and Ge are low because ionization does not occur from an atomic level to the continuum but from the valence band to the conduction band. Note also that w exceeds the energy-band gap by an amount corresponding to the average energy expended by a hole or an electron in nonionizing processes (Gou74). Since n is a large number, the statistical distribution is well approximated by a Gaussian distribution with a FWHM $\Delta n = 2.35 n^{1/2}$. Since our measurement of the energy is proportional to n, we find for the fractional energy resolution

$$\frac{\Delta E}{E} = \frac{\Delta n}{n} = \frac{2.35}{n^{1/2}} = 2.35 \left(\frac{w}{E}\right)^{1/2}, \tag{5.15}$$

and thus for a given energy E the resolution is better the smaller the quantity w. From a more detailed investigation of the ionization shower process, Fano (Fan46) has shown that the actual statistical fluctuation of the number of ion pairs produced is considerably smaller than would be expected for completely random ionizing interactions, and a more correct form of the above expression is

$$\frac{\Delta E}{E} = 2.35 \left(F \frac{w}{E}\right)^{1/2}, \tag{5.16}$$

where F is now known as the Fano factor, with $F \leq 1$. Experimental and theoretical studies of the Fano factor yield values of $F \simeq 0.4$ for gases such as Ar, and $F \simeq 0.1$ for semiconductors such as Si and Ge (Eng74, Gou74). It is the smallness of w and of the Fano factor, as well as the effective collection of all charge carriers produced in the ionization, which give semiconductor detectors excellent energy resolution (i.e., $\Delta E = 11$ keV for 5 MeV α-particles).

In summary, both types of charge-collection detectors—the ionization chamber and the semiconductor—have a 100% detection efficiency. The most commonly used Si detectors have superior energy resolution compared with gridded ionization chambers. They also can be much smaller physically, because the range of the charged particles to be detected is several orders of magnitude shorter. This feature is of great value in the design of target chambers and in setups including detectors in close geometries. Silicon detectors also have good time resolution and short dead time, of the order of ns and μs, respectively, and are thus well suited for coincidence experiments and when high count rates are involved. The elimination of window foils also is an advantage over gas-filled counters. Si detectors are relatively insensitive to neutron-induced background signals but neutrons will severely damage them. Because of their relatively small volume and low Z, they also have a low detection efficiency for γ-ray background radiation. However, if the γ-ray intensity is several orders of magnitude higher than the charged-particle flux, the γ-ray–induced background can be troublesome (Tra84). Finally, it should be cautioned that Si detectors do not last indefinitely but are damaged by radiation (in particular by neutrons), resulting in an appreciable degradation in pulse shape, depletion depth, and energy resolution, caused in part by carrier-trapping centers generated at damage sites in the crystal lattice and by removing the atoms from their normal lattice sites. Generally, significant deterioration appears to occur after exposures of about 10^{12} fast neutrons per square centimeter (or 10^{10} protons cm^{-2}). Silicon detectors have been one of the most versatile detectors for investigations of nuclear reactions pertinent to nuclear astrophysics, partly because the available thicknesses are sufficient to stop and thus to measure the energy of the most energetic charged particles emitted in reactions ($E \leq 20$ MeV), because of their large solid angles, and because of their other characteristics mentioned above.

5.3.2. Neutron Detectors

As pointed out earlier, it is generally not possible to detect neutrons directly because they are uncharged and do not cause ionization, emission of light, or any direct effects used to detect charged particles. The most commonly used neutron detection systems rely on the detection of secondary charged particles produced in neutron-induced nuclear reactions. The classical example of reaction-product detectors is the boron counter for thermal and slow neutrons. This is commonly a proportional counter filled with BF_3 gas preferably enriched in ^{10}B. The operative reaction is $^{10}B(n, \alpha)^7Li$ ($Q = +2.79$ MeV), which has a cross section of about 3800 b at a neutron energy of 0.25 eV. The α-particle and the ^7Li recoil nucleus both have short ranges, so that they usually come to rest in the gas. Since the BF_3 counter is such an efficient and simple counter for slow neutrons, one natural way of making a detector for fast neutrons is to surround a BF_3 counter with paraffin, in which the fast neutrons, slowed down by collisions primarily with the protons in the paraffin, can diffuse into the counter as thermal neutrons. Such counters typically have lengths around 30 cm and are therefore often referred to as "long counters." They are characterized by a uniform detection efficiency (to within about 10%) for neutron energies ranging from, say, 10 keV to 3 MeV (Han47). Because of this constant efficiency, when coupled to a pulsed beam and a time-of-flight system (sec. 5.4.4), the long counter can provide detailed information about the energy spectrum in the neutron flux. Other examples of reactions that have been used for slow-neutron detectors are $^6Li(n, t)^4He$ ($Q = +4.79$ MeV), $^3He(n, p)t$ ($Q = +0.76$ MeV) and neutron-induced fission on U, Pu, or Cf

$(Q = +220$ MeV). Other variations of the gas-filled reaction-product detectors are the solid, plastic, or liquid scintillation detectors (sec. 5.3.3), doped, for example, with ^{6}Li. In this case the detectors are referred to as Li-glass detectors. In these mostly inorganic scintillators the pulses due to the heavy reaction product can be separated from pulses due to electrons (produced, e.g., by γ-rays) by using pulse-shape discrimination techniques (Kin69). The specific ionization of these two groups of charged particles is different, leading to different shapes of the output light pulse, which can be enhanced by using specially developed scintillators. Because of the excellent timing characteristics, these scintillation detectors in combination with time-of-flight methods are virtually unrivaled for measurements of the neutron energy spectrum. Semiconductor devices have also been developed for neutron detection (Mil67, Bis68).

Another type of neutron detection mechanism is the elastic scattering of the neutron from a light nucleus followed by the detection of the fast, charged, recoiling nucleus (Eng74). Reactions producing a recoil proton are used in most scintillation detectors for fast neutrons. Those producing a recoil α-particle are used in gas and liquid helium scintillators and in high-pressure helium-filled proportional detectors. Proportional detectors filled with pure hydrogen can be used for neutron detection, and they have distinct advantages where high background levels of γ-rays are present. The scintillation detectors (e.g., NE 213) in combination with pulse-shape discrimination are used most widely for the measurement of neutron energies by time-of-flight methods.

Finally, the capture of slow neutrons by certain stable isotopes converts these isotopes into radioactive isotopes emitting β- and γ-rays. The intensity of these rays together with the lifetime of the radioactive isotope, the capture cross section, and the number of stable isotopes allows a measurement of the neutron flux. Gold is used most widely in the detection of thermal and higher energy neutrons via this activation method (sec. 5.4.3).

5.3.3. Gamma-Ray Detectors

As discussed previously, the interaction with γ-rays in matter produces electrons (and positrons) with various amounts of kinetic energies leading to relatively complicated response functions of the detector. In the case of photoelectric absorption, all of the energy E_γ of the incident monoenergetic photon is absorbed by a bound electron of an atom. As a result, the atom is left with an electron vacancy, resulting in the emission of X-rays or Auger electrons. The low-energy X-rays are generally absorbed in a second photoelectric event. The electrons produced lose their energy in the detector material by the energy-loss processes of excitation and ionization as discussed earlier. If the photoejected electrons and the Auger electrons are stopped in the material, the total energy E_γ is absorbed within the detector and thus the resulting pulse of the detector will correspond to the full energy of the incident γ-ray. The number of recorded events corresponding to the total energy of the incident

γ-ray have (Fig. 5.25a) a distribution of finite width, which is referred to as the full-energy peak (photopeak is also often used). In the Compton scattering event the photons are scattered by the electrons with a partial energy transfer, and a Compton electron energy spectrum will result (Fig. 5.25a) which extends continuously from zero energy up to a maximum energy of $E_e(\text{max}) \simeq E_\gamma - 0.25$ MeV. The pulses corresponding to this maximum energy form the well-known Compton edge. For a Compton scattering event, a full-energy pulse will often result if the Compton-scattered electron stops in the detector material and if the scattered γ-ray also deposits the rest of its energy in the

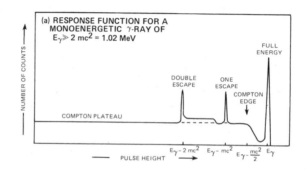

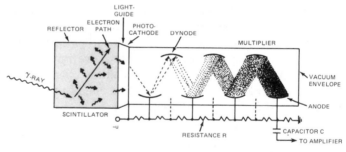

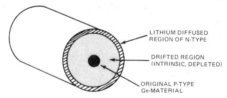

FIGURE 5.25. (a) Schematic response function of a γ-ray detector for a monoenergetic γ-ray with $E_\gamma \gg 2m_e c^2$. (b) Cross section through a scintillation detector is shown schematically, illustrating the different stages in the detection process. (c) Geometry of a coaxial-drifted Ge(Li) detector forming a P-I-N semiconductor structure. The drifted intrinsic region is depleted of free-charge carriers and is used to detect the incident nuclear radiation.

material through a series of subsequent Compton scatterings and/or photo-electric absorption. If the energy of the γ-ray is above the pair-production threshold ($2m_e c^2 = 1022$ keV), the photon can also interact in the material to create an electron-positron pair. If the volume of the detector material is sufficient, both members of the pair will stop in the material and the positron will annihilate with an atomic electron to form a pair of 511 keV γ-rays traveling in opposite directions. A full-energy pulse will result from such an event only if both members of the pair are stopped in the detector material and if both of the subsequent annihilation quanta also interact via Compton scattering or photoelectric events to deposit all of their energy in the material. Because of the fixed energy of the 511 keV annihilation radiation, if one or both of these quanta escape from the detector, the output pulse will correspond to the γ-ray energy minus 511 keV or minus 1022 keV, and the associated peaks are called single-escape peaks or double-escape peaks (Fig. 5.25a). As a consequence, the response function of such detectors to a monoenergetic γ-ray will include superposed contributions from all these interactions, with peaks corresponding to the full photon energy and (for $E_\gamma > 1022$ keV) to the escape of one or two of the annihilation quanta, plus a broad continuum corresponding to interactions where the electrons do not stop in the detector material and to Compton events in which the scattered photons escape. The magnitude of all these contributions to the response function depends on the type of detector material, on its size and its form, and on the energy of the γ-ray, and thus the calculation of response functions can become complicated, often requiring experimental calibration. Since the spectrum of a monoenergetic γ-ray is already complex, the advantages of good energy resolution become obvious when the spectrum contains contributions from several γ-rays of differing energies (Fig. 5.27). In γ-ray spectroscopic studies the work load is shared by NaI(Tl) and Ge(Li) detectors, which are discussed below.

As noted in section 5.3.1, a considerable fraction of the specific energy loss of a charged particle (here electrons created by γ-rays) goes into processes which do not directly cause ionization but which excite the neutral atoms or residual positive ions of the detector material. This excitation energy eventually appears as photons of various wavelengths which in most materials are reabsorbed within the material itself, increasing the thermal energy of the material. However, there exists a group of dielectric materials (organic and inorganic single crystals, organic liquids) which are transparent to some part of the wavelength spectrum of the photons emitted by the excited atoms and ions along the path of the charged particle. These photons can be detected outside the transparent material. It is found that their number is proportional to the energy lost within the material. Materials which exhibit this property are known as scintillators, and the process is known as the scintillation process. In the case of the inorganic crystal scintillators such as NaI the scintillation process is activated by impurities, and their luminescence is due mainly to the presence of small traces of very specific impurities such as thallium (Tl). They are then characterized as, for example, NaI(Tl) detectors, where Tl is the impurity. Note that the high atomic number of I ($Z = 53$) and the high density (3.67 g cm^{-3}) of the NaI(Tl) material gives these detectors a high efficiency for γ-ray absorption.

The principle of the operation of a scintillation detector is shown in Figure 5.25b. An incident γ-ray creates an electron in the scintillator material, which dissipates its energy E in the scintillator. A fraction ϵ_1 of this energy is converted into N photons, which are radiated in all directions:

$$N = \frac{E}{w} \epsilon_1 ,$$

where w is the energy of the scintillating photons [for NaI(Tl) detectors, $\epsilon_1 \simeq 0.1$ and $w \simeq 3$ eV]. The scintillator is surrounded by a reflector which increases the number of photons falling either on the light guide connecting the scintillator to the photomultiplier entrance window or directly on the window itself. The number of photons reaching the photocathode of the multiplier is thus reduced to

$$N_{cat} = \epsilon_2 N = \frac{E}{w} \epsilon_1 \epsilon_2 ,$$

with $\epsilon_2 \simeq 0.4$. Of the photons which reach the photocathode, a fraction $\epsilon_3 \simeq 0.2$ cause the emission of photoelectrons N_e from this photocathode:

$$N_e = \frac{E}{w} \epsilon_1 \epsilon_2 \epsilon_3 .$$

These photoelectrons are accelerated to the first electrode, or dynode, by the voltage applied between the photocathode and the first dynode, and cause the emission of S secondary electrons from this first dynode ($S \simeq 2$–5). This electron multiplication process is repeated at the succeeding dynodes, and thus, if there are n dynodes (usually $n = 6$–14), the number of electrons collected at the anode, or, equivalently, the total collected charge, will be

$$Q = N_e S^n e ,$$

where the total electron amplification S^n usually lies within the range 10^5–10^9. The charge can then be detected and the resulting signal amplified. The energy resolution at FWHM of such scintillation detectors is again limited by the statistical fluctuations in photon production, variations in light reflection, internal light absorption, and photocathode emission. It is thus given by (eq. [5.15])

$$\frac{\Delta E}{E} = \frac{2.35}{N_e^{1/2}} = 2.35 \left(\frac{w}{E \epsilon_1 \epsilon_2 \epsilon_3} \right)^{1/2} ,$$

which leads to $\Delta E/E = 4.0\%$ for a 1.3 MeV γ-ray (^{60}Co source) absorbed in a NaI(Tl) detector. In practice, one observes resolutions of 6–8% owing to additional contributions from temperature, crystal inhomogeneity, and photomultiplier effects (Eng74).

By comparison, Ge semiconductor detectors, having fewer statistical processes and more effective collection of all charge carriers produced, give an energy resolution of $\Delta E/E = 0.1\%$ for a 1.3 MeV γ-ray (eq. [5.16] with $w = 3.0$ eV and $F = 0.1$), in fairly good agreement with experimentally observed values. The high atomic number $Z = 32$ and the high density 5.3 g cm^{-3} also increase the γ-ray absorption. However, since the mean free path of a 1 MeV γ-ray in Ge is about 3 cm, such Ge semiconductor detectors must have large, sensitive volumes. Their manufacture could not be achieved by the formation of a simple P-N junction (Fig. 5.24c). In 1960 Pell (Pel60) developed a new method involving the production of a region in a Ge (or Si) single crystal, in which there is a detailed compensation of the base impurity type, usually acceptors N_a, with the opposite impurity type, and usually donors N_d such as Li. Within this region $N_a = N_d$, and the region is called intrinsic, i.e., depleted of free charge carriers (Bro79).

To maintain this condition, the detector has to be kept at LN$_2$ temperature. Another reason for cooling is to minimize the thermal leakage current. Thus, the result is a P-I-N structure (Fig. 5.25c), referred to as a coaxial Ge(Li) detector, in which the intrinsic region is used to detect nuclear radiation (such as γ-rays) in the same manner as the ionization chamber. Ge(Li) detectors with large volumes up to about 200 cm^3 are today commercially available, approaching to within a factor of 2 the volume of a 3 × 3 inch NaI(Tl) crystal [although NaI(Tl) crystals can be made much larger]. In recent years high-purity germanium crystals became available with impurity concentrations in the range 10^9–10^{11} atoms cm^{-3}. Thus, detectors with sensitive volumes similar to Ge(Li) detectors could be fabricated without requiring lithium compensation. They are referred to as intrinsic Ge detectors (Gou74). Their biggest advantage is that they can be stored at room temperature and need cooling to LN$_2$ temperature only during use.

In addition to energy resolution, another important characteristic of γ-ray detectors is their efficiency in absorbing the γ-rays. In determining absolute cross sections from a γ-ray spectrum, one of the problems is that γ-rays which were originally outside the solid angle subtended by the detector can subsequently be scattered into the detector by the material in the target chamber and the shielding material near the detector, thus adding extra counts to the spectrum. Because these γ-rays must lose energy in the scattering process, they cannot possibly contribute counts to the full-energy peak. In fact, this is the only region in the spectrum where such scattered γ-rays cannot contribute, and for this reason, quantitative γ-ray experiments are normally carried out using measurements of the full-energy peaks. Occasionally, the single- and double-escape peaks are also used for high-energy γ-rays. Therefore, the efficiencies of a γ-ray detector refer in most cases to the efficiency of the full-energy peak.

Standard techniques have been used to calculate the absolute efficiency of the full-energy peak for γ-ray detection by cylindrical NaI(Tl) crystals, including all the interactions noted above (Mar68). The results of such calculations for a 5 × 5 inch NaI(Tl) crystal are shown in Figure 5.26, where the source is assumed to be on the cylindrical axis of the crystal. The efficiency of the crystal depends on both energy and distance. In an experimental situation, corrections must be made to take into account the absorption by any material which is between the source and the detector. This absorption can also be

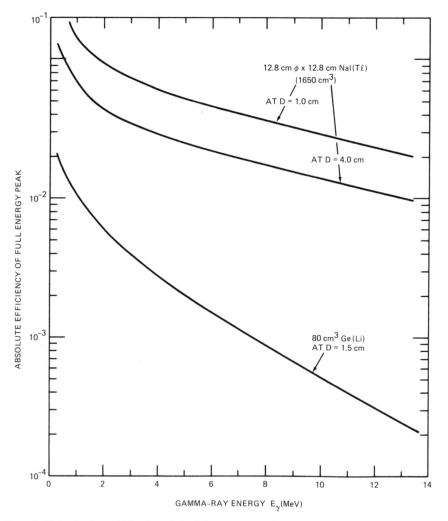

FIGURE 5.26. Absolute efficiencies of the full-energy peak for a NaI(Tl) detector and a coaxial Ge(Li) detector are compared as a function of γ-ray energy. The distances from the source to the front of the detectors are given.

calculated. However, it is useful for a particular experimental setup to make measurements using calibrated γ-ray sources placed at the location of the target. Standard sources such as ^{22}Na, ^{56}Co, ^{60}Co, ^{88}Y, ^{133}Ba, ^{137}Cs, ^{152}Eu, and ^{228}Th are commercially available and span an energy range $E_\gamma = 0.08$–3.55 MeV (Mar68, Yos80). Of course, one has to be sure that the sources were accurately calibrated. A convenient check on the deduced absolute efficiency ϵ is provided (Sch76) by sources such as ^{60}Co, which emit cascade γ-rays of equal intensity (^{60}Co: $E_{\gamma 1} = 1.17$ MeV and $E_{\gamma 2} = 1.33$ MeV). The observed number of counts in the full-energy peaks of both γ-ray transitions over an accumulating time t is given by the relations

$$N_1 = \epsilon_1 N_0 t , \qquad N_2 = \epsilon_2 N_0 t ,$$

where N_0 is the source strength and ϵ_1 and ϵ_2 are the absolute efficiencies of the detector for both γ-rays. There is a finite probability that both γ-rays are simultaneously incident on the detector, leading to a peak corresponding to the sum of their energies, i.e., $E_\gamma = 1.17 + 1.33 = 2.50$ MeV. The number of counts N_s in this summed peak is given by

$$N_s = \epsilon_1 \epsilon_2 N_0 W(\vartheta) t ,$$

where $W(\vartheta)$ describes the angular correlation between the coincident γ-rays. For the sum peak of a ^{60}Co source ($\vartheta = 0°$), one finds $W(0°) = 1.09$ for a distance of $D = 1.5$ cm from the source. The ratio of the number of counts N_s and, say, N_1 then leads to

$$\frac{N_s}{N_1} = \epsilon_2 W(0°) ;$$

thus the absolute efficiency ϵ_2 is obtained independently of the source strength N_0. Measurements of the efficiency curve can be extended (Sin71, Sch76) to higher γ-ray energies by using the isotropic (or nearly isotropic) emission of capture γ-rays in resonance reactions such as ^{11}B$(p, \gamma)^{12}$C ($E_R = 163$ keV), ^{14}N$(p, \gamma)^{15}$O ($E_R = 278$ keV), ^{24}Mg$(p, \gamma)^{25}$Al ($E_R = 230$ keV), and ^{27}Al$(p, \gamma)^{28}$Si ($E_R = 992$ keV). These reactions emit γ-rays in the energy range $E_\gamma = 0.4$–11.7 MeV, where the deduced relative efficiency of the low-energy γ-rays is matched to the results obtained with the calibrated radioactive sources. It turns out in general that the calculated efficiency curve for NaI(Tl) crystals is in good agreement with observation.

The above discussion can be applied equally well to Ge(Li) detectors. In practice, however, because of their frequently irregular or poorly defined shapes (in particular, the size and shape of the uncompensated P-type core, Fig. 5.25c), a calculation of the absolute efficiency is not usually satisfactory (Sch76). Thus an experimental efficiency determination must be made separately for each detector. The efficiency curve for an 80 cm^3 Ge(Li) detector shown in Figure 5.26 was obtained (Sch76) as described above and in the setup of Figure 5.18a, including a 3 mm thick lead sheet placed in front of the detector for X-ray reduction from the target.

Both NaI(Tl) and Ge(Li) detectors are presently used for nuclear astrophysics studies. For the detection of low γ-ray yields, e.g., at subbarrier energies, the large NaI(Tl) detectors seem to be preferable. However, in such a comparison consideration must be given to resolution, since fewer counts are needed for the higher resolution detector to show a peak above the background. There are no simple criteria for deciding which detector to use in a given situation, and it is best to have both available. The data shown in Figure 5.27 illustrate the value of the Ge(Li) detector in resolving complex γ-ray spectra. However, the higher efficiency of the NaI(Tl) detector is useful when studying reactions with very low cross sections (chap. 7), when ion beams of high intensity are not available, when γ-rays of energy higher than about 12 MeV are to be studied, and when high time resolution is required. The complexity of the γ-ray spectra can be considerably reduced by using diffraction, magnetic pair, Compton-suppressed, or pair (crystal) spectrometers (Alb60). However, because of their low efficiency, these spectrometers are not frequently used in studies of astrophysically relevant reactions. The NaI(Tl) and Ge(Li) detectors are also characterized by good time resolutions, in the range of a few ns, so that they are well suited for coincidence experiments. Similarly, their count rate limitation is in the range of 10^5 events s^{-1}. The presence of neutrons has two effects on Ge(Li) detectors. One effect is neutron

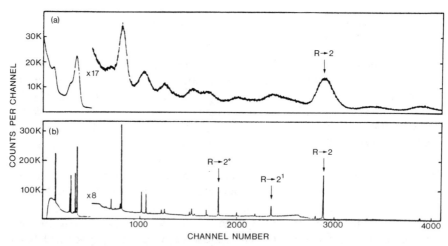

FIGURE 5.27. Pulse-height spectrum from an 1840 cm^3 NaI(Tl) detector (a) is compared with the spectrum from a 40 cm^3 Ge(Li) detector (b). In both cases the γ-rays from the 1660 keV resonance of ^{24}Mg$(p, \gamma)^{25}$Al are shown (Dwo72). The label $R \to 2$ refers to the γ-ray transition from the resonance state to the second excited state in ^{25}Al. The unprimed, single-primed, and double-primed peaks denote full-energy and single and double-escape peaks, respectively. Note that the resolution is over 50 times better for the $R \to 2$ transition in (b) than that in (a).

inelastic scattering which contributes to the γ-ray spectrum (Cha65). The other is radiation damage similar to that previously discussed for Si detectors. Severe damage occurs (Kra68) at an integrated neutron dosage of about 10^{11} neutrons cm^{-2}. By a process of redrifting and re-etching, the damaged detector can be rejuvenated, but often at the price of a reduced volume, i.e., a smaller absolute efficiency. For NaI(Tl) crystals, the primary neutron effect is the (n, γ) activation of the NaI to form ^{24}Na and ^{128}I. This capture reaction produces high-energy γ-rays in the crystal and also delayed activities from the beta-decay of ^{128}I and ^{24}Na. Both of these activities can cause serious background problems for low-yield measurements, since they occur inside the crystal and thus are detected with high efficiency.

In recent years, detector arrangements covering nearly a 4π solid angle have been built. They are referred to as crystal balls and have a number of segments ranging from 6 to 150. Many of these units use NaI(Tl) crystals. Also used are bismuth germanate crystals $(Bi_4Ge_3O_{12})$, usually abbreviated as BGO. Because of the high atomic number of bismuth ($Z = 83$) and a high density (7.13 g cm^{-3}), its absorption power is about 2.5 times higher than NaI(Tl), leading to a 16-fold reduction in volume of the finished geometry. At present several laboratories are building compact 4π crystal balls composed of BGO (or BaF_2) segments. The price to be paid for this compact spectrometer is an energy resolution about a factor of 1.7 times that of NaI(Tl). For higher resolution experiments, crystal balls composed of intrinsic Ge detectors (surrounded by BGO Compton suppressors) are in preparation. These new detector arrangements may also be used for nuclear astrophysics studies.

Finally, mention should be made of some special γ-ray detectors, which have been used in neutron experiments (chap. 9). The most common type is the Moxon-Rae detector (Mox63) for the detection of the prompt γ-rays emitted following thermal neutron capture. Moxon and Rae pointed out that the number of electrons emerging from a thick slab of low-Z absorber was almost linearly dependent on the γ-ray energy. Thus a detector based on the observation of these emerging electrons, produced mainly by Compton interaction (Fig. 5.23c), would have a detection efficiency linearly proportional to the γ-ray energy and therefore independent of the branching scheme in the γ-decay cascade following neutron capture. The absorber should itself be relatively free from neutron capture, and Moxon and Rae chose carbon in the form of graphite as the most readily available material to satisfy this criterion. The electrons are detected with a very thin sheet (about 0.5 mm thick) of plastic scintillator, which—because of its small volume—makes it relatively insensitive to direct γ-ray or neutron irradiation. These detectors have been widely used for studies of (n, γ) capture reactions at neutron energies between about 5 eV and 100 keV (Mac63).

5.4. Detection Techniques

In the preceding section we discussed various detecting devices. These devices, together with a wide variety of electronic apparatus, are collectively called nuclear instrumentation. Detailed description of such instrumentation is outside the scope of this book (Gou74a, Bor74, Bro79). In what follows we shall confine ourselves to a discussion of some specific examples, including limited performance data. Many of the devices described are commercially available. In particular, we will discuss the application of detectors and auxiliary electronics to various techniques aimed at detecting radiation resulting from nuclear reactions of astrophysical interest.

5.4.1. Electronics and Data Acquisition

The electronic components of a typical spectrometer are shown in Figure 5.28. The first element needed is a detector bias supply adjusted to suit the particular detector. The detector converts the energy deposited by the incident radiation into an electric signal—a current pulse. The current pulse charges a capacitor, producing a voltage signal. The voltage rises fast, asymptotically reaching a value proportional to the absorbed energy. The time dependence of the pulse is given by the relation $V(t) \simeq 1 - \exp(-\lambda t)$, where the time constant λ^{-1} is determined, in semiconductors, by the collection time for the free-charge carriers and, in scintillators, by the decay time of the excited atoms.

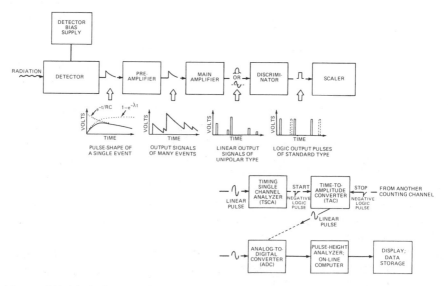

FIGURE 5.28. Block diagram of typical electronic components of a detector system. Note that linear signals are defined as those in which the output-signal amplitude is proportional to that of the input signal, such as the radiation energy. By contrast, logic signals have a fixed shape and amplitude indicating only that a pulse is present, absent, or related to time.

For example, in a Si detector a 1 MeV charged particle releases about 5×10^{-14} C of charge, producing a signal of 10 mV in a 5 pF capacitor. The capacitor (C) eventually discharges through a resistor (R) with a time constant proportional to the resistance and the capacitance (RC). The output pulse, shown in the "pulse shape of a single event" segment of Figure 5.28 by the solid curve, is the product of the two dotted curves. It is characterized by a short rise time, usually 50 ns or less, followed by a slow exponential decay, usually several microseconds or milliseconds, depending on the detector. This clipping of the pulse shape is important in minimizing the overlap of consecutive pulses at high count rates.

The electric signal from the detector is fed into a preamplifier, which is in close proximity to it. This low-noise amplifying element is specially designed to accept the signal from the detector and to amplify slightly (by a factor of 5–10). It also provides a minimum of shaping to keep the signal-to-noise ratio as high as possible. Note that electronic noise is added to the signal as a result of current fluctuations in the signal-amplifying elements. In normal operation at ordinary count rates the steep rise of the detector signal overlaps the exponential decay (tail) of the previous one, so that the output never returns to the baseline. Since the amplitude of the signal from the detector is variable (owing to differing energies of the incident radiation) and random in time, the preamplifier output signal is irregular as shown in Figure 5.28. Preamplifiers are usually designed with a low output impedance, since the output signal must go through a long coaxial cable to the main amplifier—often located some distance away.

All the useful information in the preamplifier output signal is included in the rise time and amplitude of each pulse, which reflect the production time of the signal and the energy of the incident radiation. Thus, the pulse-shaping circuits in the main amplifier (Fig. 5.28), operating with time constants much shorter than the decay of the preamplifier signal and much longer than its rise time, effectively remove the slow component of the preamplifier signal, producing individual pulses, whose amplitudes are proportional to the quantity of interest, usually the energy. The pulses shown in Figure 5.28 are of the unipolar type, and ideally they do not overlap. In practice, when the count rate is high, the amplifier output pulses overlap. Hence, a further degradation in resolution occurs at high count rates because of baseline fluctuations and pulse pileup. This problem can be partially solved by using pileup-rejecting devices (Gou74a), which sense, using fast (several ns) signals, when two signals arrive within the required shaping time. All amplifiers use shaping as a method of minimizing the sensitivity of the analyzed signal to noise from various sources. Pulse shaping in amplifiers involves some form of differentiation and integration. Differentiation serves the major purpose of limiting the duration of any signal to a reasonable time to prevent its interfering with later signals, and integration is used to average out noise fluctuations.

It follows that high count rates require short shaping times (usually 0.2–

0.5 μs), while optimal energy resolution demands long shaping times (usually 2–6 μs). The primary function of an amplifier is to amplify the signal (usually a fraction of a volt) received from the detector-preamplifier combination to a level (commonly 10 V maximum) such that subsequent electronic devices can perform accurate amplitude analysis while shaping the signal pulses to yield the best possible energy resolution. Linearity of the input/output characteristic is required to simplify interpretation of spectra recorded in the analyzer, and the term "linear amplifier" is often used in nuclear spectrometry. The gain of an amplifier must also be stable against temperature changes and as a function of time. Otherwise, significant spectrum distortion and degradation of energy resolution of the system result. This is particularly bad in the measurement of low cross sections, where the highest signal-to-noise ratio above background level is of the utmost importance. Unipolar pulse shapers are commonly preferred for good noise performance, but bipolar pulses, generated by two short time-constant differentiators, are also useful in many applications. The balanced waveform (i.e., extending both above and below the baseline in equal amounts) avoids any effects of baseline shifting between pulses, thereby maintaining the operating points of the late stages of the spectrometer at their design values, independent of count rate. Further advantages include reduction of low-frequency noise, hum, and microphony and the availability of a zero crossing point in the output waveform to use for timing purposes. Bipolar shaping is used in applications of high count rates and where the ultimate in energy resolution is not essential.

The output signals of the main amplifier might be fed into a voltage discriminator, which marks the occurrence of pulses that exceed a predetermined minimum amplitude level. If an event does not reach the discriminator level, no output is generated. When the level is reached, a logic signal (usually a step function with a pulse height of + 10 V and a width of 0.5 μs) is produced. This discrimination is illustrated in Figure 5.28. Here, of five input signals, only three have sufficient amplitude to exceed the adjusted discriminator threshold required to generate the standard output pulses, to be counted in a scaler. An extension of the discriminator is the single-channel analyzer (SCA), which has both lower-level and upper-level discriminators that can be set to define a certain acceptable amplitude range in the input signals. The region between the lower-level and upper-level discriminator settings is called the SCA window, and only linear pulses with amplitudes falling within this window produce output pulses. SCAs are generally used to select pulses of interest in one analysis channel, the output being used to gate another channel, so that the only signals processed are those coincident with the SCA output pulses. If in such applications a high time resolution is essential, fast discriminators, fast amplifiers, timing filter amplifiers, and the like (Gou74a), are frequently used to give precise time information on the input pulses either from their rise time or—in the case of bipolar pulses—from the crossover point of the pulses (i.e., when the pulse shape crosses the baseline).

For coincidence or time-of-flight work a time-to-amplitude converter (TAC) is commonly used. The device produces linear output pulses whose amplitude is linearly proportional to the time interval between a start and a subsequent stop input pulse. No output signal occurs unless a stop pulse follows a start pulse within a preselected time. The input pulses are typically fast negative logic signals (about 2 ns rise time and 1 V pulse height) generated by timing discriminators or timing SCAs (TSCAs). The basic time-to-amplitude conversion is carried out by using the start pulse to switch a constant current into an integrating capacitor and by stopping the current flow when a stop pulse occurs. An output signal occurs immediately after the stop pulse, while the integrating capacitor remembers the charge stored on it during the timing period. The system is reset immediately after the end of the linear output pulse. These and other electronic devices are commercially available as nuclear instrument modules (NIMs) in order to provide maximum compatibility among instruments produced by various manufacturers.

The final step in the signal-processing system of any nuclear spectrometer is to convert the amplitude of the signal from the main amplifier (or TAC) into digital information. This is achieved with an analog-to-digital converter (ADC), which measures the amplitude of the analog (linear) pulse and provides a digital output with a numerical value proportional to the pulse amplitude. The ADC generates a voltage rising linearly with time until it reaches the height of the input pulse. The number of cycles an auxiliary electronic oscillator (called the clock) goes through during this voltage rise is counted. This total count is a digital number whose size is proportional to the input pulse height. Once the pulse height is digitized, it can be transferred directly to a computer and handled by various computer techniques. Most commonly, a location in a magnetic core memory is reserved for each digital pulse of a given height. After each pulse is digitized in the ADC, the number in the corresponding memory location is increased by 1. The cumulative numbers in each memory location give the number of pulses of a given height, which is just the number of original particles with the corresponding energy. These numbers can be retrieved by one of several computer devices, which displays them on an oscilloscope, analyzes them, processes them in connection with the coincident pulses from other detectors (called multiparameter analysis), and transfers them to a magnetic tape or a disk for data storage. A complex instrument which performs all these functions is known as a multichannel pulse-height analyzer. The ADC divides input pulse amplitudes into as many as 8192 increments (called channels) and thus is particularly well suited for high-resolution spectroscopy, as for example when using a Ge(Li) detector. In the case of a pulse-height analyzer, the digital information from the ADC determines the memory address (channel) to which a count must be added, and thus the resulting pulse-height spectrum provides at once information on the intensity and energy of all emissions from the source. The major characteristic of an ADC is the repetition rate of the internal clock used in the digitizing and registering process. For example, if the clock rate is 100 MHz and a coding is in 4096 channels, the coding time of the ADC will be over 40 μs per event. This coding time is the major factor in determining the rate at which events can be handled by the ADC (here about 20,000 events per second).

There are many experimental situations where it is desirable to know the time relationship between two (or more) detected events. Perhaps the most common case is in studying processes in which an excited nucleus emits cascade γ-rays in its decay to the ground state, for example, $^{12}C(\alpha, \gamma\gamma)^{16}O$ (chap. 7), or emits particles as in the reaction $B(p, \alpha)2\alpha$ (Bec84). If one wants to study energy and/or angular correlations between these simultaneously emitted reaction products, one must ascertain that they originated in the same process. This is done by ensuring that they were emitted in coincidence. Such studies are referred to as coincidence experiments. A block diagram of the typical electronic devices needed in such coincidence experiments is shown in Figure 5.29 for a study involving two detectors. The signal from the pre-amplifier of each detector is fed into a normal main amplifier as well as into a fast amplifier, which is suitably designed for enhancing the fast rise time component of the preamplifier output signals. The signals from the fast amplifiers are then fed into fast discriminators, which provide fast negative logic pulses to be used as start and stop input signals for a TAC. Usually an extra delay device is inserted in the line carrying the pulse from one of the detectors, so that the time interval corresponding to coincident input signals is not zero, thus allowing one to observe the true (real) coincidences and accidental (random) coincidences simultaneously. In Figure 5.29 the real coincidences form a peak with a time width τ characterizing the time resolution of the experimental apparatus. However, as seen in the figure, this peak is super-

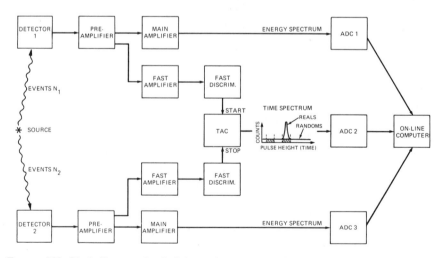

FIGURE 5.29. Block diagram of typical electronic components for coincidence spectrometry.

posed on a flat background arising from random coincidences, which contribute signals in the region of the peak as well as at all other times. To understand the relation between real and random coincidences, let us assume that N_0 processes occur per second and that the detectors have 100% detection efficiencies and subtend solid angles of Ω_1 and Ω_2. If the angular distribution of the coincident radiations is isotropic, the number of real coincidences detected per second, N_{real}, is then

$$N_{real} = N_0 \Omega_1 \Omega_2 \,,$$

leading to the peak shown in Figure 5.29. If all radiations detected come only from processes such as γ-γ cascades, the total count rates in the two detectors N_1 and N_2 are

$$N_1 = N_0 \Omega_1 \,, \qquad N_2 = N_0 \Omega_2 \,.$$

The chance that the two detectors register within the time window τ accidental coincidences per second, N_{rand}, is then given by

$$N_{rand} = \tau N_1 N_2 = N_0^2 \Omega_1 \Omega_2 \tau \,,$$

One then finds the ratio of real to random coincidences to be

$$\frac{N_{real}}{N_{rand}} = \frac{1}{N_0 \tau} \,.$$

For example, for $N_0 = 10^6$ events s^{-1} and $\tau = 1$ μs, the ratio is unity. It is clear that the ratio must be kept reasonably large. It can be increased by reducing N_0 (i.e., by reducing the beam current), but this reduces N_{real} and hence the rate of data accumulation. It is therefore important to make the time resolution τ as small as possible. With suitable fast timing techniques (Gou74a, Eng74) time resolutions in the range of a few nanoseconds have been achieved.

When a time spectrum as obtained from a TAC is used (Fig. 5.29), the contribution of random coincidences to the region of the real coincidence peak can be measured and corrected for, using data from regions in the TAC spectrum outside the peak region. Such data handling and data manipulation are possible using an on-line computer (Bor74, Eng74, Bro79), in which not only the time spectrum but also the energy spectra from both detectors are processed after digitization through the respective ADCs. In this case, processors are designed for various tasks in the handling of the multiparameter data (here energy and time). One processor might operate a selected number of ADCs independently, thus displaying, analyzing, and storing their singles spectra as is usual in a pulse-height analyzer. Another processor might operate a set of ADCs in coincidence. In the example shown in Figure 5.29 such a processor will accept input data only if all three ADCs provide an input signal within usually a few microseconds, thus representing a type of slow coincidence selection of the multiparameter input data. The accepted coincidence data can be stored in a buffer or transferred event by event to a magnetic tape, which is referred to as list-mode data acquisition. In this case the actual sorting of the data is carried out later, after the experiment, in what is called a playback analysis. However, a processor can also be used to analyze, sort, and reduce the data on line, with additional constraints imposed on the input spectra, such as energy range (data from ADC1 and ADC3) and equal time windows set on the coincidence peak and the flat random distribution (data from ADC2). The data from the ADCs might also be mathematically manipulated. The production of a mass spectrum from the detector signals in a ΔE-E telescope (sec. 5.4.2) is one example of this. Such on-line reduction of the data is one of the main reasons for using on-line computers.

When a large number of detectors and their electronics are to be controlled, as in crystal ball arrangements, CAMAC (computer assisted measurement and control) electronic modules (Bro79) are used to interface the multiple data inputs directly to the computer for processing. These CAMAC modules may also be used to control and adjust experimental equipment such as the electronics, the spectrometers, the target chamber, and even the accelerator (Eng74).

5.4.2. Detection of Reaction Products

Capture reactions A(x, γ)B play an important role in nuclear astrophysics. In this case a projectile x (usually a proton or an α-particle) incident on target nuclei A causes a reaction leading to γ-rays and a residual nucleus B, whose mass is in first order equal to the sum of A and x. Most commonly, the reaction is studied by detecting the emitted γ-rays using Ge(Li) detectors. However, in some circumstances it is possible and advantageous to detect the recoiling residual nucleus B with the same momentum as the projectile and traveling essentially in the same direction (sec. 5.6.1). If the residual nucleus is unstable, its radioactivity can be detected (sec. 5.4.3).

The $^{12}C(p, \gamma)^{13}N$ reaction (chap. 6) may be taken as an example of a reaction in which γ-rays are detected (Rol74a). In this case, the ^{12}C targets of 10 μg cm^{-2} thickness (corresponding to a 4 keV energy loss for 500 keV protons) were deposited on 0.25 mm thick Ta backings. The γ-rays were observed with a 45 cm^3 Ge(Li) detector (2.0 keV energy resolution at $E_\gamma = 1.3$ MeV). A γ-ray spectrum obtained at $E_p = 1.60$ MeV and $\theta_\gamma = 90°$ is shown in Figure 5.30. Two γ-rays, one from a transition to the ground state (DC $\rightarrow$ 0) and one from a transition to the unbound state at 2366 keV (DC $\rightarrow$ 2366), are observed. A contaminant γ-ray from the $^{23}Na(p, \alpha\gamma)^{20}Ne$ reaction is also observed. The capture γ-ray transitions are identified by their energies. If the Q-value of the reaction is known (here $Q = 1944$ keV), the energy of a γ-ray transition to a state E_x in the residual nucleus is given by the relation

$$E_\gamma = Q + \frac{M}{M + m} E_p - E_x - \Delta E_{Rec} - \Delta E_{Dop} , \qquad (5.17)$$

where M and m are the respective target and projectile masses (in exact mass units) and E_p is the projectile energy in the laboratory system. For $E_p = 1603$ keV, the first three terms in equation (5.17) lead to E_γ (DC → 0) = 3422.8 keV. The fourth term reflects a correction for the recoil energy of the emitting nucleus,

$$\Delta E_{\text{Rec}} = \frac{E_\gamma^2}{2M_B c^2} , \tag{5.18}$$

which amounts here to 0.48 keV. Finally, the Doppler shift of the γ-rays emitted by a nucleus moving at a velocity v affects the observed γ-ray energy:

$$\Delta E_{\text{Dop}} = \frac{v}{c} E_\gamma \cos \theta , \tag{5.19}$$

where θ is the angle between the recoiling nucleus and the γ-ray detector (here $\theta = 90°$, $\Delta E_{\text{Dop}} = 0$). The calculated energy of 3422.3 keV for the DC → 0 transition is in good agreement with the observed value (Fig. 5.30). A similar procedure ensured the correct identification of the 1056 keV γ-ray with the DC → 2366 keV transition.

Normally, the primary γ-ray transition to an excited state is followed by secondary emissions from the cascading to the ground state of the residual nucleus (Fig. 5.31). The correct energy and intensity of these secondary transitions provide additional criteria for ensuring the correctness of the identifica-

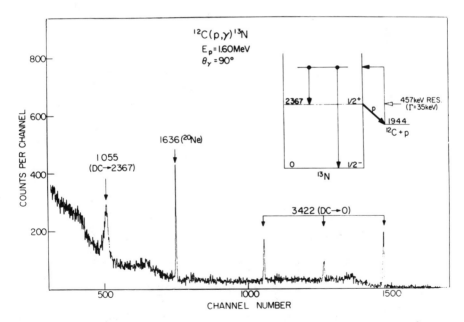

FIGURE 5.30. Gamma-ray spectrum of the $^{12}C(p, \gamma)^{13}N$ reaction obtained with a 45 cm³ Ge(Li) detector at $E_p = 1.60$ MeV and $\theta_\gamma = 90°$ (Rol74a).

tions. In the case of the DC → 2366 keV primary transition, the 2366 keV state decays mainly by proton emission because of its large (39 keV) proton (total) width. The observed width of the DC → 2366 keV γ-ray peak (Fig. 5.30) is consistent with the above width of the state. In this case the existence of the DC → 2366 keV transition could also be detected via this secondary proton emission. It should be pointed out that such transitions to unbound states with $\Gamma \gg \Gamma_\gamma$ are not relevant to nuclear astrophysics, since they do not contribute to the formation (synthesis) of the compound nucleus B. It should also be noted that the energies of the primary γ-ray transitions follow the projectile energy according to equation (5.17), and thus in performing a precise energy calibration of the Ge(Li) detector (Mar68, Alk82), one gets the beam energy E_p as a bonus (sec. 5.5). Of course, the energies of the secondary γ-ray transitions remain constant except possibly for small Doppler shifts. For nonresonant capture reactions or broad resonances, the observed peak width of the primary

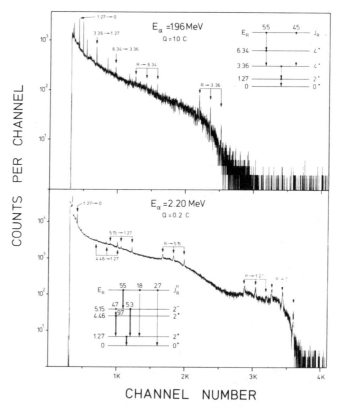

FIGURE 5.31. Gamma-ray spectra obtained at two resonances of the $^{18}O(\alpha, \gamma)^{22}Ne$ reaction are shown together with the identification of the observed γ-ray transitions (Tra78). These transitions (peaks) are superposed on an intense, neutron-induced background dominating the spectrum.

transitions to particle bound states reflects the target thickness (Fig. 5.30), if it is comparable to or larger than the resolution of the detector.

In all γ-ray measurements an accurate subtraction of background from the raw data is important. This background will normally include the ambient room and cosmic-ray background, the background associated with the operation of the accelerator, and the background due to contaminant reactions in the target. In the case of narrow resonances, backgrounds can be measured by running at energies just above or just below the resonant energy. Where this is not possible because the resonance is either not narrow or not well isolated from other resonances, the background at the resonant energy can be estimated by substituting a chemically similar target for the isotope under study. This procedure is particularly useful when gaseous targets are used. Of course, the best solution to contaminant background problems is to have targets as clean as possible. However, there are cases where the isotopic target itself causes a background. For example, studies of the capture reaction $^{18}O(\alpha, \gamma)^{22}Ne$ ($Q = +9.67$ MeV) are hampered by the reaction $^{18}O(\alpha, n)^{21}Ne$ ($Q = -0.70$ MeV), so that at all bombarding energies above $E_\alpha(\text{lab}) = 0.86$ there is a neutron-induced γ-ray background which is far greater than the number of capture γ-rays, as demonstrated in Figure 5.31 (Tra78). As can be seen in the figure, the high signal-to-noise ratio of a Ge(Li) detector allows one to observe and clearly identify the capture γ-ray transitions superposed on an intense and continuous neutron-induced γ-ray background. The price to be paid is neutron damage of the Ge(Li) detector. It follows from Figure 5.31 that the advent of the Ge(Li) detector made an enormous improvement in the signal-to-noise ratio in such studies where unavoidable backgrounds are present as well as for all studies of capture reactions with complicated γ-ray decay schemes and/or with low count rates (Buc84).

Gamma-ray spectroscopy has also been used extensively to provide information in studies of reactions such as (p, α), (p, n), (α, n), and (α, p) as well as in studies of heavy-ion–induced reactions such as $^{12}C + ^{12}C$ and $^{16}O + ^{16}O$. For example, the $^{15}N(p, \alpha_1)^{12}C^*$ reaction ($Q = +0.53$ MeV) proceeding to the 4.43 MeV first excited state in ^{12}C can be studied by detecting either the α_1-particle group or the 4.43 MeV γ-ray emitted from the product nucleus $^{12}C^*$. Because of the relatively low energies of the α_1-particles at sub-Coulomb energies, their observation in Si detectors is hampered by the intense elastic scattering yield, which increases with decreasing beam energy. It is therefore more convenient to study this reaction channel via the 4.43 MeV γ-ray intensity. Except for the direct production of residual nuclei in their ground state, the γ-ray technique provides information on the partial cross sections of a nuclear reaction nearly identical with that obtained by the light and heavy particle detection methods. Using high-resolution Ge(Li) detectors, a heavy product nucleus can be identified by the well-known energies of its characteristic γ-ray transitions (Hul80 and references quoted therein). If the residual nucleus is radioactive, measurement of its activity provides information on the total cross section, including that for direct population of the ground state (sec. 5.4.3). An important problem with all these γ-ray techniques is that, since the kinematic information specifying a given reaction (such as eq. [5.17]) is not available from these secondary reaction products, the detection and identification of characteristic γ-ray lines from the heavy residual nuclei cannot be associated uniquely with a given reaction, unless the target is extremely pure and/or the ion beam does not contain contaminant ions. For example, in the $^{16}O + {}^{16}O$ reaction studies of residual nuclei such as ^{24}Mg, ^{27}Al, and ^{27}Si (where α-particle emission is involved), traces of ^{12}C nuclei in the targets will also contribute to the observed γ-ray yields of these residual nuclei via the $^{12}C + {}^{16}O$ contaminant reaction. These additional contaminant yields can be significant, in particular at sub-Coulomb energies, because of the lower Coulomb barrier of the $^{12}C + {}^{16}O$ system. Therefore, it is of the utmost importance to use very pure targets and to take precautions to keep out any contaminants during the course of an experiment. These requirements can be met best with windowless gas targets (Hul80).

The detection of light reaction products such as neutrons, protons, and α-particles is of the same importance in studies of astrophysical reactions as γ-ray detection, and often the detection methods complement each other. This is illustrated for the $^{15}N(p, \alpha)^{12}C$ reaction, where the study of the α_1 channel is facilitated—as mentioned above—by γ-ray spectroscopic measurements, while that of the α_0 channel ($Q = +4.96$ MeV) requires a charged-particle detector. Figure 5.32 shows a spectrum of the $^{15}N(p, \alpha_0)^{12}C$ reaction obtained with a Si detector and a supersonic jet gas target (Red82). As can be seen in the figure,

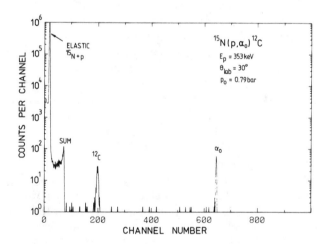

FIGURE 5.32. Particle spectrum obtained with a Si detector placed at $30°$ (Red82). A proton beam of 353 keV was incident on a supersonic jet gas target (highly enriched in ^{15}N) at an inlet pressure of the Laval nozzle of $p_0 = 0.79$ bars (Fig. 5.21). The light and heavy reaction products of the $^{15}N(p, \alpha_0)^{12}C$ reaction are seen together with the intense $^{15}N + p$ elastic scattering peak. Occasional electronic summing of pulses from this peak leads to the edge structure labeled "sum."

both the light (α_0) and heavy (^{12}C) reaction products of this reaction are observed together with the elastically scattered protons. The observed energies of these groups are consistent with the expected values from kinematic calculations. When angular distribution effects are taken into account, the relative number of counts in the α_0 and ^{12}C peaks is also consistent with expectation. Such kinematic and other criteria are usually applied in the identification of the observed charged particles. For measurements far below the Coulomb barrier, it is essential to mount the detectors close to the target to get a large solid angle of acceptance. In the measurements of ^{15}N(p, α_0)^{12}C six Si detectors were positioned in the angular range from 22° to 150° (Fig. 5.21), each detector having a solid angle of about 40 msr. Nickel foils of 1.2 μm thickness were placed in front of these detectors to stop the intense yield of elastically scattered protons. Of course, these foils also stopped the ^{12}C recoil nuclei, so that the spectra contained only the α_0-particles shifted somewhat in energy by the energy loss in the Ni foils.

In cases where high energy resolution is desired or where the reaction products have energies similar to that of the elastically scattered projectiles (Kie79), the best technique is to measure the deflection of the particles in a magnetic field (Hen74, Bro79). Ignoring relativistic corrections, the radius of curvature r of a particle with charge Ze, mass M, and kinetic energy E in a magnetic field of strength B is (eq. [5.6])

$$r = \frac{(2ME)^{1/2}}{ZeB} \, . \tag{5.20}$$

There are many designs for magnetic field shapes which focus the particles to positions that are functions of r, so a measurement of r involves only a measurement of position. Any of the detectors discussed in section 5.3.1 can be used to detect the particles.

Most detectors measure only the energies of particles with no information about the nature of the particles being observed. When only the particle of interest is present (Fig. 5.32), identification is often not necessary. However, if the reaction has many exit channels, such as in heavy-ion–induced reactions, it is generally important to identify the particles being detected. Any particle is uniquely characterized by the quantities E, Z, and M. The energy E is usually measured in a detector, where the particle is stopped. In a ΔE-E telescope (Fig. 5.33) a thin transmission detector is placed in front of a thick detector. The particle loses an energy ΔE in the thin detector, and the remainder, $E_r = E - \Delta E$, in the thick detector. By adding the coincident outputs from both detectors, one obtains a pulse proportional to the total energy E. The energy loss ΔE is given to first order by the relation

$$\Delta E \propto MZ^2/E \tag{5.21}$$

and thus is a measure of the product MZ^2, which is 1, 2, 3, and 16 for protons, deuterons, tritons, and α-particles, respectively. Separate energy spectra of all

ΔE-E TELESCOPE

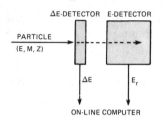

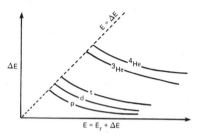

FIGURE 5.33. Experimental arrangement of a ΔE-E telescope and a plot of ΔE versus $E_r + \Delta E$ signals produced by the telescope are shown schematically. The pattern provides a unique isotopic fingerprint for light ions. The dotted line limits the low-energy range of the telescope, since these particles are fully stopped in the ΔE detector.

particles can therefore be measured simultaneously. A plot of the ΔE versus the E signals forms hyperbolae, displaced according to the quantity MZ^2 (Fig. 5.33). It should be pointed out, however, that the energy loss ΔE does not depend directly on the nuclear charge Z but depends on an effective charge $Z_{\text{eff}} \leq Z$, in particular for heavy ions at relatively low energies. Hence, the MZ^2 measurement with a ΔE-E telescope in general does not uniquely identify the particles. If the ΔE and E detectors are placed at a distance l, the time-of-flight t_f of the particles traveling from one detector to the other can be measured to yield

$$t_f = l\left(\frac{M}{2E}\right)^{1/2}. \tag{5.22}$$

With these three quantities (E, ΔE, and t_f), the particle is uniquely identified. Alternatively, a ΔE-E telescope placed at the focal plane of a deflecting magnet provides the same identification (Bro79 and references therein).

Finally, many-body reactions such as $^{11}\text{B}(p, \alpha)2\alpha$ and $^3\text{He}(^3\text{He}, 2p)^4\text{He}$ must be treated as special cases. In the three-body reaction $^{11}\text{B} + p \rightarrow 3\alpha$ ($Q = +8.68$ MeV) three identical particles are produced as reaction products. This situation creates problems in the analysis and interpretation of the α-particle spectra. As shown in Figure 5.34a the reaction can proceed—in a

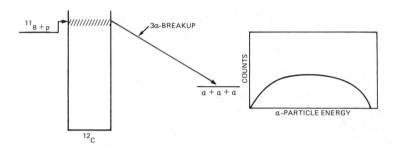

(a) DIRECT DECAY ^{11}B + p $\longrightarrow 3\alpha$ (SINGLE STEP)

^{11}B + p

3α-BREAKUP

$\overline{a + a + a}$

COUNTS

a-PARTICLE ENERGY

^{12}C

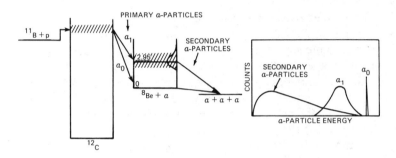

(b) SEQUENTIAL DECAY ^{11}B + p $\longrightarrow$ ^{8}Be + a $\longrightarrow 3\alpha$ (TWO STEP)

^{11}B + p

PRIMARY a-PARTICLES

a_1

a_0

2.95

0

^{8}Be + a

SECONDARY a-PARTICLES

$a + a + a$

COUNTS

SECONDARY a-PARTICLES

a_1

a_0

a-PARTICLE ENERGY

^{12}C

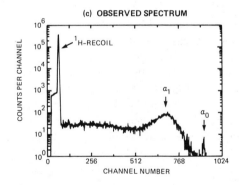

(c) OBSERVED SPECTRUM

^{1}H-RECOIL

a_1

a_0

COUNTS PER CHANNEL

CHANNEL NUMBER

FIGURE 5.34. Schematic diagram of the relevant level scheme and expected α-particle spectrum of the ^{11}B(p, α)2α reaction for the mechanisms of a direct breakup (a) and of a sequential decay (b). The observed spectrum (c) was obtained (Bec84) with a Si detector placed at $37°.5$ and a 3.5 MeV ^{11}B ion beam incident on a hydrogen jet gas target. The narrow low-energy peak arises from elastically scattered hydrogen recoil nuclei.

single step—directly into three α-particles, leading to a continuous energy distribution of the α-particles as determined by phase-space features (chap. 6). Alternatively, the reaction could be thought of as a sequential decay via the states in ^{8}Be (Fig. 5.34b), whereby the primary α-particles populating these states have well-defined energies (as in a normal two-body reaction) and their peak shapes reflect the total widths of these states. The subsequent breakup of the ^{8}Be recoil nuclei into two secondary α-particles is characterized by a continuous energy distribution. The observed spectrum could then be a superposition of both types of spectra. In order to observe the α-particle spectra over their full energy range, the reaction has been studied recently (Bec84) with a ^{11}B heavy-ion beam incident on a hydrogen gas target of the supersonic jet type (Fig. 5.21). A sample spectrum is shown in Figure 5.34c, indicating the presence of the sequential decay mechanism. If the α-particles are detected in a 4π detector over their full range of energies, the total cross section of the reaction is obtained by dividing the total number of counts by 3, since for each ^{11}B + p interaction three α-particles are emitted. In practice one has to use detectors with finite solid angles, and thus this correction factor depends critically on the type of reaction mechanisms involved (for example, two of the particles can be emitted with a very small angle separating them). Such information can be obtained by using several detectors placed around the target and operated under coincidence conditions. Such coincidence experiments have shown (Bec84) that the direct three α-particle breakup mechanism contributes at most a few percent to the ^{11}B$(p, \alpha)2\alpha$ reaction cross section, and that for detectors with finite solid angles the above correction factor is near 2.

5.4.3. Activity Method

Consider the problem of measuring the cross section for the reaction ^{14}N$(\alpha, \gamma)^{18}$F. One method would be to bombard a ^{14}N target with α-particles and detect the prompt capture γ-ray transitions. Their yield is then a measure of the total cross section (sec. 5.5). However, since the residual ^{18}F nucleus is unstable ($T_{1/2} = 110$ minutes), the cross section can be determined more easily by measuring the number of ^{18}F nuclei produced via the specific residual radioactivity (here positron emission). This is referred to as the activity method. The method involves two stages: (1) irradiation of the target by projectiles and (2) measurement of the residual radioactivity of the long-lived isotopes after the irradiation is stopped.

The net rate of increase in the number of nuclei of a radioactive isotope is given by the relation

$$\frac{dN(t)}{dt} = P(t) - \lambda N(t) . \tag{5.23}$$

The first term represents the rate of production via the nuclear reaction, and —since radioactive nuclei begin to decay as soon as they are formed—the second term is the rate of loss due to normal radioactive decay with the decay constant λ ($\lambda = \ln 2/T_{1/2}$). If the production rate $P(t)$ is constant with time, and for $N(t = 0) = 0$, the solution of the differential equation (5.23) is

$$N(t) = \frac{P}{\lambda}(1 - e^{-\lambda t}) \simeq \frac{P}{\lambda} \qquad \text{for } \lambda t \gg 1 . \qquad (5.24)$$

As shown in Figure 5.35, the growth of the function $N(t)$ approaches—for irradiation periods exceeding five half-lives—a saturation level of $N(t) = P/\lambda$. This situation is reached when the production rate P is equal to the spontaneous decay rate λN ($dN/dt = 0$ in eq. [5.23]).

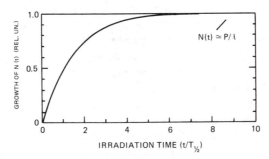

FIGURE 5.35. The growth of the number of nuclei of a radioactive isotope is shown as a function of irradiation time, expressed here in units of the half-life $T_{1/2}$. The calculation assumes a constant production rate P.

The production rate $P(t)$ is given by the relation (sec. 5.5)

$$P(t) = \int \sigma(E)I(t)y(t)dx , \qquad (5.25)$$

where $I(t)$ and $y(t)$ are the beam current (in particles per second) and the number of active target atoms (in atoms cm^{-3}), respectively. Note that the quantities in the integrand are implicit functions of x. These quantities could change during irradiation as a result of current fluctuations in the ion beam and instability of the target material (i.e., target deterioration), respectively. Due to the energy loss of the particles in the target and the energy dependence of the cross section $\sigma(E)$, one has to integrate over the target thickness. If thin targets are used, the cross section is often nearly constant over the target thickness, and equation (5.25) can be approximated by

$$P(t) = \sigma(E)I(t)n(t) , \qquad (5.26)$$

where $n(t)$ is now the total number of active target atoms (in atoms cm^{-2}). If the beam intensity $I(t)$ and target density $n(t)$ are recorded during the irradiation period, i.e., between $t = 0$ and $t = t_i$, the number of radioactive isotopes N_r present in the target after bombardment for a time period t_i is given by the relation

$$N_r = \sigma(E)e^{-\lambda t_i} \int_0^{t_i} I(t)n(t)e^{\lambda t}dt , \qquad (5.27)$$

where the numerical value of the integral is often obtained using an on-line computer.

After irradiation, the radioactive nuclei N_r will decay at their natural rate:

$$\frac{dN}{dt} = -N_r \lambda e^{-\lambda t} , \qquad (5.28)$$

with the time scale now starting at time t_i. From a measurement of the activity dN/dt, one can determine the number of radioactive nuclei present in the target and hence the absolute cross section $\sigma(E)$ using equation (5.27). From the energies of the radioactive decay products (electrons, positrons, γ-rays) as well as from the time dependence of the activity dN/dt (eq. [5.28]), one can determine the identity of the radioactive nuclei.

The activity method has been used in studies of capture reactions (chaps. 6 and 9). In this case all primary γ-ray transitions to particle bound states will eventually lead to the formation of the radioactive nucleus in its ground state, while transitions to particle unbound states do not (Fig. 5.30). Therefore, the method automatically provides a total cross section only for the former transitions, which is the quantity of astrophysical interest. The method is independent of details of the γ-ray decay scheme and of effects of γ-ray angular distributions (sec. 5.5.5). For the determination of very small cross sections this technique has the advantage of allowing measurement of the delayed activity at later times or at places removed from the intense, prompt, beam-induced radiations, and often in better geometry and with better shielding than would have been practical at the target location (Fig. 6.17). In practice, the activity technique is limited to radioactive nuclei with half-lives longer than a few seconds. In some applications such as the detection of solar neutrinos (chap. 10), the signal-to-noise ratio of the activity method is significantly enhanced by using radiochemistry techniques to chemically separate the radioactive nuclei from other elements present.

There are two pitfalls to watch for when using the activity method. First, in the above treatment it was assumed that all radioactive nuclei produced in the reactions remain in the target. If targets with thick backings are used, the nuclei will move toward the backing and be stopped there. Through the processes of diffusion or sputtering by the ion beam, a fraction of the nuclei could be lost to the target. By surrounding the target almost completely with a catcher foil, the loss rate can be measured. In practice, the loss rate is frequently found to be very small (of the order of a few percent). Second, impurities in the ion beam and in the target could cause additional production of the radioactive nuclei being studied (a similar problem has been discussed in sec. 5.4.2). In the $^{14}N(\alpha, \gamma)^{18}F$ example quoted above, ^{18}F nuclei could also be produced via reactions such as $^{16}O(^3He, p)^{18}F$, and $^{17}O(d, n)^{18}F$, and $^{18}O(p, n)^{18}F$.

Since the kinematic information and history of the formation of ^{18}F nuclei is lost in the activity method, the possibility of such contributions demands special attention. A typical capture cross section is of the order of 1 μb, while the cross sections of contaminant reactions such as those quoted above could be as high as about 100 mb. Hence, an equal amount of radioactive nuclei would be produced by the contaminant reactions if the product $I(t)n(t)$ is only 10^{-5} times that of the reaction of interest. Quantitative measurement and control of such small impurities in both the ion beam and the target material is extremely difficult. However, if such measurements are not carried out in the course of an irradiation, the value of the absolute cross section as deduced from the activity method might represent only an upper limit on the true value.

5.4.4. Time-of-Flight Techniques

In the time-of-flight (TOF) technique, the flight time t of a particle over a known path length l is measured (Fig. 5.36). From this measurement the velocity ($v = l/t$) of the particle can be obtained. If the mass of the particle is known, the energy of the particle can be determined:

$$E = \frac{1}{2} M \left(\frac{l}{t}\right)^2 . \qquad (5.29)$$

(a) NEUTRON TIME-OF-FLIGHT TECHNIQUE

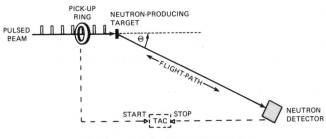

(b) CHARGED PARTICLE TIME-OF-FLIGHT TECHNIQUE

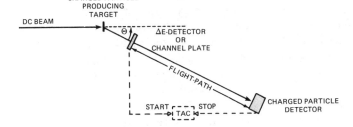

FIGURE 5.36. Schematic arrangements used for (a) neutron and (b) charged-particle time-of-flight measurements.

The energy resolution obtainable is then

$$\frac{\Delta E}{E} = \left[\left(\frac{2\Delta l}{l} \right)^2 + \left(\frac{2\Delta t}{t} \right)^2 \right]^{1/2} , \qquad (5.30)$$

where $\Delta l/l$ and $\Delta t/t$ are the fractional uncertainties in the measurements of the length of the flight path and the time. The $\Delta l/l$ uncertainty is usually small, but care must be taken in the geometrical arrangement of the timing devices, especially when short flight paths are being used. The uncertainty in the time measurement usually dominates, but it can be reduced by increasing the length of the flight path, albeit at the expense of solid angle. If the energy of the particle is known, the TOF technique can be used to determine the particle mass. This helps in identifying the particle.

The earliest applications of the TOF technique were in measurements of neutron energies (chap. 9) and in reducing background due to (α, n) reactions in studies of (α, γ) capture reactions (chap. 7 and Die73). In both applications a pulsed charged-particle beam (electrons or ions) from an accelerator is incident on a target, and the produced neutrons (or γ-rays) are observed in a detector at a distance l (Fig. 5.36a). In cyclic or linear accelerators the time structure of the pulsed beam is characteristic of the accelerator. In accelerators such as the Van de Graaff, beam bunching might be carried out usually at the ion source. The arrival time of the pulsed beam at the target (start signal) is obtained using a nonintercepting device such as a pickup ring placed close to the target. The stop signal is provided by the detector. For a flight path of 1 m and a neutron energy of 1 MeV, the neutron flight time is about 70 ns, while that of capture γ-rays is about 3 ns. Such differences in flight times are used in studies of (α, γ) capture reactions to discriminate between the prompt capture γ-rays and delayed γ-rays induced by neutron interactions in the detector and other places (chap. 7). For energy or mass measurements of charged particles emitted from a target, and experimental arrangement such as that shown in Figure 5.36b might be used (Eng74, Bro79).

The above example illustrates that good time resolution is essential when using TOF techniques. In charged-particle TOF measurements (Fig. 5.36b), a ΔE detector might be used to provide the start time and the E detector the stop signal. For light medium-energy particles this technique works well, but for heavy particles, or for low-energy light particles, multiple scattering in the first detector usually causes serious reduction in the number of particles arriving at the second detector, unless the area of this second detector is very large or the flight path relatively short. This problem can be significantly reduced if the start signal is derived from secondary electrons ejected from a thin carbon foil, which is biased at a high negative potential with respect to an electron multiplier such as a channel plate (Bro79). The carbon foil can be made as thin as 0.04 μm, which is thinner than most transmission ΔE detectors currently available. With such channel-plate devices time resolutions of the order of 1 ns or better can be obtained.

5.5. Experimental Procedures and Data Reduction

5.5.1. Cross Section and Yield of Nuclear Reactions

As discussed in chapter 3, the probability for the reaction A(x, y)B to occur is expressed in terms of a reaction cross section σ. Each target nucleus has associated with it an effective area σ (Fig. 5.37a), such that a projectile coming within this area will cause the reaction to proceed with a probability of unity. If a target of thickness Δx and area F contains n_0 active target nuclei (nuclei under study) and if the thickness Δx is small enough so that the individual areas do not overlap, the total active target area is given by $F_{act} = n_0 \sigma$. Dividing by the geometrical area F leads, then, to the reaction yield Y per incident particle:

$$Y = F_{act}/F = n_0 \sigma/F .$$

With $n = n_0/F\Delta x$ as the number of active target nuclei per unit volume, the yield for thin targets is

$$Y = \sigma n \Delta x . \tag{5.31}$$

For a solid target containing only active nuclei, n is given by

$$n = v\rho N_A/A , \tag{5.32}$$

where v, ρ, N_A, and A are respectively the number of atoms per molecule, the density of the target material, Avogadro's number ($N_A = 6.023 \times 10^{23}$), and the atomic or molecular weight of the target sample. For example, for an ^{27}Al target ($v = 1$, $\rho = 2.69$ g cm^{-3}, $A = 27$), one finds $n = 6.00 \times 10^{22}$ atoms cm^{-3}. For gas targets, the number of target atoms is given by

$$n = vL ,$$

with the Loschmidt number $L = 2.69 \times 10^{19}$ atoms cm^{-3} for a gas with a temperature of 0 °C and a pressure of 760 torr. This has to be modified for other values of temperature and pressure: $n = 9.66 \times 10^{18}$ vp/T, where T is in K and p is in torr. For example, nitrogen gas ($v = 2$) at 0 °C ($= 273$ K) and 0.76 torr contains $n = 5.38 \times 10^{16}$ atoms cm^{-3}.

Using the stopping cross section ϵ (sec. 5.2.1),

$$\epsilon = \frac{dE}{d\rho} = \frac{1}{n}\frac{dE}{dx} ,$$

the thin-target yield can be expressed alternatively as

$$Y = \sigma \frac{dE}{\epsilon} = \sigma \frac{\Delta}{\epsilon} , \tag{5.33}$$

where Δ is the energy loss of the projectiles in the target, i.e., the target thickness in units of electron volts. Since ϵ is tabulated in the laboratory system, the thickness Δ must be in the same system. If the target contains N_a active atoms

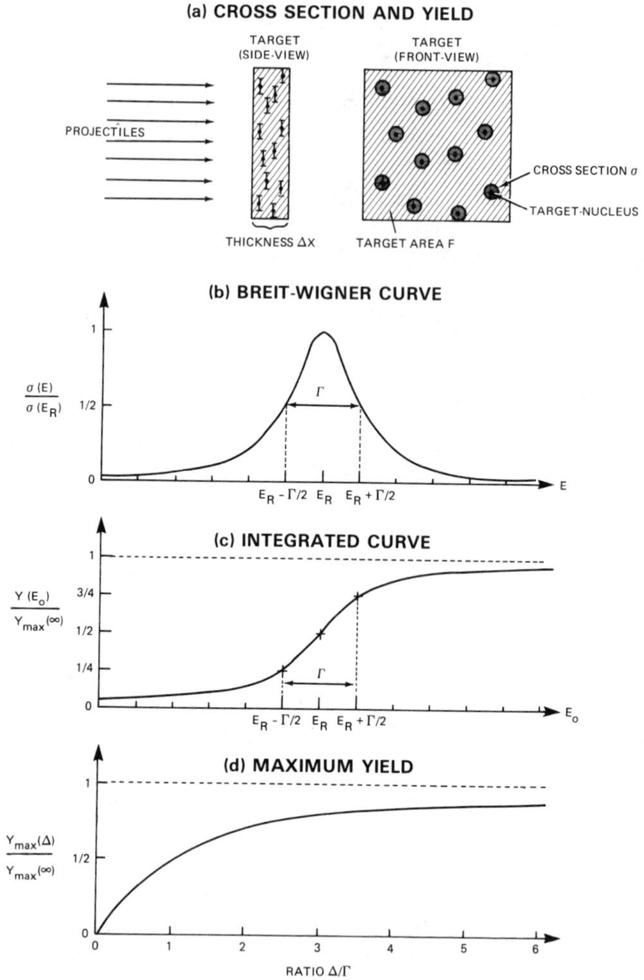

(a) CROSS SECTION AND YIELD

TARGET (SIDE-VIEW)

TARGET (FRONT-VIEW)

PROJECTILES

THICKNESS Δx

TARGET AREA F

CROSS SECTION σ

TARGET-NUCLEUS

(b) BREIT-WIGNER CURVE

$\dfrac{\sigma(E)}{\sigma(E_R)}$

Γ

$E_R - \Gamma/2 \quad E_R \quad E_R + \Gamma/2$

E

(c) INTEGRATED CURVE

$\dfrac{Y(E_0)}{Y_{max}(\infty)}$

Γ

$E_R - \Gamma/2 \quad E_R \quad E_R + \Gamma/2$

E_0

(d) MAXIMUM YIELD

$\dfrac{Y_{max}(\Delta)}{Y_{max}(\infty)}$

RATIO Δ/Γ

FIGURE 5.37. (*a*) Cross section for a nuclear reaction and active target area used to calculate the yield per incident projectile. (*b*) For a thin target of thickness Δx (in cm) or Δ (in eV) the reaction yield of a narrow resonance exhibits the Breit-Wigner curve, which has a maximum $\sigma(E_R)$ at $E = E_R$ and a FWHM of Γ. By integration of the Breit-Wigner cross section for a thick target, one obtains the curve shown in (*c*), which is a smoothed step function reaching $Y_{max}(\infty)$ asymptotically. The energy interval limited by the FWHM of the resonance curve corresponds to the integrated curve lying between $\frac{1}{2}$ and $\frac{3}{4}$ of $Y(E_0)/Y_{max}(\infty)$. For interim situations the maximum integrated resonance yield $Y_{max}(\Delta)$ compared with that for an infinite thick target $Y_{max}(\infty)$ is shown in (*d*) as a function of Δ/Γ.

per square centimeter ($N_a = n_a \Delta x$) and N_i inactive atoms per square centimeter of species i, the thin-target yield is given by

$$Y = \sigma \frac{\Delta_{\text{eff}}}{\epsilon_{\text{eff}}} , \tag{5.34}$$

where the effective target thickness Δ_{eff} and the effective stopping cross section ϵ_{eff} are determined by the relations

$$\Delta_{\text{eff}} = N_a \epsilon_a + \Sigma_i N_i \epsilon_i , \tag{5.35}$$

$$\epsilon_{\text{eff}} = \epsilon_1 + \frac{1}{N_a} \Sigma_i N_i \epsilon_i = \frac{\Delta_{\text{eff}}}{N_a} . \tag{5.36}$$

For example, in the diatomic compound Ta_2O_5 with oxygen as the active nuclei, one finds $\epsilon_{\text{eff}} = \epsilon_O + \frac{2}{5}\epsilon_{Ta}$. As noted previously, when chemical compounds are evaporated, one should not assume that the composition of the target will be the same as that of the original compound (Bec82a, Buc84a).

If the condition for a thin target is not fulfilled, the reaction yield per incident projectile is obtained by integrating the thin-target yield over the target thickness Δ:

$$Y(E_0) = \int \sigma(E) n \, dx = \int_{E_0 - \Delta}^{E_0} \frac{\sigma(E)}{\epsilon(E)} \, dE . \tag{5.37}$$

For an infinitely thick target the interaction extends from zero to E_0 (the incident energy of the projectiles). For a narrow resonance ($\Gamma \ll E_R$), the energy dependence of the cross section is decribed by the Breit-Wigner formula (chap. 4):

$$\sigma_{\text{BW}}(E, E_R) = \pi \lambdabar^2 \omega \frac{\Gamma_a \Gamma_b}{(E - E_R)^2 + (\Gamma/2)^2} . \tag{5.38}$$

If $\Delta \ll \Gamma$, the yield Y is proportional to the profile of the resonance (Fig. 5.37b). If $\Delta \gg \Gamma$ and if the energy dependence of the quantities $\lambdabar^2$, Γ_a, Γ_b, Δ, and ϵ is negligibly small over the region of the resonance, the yield is given by

$$Y(E_0) = \frac{\lambda^2}{2\pi} \omega\gamma \frac{M+m}{M} \frac{1}{\epsilon} \left[\arctan\left(\frac{E_0 - E_R}{\Gamma/2}\right) \right.$$

$$\left. - \arctan\left(\frac{E_0 - E_R - \Delta}{\Gamma/2}\right) \right]. \tag{5.39}$$

The mass ratio in this equation (M = mass of target nuclei, m = mass of projectile) takes into account that ϵ is usually given in the laboratory system. For a thick target, the yield reaches a maximum $Y_{\text{max}}(\infty)$ for $E_0 \gg E_R$, which represents the integral over the entire resonance region:

$$Y_{max}(\infty) = \frac{\lambda^2}{2} \, \omega\gamma \, \frac{M+m}{M} \frac{1}{\epsilon} . \tag{5.40}$$

The excitation function for a thick-target yield curve is shown in Figure 5.37c, where the 50% yield point corresponds to the resonance energy E_R and the energy interval between the 25% and 75% yield points is the resonance width Γ (assuming a negligible beam energy spread). For a finite target thickness Δ, equation (5.39) has a maximum at $E_0 = E_R + \Delta/2$ given by (Fow48)

$$\frac{Y_{max}(\Delta)}{Y_{max}(\infty)} = \frac{2}{\pi} \arctan\left(\frac{\Delta}{\Gamma}\right) , \tag{5.41}$$

which is illustrated in Figure 5.37d. From this figure one sees that, if the target thickness Δ is at least 6 times the natural resonance width, the maximum yield is essentially (89%) the thick-target yield.

5.5.2. Factors Affecting the Reaction Yield

As noted in section 5.1, charged-particle beams from accelerators are not monoenergetic. Their energy distribution is given by $g(E, E_0)$, where E_0 is the mean incident energy and $g(E, E_0)$ has the usual normalization

$$\int_0^\infty g(E, E_0) dE = 1 .$$

In many cases it is sufficient to approximate this distribution by a Gaussian function (Bon63, Rob84)

$$g(E, E_0) = \frac{1}{(2\pi)^{1/2}\delta_b} \exp\left[\frac{(E-E_0)^2}{2\delta_b^2}\right], \tag{5.42}$$

where the FWHM of the distribution is given by

$$\Delta_b = 2(2 \ln 2)^{1/2}\delta_b = 2.355\delta_b . \tag{5.43}$$

Neglecting the effects of beam energy straggling in the target, the reaction yield $Y(E_0)$ (eq. [5.37]) has to be folded with this beam energy distribution,

$$Y(E_0) = \int_{E_0-\Delta}^{E_0} \int_0^\infty \frac{\sigma(E')}{\epsilon(E')} g(E', E_0) dE' dE , \tag{5.44}$$

which must be evaluated numerically.

In the case of nonresonant reactions, i.e., $\sigma(E) =$ constant over the energy regions Δ and Δ_b, equation (5.44) is well approximated by equation (5.33) (thin-target yield). Similarly, for narrow resonances with $\sigma(E) = \sigma_{BW}(E, E_R)$ and for the condition $\Delta \gg \Gamma \gg \Delta_b$ (i.e., a negligible energy spread Δ_b in comparison with the resonance width Γ), all projectiles in the beam energy distribution contribute to the resonance yield, and the thick-target yield $Y_{max}(\infty)$ is given by equation (5.40). This conclusion is also valid for the condition

$\Delta \gg \Delta_b \gg \Gamma$. In the case $\Delta_b \gg \Delta \gg \Gamma$, the resonance yield is given approximately (Rob84) by

$$Y(E_0) \simeq Y_{\max}(\infty) \frac{\Delta}{\Delta_b} \exp\left[-\frac{(E_0 - E_R - \Delta/2)^2}{2\delta_b^2} \right], \tag{5.45}$$

i.e., at $E_0 = E_R + \Delta/2$ the reaction yield is reduced by the ratio Δ/Δ_b. Clearly, there is no advantage in choosing a target thickness Δ smaller than the beam energy spread Δ_b.

There is a further smearing of the effective beam energy distribution, referred to as the Doppler effect, which is a result of the thermal motion of the target atoms. The kinematic equations for the Doppler effect in nuclear reactions induced by nucleon bombardment were worked out by Bethe (Bet37) and Bethe and Placzek (Bet37a). In this case the velocities of the target nuclei are characterized by Maxwell-Boltzmann statistics, and the mean energy per degree of freedom is taken to be $\bar{\epsilon} \simeq kT$. The energy distribution of such gaseous target nuclei has a Gaussian shape, with the Doppler width given by

$$\delta_D = \left(2\,\frac{m}{M}\, EkT \right)^{1/2}$$

or a FWHM width of

$$\Delta_D = 4(\ln 2)^{1/2} \left(\frac{m}{M}\, EkT \right)^{1/2}. \tag{5.46}$$

For solid targets (crystalline), Lamb (Lam39) replaced the classical $\bar{\epsilon} \simeq kT$ approximation with the energy per mode of vibration of a Planck oscillator. For the reaction ^{27}Al$(p, \gamma)^{28}$Si it was found (Don67) that the improved calculations (using various theories of the specific heat such as the Einstein and Debye models) resulted in 40% larger Doppler widths compared with the classical approximation ($\Delta_D \simeq 40$–53 eV at $E_p = 0.77$–1.36 MeV; see also Goe85).

The beam energy spread δ_b and the Doppler broadening δ_D combine to give an effective beam spread that is Gaussian in shape with a width equal to the quadrature of the beam and Doppler standard widths,

$$\delta = (\delta_b^2 + \delta_D^2)^{1/2}.$$

In many practical situations $\delta_D \ll \delta_b$ and thus $\delta \simeq \delta_b$ (Fig. 5.12). The Doppler width can be reduced either by lowering the target temperature or by forcing the incident and target particles to add their velocities like energies, i.e., quadratically instead of linearly. Quadratic addition applies when two beams of particles collide at right angles. This crossed-beam method has been used in the $\alpha + \alpha$ elastic scattering studies to the ^{8}Be ground state (Ben68).

When a beam of charged particles penetrates matter, the slowing down is accompanied by a spreading of the beam energy owing to statistical fluctuations in the number of collisions, the energy straggling. For a homogeneous target the reaction yield $Y(E_0)$ is then given by the triple integral

$$Y(E_0) = \int_{E_0-\Delta}^{E_0} \int_0^\infty \int_0^\infty \frac{\sigma(E'')}{\epsilon(E'')} g(E', E_0) W(E, E', E'') dE'' dE' dE , \qquad (5.47)$$

where $W(E, E', E'')$ is the probability that a projectile incident on the target at an energy E'' has an energy between E' and $E' + dE'$ at a depth inside the target corresponding to the energy E ($E = E_0 - \Delta$ to $E = E_0$). In many cases of practical interest, the distribution in energy loss $W(E, E', E'')$ is sufficiently close to a Gaussian that the spreading around the average value is completely characterized by the average squared fluctuation (squared standard deviation) in energy loss Ω^2 (Bes80). For qualitative estimates of energy straggling it is often sufficient to use Bohr's basic formula (Boh48, Bes80)

$$\Omega_B^2 = 4\pi e^2 Z_1^2 Z_2 N = 2.60 \times 10^{-19} Z_1^2 Z_2 N \text{ keV}^2 , \qquad (5.48)$$

where Z_1 and Z_2 are the atomic numbers of the projectile and target atoms, respectively, and N is the target density (atoms cm^{-2}). Note that this expression is independent of projectile energy and independent of the distribution of the target electrons in velocity and space. For example, straggling of protons incident on neon target nuclei with $N = 1.8 \times 10^{15}$ atoms cm^{-2} is $\Omega_B \simeq 70$ eV (FWHM $\simeq 165$ eV). In comparison, for a proton energy of $E_p = 300$ keV the mean energy loss in the target (i.e., the target thickness Δ) is 12 eV. Thus the energy straggling in a target can be larger than the target thickness. Since $\Omega_B \propto N^{1/2}$ and $\Delta \propto N$, the ratio $\Omega_B/N \propto 1/N^{1/2}$ decreases with increasing target density N at a given projectile energy. For a homogeneous target the thick-target yield curve for a narrow resonance is expected to be a step function with a flat plateau (Fig. 5.17a). The width of the low-energy edge is due to the combination of beam energy resolution Δ_b, Doppler width Δ_D, and resonance width Γ. For the case shown in Figure 5.17a the observed width is consistent with the known values of 1.1 keV for Γ and 0.5 keV for Δ_b. The high-energy edge of the yield curve (Fig. 5.17a) has a much shallower slope than the front edge because of energy straggling. (For a Gaussian distribution the FWHM width corresponds to the 12% and 88% points of the integrated distribution.) The extracted value of $\Omega \simeq 17$ keV agrees fairly well with the Bohr model prediction when the range straggling of the implanted ^{14}N ions is included.

Finally, if the target is nonhomogeneous because of contaminants in the target or nonuniformities in the thickness, the effective stopping power $\epsilon_{\text{eff}}(E'', x)$ varies over the target thickness, which has to be taken into account in calculating the reaction yield curve. Such target contaminants critically influence the observation of an effect in resonance yield curves called the Lewis effect (Lew62, Don67). This effect is due to the discrete nature of energy losses suffered by the projectiles in the target and leads to a peak in the thick-target yield curve slightly above the resonance energy.

5.5.3. Reaction Yield and Experimental Observation

In the above sections we have discussed the reaction yield per incident projectile for a nuclear reaction A(x, y)B. If N_p projectiles are impinging on the target over a time period t, the resulting yield is $N_p Y$, which is in general emitted in all directions. In an actual experiment a detector of finite solid angle $d\Omega$ is placed at a detection angle θ (Fig. 5.38a), and thus only a fraction of the yield $N_p Y$ is detected. Often the light reaction products y are detected and the number of events $N_y(\theta)$ registered in the detector over the time period t is given by the equation

$$N_y(\theta) = N_p Y \epsilon_y d\Omega_y W_y(\theta) , \qquad (5.49)$$

where ϵ_y and $d\Omega_y$ are the absolute detection efficiency and the solid angle of the detector, respectively, and $W_y(\theta)$ represents the angular distribution of the

(a) DETECTION OF REACTION PRODUCTS

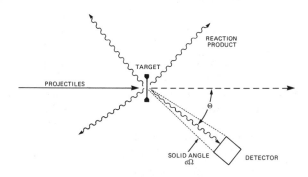

(b) EXPERIMENTAL SET-UP FOR STUDIES OF $^{18}O(p, a_0)^{15}N$

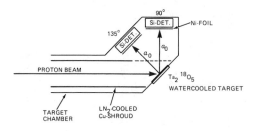

FIGURE 5.38. (a) The reaction products of a nuclear reaction are observed in a detector placed at an angle θ with respect to the beam direction. The detector subtends a solid angle $d\Omega$. (b) Experimental setup used in the studies of $^{18}O(p, \alpha_0)^{15}N$ (Lor79).

reaction products y. For unpolarized projectiles and target nuclei this distribution is symmetrical around the beam axis, depending only on the angle θ with respect to the beam direction. This angular distribution $W_y(\theta)$, as well as the quantities N_p, ϵ_y, and $d\Omega_y$, must be known to extract the reaction yield Y from the observed number of events $N_y(\theta)$. For astrophysical applications the reaction yield $Y(E)$ has to be known over a wide range of projectile energies E, and thus the energy dependence of the detection efficiency $\epsilon_y(E)$ and of the angular distribution $W_y(E, \theta)$ also must be known. The observation of low reaction yields is facilitated by high beam currents and detectors of high efficiency and large solid angles. For example, if the detector is placed at $\theta = 90°$ with a solid angle approaching a 2π geometry, the reaction products y are observed over the angular range $\theta \simeq 0°–180°$. Consequently, the registered events $N_y(\theta)$ represent angle-integrated information, and no angular distributions $W_y(E, \theta)$ need to be measured (Ket82). A similar situation prevails if the heavy reaction product B is radioactive and its production yield is measured via the activity method (sec. 5.4.3).

In the laboratory the energy dependence of the reaction yield $Y(E)$ is usually obtained by combining the following steps: (1) measurement of detailed excitation functions over a wide range of beam energies using a few detectors in close geometry, (2) measurement of angular distributions at selected beam energies over this energy range, and (3) determination of the absolute yield (i.e., cross section) at one beam energy. The experimental procedures depend also in part on the type of reaction mechanisms involved. Details of these steps and procedures are discussed in section 5.5.4.

5.5.4. Measurement of Excitation Functions

To obtain information on the reaction yield $Y(E)$ over a wide range of beam energies, in particular at as low an energy as possible, an experimental setup has to be devised which allows the use of high beam currents and detectors in close geometry. The number of counts $N_y(\theta, E)$ observed in a detector corrected for the number of incident projectiles $N_p(E)$, the detection efficiency $\epsilon_y(E)$, and the solid angle $d\Omega_y(E)$ yielded

$$\frac{N_y(\theta, E)}{N_p(E)\epsilon_y(E)d\Omega_y(E)} = Y(E)W_y^*(\theta, E) , \qquad (5.50)$$

where $W_y^*(\theta, E)$ represents the angular distribution averaged over the solid angle $d\Omega^y(E)$. In such studies it is sufficient to know the energy dependence of the quantities $N_p(E)$, $\epsilon_y(E)$, $d\Omega_y(E)$ to high accuracy. Their absolute values are needed only for the determination of the absolute yields (sec. 5.5.6). Thus equation (5.50) can be written as

$$\frac{N_y(\theta, E)}{[N_p(E)\epsilon_y(E)d\Omega_y(E)]_{rel}} = CY(E)W_y^*(\theta, E) , \qquad (5.51)$$

where the denominator contains only the relative values of the quantities $N_p(E)$, $\epsilon_y(E)$, and $d\Omega_y(E)$, i.e., their energy dependence. Their absolute magnitudes are incorporated in the constant C. Such measurements of relative excitation functions should be carried out in beam energy steps ΔE nearly equal to the target thickness Δ, in order to obtain complete information on the reaction mechanisms involved (e.g., the existence of narrow resonances).

The above procedures are illustrated using the $^{18}O(p, \alpha)^{15}N$ reaction (chap. 6). From the Q-value of this reaction ($Q = +3.797$ MeV) and the level structure of ^{15}N it is clear that at low energies the reaction can proceed only to the ^{15}N ground state. The emitted light reaction products, the α_0-particles, can be detected conveniently using Si detectors (Fig. 5.38b). To allow the use of high beam currents (up to 300 μA), in particular at the lowest beam energies, solidly backed $Ta_2{}^{18}O_5$ targets were used. These were produced by anodizing 0.2 mm thick Ta sheets in 99.9% ^{18}O–enriched water. The targets (3–20 keV thick at $E_p = 150$ keV) were mounted in a target holder at an angle of 45° with respect to the beam direction. Direct water cooling was applied to the backing. The targets were able to withstand intense beams for periods greater than several days without noticeable deterioration. A LN_2-cooled copper tube (30 cm in length) coming to within 5 mm of the target, together with the target, constituted the Faraday cup for beam integration. This target setup allowed only two Si detectors to be installed (with an active area of 300 mm^2) at angles of 90° and 135° to the beam direction and at a distance of 7 cm from the target. The detectors were insulated electrically from the target chamber. The effective solid angles $d\Omega_y$ of both Si detectors were determined by placing an Am source at the target position. Ni foils of 2 μm thickness were placed in front of the detectors to stop the elastically scattered protons. With these foils the energy resolution for particle detection went from 20 to 70 keV.

The measurements of excitation functions at $\theta_\alpha = 90°$ and 135° were carried out over a range of beam energies from $E_p = 72$ to $E_p = 935$ keV in steps of $\Delta E = 0.5$–10 keV. Since the solid angles and detection efficiencies of the Si detectors are energy-independent, the observed counts of the α_0-particles had to be corrected only for the variation in accumulated charge at each beam energy, i.e., the variation in $N_p(E)$. The resulting excitation functions are illustrated in Figure 5.39. The data reveal a few narrow resonances superposed on a smoothly varying yield, which drops by more than 10 orders of magnitude over the energy region investigated. The gross energy dependence of the data is largely due to the Coulomb barrier, as can be seen from the conversion of the data into the astrophysical $S(E)$ factor (lower part of Fig. 5.39). Four new resonances at low energies (inset in Fig. 5.39) were observed.

To calculate the reaction rate $\langle \sigma v \rangle$ for narrow resonances, values for their resonance energies E_R and resonance strengths $\omega\gamma$ are needed (chap. 4). Of course, the total resonance width Γ should also be known, to ensure that the condition of a narrow resonance prevails. Information on E_R and Γ can be obtained from thick-target yield curves. For example, from detailed excitation

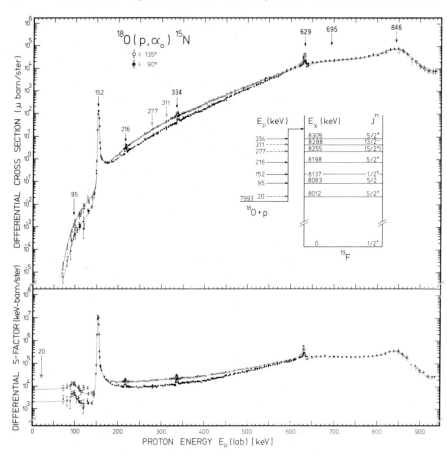

FIGURE 5.39. Excitation functions for $^{18}O(p, \alpha_0)^{15}N$ obtained at 90° and 135° are shown in the top figure (Lor79). The reaction yield has already been converted into an absolute scale and represents—in the nonresonant energy region—the differential cross section (sec. 5.5.7). The data have been corrected for effects of finite target thickness, i.e., effective beam energy within the target. The data are also shown in the form of the astrophysical $S(E)$ factor (*bottom figure*). The lines through the data points are to guide the eye.

functions obtained near the 152 keV resonance of $^{18}O(p, \alpha)^{15}N$ (in energy steps of 0.5 keV), values of $E_R = 152 \pm 1$ keV and $\Gamma \leq 0.5$ keV have been deduced (Lor79). From the observed yield on the plateau of the thick-target yield curve corrected for effects of the angular distribution and finite target thickness (eq. [5.41]), information is obtained on the relative strengths $(\omega\gamma)_{rel}$ of the observed resonances, which finally has to be converted into absolute values (sec. 5.5.6). Note that the resonance energy E_R enters exponentially in the calculation of the reaction rate $[\langle\sigma v\rangle_{res} \propto \exp(-11.605 E_R/T_9)$ with E_R(center of mass) in MeV] and therefore should be known absolutely to high accuracy. For example, an error of ± 5 keV for the $E_R(\text{lab}) = 152$ keV resonance translates at $T_9 = 0.1$ into an error in the reaction rate of -42% to $+73\%$.

Calculation of the reaction rate $\langle \sigma v \rangle$ for the nonresonant mechanism of a given reaction requires (chap. 4) a knowledge of the energy dependence of the cross section $\sigma(E)$. The observed yield $Y(E_0)$ at an incident projectile energy E_0 represents the cross section $\sigma(E)$ integrated over the target thickness Δ (eq. [5.37]):

$$Y(E_0) = \int_{E_0 - \Delta}^{E_0} \frac{\sigma(E)}{\epsilon(E)} \, dE .$$

The stopping power $\epsilon(E)$ is, in most practical situations ($\Delta \ll E_0$), nearly constant over the energy interval $E_0 - \Delta$ to E_0 and thus can be removed from the integral:

$$Y(E_0) = \frac{1}{\epsilon(E_0)} \int_{E_0 - \Delta}^{E_0} \sigma(E) dE . \tag{5.52}$$

If the cross section is also energy-independent over this interval, the mean effective energy in the target is $E_{\text{eff}} = E_0 - \Delta/2$, and the cross section at this energy is given by

$$\sigma(E_{\text{eff}}) = Y(E_0)\epsilon(E_0)/\Delta . \tag{5.53}$$

If the cross section is energy-dependent over the target thickness, determination of the effective beam energy requires a knowledge of $\sigma(E)$. In this case the effective beam energy is defined by the equation

$$Y(E_0) = \frac{1}{\epsilon(E_0)} \int_{E_0 - \Delta}^{E_0} \sigma(E) dE = \frac{2}{\epsilon(E_0)} \int_{E_{\text{eff}}}^{E_0} \sigma(E) dE , \tag{5.54}$$

where the effective energy E_{eff} corresponds to a beam energy in the target at which one-half of the yield for the full target thickness is obtained (Fig. 5.40). The cross section deduced from the observed yield $Y(E_0)$ via equation (5.53) corresponds, then, to the cross section at E_{eff}. At sub-Coulomb energies the cross section $\sigma(E)$ drops steeply with decreasing beam energy, mainly because of the Coulomb barrier (chap. 4):

$$\sigma(E) = S(E) \frac{1}{E} \exp(-2\pi\eta) .$$

For non-resonant reactions the astrophysical $S(E)$ factor varies more gently with beam energy and is often nearly constant over the relatively small energy interval of the target thickness (e.g., Fig. 5.39). In this case the integral in equation (5.54) reduces to

$$\int_{E_0 - \Delta}^{E_0} \frac{1}{E} \exp(-2\pi\eta) dE = 2 \int_{E_{\text{eff}}}^{E_0} \frac{1}{E} \exp(-2\pi\eta) dE ,$$

which must be evaluated numerically. Assuming a linear decrease in cross section from σ_1 at E_0 to σ_2 at $E_0 - \Delta$, the effective beam energy can be calculated from the expression

$$E_{eff} = E_0 - \Delta + \Delta\left\{-\frac{\sigma_2}{\sigma_1 - \sigma_2} + \left[\frac{\sigma_1^2 + \sigma_1^2}{2(\sigma_1 - \sigma_2)^2}\right]^{1/2}\right\}, \tag{5.55}$$

which is a good approximation (to better than 6%) for ratios $\sigma_1/\sigma_2 \leq 10$. For example, for the $^{18}O(p, \alpha)^{15}N$ reaction with $E_p(\text{lab}) = 70$ keV and a target thickness of $\Delta(E_p = 70 \text{ keV}) = 10$ keV, the cross section drops by a factor of $\sigma_1/\sigma_2 = 9.5$ [assuming $S(E) = $ constant] and thus $E_{eff} = 66.0$ keV. Numerical integration yields $E_{eff} = 67.2$ keV. Neglecting any target thickness correction, i.e., $E_{eff} \equiv E_0 = 70$ keV, the deduced cross section would be in error by a factor of 1.8. For $E_{eff} = E_0 - \Delta/2 = 65$ keV, the error would be 62%. These discussions illustrate again that an accurate knowledge of the absolute effective beam energy associated with an observed reaction yield is as important at low beam energies as the yield measurements themselves. The reliability and accuracy of the above procedures can be tested sensitively by analyzing data obtained with targets of various thicknesses (Ket77, Ket80). If the energies of the reaction products can be measured to high accuracy, they can provide independently accurate information on the effective beam energy (Krä82 and sec. 5.4.2). Note that if the cross section drops steeply with decreasing beam energy, there is no advantage in using very thick targets, since only the first few atomic layers in the target contribute to the observed reaction yield.

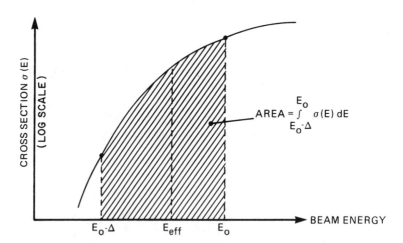

FIGURE 5.40. Shown is an energy-dependent cross section $\sigma(E)$ of a nuclear reaction. If the reaction is investigated with projectiles of incident energy E_0 and a target of thickness Δ, the observed reaction yield $Y(E_0)$ corresponds to the cross section integrated over the thickness Δ (*shaded area*). The mean effective beam energy E_{eff} is defined as that energy which divides the shaded area into two equal areas.

Usually the relative number of incident projectiles $N_p(E)$ is determined either by charge integration of the ion beam (solid targets) or by beam power integration using a calorimeter (gas targets). In the case of gas targets the elastic scattering of the projectiles on the gas target nuclei has also been observed in particle detectors and has been used to determine both the beam intensity and the target density. This method requires a priori knowledge of the energy dependence of the elastic scattering cross section $\sigma_{sc}(E, \theta)$. If the gas target density $N_t(E)$ is known from other sources [e.g., from the gas pressure measurement $p(E)$ in the target chamber: $N_t(E) \propto p(E)$] and if the elastic scattering cross section for nonidentical interacting particles is shown to follow the Rutherford scattering law (at least at low projectile energies) the relative numbers of projectiles at energies E_1 and E_2 are related to the respective numbers of observed scattering counts $N_{sc}(E)$ in the detector by the equation

$$\frac{N_p(E_1)}{N_p(E_2)} = \frac{\sigma_{sc}(E_2, \theta)}{\sigma_{sc}(E_1, E)} \frac{p(E_2)}{p(E_1)} \frac{N_{sc}(E_1)}{N_{sc}(E_2)} = \left(\frac{E_1}{E_2}\right)^2 \frac{p(E_2)}{p(E_1)} \frac{N_{sc}(E_1)}{N_{sc}(E_2)}.$$

5.5.5. Measurement of Angular Distributions

To obtain angle-independent information on the reaction yield $Y(E)$, equation (5.51) has to be corrected for the effects of angular distributions $W_y(\theta, E)$, which are described by a sum of Legendre polynomials $P_k(\theta)$:

$$W_y(\theta, E) = \Sigma_k a_k(E) Q_k P_k(\theta) \qquad (k = 0, 1, 2, \ldots), \tag{5.56}$$

where $a_k(E)$ are the energy-dependent coefficients and Q_k the attenuation coefficients due to the finite size of the detectors. The latter coefficients can be calculated from the geometry of the setup and tested experimentally using known distributions from other reactions.

In the $^{18}O(p, \alpha_0)^{15}N$ studies measurements of α_0 angular distributions were carried out in a scattering chamber, where a $Ta_2{}^{18}O_5$ target was positioned in the center of the chamber at an angle of 15° with respect to the beam direction. The beam direction was defined by 3 and 5 mm diameter Ta collimators placed near the entrance of the chamber. The α_0-particles were observed with a fixed Si detector (monitor) at $\theta = 135°$ ($d = 15$ cm) as well as with a Si detector, which could be rotated around the target at the angular range $\theta = 20°-160°$ ($d = 8$ cm). The resulting distributions obtained at selected energies are illustrated in Figure 5.41. The isotropic distribution observed at the $E_R = 152$ keV resonance is consistent with the $J_R^\pi = \frac{1}{2}^+$ assignment to this resonance. At energies away from this resonance the angular distributions exhibit pronounced anisotropies asymmetric around $\theta = 90°$, which can be explained in terms of amplitudes of different parities (Lor79).

The above setup was also involved in the studies of the $^{15}N(p, \alpha_0)^{12}C$ reaction using $Ti^{15}N$ solid targets. In addition, a quasi-point jet gas target (^{15}N enriched gas) was used with six Si detectors placed around the jet target in the

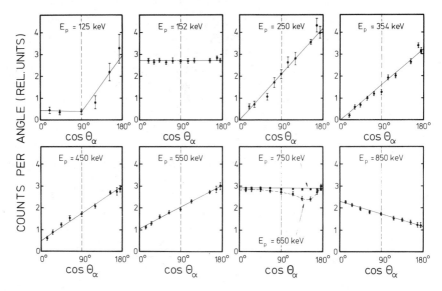

FIGURE 5.41. Results of α_0 angular distribution measurements for $^{18}O(p, \alpha_0)^{15}N$ at several beam energies (Lor79). The lines through the data points are to guide the eye.

angular range of $\theta = 22°.5–150°$ (Fig. 5.21). The observed angular distributions exhibit anisotropies asymmetric around 90°, which can be described by the expression

$$W(\theta, E) = 1 + a_1(E)P_1(\theta) .$$

The deduced a_1 coefficients as a function of proton energy are shown in Figure 5.42. Again interfering amplitudes of differing parities are required to explain these data (Red82).

Both examples emphasize that at energies far below the Coulomb barrier not only do s–partial waves in the incoming channel (and therefore leading to isotropic angular distributions) need to be considered, but also higher partial waves are required for interpretation of the data. The data caution against assuming angular distributions symmetric around 90° for nonresonant energy regions.

5.5.6. Absolute Cross Section and Resonance Strength

Measurements of excitation functions at selected detection angles and of angular distributions at appropriate beam energies provide information on the angle-integrated yield curve for a given reaction. The absolute scale of this yield curve is of critical importance to nuclear astrophysics. The relevant quantities needed are the absolute cross section σ for the nonresonant part of the reaction mechanism and the absolute resonance strength $\omega\gamma$ of the narrow resonances (chap. 4). If one of these quantities is known, the other can be determined via the expression (eqs. [5.33] and [5.40])

$$\frac{Y_r(\infty)}{Y_{nr}(E)} = \frac{\lambda^2(E_R)}{2} \frac{M + m}{M} \frac{\epsilon_{lab}(E)}{\epsilon_{lab}(E_R)} \frac{\omega\gamma(E_R)}{\sigma(W)\Delta_{lab}(E)}, \tag{5.57}$$

where the resonance yield $Y_r(\infty)$ (for a thick target) is obtained at the resonance energy E_R and the nonresonant yield $Y_{nr}(E)$ at the energy E. For the condition $E \simeq E_R$, the ratio of stopping powers cancels and the target thickness $\Delta_{lab}(E)$ can be deduced from the thick-target yield curve at the narrow resonance.

Combining equations (5.33) and (5.49), the absolute cross section $\sigma(E)$ is related to the observed number of events $N_y(\theta, E)$ in the detector by the relation

$$\sigma(E) = \frac{N_y(E, \theta)\epsilon_{lab}(E)}{N_p(E)\epsilon_y(E)d\Omega_y(E)W_y(E, \theta)\Delta_{lab}(E)}. \tag{5.58}$$

Thus, accurate information must be available on the angular distribution $W_y(E, \theta)$, the absolute number of projectiles $N_p(E)$, the detector properties $\epsilon_y(E)d\Omega_y(E)$, and the target features $\epsilon_{lab}(E)/\Delta_{lab}(E)$. Their determination is often quite time-consuming and frequently involves novel techniques.

If the reaction products y can be observed in a detector concurrently with

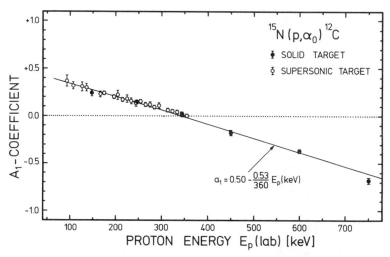

FIGURE 5.42. Observed angular distributions of the $^{15}N(p, \alpha_0)^{12}C$ reaction can be described by the expression $W(\theta) = 1 + a_1(E)P_1(\theta)$, where the deduced $a_1(E)$ coefficients are shown as a function of proton energy. The data have been obtained using Ti^{15}N solid targets as well as a ^{15}N-enriched supersonic jet gas target (Red82). The solid line through the data points assumes a relationship as given in the figure.

the elastic scattering yield, the above procedure is extremely simplified. For example, in the $^{18}O(p, \alpha_0)^{15}N$ studies (Lor79) the elastically scattered protons from the ^{18}O gas target nuclei were observed concurrently with the α_0-particles in a Si counter (Fig. 5.43). The ratio of observed counts in the two peaks depends only on the ratio of differential cross sections and the correction of solid angles from the center-of-mass system to the laboratory system (Mar68):

$$\frac{N_\alpha(E, \theta)}{N_{el}(E, \theta)} = \left\{ \left[\frac{d\sigma(E, \theta)}{d\Omega} \right]_\alpha \Big/ \left[\frac{d\sigma(E, \theta)}{d\Omega} \right]_{el} \right\} \left(\frac{d\Omega_\alpha}{d\Omega_{el}} \right)_{lab/c.m.} \tag{5.59}$$

It was also shown that the elastic scattering from ^{18}O followed the Rutherford scattering law. The deduced value for $d\sigma/d\Omega(E, \theta)_\alpha$ together with the available information on the α_0 angular distribution (Fig. 5.41) was then used to arrive at the total cross section of $\sigma = 0.68 \pm 0.06$ mb for $E_p = 345$ keV. This cross section was then used as the standard to scale the absolute rates of the non-resonant parts of the excitation functions (Fig. 5.39) as well as to deduce the resonance strengths $\omega\gamma$ of the narrow resonances using equation (5.57) (e.g., $\omega\gamma = 0.17 \pm 0.02$ eV at $E_R = 152$ keV).

Similarly, in γ-ray spectroscopic studies of a given nuclear reaction the Coulomb excitation process for the target-projectile system has been used (Hul80) as an intrinsic calibration standard in determining the absolute cross sections.

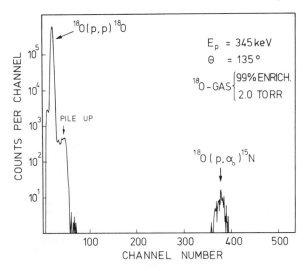

FIGURE 5.43. Particle spectrum obtained with a well-collimated Si detector (without Ni foil in front of the detector) at an angle of 135° by bombarding a 99% enriched ^{18}O gas target (2.0 torr) with protons of incident energy $E_p = 345$ keV (Lor79).

In order to avoid in part some of the above requirements for an absolute determination of resonance strengths (or cross sections), the resonance strengths for a given reaction of interest, A(x, y)B, are often determined relative to the resonance strength of a standard reaction (Eng66), C(z, w)D, in the same experimental setup (eqs. [5.40] and [5.49]):

$$\frac{N_y(\theta)}{N_w(\theta)} = \frac{N_p(x)}{N_p(z)} \frac{\epsilon_y}{\epsilon_w} \frac{W_y(\theta)}{W_w(\theta)} \frac{\lambda_x^2}{\lambda_z^2} \frac{M_A + m_x}{M_A} \frac{M_C}{M_C + m_z} \frac{\epsilon_{lab}(A)}{\epsilon_{lab}(C)} \frac{\omega\gamma(y)}{\omega\gamma(w)}. \tag{5.60}$$

Since only the ratio of the various quantities needs to be known, measurements can be made with higher precision than absolute measurements, if the standard strength $\omega\gamma(w)$ is well known (Wie80, Buc84).

Discrepancies of a factor of 2 and more have been noted in some of these standards, in part because of the assumed target stoichiometries, which could be different from the stoichiometry of the original target material. This situation prevails in particular for the nitrogen, oxygen, and fluorine nuclei, where some chemical compound is usually used to make solid targets. Recently (Bec82a) measurements of resonance strengths for such target nuclei have been made using a windowless gas target and heavy-ion beams. The strong resonances in $^{19}F(p, \alpha\gamma)^{16}O$ at $E_{c.m.} = 324$ and $E_{c.m.} = 829$ keV proceed predominantly via the 6.13 MeV state in ^{16}O, and hence the α_2-particle group feeding this state and the subsequent 6.13 MeV γ-ray decay of this state are produced in equal numbers. In the first step of the method, the resonance strengths of the α_2-particle group were determined relative to the $p + {}^{19}F$ Rutherford elastic scattering cross section, in a manner similar to that described above for the $^{18}O(p, \alpha_0)^{15}N$ measurements. The studies were carried out by interchanging the roles of projectiles and target nuclei. ^{19}F ions of $E_{lab} = 6.42$ and $E_{lab} = 16.44$ MeV were directed onto a supersonic jet hydrogen gas target (Fig. 5.21), where the deuterium in the gas was depleted to 0.001%. A sample particle spectrum obtained at the $E_{c.m.} = 324$ keV resonance is shown in Figure 5.44. The results are $\omega\gamma(\alpha_2) = 22.3 \pm 0.8$ eV and 570 ± 30 eV for the 324 and 829 keV resonances, respectively. Both resonances provided a γ-ray strength standard via the 6.13 MeV transition. In the second step of the method, both resonances were studied via γ-ray spectroscopy using an extended gas target system (Fig. 5.19b) and a Ge(Li) detector placed at 90°. With the above standards for the 6.13 MeV γ-ray yield, the γ-ray strength of several (p, γ) and (α, γ) induced resonance reactions could be related to this standard simply by changing the ^{19}F ion beam to the heavy-ion beams of interest. The experimental setup remained identical, except that for the (α, γ) induced reactions chemically pure helium gas was used. At the least, this method avoided the usual problems of target stoichiometries.

When the residual nucleus B of a reaction A(x, y)B is radioactive, the activity method (sec. 5.4.3) can also be used in determining the absolute cross section.

The above discussions assumed that the target and thus the interaction

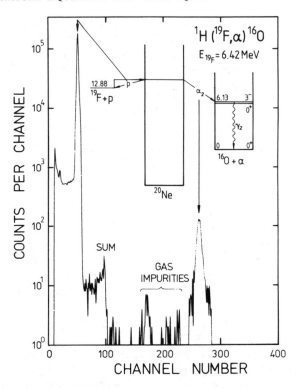

FIGURE 5.44. Sample particle spectrum obtained at $\theta_{\text{lab}} = 37°5$ for ^{19}F ions of $E_{\text{lab}} = 6.42$ MeV incident on a hydrogen gas target (Bec82a). The indicated gas impurities are less than one part in 10^4.

zone of the reaction is well defined. If an extended gas target is used, the analysis is more complicated, since both the detector efficiency and the cross section are functions of position along the target axis (Fig. 5.45). The observed count rate is then expressed as an integral over the length of the target:

$$N_y(E_0) = \int \sigma(E_{\text{c.m.}}) N_p(x) N_t(x) W_y(E_{\text{c.m.}}, \theta) \epsilon_y(E_{\text{c.m.}}, x) d\Omega_y(E_{\text{c.m.}}, x) dx .$$

For thin targets the number of projectiles $N_p(x)$ along the target axis is constant and can be replaced by the total number of beam particles, N_p. If heating effects of the ion beam on the target gas along the beam axis can be neglected, the target density $N_t(x)$ is also constant and can be determined from the pressure and temperature measurement of the target gas. For nonresonant cross sections varying slowly over the target length, the cross section can be approximated by the value at the effective beam energy $\sigma(E_{\text{eff}})$ where $E_{\text{eff}} \simeq E_0 - \Delta/2$. Similarly, the angular distribution $W_y(E_{\text{c.m.}}, \theta)$, the detection efficiency $\epsilon_y(E_{\text{c.m.}}, x)$ and the solid angle $d\Omega_y(E_{\text{c.m.}}, x)$ are slowly varying functions of

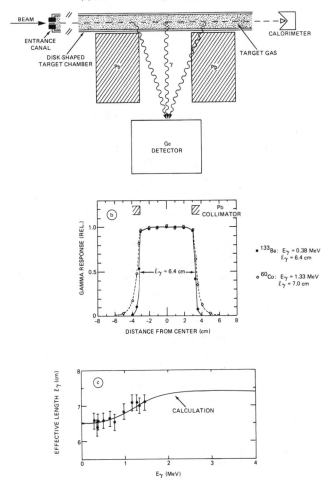

FIGURE 5.45. (a) Shown schematically is an arrangement used for γ-ray spectroscopy with an extended gas target (Tra84). The beam enters a disk-shaped target chamber through a 3 mm diameter canal and is stopped in a beam calorimeter. The effective target length l_γ seen by the 80 cm^3 Ge(Li) detector (at a distance of 133 mm from the center of the chamber) is defined here by the lead collimators (90 mm long with a cylindrical hole of 60 mm diameter). (b) Response function of the Ge(Li) detector for two γ-rays in this setup. (c) The effective target length l_γ in this setup varies with γ-ray energy. The observed and calculated values are in good agreement.

energy and can be replaced by their values at E_{eff}. The above equation then reduces to the expression

$$N_y(E_0) = \sigma(E_{eff})N_p N_t \int W_y(\theta)\epsilon_y(x)d\Omega_y(x)dx$$

$$= \sigma(E_{eff})N_p N_t W_y^*(\theta, l_y)\eta_y l_y , \tag{5.61}$$

where $W_y^*(\theta, l_y)$ represents the angular distribution integrated over the effective target length l_y and η_y is the effective detection efficiency for the reaction products y over the target length l_y. In γ-ray spectroscopy, the effective target length is also a function of γ-ray energy (Fig. 5.45).

In order to find out what the achievable precision of absolute cross-section determinations with extended gas targets is, the arrangement shown in Figure 5.45a has been used (Tra84). The response function of the Ge(Li) detector in this setup was determined with radioactive sources, which were moved along the beam axis within the target chamber (Fig. 5.45b). The observed and calculated target lengths l_y are shown in Figure 5.45c as a function of γ-ray energy. One sees that l_y has a lower limit as defined by the geometry of the setup and approaches a maximum value for $E_\gamma \geq 2$ MeV. With the knowledge of the effective target length l_y to a precision of $\Delta l_y/l_y = 3\%$, the absolute detection efficiency η_y was determined (sec. 5.3.3) with calibrated γ-ray sources placed at the center of the target chamber to an accuracy of $\Delta\eta_y/\eta_y = 4\%$; thus $\Delta(l_y\eta_y)/l_y\eta_y = 5\%$. If the number of projectiles N_p is determined with a beam calorimeter to an accuracy of 3% and if the number of target atoms N_t is known from pressure and temperature measurements of the target gas to a precision of 5%, the product $N_p N_t l_y \eta_y$ is known to an accuracy of 8%. Furthermore, if the angular distribution $W_\gamma^*(\theta, l_y)$ is known to 5% and if the statistical error in the observed number of counts $N_y(E_0)$ is 5%, the total error in the absolute cross section will be 11%.

In charged-particle spectroscopy using an extended gas target, the product $\eta_y l_y$ is determined by the selected collimator system between the target and the detector. In the disk-shaped target chamber (Figs. 5.19b and 5.45) a Si detector was placed (Tra84) at an angle $\psi = 45°$ to the beam axis and collimated by a slit-hole combination (Fig. 5.46). The effective target length l_y is here given by the relation (for notation see Fig. 5.46)

$$l_y = \frac{1}{\sin\psi}\frac{sd}{f} ,$$

where the distance d varies over the length l_y. However, for the condition $l_y/d \ll 1$ (here $l_y/d = 7 \times 10^{-3}$) this variation is negligible. As a consequence, the solid angle $d\Omega$ is identical for each point along the length l_y ($d\Omega_y = \pi r^2/d^2$), and the product $\eta_y l_y = d\Omega_y l_y$ is given by the relation

$$d\Omega_y l_y = \frac{\pi}{\sin\psi}\frac{sr^2}{fd} .$$

Quadratic addition of the errors in the geometric quantities (here 1.5%) and in the angle ψ (here $\pm 1° \cong 1.7\%$) led to an error of 3.2% for $d\Omega_y l_y$.

The uncertainty in the detection angle ψ enters more critically, if the reaction yield is determined relative to the elastic scattering yield (eq. [5.59]), where the elastic scattering cross section follows the Rutherford law. For proton scattering on heavy target nuclei the laboratory angle ψ is nearly identical with the center-of-mass angle θ. An error of $\Delta\psi = \pm 1°$ leads to an uncertainty in the Rutherford scattering cross section of $\Delta\sigma_R/\sigma_R = 9\%$ at $\psi = 45°$. In this case the angle ψ has to be known to higher accuracy. If the intensity of the elastic scattered projectiles (I_{sc}) and that of the recoil target nuclei (I_{rec}) can be measured in the same detector, the intensity ratio depends sensitively on the detection angle ψ (Mar68):

$$\frac{I_{sc}}{I_{rec}} = \left(\frac{\sin\theta}{\sin\psi}\right)^2 \left(\frac{\cos\psi}{\sin\theta/2}\right)^4 \frac{1}{4\cos\psi\cos(\theta-\psi)}.$$

This expression is valid for the Rutherford scattering law. The $^3\text{He} + {}^4\text{He}$ elastic scattering at $E_{c.m.} = 0.34\text{–}0.68$ MeV was used in the setup of Figures 5.45 and 5.46 to yield (Tra84) $I_{sc}/I_{rec} = 0.893 \pm 0.008$ and thus $\psi = 44°4 \pm 0°1$.

This setup was finally used (Tra84) to determine the absolute cross section of the broad ($\Gamma = 68$ keV) resonance at $E_p = 1.64$ MeV in the $^{15}\text{N}(p,\alpha_1)^{12}\text{C}$ reaction (Ajz82). With a ^{15}N gas pressure of 0.5 torr, the target is 1 keV thick over the extended region, fulfilling the condition for a thin target ($\Delta \ll \Gamma$). The known α_1 angular distribution (Bas59) led to a 5% correction for the observed α_1 intensity at $\psi = 44°4$ ($\theta = 49°6$). The absolute determination gave $\sigma_{\alpha_1} = 160 \pm 10$ mb, and the determination relative to the proton elastic scattering resulted in $\sigma_{\alpha_1} = 156 \pm 9$ mb. Furthermore, the 4.43 MeV secondary γ-ray

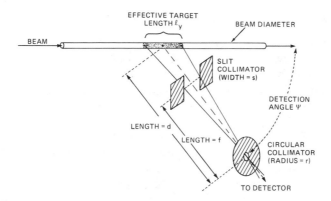

FIGURE 5.46. Schematic diagram of the setup for charged-particle spectroscopy using an extended gas target (Tra84). The beam has a maximum diameter of about 3 mm. The reaction products formed along the effective target length l_y are observed in a detector through a combination of slit-collimator ($s = 1.020$ mm, $f = 148.6$ mm) and circular aperture ($r = 1.515$ mm, $d = 175.6$ mm).

radiation from the ^{15}N$(p, \alpha_1\gamma)^{12}$C reaction was observed concurrently with the elastic scattering. In this case the product $N_p N_t$ cancels in the ratio of yields, and one obtains

$$\frac{N_\gamma}{I_{el}} = \frac{\sigma_\gamma}{(d\sigma/d\Omega)_{el}} \frac{l_\gamma \eta_\gamma}{l_\gamma \, d\Omega_\gamma} \, W^*_\gamma(\theta_\gamma, l_\gamma) \ .$$

The known angular distribution (Bas57) led to a correction factor of $W^*_\gamma(\theta_\gamma, l_\gamma) = 0.91 \pm 0.01$. The geometric ratio in the above equation,

$$R(E_\gamma) = R_{4.43} = \frac{l_\gamma \eta_\gamma}{l_\gamma \, d\Omega_\gamma} \ ,$$

is determined to be $R = 38.5 \pm 1.8$ from the geometry and absolute efficiency of the Ge detector (Fig. 5.45). The ratio can be determined independently using the observed intensity $I_{\alpha 1}$ of the α_1 particles (corrected for effects of lab–c.m. transformation), and the intensity $I_{4.43}$ of the 4.43 MeV γ-radiation (corrected for effects of angular distribution):

$$R_{4.43} = 4\pi \frac{I_{\alpha 1}}{I_{4.43}} = 36.1 \pm 0.8 \ .$$

The weighted average of both values is $R_{4.43} = 36.6 \pm 0.7$. With this result, the deduced value σ_γ from the observed 4.43 MeV γ-ray yield is 156 ± 13 mb, in good agreement with the above values, leading to a weighted average of $\sigma = 157 \pm 6$ mb. This result agrees well with the reported value of 170 ± 19 mb (Hag57) and fairly well with 190 ± 15 mb (Bas59). The weighted average of $\sigma = 163 \pm 5$ mb at the $E_p = 1.64$ MeV resonance can be used as another standard in relative measurements.

5.6. Some Future Techniques

In this section new experimental techniques, which have been investigated in recent years in several laboratories and which might be of value for future experiments in the field of nuclear astrophysics, will be discussed. Some of these techniques have already been applied to astrophysical questions, while others are in the technical development stage.

5.6.1. Detection of Recoil Nuclei in Capture Reactions

Radiative capture reactions A(x, γ)B are among the most important reactions for the formation of some elements. They are usually studied in the laboratory by detecting the emitted γ-rays. If the capture cross section is small and/or competing reactions produce a high γ-ray background, the measurements require high-resolution Ge detectors to improve the signal-to-noise ratio. Even close to the target, Ge detectors have an absolute detection efficiency of the order of 10^{-2} to 10^{-3} for $E_\gamma = 1$–10 MeV (Fig. 5.26). Since the recoiling nuclei

B travel in essentially the same direction as the beam (Fig. 5.47*a*), their direct detection with detectors having 100% efficiency would greatly improve the experimental sensitivity. There are, however, some obvious problems if a detector is placed in the beam direction ($\theta = 0°$): one observes not only the capture products but also the incident beam (including beam contaminants), the elastic scattering products, and background events (e.g., from multiple

(a) RECOILS IN RADIATIVE CAPTURE REACTIONS

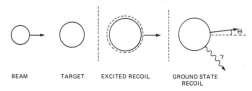

BEAM   TARGET   EXCITED RECOIL   GROUND STATE RECOIL

(b) DETECTION OF RECOILS

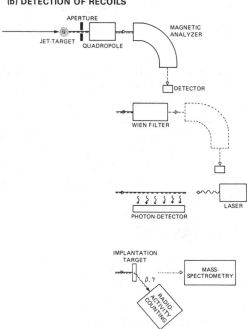

FIGURE 5.47. (*a*) In a capture reaction the projectile and the target nucleus combine to form an excited compound nucleus, which moves in the direction of the projectile. The direction is slightly changed after emission of γ-rays in the decay to the ground state. (*b*) The recoil nuclei in capture reactions can be detected with the help of a magnetic analyzer, a Wien filter, a tunable laser, radioactive counting, or mass spectrometry.

scattering processes). These problems are solved using different detection techniques (Fig. 5.47b), each of which—as we will see—has limitations.

In general, the beam direction has to be well defined by a set of collimators. The target should be fairly well localized and rather thin in order to avoid a significant energy loss (and energy straggling) of the recoil nuclei in the target. These requirements can be fulfilled by using either a thin-foil target or a supersonic jet gas target. Because of the momentum $p_\gamma = E_\gamma/c$ of the capture γ-rays, the recoil nuclei are emitted in a cone of half-angle $\theta/2 \simeq E_\gamma/p_1 c$, where p_1 is the momentum of the projectile. The recoil nuclei have on the average the same momentum as the projectiles, with a spread of

$$\frac{\Delta p}{p} = 1 \pm \frac{E_\gamma}{(2m_1 c^2 E_1)^{1/2}} \tag{5.62}$$

owing to the γ-ray recoil. In this expression m_1 and E_1 are the mass and laboratory energy of the projectile. An aperture placed behind the jet target (Fig. 5.47b) should then have a hole with a diameter matched appropriately to the cone of the recoil nuclei. A quadrupole lens should focus the divergent beam of recoil nuclei to a narrow spot near the position of the detector.

If a magnetic analyzer is used between the quadrupole lens and the detector (Fig. 5.47b), its analyzing power is determined by the rigidity

$$R_m = p/q , \tag{5.63}$$

where p and q are the momentum and charge state of the particles. Since the capture products have the same momentum as the beam, discrimination between the intense beam and the few capture products is based on the fact that the capture products can have a higher charge state than the maximum charge state of the projectiles ($q_{cap} > q_{beam}$). In addition, elastic scattering events will also be very intense in the beam direction ($\theta = 0°$) and must be taken into account. If the mass of the target nuclei is m_2, the impact of the elastically scattered projectiles p_1^* (for $m_1 > m_2$) and of the recoil nuclei p_2 are given by the relations (Mar68)

$$p_1^* = \frac{m_1 - m_2}{m_1 + m_2} p_1 \qquad (m_1 > m_2) , \tag{5.64}$$

$$p_2 = \frac{2m_2}{m_1 + m_2} p_1 . \tag{5.65}$$

The magnetic rigidity of the capture products, the beam, and the elastically scattered particles are shown in Figure 5.48 for a few capture reactions. The results show that the momentum spread of the recoil nuclei is significantly reduced if the capture reaction is studied for the condition $m_1 > m_2$, as expected from equation (5.62). This spread is high for reactions with high Q-values (i.e., high γ-ray energies). The figure also verifies the statement above that for a unique measurement of the capture products with a magnetic

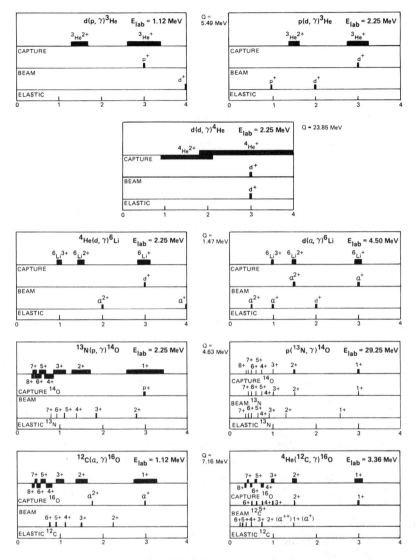

FIGURE 5.48. Relative magnetic rigidity of capture products, projectiles, and elastic scattering products for several radiative capture reactions. These relationships are independent of beam energy. The exception is the momentum distribution of the capture products, which depends on E_γ and thus on the beam energy. The various charge states of each ion can be produced in the target.

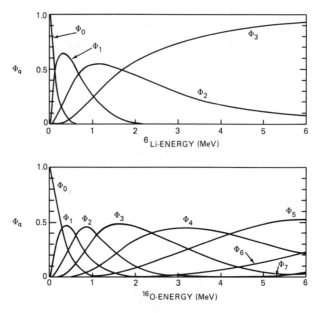

FIGURE 5.49. When charged particles pass through matter, they capture and lose orbital electrons until an equilibrium charge distribution is attained for the moving ions. Shown are the fractions Φ_q of the total beam of ^{6}Li and ^{16}O ions with charge q as a function of ion energy (Mar68).

analyzer, one has to measure the recoil nuclei in their highest charge states, so that the conditions $q_{\text{cap}} > q_{\text{beam}}$ as well as $q_{\text{cap}} > q_{\text{target}}$ are fulfilled. It follows from the figure that these conditions can be fulfilled only for $m_1 < m_2$, except for very light recoil nuclei. In the reaction ^{4}He$(d, \gamma)^6$Li, one has a unique signature for the charge states $q_{\text{cap}} = 2^+$ and 3^+ and in ^{12}C$(\alpha, \gamma)^{16}$O for $q_{\text{cap}} = 5^+$ to 8^+. The charge-state distribution for the ^{6}Li and ^{16}O ions as a function of ion energy is shown in Figure 5.49. For $E_{\text{lab}}(d) = 2.25$ MeV the ^{6}Li recoil nuclei have an energy of $E_{\text{lab}}(^6\text{Li}) = 1.13$ MeV, and thus 83% of the ^{6}Li recoil nuclei are in the $q_{\text{cap}} = 2^+$ and 3^+ charge states. By comparison, for $E_{\text{lab}}(\alpha) = 1.12$ (4.0) MeV, the ^{16}O recoil ions have $E_{\text{lab}}(^{16}\text{O}) = 0.28$ (1.0) MeV, and thus less than 0.1% (1%) of the ^{16}O ions are in the charge states $q_{\text{cap}} \geq 5^+$. Thus, a magnetic analyzer is a useful technique for detection of capture products only for reactions involving light projectiles and light target nuclei. This technique has been used in studies of the reactions $p(d, \gamma)^3$He and $d(\alpha, \gamma)^6$Li (Bel70, Rob84). Since the momentum distribution of the recoil ions (eq. [5.62]) depends on the angular distribution of the capture γ-rays, the observed data yielded information on these distributions for $\theta_\gamma = 0°-180°$. It turned out that they were asymmetric around 90°, clearly indicating the presence of multipole transitions of opposite parity.

As an alternative technique, a Wien filter can be placed between the quadrupole lens and the detector (Fig. 5.47b). Note that all ions of a given velocity

v pass the Wien filter independent of their charge state. Of course, neutral particles will be unaffected, and their discrimination requires additional deflecting devices. If the velocity of the recoil nuclei is v_r, the beam has a relative velocity of

$$\frac{v_1}{v_r} = \frac{m_1 + m_2}{m_1} , \qquad (5.66)$$

and the elastically scattered particles have relative velocities of

$$\frac{v_1}{v_r} = \frac{m_1 - m_2}{m_1} \qquad (m_1 > m_2) , \qquad (5.67)$$

$$\frac{v_2}{v_r} = 2 . \qquad (5.68)$$

These relative velocities are shown in Figure 5.50 for three capture reactions. As seen in the figure there is no interference between the capture products and

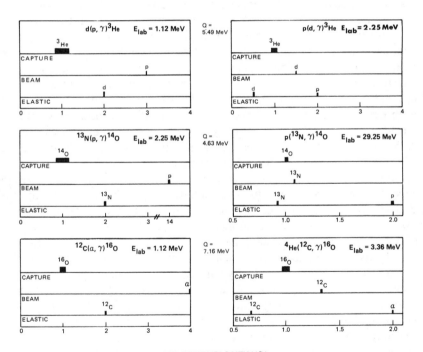

RELATIVE VELOCITY V/V$_r$

FIGURE 5.50. Shown is the relative velocity of capture products, projectiles, and elastic scattering products for several capture reactions. These relationships are independent of beam energy. The exception is the velocity distribution of the capture products, which depends on E_γ and thus on the beam energy.

the projectiles or the elastic scattering products. As the difference in mass between projectiles and capture products decreases, the velocity resolution $\Delta v/v$ of the Wien filter must also decrease. For example, the detection of the ^{14}O capture products in the ^{13}N$(p, \gamma)^{14}$O reaction requires $\Delta v/v \leq 1$, while in the interchange of the roles of projectiles and target nuclei, $p(^{13}$N$, \gamma)^{14}$O, a resolution of $\Delta v/v \leq 0.08$ is needed. Clearly, from this point of view the choice would be a proton beam and a ^{13}N target. However, the energy of the capture products E_r depends sensitively on the masses in the entrance channel $\{E_r = [m_1/(m_1 + m_2)]E_1\}$: for ^{13}N$(p, \gamma)^{14}$O with $E_1 = 2.25$ MeV one finds $E_r = 0.16$ MeV, and for $p(^{13}$N$, \gamma)^{14}$O with $E_1 = 29.25$ MeV (same center-of-mass energy) the recoil energy is $E_r = 27.16$ MeV. The low energy of the ^{14}O capture products in the first case has problems associated with the ^{14}O charge-state distribution (about 70% neutral ions; Fig. 5.49) as well as with the ^{14}O energy loss in the target, requiring unusually thin targets. Both problems are nearly absent in the second case. Thus, the study of this capture reaction is facilitated by using the $p(^{13}$N$, \gamma)^{14}$O reaction, which requires a Wien filter with a velocity resolution of at least 0.08.

As noted above, neutral particles will be unaffected by the Wien filter, emerging at the exit slit of the Wien filter along with the capture products. Such neutral particles can originate from the neutralization of the projectiles in the target zone and will usually be orders of magnitude more intense than the capture products. Using a magnetic analyzer (or an electrostatic analyzer) the charged capture products can be separated from these neutral particles (Fig. 5.47b) and subsequently observed in detectors placed appropriately at the exit (or exits) of the analyzer. Of course, the neutralized capture products also escape detection, but they are often negligible at high recoil energies. Detection techniques involving a Wien filter plus an additional deflecting device are presently being exploited in several laboratories.

Atomic methods could also be used in detecting capture products, since the atomic numbers of these products are different from those of the projectiles and elastic scattering products. A tunable laser could excite specific states of the orbital electrons and the resulting photons could be observed in a detector (Fig. 5.47b). A capture product could be excited several times along its flight path, significantly increasing its detection efficiency. This atomic method (Hur76) is hampered by the velocity spread as well as the angular divergence of the capture products, leading to a relatively broad Doppler spectrum for excitation (and emission) of specific states in the atomic cloud. A tunable laser could be matched to the mean velocity of the capture products, but it would excite only a small fraction of the atoms. It is not yet clear whether the final efficiency is significantly higher than that for capture γ-rays using Ge detectors. This approach will need much research in the future.

Finally, the capture products could be implanted in a solid target placed behind the primary jet gas target (Fig. 5.47b). When the primary target is a solid, the implantation takes place in the target material itself. If the capture products are radioactive with short half-lives, they are detected using standard radioactive counting methods (Hin62, Oes75). If they are long-lived or stable, ultrasensitive mass spectrometry based on accelerators (sec. 5.6.2) can be used. In both methods the kinematic information on the capture products (Figs. 5.48 and 5.50) is lost. One has therefore to ensure that the implanted species were not produced by other contaminant reactions or that they were not already present in the implantation target as impurities.

5.6.2. Accelerator Mass Spectrometry

Let us start with a reaction A(x, y)B having a cross section of 1 μb, which is initiated in a target having 10^{18} atoms cm^{-2} using a beam of 200 μA. The reaction proceeds at a rate of 72,000 reactions per minute, i.e., 72,000 recoil nuclei B are implanted in the target per minute (assuming a negligible loss via kinematic or sputtering effects). If the recoil nuclei are radioactive with relatively short lifetimes, then efficient apparatus for the detection of the decay particles (α-, β-, γ-radiation) permits detection of all or nearly all of a small number of radioactive atoms, in the presence of a large number of nonradioactive atoms. For example, if the recoil nuclei B in the above example have a half-life of 10 minutes and if their production is stopped after 1 minute, the initial decay rate is 5000 events per minute, and 90% of the radioactive species are detected in 33 minutes. As the half-life increases, the time required to carry out an experiment with a small number of radioactive atoms increases. For a half-life of, say, 1 year the decay rate of the recoil nuclei B initially would be 0.1 events per minute, and detection of 90% of all species would take 3.3 years. Efficient detection of the radioactive-decay products becomes impossible for long-lived species unless the experiment can be continued over long time periods. Consequently, studies of long-lived radioactivities invariably use very large numbers of atoms, the apparatus detecting the decay of only a small fraction of the total. For example, ^{26}Al has a half-life of 0.72×10^6 years, and a decay rate of 10 events per minute requires 5×10^{12} atoms of ^{26}Al.

If such long-lived atoms could be counted efficiently using mass spectrometry (Oes75, DeB78, Lit80), it would be possible to determine the presence of very small quantities in the sample. Assuming an efficiency of only 1% for converting atoms in the sample to ^{26}Al ions, one would have—for the above example—a beam of 5×10^{10} ^{26}Al ions. Under these circumstances mass-spectrometric detection sensitivity surpasses by far the sensitivity of radioactive counting methods.

The detection of rare stable or long-lived radioactive atoms, such as ^{14}C (half-life = 5730 y), by conventional mass spectrometry is complicated by the presence of much more abundant molecules and atoms, such as $^{12}CH_2$ and ^{14}N, with nearly the same mass. Their masses can be compared with the mass of ^{14}C by evaluating the average masses divided by the difference in masses ($M/\Delta M$), which are 1134 and 84,000, respectively. Such very high mass resolutions are required to eliminate molecular and atomic interferences, but can only be achieved at the expense of low efficiency for the already rare atoms. Since building a very large and expensive magnetic spectrometer does not seem attractive, the problems are reduced by using higher energy tandem electrostatic accelerators or cyclotrons. Such devices have increased the sensitivity of mass spectrometry by several orders of magnitude, giving rise to the term "ultrasensitive mass spectrometry" (Lit80).

Such an ultrasensitive mass spectrometer using a tandem accelerator is shown schematically in Figure 5.51. A sputter ion source produces negative ions. After magnetic analysis, the ions are accelerated and focused onto the first aperture, AP1. The negative ions are then accelerated toward the positive high-voltage terminal of the tandem, where several electrons are stripped off by a gas (or foil) stripper producing positive ions in various charge states q. These ions are then further accelerated, focused onto the next aperture AP2, and analyzed by a magnet, which specifies the product $(M/q)(E/q)$. A subsequent electrostatic analyzer then measures the quantity (E/q), and a heavy-ion counter determines the energy E. In the particle identification, ΔE-E telescopes and/or time-of-flight techniques are also often used. The basic ideas of the mass spectrometer technique are the discrimination between atomic species at the ion source and further discrimination by elimination of charged molecules in the stripping process. For example, nitrogen does not produce long-lived negative ions to interfere with the detection of ^{14}C, and while several stable or metastable doubly charged molecules are known, no triply charged molecules have been observed. If only positive ions with charge states $q \geq 3^+$ are subsequently analyzed, the interfering molecular species are elimi-

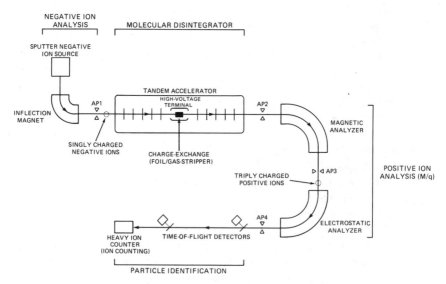

FIGURE 5.51. Schematic diagram of an ultrasensitive mass spectrometer involving a tandem accelerator (Lit80).

nated from the mass spectrum. The magnetic and electrostatic analyzers specify the quantity M/q, and, since in this approach M can be regarded as an integer, only modest mass resolution is needed. Ambiguities can arise, however, if M and q have common factors. If it is possible also to measure the energy of the ion, the charge state q and thus the mass M are uniquely determined. The use of energy, mass, and charge signatures, at ion energies such that charge state 3^+ or higher is the dominant product in the terminal gas stripper (Fig. 5.51), is the basis for much of the ultrasensitive mass spectrometry.

Radioactive isotopes ultimately decay to stable isotopes. When only beta-decay is involved, the mass difference between a stable and radioactive isotope pair is often so small that they can be regarded as having the same mass. A pair of such isotopes (e.g., ^{26}Al and ^{26}Mg) are referred to as isobars. The separation of isobars with high efficiency and the elimination of molecules are major features of ultrasensitive mass spectrometry of radioisotopes. Of course, any difference between the isobars can be exploited to separate them. However, even though they are different chemical elements, separation by conventional chemical methods is very difficult because the element of interest is often less than 10^{-9} times as abundant as the other element. A general method for discriminating between isobars does not exist, and each isobaric pair has to be treated as a special case (Lit80). For example, when the rare isobar has a larger nuclear charge than the abundant isobar, it can be identified uniquely if it is fully stripped of atomic electrons. Differences in the range or the rate of energy loss in matter also can be used to separate light or perhaps even heavy isobars from one another.

Accelerator mass spectrometry is a very sensitive tool for the solution of various problems in nuclear astrophysics. As techniques are improved it will find further applications, such as the measurement of low reaction cross sections, isotopic anomalies, or solar neutrino absorption in minerals (chap. 10).

5.6.3. Radioactive Ion Beams

In the explosive burning phase of stellar evolution (chap. 8), nuclear burning times can be measured in seconds or less. If the lifetime of a radioactive nucleus is longer than or the same as the burning time, then that nucleus will become involved in the nuclear-burning processes. At the high stellar temperatures $T_9 \simeq 1$, a significant fraction of nuclei are in excited states, and hence reactions involving excited nuclei can also take place. Such a situation is of special interest if an isomeric state close to the ground state exists. For example, the ^{26}Al nucleus ($T_{1/2} = 0.72 \times 10^6$ y) has an isomeric state at 229 keV ($T_{1/2} = 6.3$ s), which is significantly populated if thermodynamic equilibrium is established (chaps. 3 and 6). The most important reactions involving these unstable (radioactive) nuclei are those induced by protons, α-particles, and neutrons. In what follows we will discuss various laboratory approaches to the study of such reactions. Since protons and α-particles are naturally available, while neutrons have to be produced artificially, we will discuss mainly p- and α-induced reactions. Note that because of the high stellar temperature the cross sections need to be known only at energies near the Coulomb barriers ($E_{c.m.} = 0.5$–5.0 MeV). At these energies the cross sections are often in the range of μb to mb.

An important question in this field is the relative merit of radioactive targets (RT) and radioactive ion beams (RIB). Suppose, for example, that we have a separated beam of radioactive nuclei A and that we want to study the reaction $A + x \rightarrow B + y$, where x is a stable nucleus (here protons or α-particles). Should we (Fig. 5.52) collect nuclei A to form a radioactive target and bombard this target with the projectiles x, or should we accelerate the nuclei A and let them impinge on a target of x-type nuclei? If the radioactive nuclei with half-life $T_{1/2}$ [$= (\ln 2)\tau$] are produced with a rate N_A and form a target of area F (Fig. 5.52b), the total number of reactions A(x, y)B over the bombarding time T are (secs. 5.4.3 and 5.5.1)

$$N(\text{RT}) = \frac{\sigma N_A N_x(b)\tau}{F}\left[T - \tau + \tau \exp\left(-\frac{T}{\tau}\right)\right],$$ (5.69)

where σ is the reaction cross section and $N_x(b)$ the beam current. If the radioactive nuclei are accelerated (with a loss factor f) and impinge on a target of $N_x(t)$ target atoms per square centimeter (Fig. 5.52c), the total number of reactions over the time T is

$$N(\text{RIB}) = \sigma N_A N_x(t) f T .$$ (5.70)

If the half-life $T_{1/2}$ is much longer than the necessary running time T, both approaches provide the same number of reactions, $N(\text{RT}) = N(\text{RIB})$, for the condition

$$T = 2Ff \frac{N_x(t)}{N_x(b)} .$$ (5.71)

Note that this condition is independent of σ, N_A, and $T_{1/2}$. If one takes, as an order-of-magnitude estimate, the values $F = 1$ cm^2, $f = 10\%$, $N_x(t) = 10^{19}$ atoms cm^{-2}, and $N_x(b) = 100$ μA $= 6 \times 10^{14}$ projectiles s^{-1}, one finds $T = 0.9$ h, and hence $T_{1/2} \gg 1$ h. Clearly, for radioactive nuclei with half-lives longer than a day the radioactive target approach is more favorable. Among the light nuclei are ^{3}H ($T_{1/2} = 12.3$ y), ^{7}Be (53.4 d), ^{10}Be (1.6×10^6 y), ^{14}C (5.7×10^3 y), ^{22}Na (2.6 y), ^{26}Al (7.2×10^5y), and ^{32}Si (280 y). These nuclei have sufficiently long half-lives to allow for production of an off-line radioactive target and the transport of the target to the experimental site (Fig. 5.52a). An amount of about 1 μg of radioactive material is needed. Such quantities can be produced with a reactor or with high-current light-ion reactions (Wie77, Fil83, Buc84). Sometimes the radioactive nuclei are commercially available. With the use of techniques such as clean and efficient chemistry,

(a) RADIOACTIVE TARGET (OFF-LINE)

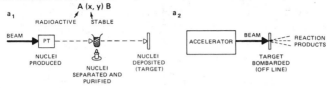

(b) RADIOACTIVE TARGET (ON-LINE)

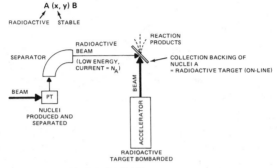

(c) RADIOACTIVE ION BEAM (RIB)

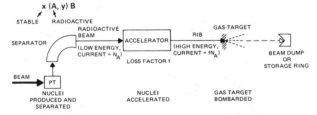

(d) RADIOACTIVE BEAM VERSUS RADIOACTIVE TARGET
(CHARGED PARTICLE REACTIONS)

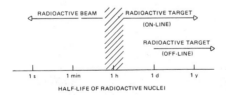

FIGURE 5.52. Schematic diagram of experimental techniques are shown (*a–c*) for the study of charged-particle–induced nuclear reactions involving a radioactive nucleus in the entrance channel. A decision on the most advantageous way of studying such a reaction is found to be a function of the half-life of the radioactive nuclei (Nit84, Cra84). If the half-life is shorter than 1 hour, then the radioactive beam experiment is more advantageous. If it is longer, then the radioactive target experiment is the way to go. It should be noted that this conclusion is independent of the reaction cross section and the intensity of the primary radioactive beam.

distillation and evaporation in vacuum, and electrodeposition, suitable targets can be produced from the primary target material.

In the other extreme case, $T_{1/2} \ll T$, equality of equations (5.69) and (5.70) leads to the relation

$$T = \frac{\tau^2}{\tau - Ff N_x(t)/N_x(b)} , \tag{5.72}$$

which requires for positive times T the condition

$$T_{1/2} \geq (\ln 2) Ff \frac{N_x(t)}{N_x(b)} . \tag{5.73}$$

Note again that both relationships are independent of σ and N_A. With the above numerical values one finds $T_{1/2} \geq 0.32$ h. With, say, $T_{1/2} = 0.5$ h, the necessary running period for the equality condition is $T = 2.1$ h. It follows, then, that for half-lives shorter than about 1 hour, radioactive ion beams are needed (Fig. 5.52c), while for half-lives longer than 1 hour the production of an on-line radioactive target is better (Fig. 5.52d). For example, take $T_{1/2} = 1$ s and $T \gg 1$ s and again use the above values for $F, f, N_x(t)$, and $N_x(b)$. Then the ratio of productivity for both approaches is $N(\text{RIB})/N(\text{RT}) \simeq 10^3$. In the time range $T_{1/2} \simeq 1$ hour to 1 day (Fig.5.52d) there are several species among the light nuclei: ^{18}F ($T_{1/2} = 1.8$ h), ^{24}Na (15.0 h), ^{28}Mg (21.1 h), and ^{31}Si (2.6 h).

When neutron-induced reactions are involved, the quantities $N_x(t)$ and $N_x(b)$ in the above expressions have to be replaced by the respective neutron flux over the running period T: $\phi_n(t)T$ and $\phi_n(b)T$. In the RIB approach the neutron target could be the thermal neutron flux $\phi_n(t)$ from a high-flux reactor such as the ILL reactor at Grenoble [$\phi_n(t) \simeq 10^{10}$ neutrons cm^{-2} s^{-1}]. For the radioactive target approach the neutron beam could have a Maxwellian energy distribution at the relevant stellar energy kT (chap. 9) produced, for example, via the ^{7}Li(p, n)^{7}Be reaction near threshold [$\phi_n(b) \simeq 10^7$ neutrons cm^{-2} s^{-1}]. With these neutron fluxes the dividing line between the radioactive beam and the radioactive target approach is found to lie near half-lives of $T_{1/2} = 1$ minute. Several reactions of the type (n, γ), (n, p), and (n, α) involving radioactive targets have already been investigated (e.g., Tra84).

The above discussions were based solely on a comparison of productivity for the two experimental approaches. However, in the radioactive target technique there are two major problems. The first problem is the large background due to the radioactive decay of the target nuclei (β- and γ-radiation), particularly for nuclei with relatively short half-lives. For example, a ^{7}Be target of 0.1 μg produces a 478 keV γ-ray flux of 1.3×10^8 events s^{-1} = 3.5 mCi, and a 0.1 μg ^{18}F target a flux of β^+-particles and 511 keV γ-rays of 3.5×10^{11} events s^{-1} = 9.5 Ci. Special detection techniques are required in the observation of reaction products bathed in such a high background. The second problem arises from chemical contaminants in the radioactive target (decay products of radioactive nuclei and impurities in the target and target backing), which can significantly reduce the signal-to-noise ratio. In the radioactive ion beam approach the beam is stopped in a beam dump far away from the target area, and high-purity gas targets can be used. Consequently, both problems are significantly reduced in the RIB approach and might shift the dividing line between approaches to half-lives somewhat longer than 1 hour (Fig. 5.52d).

In comparing the merits of different techniques in the study of nuclear reactions, a quantity called the luminosity, L, is often referred to, which is defined as the product of the number of projectiles times the number of target atoms:

$$L = N_p N_t . \tag{5.74}$$

The luminosity is related to the reaction cross section, σ, the detection efficiency ϵ of the reaction products, and the observed number of events N by the equation (sec. 5.5)

$$N = L\epsilon\sigma . \tag{5.75}$$

For example, a radioactive ion beam with $N_A = 10^{10}$ atoms s^{-1} and an accelerator loss factor $f = 10\%$ provides a beam of $N_p = fN_A = 10^9$ projectiles s^{-1}. With $N_t = 10^{19}$ target atoms cm^{-2}, the luminosity is $L = 10^9$ s$^{-1} \times 10^{19}$ cm$^{-2} = 10^{28}$ s^{-1} cm$^{-2} = 10$ mb^{-1} s^{-1}. For $\sigma = 1$ mb and $\epsilon = 10\%$, one finds $N = 1$ event s^{-1}.

Several techniques have been proposed for the production of radioactive ion beams (Rol79, Boy81, Hai83, Nit84, Cra84, Buc85). In the conversion method, primary production reactions with large cross sections and favorable kinematics are used to convert the primary heavy-ion beam into a secondary radioactive ion beam. For example, the reactions ^{1}H(^{7}Li, ^{7}Be)n and ^{3}He(^{16}O, ^{15}O)^{4}He have cross sections in the range of 10–100 mb and emit the radioactive nuclei ^{7}Be and ^{15}O into a narrow forward cone with energies similar to the primary heavy-ion beam ($E \simeq 1$ MeV amu^{-1}). The primary and secondary beams are separated in the beam transport systems by using kinematic constraints. For optimum conditions a useful radioactive beam intensity $N_A \simeq 10^6$–10^8 ions s^{-1} is expected (Boy81, Hai83, Nit84), and thus $L \simeq 10^7$ s$^{-1} \times 10^{19}$ cm$^{-2} = 0.1$ mb^{-1} s^{-1}. The conversion method suffers from the small conversion factor of 10^{-4} to 10^{-7} for the production of secondary beams.

With high-energy heavy-ion beams ($E \geq 100$ MeV amu^{-1}) it is found that many isotopes are produced with large cross sections ($\sigma \leq 100$ mb) via projectile fragmentation (Nit84, Cra84). These fragments are emitted into a very narrow cone in the forward direction and have a velocity similar to the primary heavy-ion beam. With a heavy-ion beam intensity of 100 nA $= 6 \times 10^{11}$ ions s^{-1} and a primary target of 7×10^{22} atoms cm^{-2} (e.g., 1 g cm^{-2} of ^{9}Be), the secondary beam intensity can be as high as 4×10^9 ions s^{-1}, reflecting a conversion factor of 10^{-2} for radioactive ions not too far removed from the mass of the projectile. Subsequent purification and energy spread reduction of the beam can be achieved at the cost of intensity. It could reduce the intensity significantly depending on the beam characteristics desired. For nuclear reactions of astrophysical interest the radioactive ion beams may have to be decelerated to energies around 1 MeV amu^{-1}, causing another reduction in beam intensity. Such techniques are presently being developed at several heavy-ion laboratories (Nit84, Cra84). Another approach involves relativistic heavy-ion beams ($E \geq 1$ GeV amu^{-1}), where the production mechanism is electromagnetic dissociation of the projectile (Nit84).

A combination of an on-line isotope separator and an accelerator has been proposed (Nit84, Cra84, Buc85) for the production of a radioactive beam (Fig. 5.52c). In this method a thick target ($\simeq 6 \times 10^{23}$ atoms cm^{-2}) is bombarded with an intense proton beam ($\simeq 10$ μA $= 6 \times 10^{13}$ protons s^{-1}) of about 200–600 MeV. The radioactive nuclei are produced by spallation or proton-induced fission ($\sigma \simeq 20$ mb) at a rate of about 10^{12} ions s^{-1}. Nuclei are continuously extracted from the target and separated on-line to provide a homogeneous radioactive beam at a well-defined low energy ($E \simeq 60$ keV). This beam is then accelerated to provide a radioactive ion beam with characteristics appropriate for studies of astrophysically interesting nuclear reactions. Accelerator systems such as a combination of an RFQ (radio-frequency quadrupole) and a linac accelerator (Sto79 and sec. 5.1.2) appears to be able to provide the necessary beam energies ($E \simeq 1$ MeV amu^{-1}) and acceptable energy spread (Buc85). These systems have beam transmission efficiencies of about 50%. The luminosity could therefore be as high as $L = 0.5 \times 10^{12}$ s^{-1} $\times 10^{19}$ cm^{-2} $= 5$ μb^{-1} s^{-1}, which is sufficient for most applications.

It is clear that much research and technical development are needed on all aspects of the problem (targets, ion sources, accelerators, radiation safety, and especially improved detection of reaction products; e.g., Rol86a) before nuclear reactions involving short-lived radioactive nuclei can be carried out in the laboratory. In the near future it might be possible to use heavy-ion storage rings with beam cooling (sec. 5.6.4) to store radioactive ions and hence to enhance further the luminosity for some experiments (Nit84, Cra84, Hab84, Buc85).

5.6.4. Storage Rings

In a standard setup (Fig. 5.1) the ion beam from an accelerator is used only once. Most of the beam does not cause a reaction and is subsequently stopped at a Faraday cup downstream from the target area. If the ion beam from the accelerator could be used several times to cause the nuclear reaction, the luminosity would clearly be increased. It also would represent a more efficient use of the "expensive" ion beam, particularly when radioactive ion beams are involved. Such repeated use of an ion beam can be achieved using storage rings (Fig. 5.53), in which particles are accumulated and circulate continuously, colliding repeatedly with a thin internal target ($\simeq 10$ ng cm^{-2}). A storage ring (O'Ne63, O'Ne66, Wil80a, Hab84) resembles a synchrotron (sec.

5.1.2 and Liv62) with vacuum chambers surrounded by bending and focusing magnets and with at least one radio-frequency accelerating cavity. Special magnets and electrodes are used to inject particles from an external accelerator into the ring. Usually the ring is not used to change the particle energy but rather to maintain the particles in a fixed orbit at a constant energy. The radio-frequency accelerating cavity supplies only enough power to make up for the energy lost by the passage through the target and by synchrotron radiation.

Ions from the accelerator have an initial energy spread, which is increased by their interaction with the target in the storage ring. Ions with such a wide range of energies and directions cannot be accumulated and circulated efficiently for long periods of time. In their own frame of reference the ions in the beam are like a hot gas with a temperature T_i. This temperature must be reduced substantially for successful operation of a storage ring. Otherwise a

STORAGE RING

QUADRUPOLE
FOCUSING
MAGNETS

DIPOLE
BENDING
MAGNETS

RADIO-FREQUENCY
CAVITY

INJECTION

BEAM FROM
ACCELERATOR

RF-OSCILLATOR

ELECTRON GUN

TARGET STATION

COOLING
VIA
INTENSE
MONOCHROMATIC
ELECTRON BEAM

CIRCULATING
ION BEAM

COLLECTOR

ELECTRON COOLING

ELECTRON

ENVELOPE OF
ELECTRON/ION
BEAM

ION

FIGURE 5.53. In a storage ring particles from an external accelerator are injected, accumulated, and confined to an orbit by a combination of bending and focusing magnets (Hab84). Their energy is maintained by a radio-frequency accelerating cavity. An intense beam of particles (several mA) can be stored in the ring for relatively long periods (minutes). They may circulate in the ring several million times in bunches or clusters. The interaction of the beam with thin internal targets heats the ion beam, and thus beam-cooling techniques are needed.

significant fraction of the circulating beam will be lost during the many revolutions. Such a temperature reduction (beam cooling) can be accomplished by several techniques, most notably electron cooling and stochastic cooling (Bud76, Mee72, Col81). In electron cooling an intense beam of electrons (several amperes) is passed through a long, straight section of the ring (a few meters) parallel to the ion beam and at the same average speed (Fig. 5.53). To match velocities, a 100 MeV proton beam requires an electron energy of only 50 keV. These electrons have a much lower temperature T_e (or $\Delta E_e \simeq 0.2$ eV) than the stored ion beam. While moving with the electron gas, the ions undergo Coulomb interactions (sec. 5.2.1), and momentum exchange takes place. Ions slower than the electrons are accelerated, and faster ions are decelerated. Ion momentum components transverse to the electron trajectory also are reduced. As a result, the electrons carry off the randomly directed components of the ion's momentum. In effect, the hot gas of ions gives up its heat to the relatively cold gas of electrons. At equilibrium ($T_i = T_e$) the ions have a velocity spread given by the relation

$$\frac{\Delta v_i}{v_i} = \left(\frac{m_e}{m_i}\right)^{1/2} \frac{\Delta v_e}{v_e},$$

which is significantly reduced because of the small mass ratio of $m_e/m_i \leq 1/2000$. At the end of the straight section the two kinds of particles are separated by an electric or magnetic field. In some devices the electrons strike a collector, but the ions continue to circulate and pass through the cooling region repeatedly. In other devices the electrons have their own loop. As noted above, there is also a variation in speed along the beam axis, e.g., by ramping the electron velocity the energy of the ion beam can also be adjusted.

Storage rings have been built successfully for electrons and light ions. Such a storage ring has not worked to date for heavy ions. Studies and development are underway (Hab84).

6

Hydrogen Burning

Results from earlier nuclear physics research, in particular the discovery of the enormous energy stored in nuclei (chap. 3), led astrophysicists to suspect that reactions among nuclear species were the source of the energy in stars (Edd20). This suspicion, coupled with the discovery of the tunneling effect in 1928 (Gam28, Con29; chap. 4), prompted Atkinson and Houtermans in 1929 (Atk29) to produce the first qualitative theoretical treatment of the problem. Their paper, in which they discussed "how to cook a nucleus in a pot," treated the energy production in stars, as well as the stellar nucleosynthesis of the elements. With the nuclear data available to them, they concluded that at relevant stellar temperatures only hydrogen could penetrate the Coulomb barrier with sufficient ease to induce fusion and that only hydrogen-induced fusion of light nuclei would be important. Since it was known that hydrogen is the most abundant element in the sun and the universe, it was concluded (Atk29, Atk31) that the energy-producing mechanisms involved primarily hydrogen. Atkinson and Houtermans suggested that the nucleus acts as a sort of trap and cooking pot combined, catching four protons and two electrons in such a way that finally a helium nucleus is cooked with the release of a large amount of energy (26.73 MeV). The mechanism for this cooking, in which helium is formed from hydrogen, and the quantitative aspects of the problem were first elucidated in the late 1930s by von Weizsäcker (Wei37, Wei38), Bethe and Critchfield (Bet38), and Bethe (Bet39), from whose work it was clear that two different sets of reactions could convert sufficient hydrogen into helium to provide the energy needed for a star's luminosity, namely, the proton-proton (p-p) chain and the CN cycle. This fusion of hydrogen into helium fuels the prodigious luminosity of stars for the greater part of their lives and takes place either in the cores of stars or, later, in relatively thin spherical shells surrounding the helium cores, which are the ashes of hydrogen burning.

In first-generation stars, where essentially only hydrogen is available, energy is produced predominantly by the p-p chain. In second-generation stars, which contain heavier elements derived from the ashes from burning in previous stars, carbon and nitrogen nuclei are available to act as catalysts for the operation of the CN cycle. Of course, there may also be heavier nuclei such as Ne, Na, Mg, and Al available which will react with the protons, resulting in the NeNa cycle and MgAl cycle. While these cycles may be important for the production of new elements, they do not contribute significantly to energy production in stars. The p-p chain, the CN cycle, and other cycles are discussed in more detail in what follows.

6.1. The Proton-Proton Chain

While the basic concept of the hydrogen-burning process

$$4p \rightarrow {}^4\text{He} + 2e^+ + 2v \qquad (Q = 26.73 \text{ MeV})$$

was correct, the details required further elucidation. It was clear that the probability of a simultaneous interaction of four protons in a stellar environment was essentially zero. This led, then, to the concept of a series of two-body interactions producing the same end result.

The first step, the $p + p \rightarrow {}^2\text{He}$ reaction, did not produce a stable nucleus via the strong or electromagnetic interaction, but rather brought about a statistical equilibrium between ${}^2\text{He}$ and two protons (Fig. 6.1). Assuming that subsequently the reaction $p + {}^2\text{He} \rightarrow {}^3\text{Li}$ took place did not improve the situation, since ${}^3\text{Li}$ also is an unstable nucleus. This apparent impasse was resolved when Bethe and Critchfield (Bet38) demonstrated that the weak interaction could give rise to the process (Fig. 6.1)

$$p + p \rightarrow d + e^+ + v \qquad (Q = 1.44 \text{ MeV}),$$

where $K = Q - 2m_e c^2 = 0.42$ MeV is the kinetic energy shared by the positron and the neutrino. Note that the quantity Q is the total energy released in the process, including the annihilation energy of the emitted positrons. Since this process proceeds via the weak interaction, the cross section is about 20 orders of magnitude smaller than cross sections associated with nuclear interactions. As a consequence, the cross section for this reaction has not been measured, and only a theoretical value is available.

6.1.1. Theoretical Cross Section for the $p + p$ Reaction

The total Hamiltonian for the $p + p$ reaction can be written as the sum of a nuclear term (H_n) and a weak-interaction term (H_β). Since the weak-interaction term is small compared with the nuclear term ($H_\beta \ll H_n$), first-order perturbation theory can be applied, resulting in the expression for the differential cross section (Bla62)

$$d\sigma = \frac{2\pi}{\hbar} \frac{\rho(E)}{v_i} |\langle f | H_\beta | i \rangle|^2 \,, \tag{6.1}$$

which is referred to as "Fermi's Golden Rule." In this expression the quantity $\rho(E)$ is a statistical factor representing the density of final states in the process, and v_i is the relative velocity in the incident channel. The symbol within the squared brackets represents the transition matrix element between the initial $(p + p)$ and final states $(d + e^+ + v)$ resulting from the weak interaction represented by H_β.

As stated above, the statistical factor $\rho(E)$ is equal to the number of final

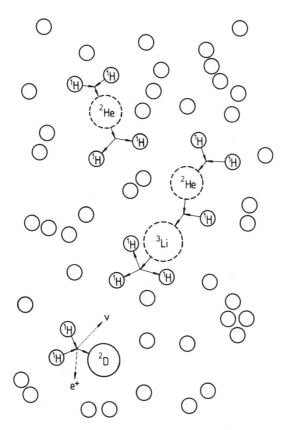

FIGURE 6.1. Schematic representation of hydrogen reacting with itself in a stellar interior. The nuclear and electromagnetic forces lead only to unstable heavier nuclei and are therefore not involved in the fusion of hydrogen into helium. It is only through the weak force that the stable isotope deuterium is synthesized. The hydrogen-burning rate is thus limited by the weakness of this force. Consequently, stars consume their nuclear fuel very slowly and still exist today.

states dN in the energy interval dE (i.e., between E and $E + dE$) of the total energy E:

$$\rho(E) = \frac{dN}{dE} .$$

One can show that the unit volume in the space defined by the conjugate variables of space and momentum has the value of h^3 (Bla62, Fra74). For a given volume V the number of states dn in the momentum range between p and $p + dp$ is then given by the total phase space divided by the unit volume:

$$dn = V \frac{4\pi p^2 dp}{h^3} .$$

Applying this equation to the present situation, one finds for the total number of states available for both the electrons and neutrinos,

$$dN = dn_e \, dn_v = \left(V \frac{4\pi p_e^2 \, dp_e}{h^3} \right)\left(V \frac{4\pi p_v^2 \, dp_v}{h^3} \right) .$$

Assuming a zero rest mass for the neutrino and neglecting the recoil energy of the deuterium, the total energy E is shared between the electron and neutrino, i.e., $E = E_e + E_v = E_e + cp_v$. We eliminate dp_v and p_v from the above equation by noting that for a constant value of p_e one has $dE = dE_v = cdp_v$. Thus the

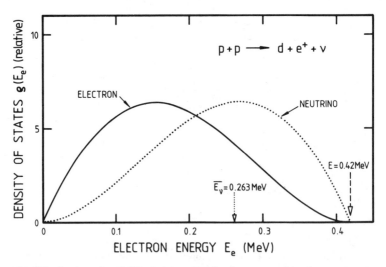

FIGURE 6.2. Distribution of available electron states for the $p + p \rightarrow d + e^+ + v$ process as a function of electron energy E_e. Note that the distribution goes to zero at $E_e = 0$ and at the maximum energy $E_e = E = 0.42$ MeV and that the highest density occurs, as one might expect, in between these extreme cases. Also shown is the corresponding distribution of neutrino states (*dotted line*), where the energy scale is now for the neutrinos. The mean neutrino energy is $\bar{E}_v = 0.263$ MeV.

above equation can be rewritten in terms of the total energy E, the electron energy E_e, and the electron momentum p_e:

$$\rho(E) = \frac{dN}{dE} = dn_e \frac{dn_v}{dE} = \frac{16\pi^2 V^2}{c^3 h^6} p_e^2 (E - E_e)^2 dp_e = \rho(E_e) dp_e .$$

This is the density of final states for which the electron has a momentum between p_e and $p_e + dp_e$ irrespective of the neutrino momentum, while the total energy is between E and $E + dE$. A graphic representation of this distribution of available states is given in Figure 6.2. The differential cross section can be written as

$$d\sigma = \frac{2\pi}{\hbar} \frac{\rho(E_e)}{v_i} |\langle f | H_\beta | i \rangle|^2 dp_e .$$

The matrix element contained in equation (6.1), $H_{if} = \langle f | H_\beta | i \rangle$, may be written in terms of wave functions as

$$H_{if} = \int \Psi_f^* H_\beta \Psi_i d\tau ,$$

where Ψ_i is the wave function of the two protons in the initial state (entrance channel) and Ψ_f describes the electron and neutrino fields (Ψ_e and Ψ_v) as well as that of the deuterium nucleus (Ψ_d):

$$H_{if} = \int [\Psi_d \Psi_e \Psi_v]^* H_\beta \Psi_i d\tau .$$

Because of the weak interaction between the nucleus and the two leptons, the fields of the leptons can be represented by plane waves:

$$\Psi_e = \frac{1}{V^{1/2}} e^{i(k_e r)} , \qquad \Psi_v = \frac{1}{V^{1/2}} e^{i(k_v r)} ,$$

where the leptons are normalized to the volume V, i.e.,

$$\int_V \Psi_i^* \Psi_i d\tau = 1 .$$

Since the radial wave function of the deuteron Ψ_d vanishes rapidly outside the nuclear domain (i.e., nuclear radius R_0), the integration need not be extended far beyond R_0. Owing to the constraints of the nuclear radius and the limited energy available for the process ($K = 0.42$ MeV), the product kR_0 for both the electron and neutrino fields is much less than unity (i.e., $R_d = 1.7$ fm, $\bar{E}_v = 0.26$ MeV, $k_v R_0 = 2.2 \times 10^{-3}$). As a consequence, the plane waves can be expanded in terms of kr (neglecting a barrier [Fermi] factor):

$$\Psi_e = \frac{1}{V^{1/2}} [1 + i(k_e \cdot r) + \cdots] ,$$

$$\Psi_v = \frac{1}{V^{1/2}} \left[1 + i(k_v \cdot r) + \cdots \right] .$$

It is sufficient to retain only the first term in this expansion. The matrix element H_{if} reduces then to the expression

$$H_{if} = \frac{1}{V} \int \Psi_d^* H_\beta \Psi_i \, d\tau .$$

The strength of the weak interaction H_β is governed by a coupling constant, g, which is analogous to the electric charge acting as the coupling constant in the electromagnetic interaction. The matrix element H_{if} can be written as

$$H_{if} = \frac{g}{V} \int \Psi_d^* \Psi_i \, d\tau$$

$$= \frac{g}{V} M_{\text{spin}} M_{\text{space}} ,$$

where the remaining overlap matrix element has been factored into the spin and space components, the values of which will be discussed below.

The differential cross section can now be written as

$$d\sigma = \frac{2\pi}{\hbar} \frac{1}{v_i} \frac{16\pi^2}{c^3 h^6} g^2 M_{\text{spin}}^2 M_{\text{space}}^2 p_e^2 (E - E_e)^2 dp_e ,$$

and the total cross section is obtained by integration over the total range of electron momenta:

$$\sigma = \frac{2\pi}{\hbar} \frac{1}{v_i} \frac{16\pi^2}{c^3 h^6} g^2 M_{\text{spin}}^2 M_{\text{space}}^2 \int_0^E p_e^2 (E - E_e)^2 dp_e .$$

Introducing the new variable

$$W = \frac{E + m_e c^2}{m^e c^2} ,$$

one arrives at an integral transformation:

$$\int_0^E p_e^2 (E - E_e)^2 dp_e = \frac{(m_e c^2)^5}{c^3} \int_1^W (W_e^2 - 1)^{1/2} (W - W_e)^2 W_e \, dW_e$$

$$= \frac{(m_e c^2)^5}{c^3} f(W) .$$

The integral $f(W)$ can be written as a sum of three integrals. Introducing additional changes of variables and carrying out the integration results in the following analytic expression:

$$f(W) = (W^2 - 1)^{1/2} \left(\frac{W^4}{30} - \frac{3W^2}{20} - \frac{2}{15} \right) + \frac{W}{4} \ln \left[W + (W^2 - 1)^{1/2} \right] .$$

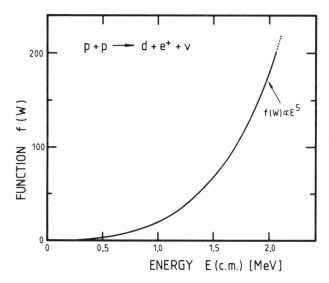

FIGURE 6.3. The function $f(W)$, which is a measure of the total phase space available for the $p + p \rightarrow d + e^+ + v$ reaction, is shown as a function of energy E. As might be expected, the available phase space increases rapidly with E. For large values of energy, the phase space and hence the probability for the process increase as the fifth power of the energy.

A graphic illustration of this function $f(W)$ is presented in Figure 6.3. For large values of W, i.e., large energies E, the function $f(W)$ approaches the form $f(W) \propto W^5 \propto E^5$.

The total cross section is given now by the equation

$$\sigma = \frac{1}{v_i} \frac{2\pi}{\hbar} \frac{16\pi^2}{c^3 h^6} \frac{(m_e c^2)^5}{c^3} f(W) g^2 M_{\text{spin}}^2 M_{\text{space}}^2$$

$$= \alpha \frac{1}{v_i} f(W) g^2 M_{\text{spin}}^2 M_{\text{space}}^2 , \qquad (6.2)$$

where α represents the constants, which, when replaced by their numerical values, result in

$$\alpha = \frac{m^5 c^4}{2\pi^3 \hbar^7} = 1.45 \times 10^{70} \text{ eV}^{-2} \text{ s}^{-1} \text{ cm}^{-6} .$$

Investigations of allowed weak-interaction phenomena indicate (Bla62, Fra74) that, among all possible types of couplings (scalar, ... , tensor), only two types are experimentally observed. When the spins of the leptons in the final states are found to be antiparallel ($s_e + s_v = s_{\text{tot}} = 0$), the process is referred to as Fermi or vector coupling with a coupling constant $g = C_V$. In the case where the spins are parallel ($s_e + s_v = s_{\text{tot}} = 1$), the process is called a Gamow-Teller or axial vector coupling with $g = C_A$. For the Fermi coupling there can be no change in the angular momentum between the initial and final states of the nuclei ($\Delta J = 0$). For the Gamow-Teller coupling, one arrives at the selection rule $\Delta J = 0$ or ± 1. However, the $\Delta J = 0$ possibility cannot proceed in this case between two states of zero angular momentum. If there are no changes in the orbital angular momentum between the initial and final states of the nuclei, the processes have maximum decay probabilities and are referred to as superallowed transitions. An example of a pure Fermi process is the decay $^{14}O(J_i^\pi = 0^+) \to {}^{14}N(J_f^\pi = 0^+)$. The decay $^6He(J_i^\pi = 0^+) \to {}^6Li(J_f^\pi = 1^+)$ is an example of a pure Gamow-Teller transition.

In the process $p + p \to d + e^+ + \nu$ the final nucleus, i.e., the ground state of deuterium ($J_f^\pi = 1^+$), is described predominantly by a relative orbital angular momentum $l_f = 0$ and $S_f = 1$ (triplet S-state). For a maximum probability of the process, i.e., a superallowed transition, the orbital angular momentum should not change in the transition. Hence, the two protons in the initial state should interact in an $l_i = 0$ orbital angular momentum state. Since one has here two identical particles (two protons), the Pauli principle requires that $S_i = 0$ so that the total wave function will be antisymmetric in (space $\times$ spin) coordinates. The process is characterized by the features $S_i = 0$ ($l_i = 0$) $\to S_f = 1$ ($l_f = 0$) and is, therefore, a Gamow-Teller transition with a coupling constant $g = C_A$. The value of this coupling constant can be extracted from measured lifetimes of Gamow-Teller transitions in the neutron, tritium, and ^{6}He decays, using an equation similar to equation (6.2):

$$\lambda = 1/\tau = \alpha f(W)g^2 M_{\text{space}}^2 M_{\text{spin}}^2 .$$

The remaining components of the transition are the spin and space matrix elements. The spin matrix element M_{spin} is obtained by summing over the final states, averaging over the initial states, and dividing the result by 2, to take into account the fact that the interaction involves identical particles (Sal52):

$$M_{\text{spin}}^2 = \tfrac{3}{2} .$$

The space matrix element M_{space} involves the following integral:

$$M_{\text{space}} = \int_0^\infty \chi_f(r)\chi_i(r)r^2 dr \ \text{cm}^{3/2} ,$$

where $\chi_i(r)$ and $\chi_f(r)$ represent the radial wave functions in the initial and final states of the process (Fig. 6.4). This integral involves Coulomb wave functions [i.e., barrier penetration $P_l(E, R_n) \ll 1$ at stellar energies] and cannot be represented in an analytic form but rather is evaluated by numerical integration. This calculation has been carried out (Fri51, Sal52, Bah69, Bah82a, Par85) and leads to a total cross section at $E_p(\text{lab}) = 1$ MeV of 10^{-47} cm^2. Even with a proton beam of 1 mA current incident on a thick hydrogen target (10^{23} atoms cm^{-2}), an interaction takes place only once in 10^6 years. Thus, it is impossible to determine this cross section experimentally with currently available techniques.

Since the $p + p \rightarrow d + e^+ + v$ reaction is governed by a nonresonant reaction mechanism, it proceeds at all energies, and, therefore, all stellar temperatures, with a cross section smoothly varying with energy [$S(0) = 3.8 \times 10^{-22}$ keV b, $dS(0)/dE = 4.2 \times 10^{-24}$ b], and the formalism for nonresonant reactions (chap. 4) may be applied in arriving at the reaction rate per particle pair, $\langle \sigma v \rangle_{pp}$. For example, at the temperature $T_6 = 15$, this rate is $\langle \sigma v \rangle_{pp} = 1.19 \times 10^{-43}$ cm^3 s^{-1}. Assuming parameters roughly like those in

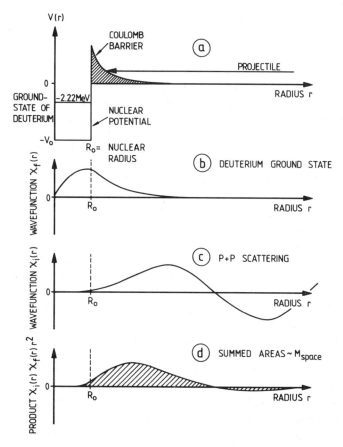

FIGURE 6.4. Shown schematically are a few ingredients used in the numerical evaluation of the space matrix element M_{space} for the $p + p \rightarrow d + e^+ + v$ reaction. The potential is shown in (a), where, for a given nuclear radius R_0, the observed binding energy of the deuterium determines the potential depth V_0. The deuterium radial wave function $\chi_f(r)$ is determined by the potential $V(r)$. Because of the loosely bound ground state, $\chi_f(r)$ extends far outside R_0 with appreciable amplitudes (b). The initial wave function $\chi_i(r)$ is obtained from $p + p$ elastic scattering data, which gives (c) a small amplitude for $r \leq R_0$ and has the usual oscillating pattern of a plane wave for $r \gg R_0$. The radial integrand in M_{space} (d) then has its major contributions in regions far outside R_0 (hatched areas).

the sun's interior, i.e., an equal mixture by mass of hydrogen and helium ($X_H = X_{He} = 0.5$) and a density $\rho = 100$ g cm^{-3}, then the mean lifetime of a proton against combustion into deuterium is (chap. 3) $\tau_H(H) = 1/N_H\langle\sigma v\rangle_{pp} = 0.9 \times 10^{10}$ y. This lifetime of a proton in a star like our sun of about 10^{10} y (Fig. 6.7) corresponds to the age of the oldest known stars. Thus, it is because of the weak interaction (and to some extent to the small Coulomb barrier penetration—about 10^{-2} in the rate), that stars consume their nuclear fuels very slowly and thus continue to exist today.

All subsequent reactions in the p-p-I chain (Fig. 6.9) involve electromagnetic and nuclear forces, and thus they proceed at a much faster rate than the initial $p + p$ reaction. Clearly, the overall rate of the conversion of four protons into a ^{4}He nucleus is governed by the $p + p$ reaction.

Using the fact that the energy produced in every fusion process $4p \rightarrow {}^4$He is 26.20 MeV (corrected for neutrino loss = 0.53 MeV; see below) and the total luminous energy radiated from the surface of the sun per second is $L_\odot = 2.4 \times 10^{39}$ MeV s^{-1} (chap 1), the rate at which such fusion processes occurs can be calculated: $N = L_\odot/26.20$ MeV $= 0.92 \times 10^{38}$ processes s^{-1}. Thus a total mass of 616 million tons of hydrogen is processed into helium every second. Since the solar mass is 2×10^{33} g (chap. 1), the upper limit on the lifetime of the sun, assuming that all the hydrogen fuel is processed at a uniform rate, is 1.0×10^{11} y. This estimate is about 20 times the past life of the sun, and at first our star seems to be in its infancy. Actually, burning takes place only in the interior of the star and involves, during the star's lifetime, about 10% of the total hydrogen. Hence, the expected lifetime of the sun is on the order of 10^{10} y. The present age of the sun is 4.5×10^9 y (chap. 10), about one-half of the expected lifetime. Thus, the sun is one of many middle-aged stars that have not yet exhausted their initial supply of hydrogen fuel.

The 26.73 MeV energy produced in every fusion process is shared among the resulting nuclei, γ-rays, and neutrinos (Fig. 6.9). Since neutrinos have essentially a unit probability of escaping from a star (chap. 10), a small fraction of the energy produced is lost to the star by the escaping neutrinos. On the average a neutrino emitted in the $p + p$ reaction takes with it an energy of 263 keV (Fig. 6.2). For the conversion $4p \rightarrow {}^4$He $+ 2e^+ + 2v$ a total energy loss results of 2×263 keV $= 526$ keV.

Since the temperature in the interior varies from star to star and the energy production depends on temperature (chap. 3), it is of interest to see how the energy production of the p-p chain changes as a function of temperature. This dependence is governed by the equation (chap. 4)

$$\epsilon \propto \langle\sigma v\rangle_{pp} \propto T^{-2/3} \exp(-\tau),$$

where τ is defined by equation (4.23). Figure 6.5 is a graphical representation of the temperature dependence for energy generation ϵ. As one might expect, ϵ increases smoothly with temperature. Since the temperature in the stellar interiors of main-sequence stars increases with the total mass of a star (chap. 2),

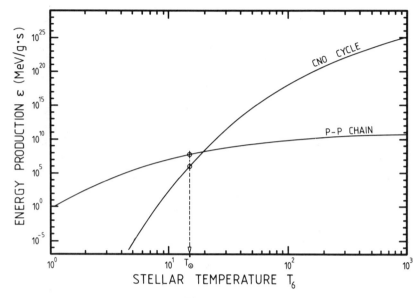

FIGURE 6.5. Energy generation as a function of central temperature for the *p-p* chain and the CNO cycle (Ibe65). Both curves drop with decreasing temperature, mainly as a result of effects of the Coulomb barriers. The changeover from the *p-p* chain to the CNO cycle is near $T_6 = 20$. Thus, the *p-p* chain is the principal energy source in the sun and stars with $M \leq M_\odot$. The changeover depends also on the abundance of C, N, and O in the star. Thus, for stars much more massive than the sun, the CNO cycle is responsible for energy production provided that at least one of the elements, C, N, or O, is present to act as a catalyst (second- or later-generation stars), and provided that their initial mass concentration is at least 1%.

more massive stars consume their available nuclear fuel more rapidly and thus have a higher energy production. Consequently, they have relatively shorter lifetimes compared with low-mass stars.

One can see from Figure 6.5 that the change in energy production ϵ is more rapid at lower temperatures. In constructing stellar models, it is important to know the temperature dependence of energy production $\epsilon(T_0)$ near any given temperature T_0. Using the above relation, one arrives at the expression

$$\epsilon(T)/\epsilon(T_0) = (T/T_0)^{(\tau_0 - 2)/3} .$$

For example, at the central solar temperature of $T_6 = 15$, one finds $\epsilon(T) \propto T^4$ and an absolute value of $\epsilon(T_0) = 5.1 \times 10^7$ MeV g^{-1} s^{-1} (sec. 6.1.5). The average in the sun is $L_\odot/M_\odot = 1.2 \times 10^6$ MeV g^{-1} s^{-1}.

It is clear from the discussions above that the $p + p$ reaction is fundamental to energy generation in stars and to stellar evolution. In fact, without this reaction no elements heavier than helium would exist at all, and the universe would be rather uninteresting. While the calculations of the $p + p$ cross section appear to be on rather firm ground, experimental verification would be comforting. At present such experimental verifications appear to be hopeless.

However, as technologies improve and beam currents increase (i.e., kilo- to mega-ampere proton-beam currents), the prospect for carrying out such experiments greatly improves, but severe background problems will always remain.

An alternative reaction $p + e^- + p \rightarrow d + v$ ($Q_s = 1.44$ MeV) resulting in the synthesis of deuterium has been proposed. However, calculations similar to those described above show (Bah69) that the cross section for this reaction is lower by 4 orders of magnitude than that of the $p + p$ reaction. Hence, this reaction, called the PEP reaction, plays no significant role in hydrogen burning. It might play a role, however, in the detection of high-energy solar neutrinos, since in this case monoenergetic neutrinos of $E_v = 1.44$ MeV are emitted (chap. 10).

TABLE 6.1 Summary of Deuterium-Burning Reactions

Reaction[a]	Q-Value (MeV)	S(0) (keV b)	References
$d(p, \gamma)^3$He	5.494	0.25×10^{-3} [b]	Gri63, Bai70
$d(d, \gamma)^4$He	23.847	$\approx 0.03 \times 10^{-3}$	Bia69
$d(d, p)t$	4.033	39	Arn54
$d(d, n)^3$He	3.269	37	Arn54
$d(^3\text{He}, p)^4$He	18.354	6240	Arn54
$d(^3\text{He}, \gamma)^5$Li	16.388	≈ 0.3	Bia68, Bus68
$d(^4\text{He}, \gamma)^6$Li	1.472	$\leq 0.03 \times 10^{-3}$	Rob81

[a] All reactions are of the nonresonant type. Deuterium as a target, is also often labeled by the symbols ^{2}H, ^{2}D, or just D.
[b] For experimental evidence see sec. 6.1.6.

6.1.2. Burning of Deuterium

Deuterium is synthesized in stars by the reaction $p + p \rightarrow d + e^+ + v$. Destruction of deuterium can occur in principle via any of the reactions listed in Table 6.1. The total reaction rate r_{12} of these processes depends on the reaction rate per particle pair $\langle \sigma v \rangle_{12}$ and the number of particles available to participate in the reactions (chap. 3):

$$r_{12} = \frac{N_1 N_2}{1 + \delta_{12}} \langle \sigma v \rangle_{12} \, .$$

Because of the overwhelmingly large number of protons in the stellar environment, processes involving protons ought to predominate. Thus, of the deuterium-burning reactions listed in Table 6.1, the $d(p, \gamma)^3$He reaction is thought to be the predominant one. If one assumes this to be the case, the time dependence of deuterium abundance in a star is described by the following differential equation:

$$\frac{dD}{dt} = r_{pp} - r_{pd}$$

$$= \frac{H^2}{2} \langle \sigma v \rangle_{pp} - HD \langle \sigma v \rangle_{pd},$$

where the first term represents the production rate of deuterium and the second term its destruction. Clearly, this equation describes a self-regulating system that eventually reaches a state of quasi-equilibrium ($dD/dt = 0$) and has an abundance ratio given by the equation

$$\left(\frac{D}{H} \right)_e = \frac{\langle \sigma v \rangle_{pp}}{2 \langle \sigma v \rangle_{pd}} \tag{6.3}$$

The $p + p$ reaction involves the weak force, and the $d(p, \gamma)^3$He reaction involves the electromagnetic force. Therefore, the abundance ratio $(D/H)_e$ must be of the same order as the ratio of the strengths of these forces; i.e., $(D/H)_e \ll 1$. Using the reaction rates per particle pair for the two reactions, one arrives at a ratio $(D/H)_e = 5.6 \times 10^{-18}$ at $T_6 = 5$. The value of this ratio decreases slowly with increasing temperature; for example, at $T_6 = 40$, $(D/H)_e = 1.7 \times 10^{-18}$.

In section 6.1.1 the lifetime of a proton against conversion to D was found to be $\tau_H(H) = 9 \times 10^9$ y. In contrast, the lifetime for deuterium in the same stellar environment is $\tau_H(D) = 1.6$ s (Fig. 6.7), again reflecting the relative magnitude of the strength of the forces involved. Clearly, the equilibrium condition is attained in approximately the same time.

The ^{3}He nuclei produced in the deuterium-destroying reaction $d(p, \gamma)^3$He can also react with the remaining deuterium via the two reactions $d(^3\text{He}, p)^4$He and $d(^3\text{He}, \gamma)^5$Li, the former reaction being faster (Table 6.1). In the same stellar environment as used above, the mean lifetime of deuterium against destruction via the $d(^3\text{He}, p)^4$He reaction can be calculated to be $\tau_{3\text{He}}(d) = 10^4$ s. Although in this calculation the reaction rate per particle pair is 7 orders of magnitude larger than for the $d(p, \gamma)^3$He reaction (Table 6.1), the mean lifetime is 4 orders of magnitude longer because of the extremely small number density ^{3}He. Similar considerations apply to the other deuterium-destroying reactions listed in Table 6.1, showing them also to be insignificant compared with the $d(p, \gamma)^3$He reaction.

The abundance ratio of deuterium relative to hydrogen resulting from nucleosynthesis in the cores of stars is on the order of $(D/H)_e \simeq 10^{-18}$ to 10^{-17} (Fig. 6.6). This ratio contrasts with the observed primordial ratio, $(D/H)_{obs} \simeq 10^{-5}$ (1.5×10^{-4} for the solar system). Thus, to explain this observed ratio, one must look elsewhere for the synthesis of deuterium—for example, in the early phase of the universe prior to the formation of stars (chap. 2).

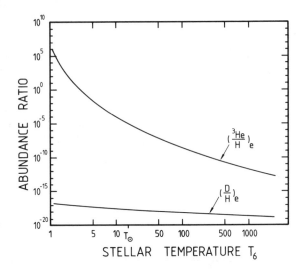

FIGURE 6.6. Shown is the temperature dependence of the abundance ratios $(^3He/H)_e$ and $(D/H)_e$ at equilibrium. The large difference in the temperature dependences for $(^3He/H)_e$ and $(D/H)_e$ is predominantly due to the higher Coulomb barrier involved in ^{3}He burning. The difference in the Coulomb barriers is also responsible for the vastly different relative abundances of ^{3}He and D in stars.

6.1.3. Burning of ^{3}He and Completion of the p-p-I Chain

The ^{3}He nuclei synthesized in the p-p chain via the sequence of reactions $p + p \rightarrow d + e^+ + \nu$ and $d(p, \gamma)^3$He are themselves largely consumed by a number of exothermic reactions ($Q \geq 0$), which are summarized in Table 6.2. Note that the largest astrophysical $S(0)$ factors occur for the reactions ^{3}He(d, p)^{4}He and ^{3}He(^{3}He, $2p$)^{4}He. This suggests that these are the key reactions and are comparable in importance. However, since the total reaction rate r depends not only on the $S(0)$ factors but also on the number of particles available to participate in the reactions, and since the abundance of deuterium in a star is extremely small, the ^{3}He(^{3}He, $2p$)^{4}He reaction should be predominantly responsible for the destruction of ^{3}He. The reaction completes the fusion of four protons into ^{4}He. The set of reactions involved in this fusion process is called the p-p-I chain (Fig. 6.9).

As a historical note, the classic paper of Bethe (Bet39) shows that the mode of completion of the p-p chain [after $p + p \rightarrow d + e^+ + \nu$ and $d(p, \gamma)^3$He] depends on the existence of ^{4}Li as a nucleus stable or almost stable to particle breakup. In this case, the reaction ^{3}He(p, γ)^{4}Li($e^+ \nu$)^{4}He would complete the chain with a rate similar to that for the $d(p, \gamma)^3$He reaction, owing to the large abundance of protons in the stellar environment. Experiments have shown, however, that ^{4}Li is not particle-stable ($p + {}^3$He $\rightarrow {}^4$Li, $Q = -2.5$ MeV) and, therefore, the ^{3}He(p, γ)^{4}Li reaction does not occur in the p-p chain (Bas59a,

TABLE 6.2 Exothermic Reactions Leading to Consumption of ^{3}He

Reaction[a]	Q-Value (MeV)	$S(0)$ (keV b)	References
^{3}He$(d, \gamma)^5$Li[b]	16.388	$\simeq 0.3$	Bia68, Bus68
^{3}He$(d, p)^4$He	18.354	6240	Arn54
^{3}He$(^3$He$, \gamma)^6$Be[c]	11.497	$\simeq 0.8$	Har67a
^{3}He$(^3$He$, 2p)^4$He	12.860	5500[d]	Wan66, Dwa71, Dwa74, Kav82
^{3}He$(\alpha, \gamma)^7$Be	1.587	0.53[d]	Par63, Nag69, Krä82, Osb82, Kav82

[a] All reactions are predominantly of the nonresonant type.

[b] ^{5}Li breaks up subsequently into ^{4}He $+ p$. This reaction plays no role relative to ^{3}He$(d, p)^4$He, owing to the much smaller $S(0)$ value.

[c] ^{6}Be breaks up subsequently into ^{4}He $+ 2p$. This reaction plays no role relative to ^{3}He$(^3$He$, 2p)^4$He, owing to the much smaller $S(0)$ value.

[d] For experimental evidence see sec. 6.1.6.

Par64a, Fia73). Thus, the completion of the p-p chain is more complicated. In 1951 Lauritsen (Lau51) and Schatzmann (Sch51) suggested the alternative completion via the ^{3}He$(^3$He$, 2p)^4$He reaction.

The time dependence of the ^{3}He abundance in a star is described by an equation similar to that given in section 6.1.2:

$$\frac{d(^3He)}{dt} = r_{pd} - r_{^3He^3He} = r_{12} - r_{33}$$

$$= HD\langle\sigma v\rangle_{12} - 2\,\frac{^3He\,^3He}{2}\,\langle\sigma v\rangle_{33} .$$

The subscripts in these and subsequent equations represent the atomic masses of the interacting nuclei, and the chemical symbols the number densities of the interacting nuclei. The first term represents the production rate of ^{3}He via the $d(p, \gamma)^3$He reaction, and the second term its destruction via the ^{3}He$(^3$He$, 2p)^4$He reaction. The factor 2 in the numerator of the second term is needed, since two $d(p, \gamma)^3$He reactions must occur to provide the two ^{3}He nuclei required for the second reaction. The factor 2 in the denominator arises from the $(1 + \delta_{ij})^{-1}$ term. Note that in the above equation the ^{3}He destruction via the ^{3}He$(\alpha, \gamma)^7$Be reaction has been neglected (see later).

Since the abundance of deuterium (D) results from an equilibrium state (sec. 6.1.2), it can be expressed as $(D/H)_e = \langle\sigma v\rangle_{11}/2\langle\sigma v\rangle_{12}$. The above differential equation can be written as

$$d(^3He)/dt = (H)^2\langle\sigma v\rangle_{11}/2 - (^3He)^2\langle\sigma v\rangle_{33} .$$

This equation describes a self-regulating system that eventually reaches a state of equilibrium $[d(^3He)/dt = 0]$ with an abundance ratio given by the equation

$$(^3He/H)_e = (\langle\sigma v\rangle_{11}/2\langle\sigma v\rangle_{33})^{1/2} . \tag{6.4}$$

Using the theoretical rate $\langle\sigma v\rangle_{11}$ for the $p + p$ reaction (sec. 6.1.1) and the rate $\langle\sigma v\rangle_{33}$ for the $^3He(^3He, 2p)^4He$ reaction based on the experimentally observed $S_{33}(0)$ factor (Table 6.2), the equilibrium ratio $(^3He/H)_e$ may be calculated as a function of stellar temperature. The results for a wide range of stellar temperatures are shown in Figure 6.6.

Because the Coulomb barrier E_C involved in the $^3He(^3He, 2p)^4He$ reaction $(E_C = 1.39 \text{ MeV})$ is high compared with that of the $d(p, \gamma)^3He$ reaction $(E_C = 0.44 \text{ MeV})$, the stellar lifetime of ^3He $[\tau_{3He}(^3He)]$ is much longer than that of deuterium $[\tau_H(D)]$. As a consequence, a greater abundance of ^3He is attained before equilibrium is established between production and destruction of ^3He (Fig. 6.6). Combining the equations $\tau_{3He}(^3He) = 1/^3He\langle\sigma v\rangle_{33}$ and $(^3He/H)_e = (\langle\sigma v\rangle_{11}/2\langle\sigma v\rangle_{33})^{1/2}$, one obtains for the lifetime for the destruction of the ^3He nuclei at equilibrium the relationship

$$\tau_{3He}(^3He)_e = (1/H)(2/\langle\sigma v\rangle_{11}\langle\sigma v\rangle_{33})^{1/2} . \tag{6.5}$$

For conditions $\rho = 100 \text{ g cm}^{-3}$ and $X_H = X_{He} = 0.5$, the resulting lifetimes as a function of stellar temperature are illustrated in Figure 6.7. For example, at $T_6 = 15$ one finds $\tau_{3He}(^3He) = 2.2 \times 10^5$ y, a time much longer than the 1.6 s found for $\tau_H(D)$ (sec. 6.1.2). Relations similar to equation (6.5) can be derived for the other reactions listed in Table 6.2. The resulting lifetimes, also plotted in Figure 6.7, confirm that the deuterium-induced reactions play no role in the burning of ^3He.

The time for ^3He to reach the equilibrium condition is also of interest. It can be determined, for conditions in which $\tau_{4He}(^3He) \gg \tau_{3He}(^3He)$ (Fig. 6.7), by combining the following equations:

$$\frac{d(^3He)}{dt} = \frac{(H)^2}{2}\langle\sigma v\rangle_{11} - (^3He)^2\langle\sigma v\rangle_{33} ,$$

$$\left(\frac{^3He}{H}\right)_e = \left(\frac{\langle\sigma v\rangle_{11}}{2\langle\sigma v\rangle_{33}}\right)^{1/2} ,$$

to arrive at the differential equation

$$\frac{d(^3He)}{dt} = (H)^2\langle\sigma v\rangle_{33}\left[\left(\frac{^3He}{H}\right)_e^2 - \left(\frac{^3He}{H}\right)_t^2\right] .$$

Using the boundary condition $^3He/H = 0$ at the time $t = 0$, the solution of this differential equation is

$$\left(\frac{^3He}{H}\right)_t = \left(\frac{^3He}{H}\right)_e \tanh\left[\frac{t}{\tau_{3He}(^3He)_e}\right] . \tag{6.6}$$

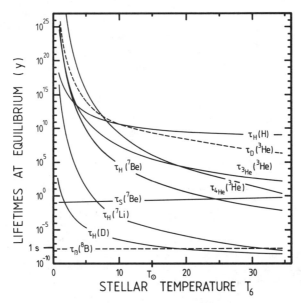

FIGURE 6.7. Plotted are the equilibrium lifetimes of ^{3}He resulting from different burning processes (Table 6.2). The ^{3}He(α, γ)^{7}Be reaction leading to the $\tau_{^4\text{He}}(^3\text{He})$–curve is important only in stars which have an appreciable amount of ^{4}He. Shown for comparison is the lifetime of hydrogen against destruction via the $p + p$ reaction and those of D, ^{7}Li, and ^{7}Be against destruction via hydrogen-burning interactions. The electron-capture lifetime of ^{7}Be in stars, $\tau_s(^7\text{Be})$, and the laboratory lifetime of the positron decay for ^{8}B are also shown. All curves assume conditions of $\rho = 100$ g cm^{-3}, $X_H = X_{He} = 0.5$.

This equation describes the time dependence of the $^3He/H$ abundance ratio at a given stellar temperature. Introducing the variable f defined as

$$f = \left(\frac{^3He}{H}\right)_t \bigg/ \left(\frac{^3He}{H}\right)_e, \qquad (6.7)$$

which is the fraction of the ^{3}He abundance at time t compared to that at equilibrium, the time t_f required to reach the fraction f is determined by the equation

$$t_f = \tau_{3He}(^3\text{He})_e \tanh^{-1} f. \qquad (6.8)$$

For example, at solar conditions (i.e., $T_6 = 15$) and $f = 0.99$, the necessary time is found to be $t_f = 5.8 \times 10^5$ y. A graphical representation of this time t_f as a function of stellar temperature is shown in Figure 6.8.

From Figure 6.8 one sees that the time t_f required to reach 99% equilibrium ($f = 0.99$) is long. Under certain conditions it may be of the same order as the lifetime of some stars. Thus, ^{3}He abundance depends on the age of a star. In our sun only after about 10^6 y would there be enough ^{3}He for the ^{3}He(^{3}He, $2p$)^{4}He reaction to operate at maximum rate. At a stellar temperature of $T_6 < 8$ ($M < M_\odot$), equilibrium would require more than 10^9 y

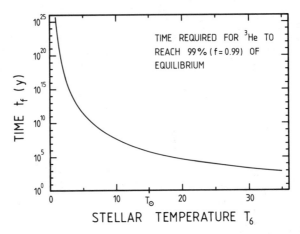

FIGURE 6.8. Temperature dependence of the time t_f required for ^{3}He to reach 99% of its equilibrium abundance (for conditions of $\rho = 100$ g cm^{-3} and $X_H = X_{He} = 0.5$). Because this time exceeds 10^9 y for $T_6 < 8$, it is reasonable to assume that ^{3}He has not achieved equilibrium at such low temperatures (or, equivalently, low stellar masses).

(Fig. 6.8); since these stars spend only about 10^{10} y on the main sequence of the H-R diagram (chap. 2), equilibrium would not be reached.

The calculation of the energy production in a star via the p-p chain depends upon whether the ^{3}He equilibrium condition has been reached. As previously noted, ^{3}He is synthesized via the reactions $p + p \to d + e^+ + v$ and $d(p, \gamma)^3$He, where the overall process $3H \to {}^3$He is limited by the $p + p$ reaction rate r_{11}. With an energy release in the $3H \to {}^3$He conversion of $Q = 6.936$ MeV and a mean energy loss of $\bar{E}_v = 0.263$ MeV from the neutrinos of the $p + p$ reaction (Fig. 6.2), the effective energy deposited in the star is $Q_{eff} = 6.673$ MeV. The rate of energy production ϵ is thus given by the equation

$$\rho\epsilon(3H \to {}^3He) = 6.673r_{11} \text{ MeV cm}^{-3} \text{ s}^{-1} .$$

For the ^{3}He(^{3}He, $2p$)^{4}He reaction, the energy release is $Q = 12.860$ MeV, leading to an energy production rate of

$$\rho\epsilon(^3He + {}^3He) = 12.860r_{33} \text{ MeV cm}^{-3} \text{ s}^{-1} .$$

The total rate of energy production is the sum of these two rates:

$$\rho\epsilon_{tot} = 6.673r_{11} + 12.860r_{33} \text{ MeV cm}^{-3} \text{ s}^{-1} . \tag{6.9}$$

Since the total energy production rate is sensitive to the ^{3}He abundance (involved in the r_{33} rate), stellar model calculations must use a continuously increasing ^{3}He abundance until the equilibrium state abundance of ^{3}He is reached. Step by step, the calculations proceed as follows:

1. The stellar model starts at time $t = 0$ with $^3He(t = 0) = 0$ (or more precisely with $(^3He/H) = (D + {}^3He/H)_{primordial} \simeq 10^{-4}$;

2. The ^{3}He production rate is given by the rate r_{11} at $t = 0$;
3. After a time interval $t = \Delta t$, another stellar model has to be constructed incorporating the ^{3}He abundance at the time $t = \Delta t$: $^3He(t = \Delta t) = r_{11}\Delta t$;
4. Similarly, the stellar model at the interval $t = 2\Delta t$ involves a ^{3}He abundance of $^3He(t = 2\Delta t) = {}^3He(t = \Delta t) + r_{11}\Delta t - 2r_{33}\Delta t$;
5. For each successive model modification, one calculates the total energy production rate $\rho\epsilon_{tot}$ (eq. [6.9]);
6. Finally, if the ^3He equilibrium is reached, calculations of the energy production rate is again simplified. With $d(^3He)/dt = 0$, one obtains $2r_{33} = r_{11}$, and

$$\rho\epsilon_{tot} = 13.103r_{11} \text{ MeV cm}^{-3} \text{ s}^{-1}.$$

Inserting the theoretical reaction rate for the $p + p$ reaction, one obtains for the above equation (chap. 3)

$$\epsilon_{tot} = 1.47 \times 10^{12} \rho X_H^2 T_6^{-2/3} \exp\left(-33.72 T_6^{-1/3}\right)$$

$$\times \left(1 + 0.012 T_6^{1/3} + 0.0107 T_6^{2/3} + 0.0009 T_6\right) \text{ MeV g}^{-1} \text{ s}^{-1}. \quad (6.10)$$

The chain of reactions $p + p \rightarrow d + e^+ + v$, $d(p, \gamma)^3$He, and ^{3}He$(^3$He, $2p)^4$He, which completes the conversion of four protons into ^{4}He with an energy release of 26.2 MeV per fusion process, is frequently called the p-p-I chain (Fig. 6.9).

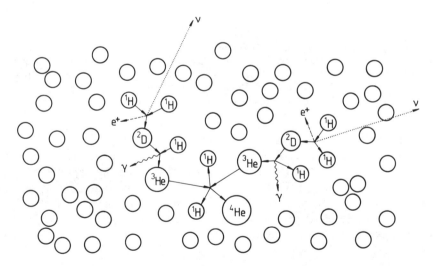

FIGURE 6.9. Schematic representation of the reactions involved in the fusion of four protons into ^{4}He by the p-p-I chain. This fusion process, with an energy release of 26.73 MeV, occurs in the cores of all first-generation stars and predominantly in the cores of second-generation stars of low mass ($M < M_\odot$) during their hydrostatic hydrogen-burning episode (main-sequence stars). The energy stored in the stellar core is reduced by the energy of the escaping neutrinos ($\bar{E}_v = 2 \times 0.26$ MeV $= 0.52$ MeV), since these neutrinos have essentially a probability of unity to leave the star.

ELECTRON CAPTURE OF ^{7}Be

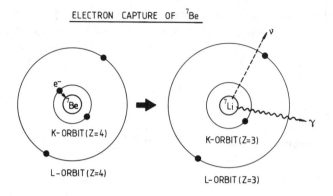

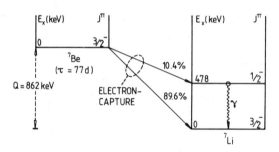

FIGURE 6.10. Schematic representation of the capture of an atomic K electron by the ^{7}Be nucleus, leading to ^{7}Li with the emission of a monoenergetic neutrino ($E_\nu = 862$ keV). In 10.4% of all cases, the capture proceeds to the 478 keV first excited state in ^{7}Li ($E_\nu = 384$ keV), which in its decay emits a 478 keV photon. The average energy of the escaping neutrinos is thus $\bar{E}_\nu = 814$ keV. Note that ^{7}Be can only decay to ^{7}Li via this electron-capture process because the alternative β^+ process $^7\text{Be} \rightarrow {}^7\text{Li} + e^+ + \nu$ is energetically forbidden ($Q = -160$ keV).

If a star has sufficient ^{4}He derived from the ashes of previous stars (or the big bang) or synthesized through the operation of the p-p-I chain, the ^{3}He nuclei can also be burned in a significant way through the capture reaction $^3\text{He}(\alpha, \gamma)^7\text{Be}$ (Table 6.2). In a star like the sun, the burning of ^{3}He by the $^3\text{He}(\alpha, \gamma)^7\text{Be}$ reaction proceeds at a stellar rate somewhat slower than the $^3\text{He}(^3\text{He}, 2p)^4\text{He}$ reaction (Fig. 6.7). Thus, subsequent reactions involving the burning of ^{7}Be are of importance as alternative ways of completing the fusion process $4\text{H} \rightarrow {}^4\text{He}$.

6.1.4. Stellar Fates of ^{7}Be

In the earth-bound laboratory, the ^{7}Be atom is radioactive, with the only energetically allowed decay being (Fig. 6.10) the capture of an atomic electron:

$$^7\text{Be}(e^-, \nu)^7\text{Li} .$$

The energy released in the reaction ($Q = 0.862$ MeV) is carried away by the escaping monoenergetic neutrinos ($E_\nu = 862$ and 384 keV). The terrestrial mean decay time is observed to be $\tau = 76.9$ d (Ajz84). This decay mode, called nuclear electron capture (EC), represents another type of decay governed by the weak force (Bam77). The decay rate $\lambda_{EC} = 1/\tau_{EC}$ is described, by analogy to that for the $p + p$ reaction (sec. 6.1.1), by Fermi's Golden Rule:

$$\lambda_{EC} = \frac{2\pi}{\hbar}\, \rho(E) |\langle f|H_\beta|i\rangle|^2 ,$$

where the phase-space factor is

$$\rho(E) = 4\pi V E_\nu^2/c^3 h^3$$

and the interaction matrix element is given by

$$H_{if} = \int [\Psi_{^7\mathrm{Li}}\Psi_\nu]^* H_\beta \Psi_{^7\mathrm{Be}}\Psi_{e^-}\, d\tau .$$

Since the radial wave functions of the ^{7}Li and ^{7}Be nuclei vanish rapidly outside the nuclear domain, the electron wave function in the nuclear domain can be approximated by

$$\Psi_{e^-}(r) \simeq \Psi_{e^-}(0)$$

and the neutrino field by a plane wave,

$$\Psi_\nu(r) = \frac{1}{V^{1/2}} \exp(i k_\nu \cdot r) \simeq \frac{1}{V^{1/2}} .$$

The interaction matrix element reduces to

$$H_{if} = \frac{\Psi_{e^-}(0)}{V^{1/2}} \int \Psi_{^7\mathrm{Li}}^* H_\beta \Psi_{^7\mathrm{Be}}\, d\tau = \frac{\Psi_{e^-}(0)}{V^{1/2}}\, g \int \Psi_{^7\mathrm{Li}}^* \Psi_{^7\mathrm{Be}}\, d\tau = \frac{\Psi_{e^-}(0)}{V^{1/2}}\, g M_n ,$$

where the symbol M_n represents the nuclear matrix element. Combining the above equations leads to the expression for the decay rate, λ_{EC}:

$$\lambda_{EC} = \frac{1}{\tau_{EC}} = \frac{g^2 M_n^2}{\pi c^3 \hbar^4} E_\nu^2 |\Psi_e(0)|^2 . \tag{6.11}$$

In the electron-capture process not only does the electron cloud provide the electrons being captured by the nucleus, but also the influence of the entire electron cloud affects the rate of the process, i.e., the probability of electrons being at the nucleus ($\lambda_{EC} \propto |\Psi_e(0)|^2$). Since the capture can occur from any of the various electron shells (K, L, M, ...), all of them contribute to the total decay rate λ_{EC}:

$$\lambda_{EC} = \lambda_K + \lambda_L + \lambda_M + \cdots$$

$$= \frac{g^2 M_n^2}{\pi c^3 \hbar^4} E_\nu^2 [|\Psi_K(0)|^2 + |\Psi_L(0)|^2 + |\Psi_M(0)|^2 + \cdots] .$$

The electron density at the nucleus of the various shells is given approximately by (Bla62, Bam77)

$$|\Psi_n(0)|^2 \simeq \frac{1}{\pi}\left(\frac{Z}{na_0}\right)^3 ,$$

where $a_0 = 0.529 \times 10^{-8}$ cm is the Bohr orbital radius and $n = 1, 2, 3, \ldots$ is the principal quantum number. It therefore follows that

$$\frac{\lambda_K}{\lambda_L} \simeq 8 , \qquad \frac{\lambda_K}{\lambda_M} \simeq 27 , \qquad \frac{\lambda_K}{\lambda_N} \simeq 256 , \ldots .$$

Clearly, the electron-capture process is dominated by K electrons ($n = 1$). Electrons from the outer atomic shells also contribute, but by a rapidly decreasing amount. A more precise treatment of the electron-capture process must also involve the wave functions of the whole atomic clouds of both the mother and daughter atoms (Odi56, Bah62, Bam77).

Since the early days of nuclear science studies, it has been accepted that the lifetime, or decay constant, of a radioactive substance is independent of the extranuclear environment. This expectation of constant nuclear decay rates was firmly established by various laboratory studies of α- and β-emitting nuclei. However, Segrè and Dandel pointed out in 1947 (Seg47, Dan47) that, in the case of electron-capture decays, the decay rate is directly related to the density of atomic electrons at the nucleus ($\lambda_{EC} \propto |\Psi_e(0)|^2$; eq. [6.11]) and that, for low-Z nuclei such as ^{7}Be, the chemical environment can measurably influence the decay rate by altering the distribution of electrons. Experimental studies, where changes of 0.1% in the ^{7}Be lifetime have been observed, confirm these effects (Seg49, Joh70, Eme72).

It is clear from the above considerations that, if a radioactive atom such as ^{7}Be, which normally decays by electron capture, has all of the orbital electrons removed, the decay rate goes to zero ($|\Psi_e(0)|^2 = 0$) and the nucleus becomes essentially stable. Such a situation is approximated in the center of stars. Here, at temperatures of $T_6 = 15$, the mean kinetic energy is $E = kT = 1.3$ keV, which is much larger than the binding energy of the most tightly bound electrons (the K-shell electrons) of many light atoms. Thus a nucleus such as ^{7}Be with a K-shell binding energy $E_K = 0.22$ keV is almost completely ionized. The lifetime of ^{7}Be in such an environment is expected to increase considerably. However, the ^{7}Be nuclei are immersed in a sea of free electrons resulting from the ionization process. Thus, in stellar environments with an appreciable density of free electrons, the rate of electron-capture decay is dominated by the capture of continuum electrons (Fig. 6.11). This proceeds in much the same manner as the capture of orbital electrons. Rates for the capture of continuum electrons were first estimated by Bethe and Bacher in 1936 (Bet36) and were improved subsequently by many authors (Bet39, Cri42, Sch53, Bah62a, Ibe67a, Bah68, Bah69a).

ELECTRON CAPTURE OF ^{7}Be IN STARS

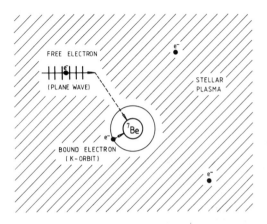

FIGURE 6.11. In the high-temperature stellar environment, atoms such as ^{7}Be are essentially stripped of their atomic electrons (stellar plasma of positive ions and free electrons). The ^{7}Be nucleus then would be stable if it were not immersed in a sea of free electrons. The decay rate for this process is proportional to the density of free (continuum) electrons surrounding the ^{7}Be nucleus. If the stellar temperature is not too high, a small probability exists for finding an electron in the K orbit of the ^{7}Be atom. This will further enhance the decay rate for electron capture.

Since all other factors in the capture of continuum electrons are approximately the same as those in the case of the atomic electron capture, the ^{7}Be lifetime in a star, τ_s, is related to that observed on earth, τ_e, by the relation:

$$\frac{\tau_s}{\tau_e} = \frac{\lambda_e}{\lambda_s} \simeq \frac{2\,|\Psi_e(0)|^2}{|\Psi_f(0)|^2} , \tag{6.12}$$

where $|\Psi_f(0)|^2$ represents the density of free electrons at the nucleus. The factor of 2 in the denominator is needed because two spin orientations of the electron have to be taken into account in calculating the decay rate λ_e. The rate λ_s is calculated by averaging over these two orientations. For a uniform density of free electrons (i.e., the electrons are described by plane waves), the density n_e of free electrons in stellar matter composed of pure hydrogen is given by the equation

$$|\Psi_f(0)|^2 = n_e = \rho/M_{\rm H} ,$$

where ρ is the stellar density (g cm^{-3}) and $M_{\rm H}$ is the mass of the hydrogen nucleus. If the stellar matter is described by the mass fraction $X_{\rm H}$ of hydrogen and the mass fraction $(1 - X_{\rm H})$ of heavier elements $M(A, Z)$, the density is given by

$$n_e = X_{\rm H}\,\frac{\rho}{M_{\rm H}} + (1 - X_{\rm H})\,\frac{Z\rho}{M(A, Z)} .$$

Since for the heavier elements $M(A, Z) \simeq 2ZM_H$, one arrives at

$$n_e \simeq \frac{\rho}{M_H}\left(\frac{1 + X_H}{2}\right).$$ (6.13)

Because of the nuclear charge Z, the above result must be corrected by using Coulomb-distorted plane waves appropriate to electron scattering from a $+Ze$ charge rather than simplified pure plane waves. The correction function $F(A, Z)$ is given approximately by the expression (Bah69a, Ibe67a)

$$F(A, Z) \simeq 2\pi\eta = 2\pi\frac{Ze^2}{hv} = 2\pi Z\frac{e^2}{\hbar c}\left(\frac{m_e c^2}{2E_e}\right)^{1/2} = 2\pi Z\alpha\left(\frac{m_e c^2}{2E_e}\right)^{1/2},$$

where η is the Sommerfeld parameter, Z is the nuclear charge, and v is the velocity of the interacting species ($v \simeq v_e$). Since the velocities of the electrons are described by a Maxwell-Boltzmann distribution for normal (non-degenerate) stellar matter, the function $F(A, Z)$ has to be averaged over the distribution. The effect of averaging can be described approximately by inserting $3kT/2$ for E_e; hence

$$\langle F(A, Z)\rangle \simeq \langle 2\pi\eta\rangle \simeq 2\pi Z\alpha\left(\frac{m_e c^2}{3kT}\right)^{1/2}.$$

Combining the above equations gives the relation

$$\tau_s = \frac{2|\Psi_e(0)|^2}{(\rho/M_H)[(1 + X_H)/2]2\pi Z\alpha(m_e c^2/3kT)^{1/2}}\tau_e,$$ (6.14)

where

$$|\Psi_e(0)|^2 \simeq \frac{1}{\pi}\left(\frac{Z}{a_0}\right)^3$$

for K-electron capture. Inserting numerical values for the ^7Be nucleus, one obtains

$$\tau_s(^7Be) = 7.06 \times 10^8 \frac{T_6^{1/2}}{\rho(1 + X_H)}\ \text{s}.$$ (6.15)

Note that $\tau_s(^7Be)$ depends on the stellar parameters ρ, X_H, and T, i.e., the stellar lifetime of ^7Be varies with stellar class and with location within the star. Note also that the temperature dependence of $\tau_s(^7Be)$ arises from the nuclear Coulomb field, which modifies the wave function of an incident electron. A more exact treatment of the process (Bah69a) provides a formula of the same form but with a numerical value of 4.72×10^8. For solar conditions ($\rho = 100$ g cm^{-3}, $X_H = 0.5$, and $T_6 = 15$), the lifetime of ^7Be against capture of continuum (free) electrons is $\tau_s(^7Be) = 140$ d compared with $\tau_e(^7Be) = 77$ d on earth. It is evident that the stellar lifetime $\tau_s(^7Be)$ has a relatively weak temperature dependence [$\tau_s(^7Be) \propto T^{1/2}$; Fig. 6.7]. This contrasts with the strong temperature dependence of the lifetime against destruction by positive charged particles, i.e., the lifetime for the competing reaction $^7\text{Be}(p, \gamma)^8\text{B}$.

It is assumed above that, under stellar conditions, the ^7Be nucleus captures electrons solely from the continuum. There is, however, a nonzero probability that some ^7Be atoms are only partially ionized, leaving one or two electrons in the K shell. In this case, the nuclear electron-capture probability of ^7Be is larger, and these electrons must be taken into account (Ibe67a, Bah69a):

$$\lambda_{\text{tot}} = \lambda_c + \lambda_K$$

where λ_c represents the decay rate due to continuum electrons and λ_K the stellar K-capture rate. For this situation one can write

$$\frac{\lambda_K}{\lambda_c} = \frac{|\Psi_K^s(0)|^2}{|\Psi_f(0)|^2} .$$

Considering only one K-shell electron (Fig. 6.11), one can use the approximation

$$|\Psi_K^s(0)|^2 \simeq \frac{1}{\pi}\left(\frac{Z}{a_0}\right)^3\left[1 - \exp\left(-\frac{E_K}{kT}\right)\right],$$

where the occupation probability of the K shell is assumed to depend on the Boltzmann factor. From the above discussion, one arrives at the equation

$$|\Psi_f(0)|^2 \simeq \frac{\rho}{M_H}\left(\frac{1 + X_H}{2}\right)\langle 2\pi\eta\rangle .$$

Inserting numerical values for ^7Be under solar conditions, one finds

$$\frac{\lambda_K}{\lambda_c} \simeq 0.21 .$$

Hence $\lambda_{\text{tot}} \simeq 1.21\lambda_c$, i.e., the K-electron–capture process increases the total decay rate by approximately 21%. This is consistent with the results of more sophisticated calculations (Ibe67a, Bah69a). Thus, the solar lifetime of ^7Be is $\tau_s(^7\text{Be}) = 120 \text{ d} = 0.33$ y.

The enhanced lifetime of ^7Be in the stellar interior increases the probability of its destruction via charged-particle–induced nuclear reactions. Since the number of protons greatly exceeds that of any other charged particle and since the Coulomb barrier is greater for particles of higher Z, the proton-induced capture reaction $^7\text{Be}(p, \gamma)^8\text{B}$ ($Q = +0.138$ MeV) is overwhelmingly dominant. This reaction has been studied in the laboratory over a wide range of energies (sec. 6.1.6). The resulting $S(E)$ factor, $S(0) = 0.024$ keV b, has been used to determine the corresponding reaction rate. For solar conditions ($\rho = 100$ g cm^{-3}, $X_H = 0.5$) the lifetime of ^7Be against hydrogen burning via the above reaction has been calculated as a function of stellar temperature. The results (Fig. 6.7) show the anticipated greater temperature dependence of charged-particle burning compared with destruction via the electron-capture process. At solar temperatures ($T_6 = 15$) one finds a mean lifetime of $\tau_H(^7Be) = 150$ y, compared with $\tau_s(^7Be) = 0.33$ yr. This means that 99.8% of the time solar destruction of ^7Be proceeds through the electron-capture process, and only 0.2% of the time via hydrogen burning. Both possible ^7Be end products, ^7Li and ^8B, require further reactions to complete the conversion of four protons into ^4He. The two branches are called the p-p-II and p-p-III chains (sec. 6.1.5).

6.1.5. The Three p-p Chains

Most of the ^7Be produced in stars like the sun becomes ^7Li via the electron-capture process

$$^7Be(e^-, \nu)^7Li .$$

Arguments based on reaction Q-values, Coulomb barriers, reaction cross sections, and abundances of interacting nuclei lead to the conclusion that ^7Li itself is predominantly and rapidly consumed by the hydrogen-burning reaction (Fig. 6.7)

$$^7Li(p, \alpha)\alpha ,$$

with $Q = 17.347$ MeV and $S(0) \simeq 100$ keV b (sec. 6.1.6). In this reaction, the conversion of four protons into an α-particle (^4He) is completed. The series of reactions

$$p(p, e^+\nu)d ,$$

$$d(p, \gamma)^3He ,$$

$$^3He(\alpha, \gamma)^7Be ,$$

$$^7Be(e^-, \nu)^7Li ,$$

$$^7Li(p, \alpha)\alpha$$

is called the p-p-II chain.

The remaining ^7Be is consumed by hydrogen burning via the $^7Be(p, \gamma)^8B$ reaction. The end product ^8B is a radioactive nucleus that decays with a lifetime of $\tau = 1.1$ s to ^8Be:

$$^8B \rightarrow {}^8Be + e^+ + \nu \quad (Q = 17.979 \text{ MeV}) .$$

Because of the selection rules involved, this β-decay goes predominantly to the first excited state ($E_x = 2.94$ MeV) in ^8Be (Fig. 6.12) with an energy release of $Q = 15.039$ MeV. The first excited state in ^8Be has a lifetime of 4×10^{-22} s. It decays quickly into two α-particles, releasing 2.94 MeV of energy. This process completes the $4p \rightarrow {}^4He$ conversion. This total set of reactions

BETA - DECAY OF ^{8}B

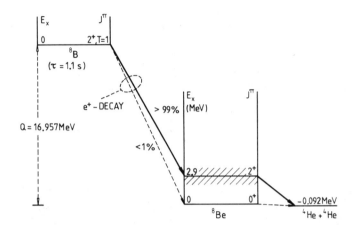

$$^8B \longrightarrow {^8Be}^* + e^+ + v$$
$$ \llcorner\;^4He + {^4}He$$

FIGURE 6.12. The positron decay of ^{8}B goes mainly to the $\Gamma = 1.6$ MeV broad first excited state in ^{8}Be, which in turn fissions into two α-particles. The average energy of the escaping neutrinos is $\bar{E}_v(^8B) = 7.30$ MeV. These neutrinos play a dominant role in solar neutrino detection experiments (chap 10).

$$p(p, e^+v)d \; ,$$
$$d(p, \gamma)^3He \; ,$$
$$^3He(\alpha, \gamma)^7Be \; ,$$
$$^7Be(p, \gamma)^8B \; ,$$
$$^8B(e^+v)^8Be^* \; ,$$
$$^8Be^*(\alpha)\alpha \; ,$$

first suggested by Fowler in 1958 (Fow58; see also Cam58), is collectively called the p-p-III chain.

Note that at the end of the p-p chains II and III there are two ^{4}He nuclei; however, one ^{4}He nucleus served only as a catalyst in the burning process. The mean lifetimes of all processes involved in the three chains are shown in Figure 6.7 as a function of stellar temperature for conditions $\rho = 100$ g cm^{-3} and $X_H = X_{He} = 0.5$. The results demonstrate again that the conversion of

$$4H \rightarrow {^4}He + 2e^+ + 2v$$

THE REACTIONS OF THE P-P CHAIN

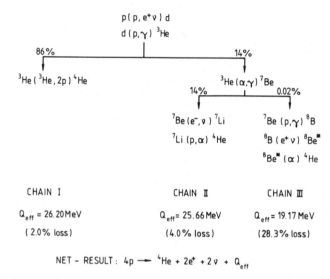

FIGURE 6.13. Shown schematically are the chains of reactions involved in the conversion of four protons into a ^{4}He nucleus. The percentages shown are calculated (Bah80, Bah82a) for solar conditions. The set of reactions does not represent a cycle, as it was sometimes called, but rather a chain of reactions, because ^{4}He is the end product of the synthesis of four protons with no cycling catalyst required. In a way, ^{4}He acts as a catalyst in chains II and III because the interaction of one α-particle leads eventually to the production of two.

is limited by the rate of the $p + p$ reaction. Figure 6.13 shows schematically the entire process. The branching percentages have been calculated (Bah80, Bah82a) for conditions prevailing in the center of the sun.

In all three chains, the total energy released is the same, $Q = 26.73$ MeV, but the amount of energy carried away by the escaping neutrinos differs in each chain. Accordingly, this difference reduces the effective energy remaining in the stellar interior. To calculate the mean energy of the escaping neutrinos, the energy spectrum of the neutrinos must be known. For the $p + p$ reaction and the electron capture of ^{7}Be, one finds $\bar{E}_\nu(p + p) = 0.26$ MeV (Fig. 6.2) and $\bar{E}_\nu(^7\text{Be}) = 0.81$ MeV (Fig. 6.10). In the case of ^{8}B decay, the calculation is complicated by the large width of the first excited state in ^{8}Be (Fig. 6.12), which leads to $\bar{E}_\nu(^8\text{B}) = 7.30$ MeV. The resulting effective energies remaining in the three p-p chains are summarized below.

Chain I:

$$\bar{E}_\nu(p + p) = 0.26 \text{ MeV} ,$$

$$Q_{\text{eff}} = Q - 2\bar{E}_\nu(p + p)$$

$$= 26.20 \text{ MeV} \qquad (2.0\% \text{ energy loss});$$

Chain II:

$$\bar{E}_\nu(p + p) = 0.26 \text{ MeV} ,$$

$$\bar{E}_\nu(^7Be) = 0.81 \text{ MeV} ,$$

$$Q_{eff} = Q - \bar{E}_\nu(p + p) - \bar{E}_\nu(^7Be)$$

$$= 25.66 \text{ MeV} \qquad (4.0\% \text{ energy loss});$$

Chain III:

$$\bar{E}_\nu(p + p) = 0.26 \text{ MeV} ,$$

$$\bar{E}_\nu(^8B) = 7.30 \text{ MeV} ,$$

$$Q_{eff} = Q - \bar{E}_\nu(p + p) - \bar{E}_\nu(^8B)$$

$$= 19.17 \text{ MeV} \qquad (28.3\% \text{ energy loss}).$$

The largest energy loss due to escaping neutrinos occurs in chain III.

In a star with sufficient ^4He to act as catalyst, all three chains (Fig. 6.13) operate concurrently. The details of their operation depend on density, temperature, and chemical composition of the star. The chemical elements involved in the three p-p chains change their abundances as a function of time owing to nuclear processing. Their time dependence is given by the following set of differential equations ($\langle \sigma v \rangle \equiv \lambda$):

$$\frac{d(H)}{dt} = -2\lambda_{11}\frac{(H)^2}{2} - \lambda_{12}HD + 2\lambda_{33}\frac{(^3He)^2}{2} - \lambda_{17}H(^7Be) - \lambda_{17}^* H(^7Li) ,$$

$$\frac{d(D)}{dt} = \lambda_{11}\frac{(H)^2}{2} - \lambda_{12}HD ,$$

$$\frac{d(^3He)}{dt} = \lambda_{12}HD - 2\lambda_{33}\frac{(^3He)^2}{2} - \lambda_{34}(^3He)(^4He) ,$$

$$\frac{d(^4He)}{dt} = \lambda_{33}\frac{(^3He)^2}{2} - \lambda_{34}(^3He)(^4He) + 2\lambda_{17}H(^7Be) + 2\lambda_{17}^* H(^7Li) ,$$

$$\frac{d(^7Be)}{dt} = \lambda_{34}(^3He)(^4He) - \lambda_{e7}n_e(^7Be) - \lambda_{17}H(^7Be) ,$$

$$\frac{d(^7Li)}{dt} = \lambda_{e7}n_e(^7Be) - \lambda_{17}^* H(^7Li) .$$

These complicated nonlinear equations are usually solved using simplifying assumptions (Cla68). The calculations reveal that the total energy production in the sun is $\epsilon_{tot} = 5.1 \times 10^7 \text{ MeV g}^{-1} \text{ s}^{-1}$. The observed energy radiated from the sun's surface is $L_\odot = 2.4 \times 10^{39} \text{ MeV s}^{-1}$. From these values it follows that the mass of the sun involved in hydrogen burning is $m_\odot = L_\odot/\epsilon_{tot} = 4.7 \times 10^{31}$ g, compared with the total mass of $M_\odot = 2.0 \times 10^{33}$ g. Therefore, only a small part of the sun, 2.4% by mass, is involved in the hydrogen-burning process $4H \rightarrow {}^4He$.

In summary, the rate of the p-p chain depends entirely on the rate of the $p + p$ reaction, and the rate of energy generation ϵ depends on how many of the subsequent reactions (Fig. 6.13) come into equilibrium with the $p + p$ reaction. The second reaction, $d(p, \gamma)^3He$, is in equilibrium at all temperatures and the equilibrium amount, by number, of deuterium nuclides is given by $(D/H)_e \simeq 10^{-17}$. Below an effective central temperature of $T_6 = 8$, the p-p chain terminates at ^3He (Fig. 6.8) with an energy generation close to one-half that of the full chain. In the temperature range about $T_6 = 8$–13, the chain is completed predominantly by the $^3He(^3He, 2p)^4He$ reaction. At higher central temperatures in stars containing amounts of ^4He comparable in mass to hydrogen, completion of the chain will be mainly via the $^3He(\alpha, \gamma)^7Be$ reaction. Since this mode of completion requires only one initial $p + p$ reaction (Fig. 6.13) rather than the two required for completion by the reaction $^3He(^3(He, 2p)^4He$, the rate of ^4He production and energy generation is doubled at high temperatures.

Note that, since the energy loss by escaping neutrinos represents on the average only a few percent of the total energy produced (Fig. 6.13), it has very little influence on the course of stellar evolution. However, the effects due to escaping neutrinos will be very significant in the late stages of stellar evolution (chap. 8). It should also be emphasized that the p-p chain provides the important mechanism by which ^4He is synthesized from hydrogen alone without heavier nuclei acting as a catalyst (secs. 6.2 and 6.3). This permits formulation of a general theory for stellar nucleosynthesis of all heavy elements, starting with hydrogen (chap. 2).

6.1.6. Laboratory Approach to p-p Chain Reactions

Measurement of cross sections for most of the reactions involved in the three p-p chains (Fig. 6.13) has been achieved in the laboratory. Experimental information has been obtained for a wide range of energies of interacting nuclei, ranging from much higher than those existing in the stellar environment to as low an energy as possible (chap. 4). This experimental information, together with theoretical guidance, then has been used to extrapolate the cross sections to stellar energies. There are two exceptions to the above statement. Presently no experimental data at any energy exist for the $p + p$ reaction and the stellar electron capture of ^7Be. The problems associated with their laboratory investigations are discussed below.

6.1.6.1. The $d(p, \gamma)^3He$ Reaction

The $d(p, \gamma)^3He$ capture reaction has been studied over a wide range of beam energies (Gri63, Bai70) and has been shown (Don67a) to be dominated by the direct-capture (nonresonant) mechanism (chap. 4). Since this reaction has the lowest Coulomb barrier of all the reactions in the p-p chain (Fig. 6.13), measurements have been extended to energies as low as 16 keV in the center-of-mass system (Fig. 6.14). This energy is not far from the calculated Gamow energy window at $E_0 = 6.5 \pm 3.3$ keV for $T_6 = 15$ (chap. 4). Since the experimental data over a wide range of energies are well described by the direct-capture model, there is considerable confidence in the validity of extrapolation over the relatively small interval between 16 and 6.5 keV. This extrapolation

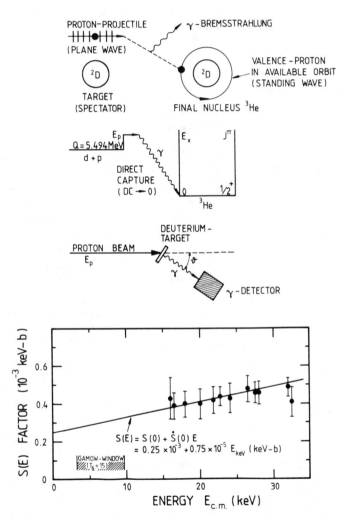

FIGURE 6.14. The $d(p, \gamma)^3$He reaction proceeds entirely by the direct-capture (DC) process into the ^{3}He ground state. With a proton beam of energy E_p incident on a deuterium target, the capture transitions were observed in γ-ray detectors at several angles θ. The angle-integrated yield (total cross section) was thus obtained as a function of beam energy. The results (Gri63, Bai70) are shown in the form of the $S(E)$ factor and are well explained by the DC model (*solid line*; Don67a). Also shown is the Gamow energy window for $T_6 = 15$.

leads to an astrophysical S-factor at zero energy of $S(0) = 0.25 \times 10^{-3}$ keV b with a slope of $(dS/dE)_0 = 0.75 \times 10^{-5}$ b (Fig. 6.14).

6.1.6.2. The $^3\mathrm{He}(^3\mathrm{He}, 2p)^4\mathrm{He}$ Reaction

In contrast to other charged-particle–induced reactions, the $^3\mathrm{He} + {}^3\mathrm{He}$ interaction gives rise to three particles in the exit channel, namely, two protons and one α-particle. The production of three particles in the exit channel for this reaction ($Q = 12.860$ MeV) can be visualized as resulting from either a direct process,

$$^3\mathrm{He} + {}^3\mathrm{He} \rightarrow 2p + {}^4\mathrm{He} \;,$$

or a sequential process via the nucleus $^5\mathrm{Li}$, which is unstable in its ground state $[^5\mathrm{Li(g.s.)} \rightarrow p + {}^4\mathrm{He}, Q = 1.965$ MeV$]$:

$$^3\mathrm{He} + {}^3\mathrm{He} \rightarrow p + {}^5\mathrm{Li(g.s.)} \rightarrow p + p + {}^4\mathrm{He} \;.$$

In the direct process, the emitted protons range in energy from zero to $E_{\max}^p = \frac{4}{6} Q = 8.6$ MeV, where the intensity distribution is governed predominantly by the availability of phase space. The associated α-particles are also distributed over an energy range up to 4.3 MeV. In contrast, the two-step process is characterized by the emission of essentially monoenergetic protons and α-particles, with $E_p = 1.6$ and 9.1 MeV and $E_\alpha = 0.4$ MeV. The energies of these monoenergetic particles are smeared out somewhat owing to the width $\Gamma = 1.5$ MeV of the $^5\mathrm{Li}$ ground state. From the particle spectra observed over a wide range of beam energies (Wan66, Bac67, Dwa71), it is clear that the sequential process prevails at the higher energies ($E_{\mathrm{c.m.}} > 1$ MeV), but the direct process is dominant at lower energies. The observed energy dependence of the $S(E)$ factor is illustrated in Figure 6.15, where the data have been fitted to a polynomial function (Dwa71):

$$S(E) = S(0) + \left(\frac{dS}{dE}\right)_0 E + \frac{1}{2}\left(\frac{d^2S}{dE^2}\right)_0 E^2$$

$$= 5.5 - 3.1E + 1.4E^2 \text{ MeV b} \;.$$

May and Clayton (May68) suggested a mechanism for this $^3\mathrm{He} + {}^3\mathrm{He}$ interaction at low beam energies in which a neutron tunnels from one $^3\mathrm{He}$ to the other, unimpeded by the Coulomb barrier, up to a radial distance where the nuclei overlap appreciably. In this model (Fig. 6.15), a diproton remains and subsequently fissions into two protons. The calculated $S(E)$ factor (dotted line in Fig. 6.15) describes the observed energy dependence of the data very well, thus providing confidence in the above extrapolation.

The magnitude of the $S(E)$ factor for this reaction is of special interest in relation to the solar neutrino problem (chap. 10). Partly on the basis of theoretical arguments, it has been suggested (Fow72, Fet72) that a low-energy resonance might exist in this reaction. If it is sufficiently low and narrow in reaction energy, it might have been unobserved in previous measurements. Such a low-energy resonance would significantly enhance the ^{3}He + ^{3}He reaction rate at solar energies and hence enhance the ^{3}He + ^{3}He route (chain I) at the expense of the alternative ^{3}He + ^{4}He branch (chains II and III; Fig. 6.13). If so, the discrepancy between predicted and observed solar neutrino fluxes might be accounted for or at least decreased. This expected low-energy reson-

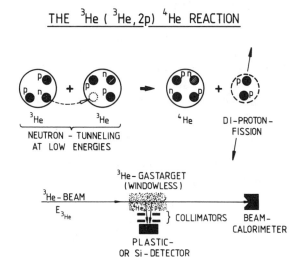

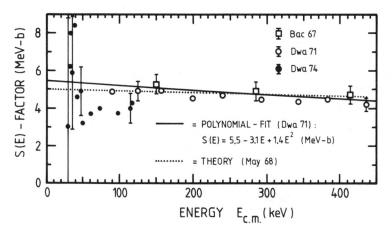

FIGURE 6.15. The ^{3}He(^{3}He, 2p)^{4}He reaction is studied by impinging a ^{3}He beam on a windowless ^{3}He gas target (chap. 5). The spectra at low beam energies reveal that a predominantly direct process is involved. This direct process can be visualized as a neutron tunneling between the two ^{3}He nuclei in the entrance channel, while the residual diproton subsequently fissions into two protons. Note that the lowest data point has an uncertainty of a factor of 2.

ance would correspond to an excited state in ^{6}Be at $E_x \simeq 11.6$ MeV. The search for this state in ^{6}Be via the reaction channels ^{6}Li$(p, n)^6$Be, ^{6}Li$(^3$He, $t)^6$Be, and ^{4}He$(^3$He, $n)^6$Be, as well as in the mirror nucleus ^{6}Li via the reactions ^{7}Li$(^3$He, ^{4}He$)^6$Li and ^{6}Li$(e, e^*)^6$Li, has not been successful (Mak72, Par73, Hal73, Won73, Fag73, Car73, Vie75, McD77). In 1974 Dwarakanath (Dwa74) carried out a search for this resonance state in a more direct way by extending the ^{3}He$(^3$He, $2p)^4$He reaction studies down to $E_{c.m.} = 30$ keV (Fig. 6.15). Although the data suggest an increase in the $S(E)$ factor at the lowest energies, the large uncertainties in the data points preclude any confirmation of the existence of this resonance at least down to $E_{c.m.} = 40$ keV. Below 40 keV the available data neither confirm nor rule out the existence of the hypothetical and "conspiratorial" resonance. Clearly, additional work is desirable to resolve this question.

Kavanagh pointed out (Kav82) that the absolute values reported by Dwarakanath (Dwa74) and by Wang and coworkers (Wan66) are systematically about 20% lower than the combined previous results of Dwarakanath and Winkler (Dwa71) and Bacher and Tombrello (Bac67). Thus the lower value of $S(0) = 5.0 \pm 0.5$ MeV b has been recommended (Kav82), while the slope (determined earlier) is retained. (For recent experimental work see Kra87).

6.1.6.3. The Capture Reaction ^{3}He$(\alpha, \gamma)^7$Be

Parker and Kavanagh (Par63) studied the capture reaction ^{3}He$(\alpha, \gamma)^7$Be over a wide range of beam energies ($E_{c.m.} = 180$–2500 keV), using a gas cell with a conventional entrance foil through which the incident beam had to pass. The γ-rays from the capture transitions were observed with NaI(Tl) crystals. Their results for the $S(E)$ factor are shown in Figure 6.16a, together with the predicted energy dependence based on the direct-capture (DC) model (Tom63). Note that the $S(E)$ data indicate a somewhat steeper rise with decreasing beam energy than is suggested in the theoretical prediction. If the DC model prediction is matched to the whole data set, one arrives at an extrapolated S-factor at zero energy of $S(0) = 0.47 \pm 0.05$ keV b. Subsequently, the measurements have been extended to lower energies ($E_{c.m.} = 164$–245 keV, Nag69) using a differentially pumped, windowless gas target to minimize energy loss and energy straggling of the beam. The capture γ-rays were detected with NaI(Tl) crystals. The combined results are fitted by a polynomial (Fig. 6.16a), leading to an extrapolated value of $S(0) = 0.61 \pm 0.07$ keV b. A reanalysis of both data sets separately, guided by the DC model predictions, led to a mean value of $S(0) = 0.52 \pm 0.05$ keV b (Kav82, Par85).

In view of the importance of this reaction to the solar neutrino question (chap. 10), it has been restudied extensively using high-resolution Ge(Li) detectors in combination with windowless gas targets, including in one case a quasi-point supersonic jet. Data have been obtained down to $E_{c.m.} = 107$ keV (Krä82). The resulting S-factors are shown in Figure 6.16b. The observed angular distributions for the capture transitions into the ^{7}Be ground state (DC → 0) and 429 keV first excited state (DC → 429), the energy dependence of their intensity ratio, as well as the energy dependence of the $S(E)$ factor data, are in excellent agreement with the DC model calculations. Using the DC model, the extrapolation of the data led to $S(0) = 0.31 \pm 0.04$ keV b. Other

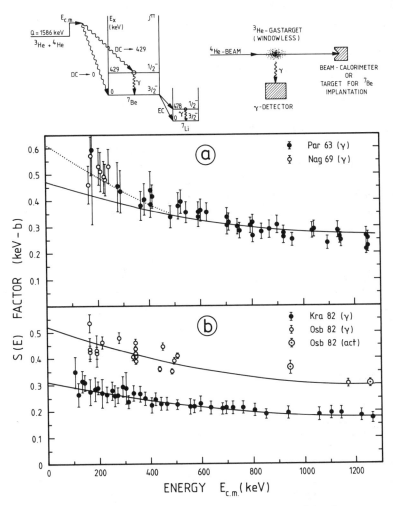

FIGURE 6.16. The ^{3}He(α, γ)^{7}Be reaction represents another example of the direct-capture (DC) mechanism, which proceeds in this case to the ground state and the 429 keV first excited state of ^{7}Be. The older results (a) are shown together with the theoretical $S(E)$ factor; the energy dependence of the data is not in good agreement with the DC model prediction (*solid line*). A polynomial fit (*dotted line*) leads to $S(0) = 0.61$ keV. (b) The energy dependence of the more recent data are in excellent agreement with the DC model calculation (*solid lines*). However, the two data sets (Krä82, Osb82) differ by about 40% in their absolute values.

measurements (Osb82) confirmed the energy dependence of the branching ratio and the $S(E)$ factor (Fig. 6.16b); however, a discrepancy of about 40% in the absolute values remains. The higher absolute values (Osb82) were confirmed for the higher energy region by several investigators (Rob83, Vol83) using an activation technique (chap. 5). The extrapolated absolute $S(0)$ factor, arrived at using only the higher values, is $S(0) = 0.52 \pm 0.02$ keV b (Par85). Efforts to resolve the existing discrepancy in the absolute values are continuing (Ale84) and have already indicated a solution (Hil87).

Although the agreement with the DC model over a wide range of energies gives credence to the extrapolation of the data to zero energy, there is still a considerable gap between the lowest measured value of $E_{c.m.} = 107$ keV and the Gamow energy window of $E_0 = 22.4 \pm 6.2$ keV (for $T_6 = 15$). In view of this, data at lower energies would be desirable.

6.1.6.4. Termination of the p-p Chains II and III

The ^{7}Be produced in the ^{3}He(α, γ)^{7}Be reaction proceeds to the termination of the p-p chain via two different paths. In the p-p-II case, the ^{7}Be decays via electron capture to ^{7}Li. The ^{7}Li is, in turn, burned in the ^{7}Li(p, α)α reaction, which terminates chain II. Since the ^{7}Li(p, α)α reaction goes much faster than the electron capture process producing the ^{7}Li, it has no influence on the relative strength of the p-p-II chain (Fig. 6.13). This reaction has been studied over a wide range of energies down to 23 keV (Saw53). The extrapolated S-factor has been determined to be $S(0) \simeq 100$ keV b (see, however, Rol86).

The alternative fate of ^{7}Be is to be burned in the ^{7}Be(p, γ)^{8}B reaction. While this is the weakest of the three branches of the p-p chain (Fig. 6.13), it is nevertheless important because the $\bar{E}_\nu = 7.30$ MeV neutrino produced in the 1.1 s positron decay of ^{8}B,

$$^8\text{B} \rightarrow {}^8\text{Be*} + e^+ + \nu \,,$$

provides most of the neutrinos for the solar detection reaction

$$^{37}\text{Cl} + \nu \rightarrow {}^{37}\text{Ar} + e^-$$

(chap. 10). The resulting ^{8}Be* fissions quickly into two α-particles (Fig. 6.12), ^{8}Be* $\rightarrow 2\alpha$, thus terminating the p-p-III branch of the p-p chain. Owing to problems associated with the fabrication of a radioactive target with a half-life of 53 days, direct measurements of the ^{7}Be(p, γ)^{8}B cross sections are extremely difficult. An additional complication is the high γ-ray background from the decay of the excited state in ^{7}Li at 478 keV (Fig. 6.10), populated in the electron capture of ^{7}Be (10.4% branch). Indirect methods, such as observing the delayed positrons emitted in the decay of ^{8}B or the delayed two α-particles resulting from the fission of ^{8}Be (Fig. 6.12), have also been used in determining the cross section. In spite of the difficulties, several investigators have studied the reaction over a wide range of energies (Kav60, Par66, Par68, Kav69,

Vau70, Kav72, Wie77, Fil83). Figure 6.17 shows the results of the most recent experiments (Fil83). The extrapolated $S(0)$ factors range from 0.016 to 0.045 keV b (Bar80, Kav82). A weighted average adopted in the most recent work is $S(0) = 0.0238 \pm 0.0023$ keV b (Fil83, Par85). Note that the discrepancy among the reported absolute $S(0)$ values is as high as a factor of 2, an uncomfortably large discrepancy, but not a surprising one in view of the difficulties involved in determining absolute cross sections (chap. 5).

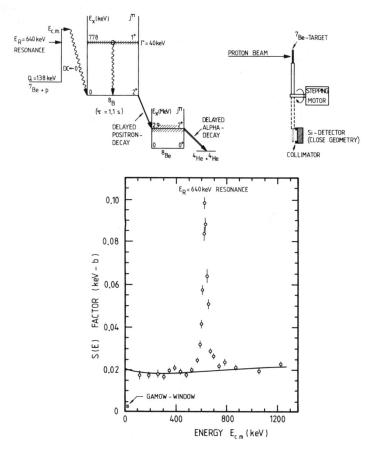

FIGURE 6.17. The ^{7}Be$(p, \gamma)^8$B reaction proceeds at energies away from the 640 keV resonance via the DC mechanism. The ^{7}Be target emits an intense γ-ray flux of 478 keV energy, which to date prohibits the direct measurement of the DC $\to$ 0 γ-ray yield. The process has, therefore, been studied indirectly by observing either the delayed positrons or the delayed α-particles emitted in the decay of the produced ^{8}B nuclei. The energy dependence of the $S(E)$ data (Fil83) is in good agreement with previous work, except for the absolute values. The solid curve is the result of a DC model calculation (Tom65) that has been normalized to the data.

6.1.6.5. Processes Involving the Weak Force

All of the charged-particle–induced reactions discussed above are governed either by the electromagnetic or the strong (nuclear) force. The two remaining reactions,

$$p + p \rightarrow d + e^+ + \nu \,,$$
$$^7\text{Be} + e^- \rightarrow {}^7\text{Li} + \nu \,,$$

proceed via the weak force. As previously mentioned, the cross sections for such weak interactions, which are on the order of 10^{-47} cm^2 at low energies, appear impossible to measure with techniques and facilities presently available. This is especially true for low-energy reactions. However, since the cross sections are strongly energy-dependent ($\sigma_{p+p} \propto E^5$), it may be possible to investigate these reactions at higher energies. For example, at a bombarding energy of 100 MeV, the $p + p$ cross section becomes about 3×10^{-15} b. If the reaction were investigated via the emitted positrons, a very pure hydrogen target would be required to avoid positron production via heavy contaminant nuclei. At still higher proton energies, meson production becomes a problem.

Accelerators producing proton beams of several hundred MeV are already available. These accelerators, combined with storage rings (chap. 5), may produce proton currents on the order of several milliamperes. If a supersonic jet gas target of hydrogen gas with high purity were installed in the beam of such a storage ring, a measurable count rate for the reaction

$$p + p \rightarrow d + e^+ + \nu$$

might be achieved.

A radioactive nucleus such as ^{7}Be, which decays only by electron capture, should be stable if all the atomic electrons were removed. This situation is approximated in a stellar environment. Experimental verification of this assertion would require (Fig. 6.18), as a first step, the production of a ^{7}Be beam with the entire atomic electron cloud removed. Such a beam could be made by bombarding a hydrogen jet target with a high-energy ^{7}Li beam (i.e., of 250 MeV), using the primary reaction ^{1}H(^{7}Li, ^{7}Be)n. If this ^{7}Be^{4+} beam could then be injected into an appropriate storage ring of extremely good vacuum, measurement of changes in the beam current of the recirculating ^{7}Be^{4+} beam would determine its stability. (Note that today large quantities of ^{7}Be also can be obtained commercially, making injection into the storage ring more feasible.)

In a stellar environment, even though the orbital electrons are absent, there is a sea of free electrons in the plasma that can be captured by the nucleus. The cross section for the capture of continuum electrons is of the order of 10^{-47} cm^2. An experimental setup simulating this stellar process might be achieved using a ^{7}Be^{4+} beam in a storage ring equipped for electron cooling.

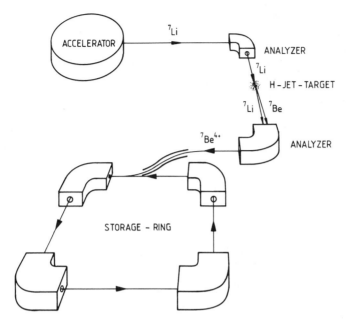

FIGURE 6.18. Conceptual experimental setup to test the assertion that $^7Be^{4+}$ completely stripped of its atomic cloud is a stable nucleus. If electron cooling were provided in the storage ring, one might also determine the probability of $^7Be^{4+}$ for electron capture by periodically passing the nuclei through the electron cloud (chap. 5).

6.2. The CNO Cycles

Since first-generation stars (Population II) consist mainly of hydrogen, their energy is produced primarily by the direct fusion of hydrogen into helium, namely, through the operation of the *p-p* chain. Those first-generation stars that are still shining—for example, stars in the globular clusters (referred to as Population II stars)—are very old. Most of the present stars are second- or third-generation stars (Population I). They were formed from material which, in addition to hydrogen, contained heavier elements synthesized in massive first-generation stars and blown into space during the violent death throes of those stars. In stars somewhat more massive than the sun, higher temperatures and densities will occur in the core before the star achieves hydrostatic equilibrium. In these stars energy can be produced by hydrogen burning of the heavier elements in another, faster chain of reactions. The favored reactions will involve those heavier elements with the smallest Coulomb barrier and the highest abundances. The lightest elements with masses greater than helium which could fulfill both of these conditions are the elements carbon and nitrogen. The elements between ^4He and ^{12}C (Li, Be, and B), while fulfilling the

Coulomb barrier condition, are extremely low in abundance (Fig. 1.23). The actual mechanism for this burning process, which involves the elements carbon and nitrogen, was originally suggested by Bethe (Bet39) and von Weizsäcker (Wei37, Wei38). This mechanism, originally referred to as the CN cycle, consists of the following sequence of reactions (Fig. 6.19):

$$^{12}C(p, \gamma)^{13}N(e^{+}v)^{13}C(p, \gamma)^{14}N(p, \gamma)^{15}O(e^{+}v)^{15}N(p, \alpha)^{12}C \ .$$

THE CN CYCLE

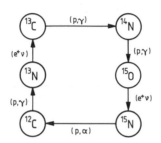

NET RESULT : $4p \longrightarrow {}^{4}He + 2e^{+} + 2v$ $(Q = 26.73 \, MeV)$

FIGURE 6.19. The C and N isotopes here serve as catalysts for the same end result as in the p-p chain, namely, the conversion of four protons into one ^{4}He.

As in the p-p chain, the net result of this CN cycle of reactions is the conversion of four protons into helium: $4p \rightarrow {}^{4}He + 2e^{+} + 2v$ $(Q = 26.73 \, MeV)$. Since the two neutrinos involved in the two β-decays are of relatively low energy, most of the energy liberated is retained in the stellar interior. Of interest in the CN cycle is the fact that, if the cycle begins, for example, with the isotope ^{12}C, the burning process also ends with ^{12}C, and thus ^{12}C can be used over and over again. Although there is not much carbon (compared with hydrogen) in a star, its repeated use helps to make the cycle an effective energy source. The carbon serves only as a catalyst. The same is true if one assumes that the cycle begins with any of the other nuclei involved (Fig. 6.19); therefore, only hydrogen nuclei are consumed.

If, at the beginning, only ^{12}C nuclei are available, the CN cycle will result in the synthesis of heavier elements, and new elements will be created. However, the total number of heavy elements will not exceed the number of original ^{12}C nuclei. As in the p-p chain, the rate of energy production in the CN cycle is governed by the slowest reaction. Since nitrogen isotopes have the highest Coulomb barrier and the $^{14}N(p, \gamma)^{15}O$ reaction goes essentially via the electromagnetic force, whereas the $^{15}N(p, \alpha)^{12}C$ reaction is governed by the nuclear force, the $^{14}N(p, \gamma)^{15}O$ reaction should be the slowest (sec. 6.2.1). Measurements of this and other reactions of the CN cycle were carried out over a period of years beginning around 1950, because it appeared at that time that this cycle would be the primary energy source in the sun. This assumption was invalid because the solar mass was too small to provide a sufficiently high temperature in its interior for this cycle to dominate. However, the information acquired was important because of the dominant role the CN cycle plays in the production of energy in more massive stars as well as in the nucleo-synthesis of the various isotopes of the light elements carbon and nitrogen. In the course of these investigations, it was discovered that the hydrogen burning involving these elements was much more complicated, having several additional cycles. The available information regarding the reactions involved and the relative importance of each of the cycles is discussed below.

6.2.1. The CN Cycle

The $^{12}C(p, \gamma)^{13}N$ and $^{13}C(p, \gamma)^{14}N$ reactions have been studied over a wide range of beam energies extending down to around 80 keV (Bai50, Hal50, Lam57, Heb60, Rol74b, Woo52, Sea52, Hes61). The observed $S(E)$ data at low energies are illustrated in Figure 6.20, indicating in both cases that the $S(E)$ factor is influenced by the low-energy tail of a broad resonance. The extrapolated $S(0)$ factor is $S(0) = 1.40$ and 5.50 keV b (Bar71) for ^{12}C and ^{13}C, respectively. Since there are no states reported near the proton thresholds in ^{13}N or ^{14}N, one has confidence in the validity of the above extrapolations.

For the $^{14}N(p, \gamma)^{15}O$ reaction ($Q = 7293$ keV) various investigators (Woo49, Dun51, Lam57, Pix57, Bai63, Hen67) have obtained data to as low as 100 keV (Fig. 6.21). However, no one has measured this reaction continuously

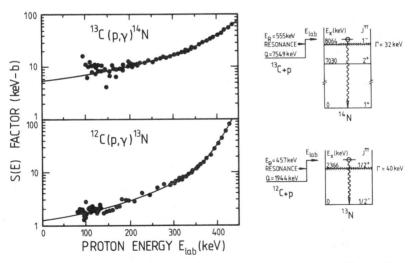

FIGURE 6.20. Energy dependence of the $S(E)$ factor for the reactions $^{12}C(p, \gamma)^{13}N$ and $^{13}C(p, \gamma)^{14}N$. The yields are dominated in each case by the effects of the low-energy tails of a broad resonance.

over the entire energy range of interest. An extrapolation of the data from the various groups led to different values for $S(0)$, ranging from about 2 to 10 keV b (Fig. 6.21). This discrepancy again illustrates the difficulty in determining absolute cross sections. Since this reaction is critical in determining the rate of energy production in this cycle, further detailed measurements over the entire energy range, as well as the determination of the influence of the subthreshold state at $E_x(^{15}O) = 7271$ keV ($E_R = -22$ keV), would be desirable. The value of $S(0) = 3.32$ keV b has been recommended (Fow75) for use in stellar model calculations. (For recent experimental work, see Sch87).

The most recent investigation of the $^{15}N(p, \alpha_0)^{12}C$ reaction (Fig. 4.7) leads

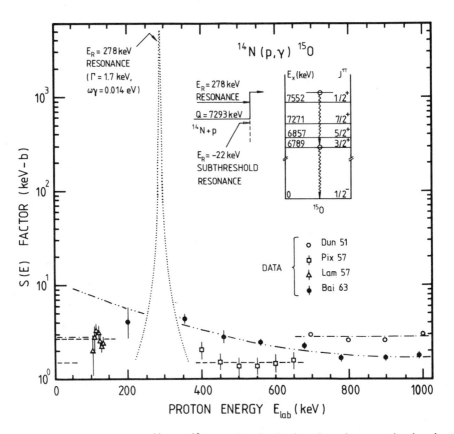

FIGURE 6.21. As shown, the $^{14}N(p, \gamma)^{15}O$ reaction is dominated at low energies by the $\Gamma = 1.7$ keV broad resonance at $E_R = 278$ keV. Data have been obtained by various investigators at energies outside this resonance. Note that where data of different investigators overlap, the absolute values differ by about a factor of 2. The observed nonresonant yields may in principle be influenced by the tails of the $E_R = -22$ and 278 keV resonances as well as by the direct-capture process, and interference effects could also be present. The available data allow no clear picture of the capture mechanisms involved.

to $S(0) = 65 \pm 4$ MeV b (Red82), which is in agreement with the results of earlier work (Sch52, Heb60a, Zys79). By comparison, the $S(0)$ factor for the $^{15}N(p, \alpha_1)^{12}C$ reaction was found to be negligible ($S(0) = 0.1$ keV b; Rol74c).

6.2.2. The CNO Bi-cycle

In discussing the reactions by which the CN cycle operates, the loss of catalytic material to the process via the $^{15}N(p, \gamma)^{16}O$ reaction was neglected. While this reaction does take place, the subsequent set of reactions,

$$^{16}O(p, \gamma)^{17}F(e^+v)^{17}O(p, \alpha)^{14}N ,$$

restores the catalytic material to the cycle. Thus a second cycle exists, and the two cycles are referred to as the CNO bi-cycle (Fig. 6.22). The importance of the second cycle (involving oxygen isotopes) relative to the CN cycle is governed by the ratio of $S(0)$ factors for the two reactions $^{15}N(p, \alpha)^{12}C$ and $^{15}N(p, \gamma)^{16}O$.

THE CNO BI-CYCLE

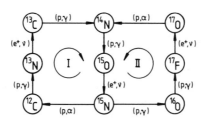

FIGURE 6.22. The interlocking sequence of reactions involved in hydrogen burning via the CNO bi-cycle. The burning of ^{17}O is assumed in this figure to proceed entirely by the $^{17}O(p, \alpha)^{14}N$ reaction.

The $^{15}N(p, \gamma)^{16}O$ reaction ($Q = 12.126$ MeV) was examined over a wide range of beam energies with modern γ-ray detectors (Rol74c). The data for the dominant ground-state transition show (Fig. 6.23) that the capture process is controlled primarily by the two known $J^\pi = 1^-$ resonances at $E_p = 338$ and 1028 keV. Therefore, the data were first analyzed in terms of them alone. If the usual energy dependence of the partial and total widths (chap. 4) as well as the interference between the two resonances are taken into account, the dashed curve in Figure 6.23 is obtained. However, the calculated yield deviates significantly from the measurements at the tails of both resonances. These discrepancies can be explained by including a DC process as part of the radiative capture mechanism (solid line in Fig. 6.23), which leads to $S(0) = 64 \pm 6$ keV b (see also Heb60a). These studies illustrate the problems involved in broad overlapping resonances, which may interfere not only with each other but also with a nonresonant process. It is interesting to note that a clarification of the

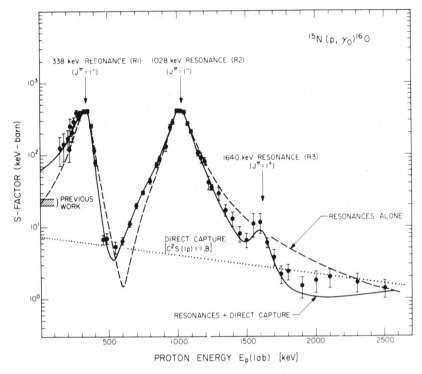

FIGURE 6.23. *S*-factor for $^{15}\text{N}(p, \gamma_0)^{16}\text{O}$. The dashed curve represents the results from an analysis in terms of the two 1^- resonances alone. The solid curve includes the results when interference effects of a direct-capture component (*dotted curve*) are included in the capture mechanism.

reaction mechanisms involved in this capture process required yield measurements over a wide and continuous range of beam energies and that a reliable extrapolation of the data to stellar energies hinged sensitively on the observed features of the tails of both $J^\pi = 1^-$ resonances (Fig. 6.23), especially at energies between the two resonances. An extrapolation of the very low energy data alone, with their relatively large errors, would have resulted in a large uncertainty in $S(0)$.

The ratio of the $S(0)$ factors for $^{15}\text{N}(p, \alpha)^{12}\text{C}$ ($= 65$ MeV b) and $^{15}\text{N}(p, \gamma)^{16}\text{O}$ ($= 64$ keV b) turns out to be about 1000:1. Thus, the second cycle operates only for every 1000 operations of the main CN cycle. Clearly, the second cycle contributes very little to the total rate of energy production. It is, however, important for the nucleosynthesis of the ^{16}O and ^{17}O isotopes.

The data obtained for the $^{16}\text{O}(p, \gamma)^{17}\text{F}$ reaction ($Q = 0.601$ MeV) over a wide range of energies ($E_{\text{c.m.}} \geq 140$ keV) show the smooth variation with energy characteristic of the direct-capture mechanism (Lau51a, War54, Lam57, Tan59, Hes58, Rol73). Extrapolating to zero energy using the DC model, one arrives at $S(0) = 7.5$ keV b (Bar71).

The reaction $^{17}O(p, \alpha)^{14}N$, which returns the lost catalytic material to the CN cycle, has been studied at energies $E_{c.m.} \geq 400$ keV (Kie79). The $S(0)$ factor was determined by summing the effects of the low-energy tails of all observed resonances, with the result $S(0) = 1.7$ keV b. It was suggested (Bro62) that the $^{17}O(p, \alpha)^{14}N$ reaction rate is dominated in the stellar energy range by a resonance at $E_p = 66$ keV through the compound state in ^{18}F at $E_x = 5668$ keV. For a calculation of this reaction rate via the Breit-Wigner expression (chap. 4), all necessary parameters were known from other experiments, except the proton partial width. The reaction rate arrived at by assuming a reduced proton width of $\theta_p^2(l = 1) = 0.007$ had been accepted for use in stellar model calculations (see below).

6.2.3. The Discovery of Additional Cycles

The calculated reaction rate for the $^{17}O(p, \alpha)^{14}N$ reaction depends critically on the assumed value of the reduced proton width for the $E_p = 66$ keV resonance. Thus, an experimental determination of this width is important. Reduced proton widths can be determined from the absolute cross sections of DC γ-ray transitions into such states (Rol73). A search for the DC transition into the 66 keV resonance state was carried out via the $^{17}O(p, \gamma)^{18}F$ reaction in the nonresonant energy region. While the results yielded only an upper limit (Rol74a, Rol75a), the value was lower than the previously assumed value by a factor of 60. As a consequence, the $^{17}O(p, \alpha)^{14}N$ reaction rate was also lowered by approximately this factor. This result, together with available information for the $^{17}O(p, \gamma)^{18}F$ reaction (Rol73), led to the conclusion that the $^{17}O(p, \gamma)^{18}F$ reaction cannot be neglected in the hydrogen burning of ^{17}O.

If the subsequent hydrogen burning of ^{18}O, resulting from the β-decay of ^{18}F, proceeds predominantly through the $^{18}O(p, \alpha)^{15}N$ reaction, one must conclude that the CNO cycle is in fact tri-cycling (Fig. 6.24). In order to quantify this hypothesis, both of the competing reactions $^{18}O(p, \alpha)^{15}N$ and $^{18}O(p, \gamma)^{19}F$ must be known. Both have been studied in detail over a wide

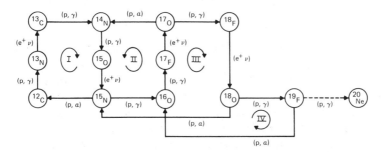

FIGURE 6.24. Illustration of the four CNO cycles involved in the conversion of hydrogen into helium. Catalytic material could be lost from the cycles via the $^{19}F(p, \gamma)^{20}Ne$ reaction, which would provide a link to the NeNa cycle (Fig. 6.27).

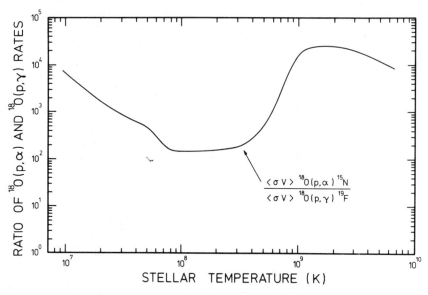

FIGURE 6.25. The ratio of the stellar reaction rates for the reactions $^{18}O(p, \alpha)^{15}N$ and $^{18}O(p, \gamma)^{19}F$ is plotted as a function of stellar temperature (Wie80).

range of energies (Lor79, Wie80), and the deduced ratio of their reaction rates is shown in Figure 6.25 as a function of stellar temperature. The results indicate that the hydrogen burning of ^{18}O is almost entirely through the (p, α) reaction at low temperatures, $T_9 \leq 0.02$, and at high temperatures, $T_9 \geq 0.7$. However, in the temperature range between these extremes, one CNO catalyst nucleus is lost from the oxygen-side cycle via the $^{18}O(p, \gamma)^{19}F$ reaction for every 150 operations of the side cycle because of a strong low-energy resonance ($E_R = 152$ keV).

If the subsequent hydrogen burning of ^{19}F proceeds predominantly through the $^{19}F(p, \alpha)^{16}O$ reaction, the CNO catalytic material will remain in the cycle, implying a fourth branch in CNO hydrogen burning (Fig. 6.24):

$$^{16}O(p, \gamma)^{17}F(e^+ v)^{17}O(p, \gamma)^{18}F(e^+ v)^{18}O(p, \gamma)^{19}F(p, \alpha)^{16}O .$$

Alternatively, if the competing $^{19}F(p, \gamma)^{20}Ne$ reaction is strong enough, CNO catalytic material will be lost through this reaction (Fig. 6.24), and the various CNO cycles could eventually cease to be an energy source in hydrogen burning. However, the ^{20}Ne produced will form the basis for further hydrogen burning through the NeNa cycle (sec. 6.3). To determine accurately the extent to which the catalytic material returns or is lost via hydrogen burning of ^{19}F, a program of further study is needed, especially experimental investigations of the ^{19}F hydrogen-burning reactions at beam energies below 0.25 MeV. Such studies also could be of interest in determining the location of the site of ^{19}F nucleosynthesis (Fig. 7.11).

Hydrogen burning at temperatures and densities far in excess of those attained in the interiors of ordinary main-sequence stars is expected to occur in a variety of astrophysical sites, such as supermassive stars ($M/M_\odot \simeq 10^5$–10^8; Hoy65, Fri73, Fri80), novae and supernovae outbursts (How71, Sta74, War76a, Gal78), and accreting neutron stars (Taa78). In these various events it is not uncommon to encounter hydrogen burning at temperatures from 10^8 to 10^9 K and perhaps higher. The CNO cycles operate, then, on a rapid enough time scale (of the order of several seconds) that β-unstable nuclei like ^{13}N will live long enough to be burned by nuclear reactions before they β-decay. In this case, like the many-headed hydra, the CNO cycle develops (Hoy65, Cau77, Aud73, Wie82) into a many cycled monster (Fig. 6.26). In this hot, or β-limited, CNO cycle the rate at which hydrogen is converted into helium is limited (Hoy65, Fri77) by the β-decay lifetimes of the proton-rich nuclei ^{14}O and ^{15}O (Fig. 6.26) rather than the proton-capture rate of ^{14}N, which characterizes lower temperatures. It has been pointed out (Wal81) that, for temperatures near 5×10^8K and higher, material of the hot CNO cycles can leak out of the cycles. This "breakout" from the hot CNO cycles can lead to a diversion of nuclear flow to heavier nuclei, such as the iron-group nuclei. While the exact path followed by the ensuing chain of proton captures depends on composition and temperture, its qualitative nature is quite analogous to the classical r-process of neutron addition (chap. 9). The hot hydrogen

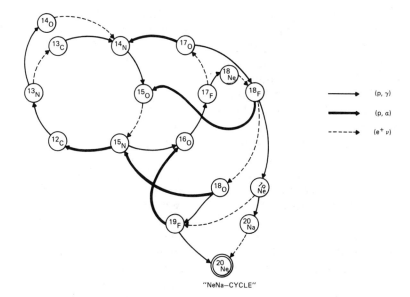

FIGURE 6.26. Hydrogen burning in the "hot" CNO cycles, which include proton-induced reactions on the unstable nuclei ^{13}N, ^{17}F, ^{18}F, and ^{19}Ne. The catalytic material transferred to ^{20}Ne could be involved in the NeNa cycle (Fig. 6.27). Such sets of interlocking reactions are also called a network of nuclear reactions.

bath converts the CNO seed nuclei into isotopes near the region of proton-unbound nuclei (the so-called proton drip line). For each neutron number, a maximum mass is reached where the proton-capture flow must wait on a weak interaction (usually β^+-decay) before the buildup of still heavier elements can continue. Because of its analogy with the r-process, this new nucleosynthetic process has been termed the "rp-process" (Wal81). Unlike the more traditional r-process, the rp-process is hindered by an ever-increasing Coulomb barrier for proton addition as progressively larger masses are formed. Thus, for heavier nuclei with large charges, the path of the rp-process may not extend all the way to the proton drip line, but only to the point where β-decay competes favorably with proton capture. Determining the stellar rates of the numerous reactions involved in the hot CNO cycles (Fig. 6.26) and the rp-process represents an enormous challenge to the experimentalist (chap. 5).

6.2.4. Consequences of the CNO Cycles

Because the branching ratios at the branch points are such that only a small amount of material goes over into the side cycles (Fig. 6.24), and also because the Coulomb barriers of the reactions in the side cycles are considerably higher than those in the main CN cycle, these cycles do not contribute significantly to energy production. However, the side cycles with their possible connection to the NeNa cycle are important to the nucleosynthesis of elements up to ^{23}Na. In particular, the isotopes ^{13}C, ^{14}N, ^{15}N, ^{17}O, ^{18}O, and possibly ^{19}F are believed to result from the operation of the CNO cycles (Cau77).

It was stated earlier that the ^{14}N$(p, \gamma)^{15}$O reaction is the slowest and, therefore, the controlling reaction in the "cold" CN cycle. Calculations of the lifetimes τ for the reactions ^{12}C$(p, \gamma)^{13}$N, ^{13}C$(p, \gamma)^{14}$N, ^{14}N$(p, \gamma)^{15}$O, and ^{15}N$(p, \alpha)^{12}$C for the conditions $\rho = 100$ g cm^{-3}, $X_H = 0.5$, and $T_6 = 10$ yield values of $\tau = 6.1 \times 10^9$, 1.1×10^9, 2.1×10^{12}, and 1.0×10^8 y, respectively. The ^{14}N burning time is over 2 orders of magnitude slower than the burning times of the other elements, thus confirming the above statement.

Assuming equilibrium conditions and ignoring the effects of all CNO side cycles, the relative abundances of CN elements are inversely related to their respective reaction rates. For stellar conditions $\rho = 100$ g cm^{-3}, $X_H = 0.5$, and $T_6 = 50$ the percentages are 5.5% (^{12}C), 0.9% (^{13}C), 93.6% (^{14}N), and 0.004% (^{15}N). The large abundance of ^{14}N is to be expected in view of the slow hydrogen-burning rate of ^{14}N. In practice, all the competing CNO cycles, as well as the time dependence of the abundance of hydrogen fuel and of the stellar temperature, must be taken into account. The time dependence of the elements is then described by a set of time-dependent coupled differential equations similar to those given in section 6.1.5 for the p-p chain. An analytical solution to these equations cannot be obtained, and only numerical approximations, requiring the use of large computers, are possible (Cau62, Cau65, Cau72, Cau77, Aud73, Wie82). The results show that, in spite of the operation of many side cycles, at the completion of hydrogen burning via the CNO cycles, the major component of the ashes is still ^{14}N. There is, thus, a transformation of the nuclear species predominantly to ^{14}N from the ^{12}C and ^{16}O seed nuclei, which were previously produced in helium burning (chap. 7) in other stars and are the most abundant isotopes present at the beginning of hydrogen burning in the CNO cycles. The survival of ^{14}N in hydrogen burning was of great importance for human evolution because, without nitrogen, life as we know it, including human life, would not exist (chap. 7).

As with the reaction rates, the rate of energy production ϵ also is temperature-dependent. In the p-p chain, the $p + p$ reaction, which proceeds via the weak force, governs the rate of energy production. For the CNO cycles, the governing ^{14}N$(p, \gamma)^{15}$O reaction proceeds via the electromagnetic force. The temperature dependence of ϵ_{pp} is shown in Figure 6.5, along with that of the energy production ϵ_{CNO} for the CNO cycles. It is clear that the principal source of energy production at low stellar temperatures is the p-p chain, while at higher temperatures the CNO cycles become increasingly important, exceeding that of the p-p chain at temperatures above $T_6 \simeq 20$. The significantly steeper rise of ϵ_{CNO} compared with that of ϵ_{pp} is predominantly due to the higher Coulomb barriers involved in the CNO reactions. For example, near $T_6 = 15$ (the sun) the p-p chain results in $\epsilon_{pp} \propto T^4$ (sec. 6.1.1), while the CNO cycle has a temperature dependence of $\epsilon_{CNO} \propto T^{18}$. Once stellar temperatures reach a level such that the effects of the Coulomb barriers no longer strongly affect the rate of energy production, both curves in Figure 6.5 have lower gradients. At a high enough temperature, the ratio of the rate of energy production via the two mechanisms should approach the ratio of the strengths of the electromagnetic and weak forces.

The temperature in the stellar interior is determined by the mass of the star (chap. 2). As a consequence, very massive stars with masses $M > M_\odot$ and significant amounts of heavy elements produce energy mainly through the operation of the CNO cycles, while low-mass stars like our sun operate mainly on the p-p chain.

6.3. Other Cycles

In second-generation stars, whose stellar temperatures are higher than those for the quiescent CNO cycle, additional cycles for fusing hydrogen into helium come into play, such as the NeNa cycle and the MgAl cycle. The sequence of reactions involved in these cycles are displayed in Figure 6.27. Because of the higher Coulomb barriers involved in these reactions, both cycles are relatively unimportant as additional energy sources in stars. They are important, however, for the nucleosynthesis of elements between ^{20}Ne and ^{27}Al. For example, the NeNa cycle may play a role in understanding Ne-E, the highly enriched ^{22}Ne (almost pure ^{22}Ne) found in meteoritic samples, while the MgAl cycle may provide the mechanism for production of ^{26}Al, the decay of which gives rise to the ^{26}Mg/^{27}Al anomaly found in some meteorites

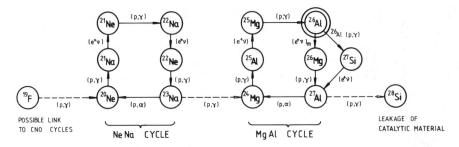

FIGURE 6.27. Sequences of nuclear reactions and β-decays involved in the low-temperature operation of the NeNa and MgAl cycles. The isomeric state in ^{26}Al (^{26}Al^m) decays quickly to ^{26}Mg, while ^{26}Al in its ground state (^{26}Al^0) lives long enough ($T_{1/2} = 0.72 \times 10^6$ y) to be involved in nuclear burning via the ^{26}Al$(p, \gamma)^{27}$Si reaction. The (p, γ) reactions on ^{19}F and ^{23}Na lead to possible links between the CNO and NeNa cycles and the NeNa and MgAl cycles, respectively. The ^{27}Al$(p, \gamma)^{28}$Si reaction leads to a leakage of nuclei out of the MgAl cycle.

(chap. 10). In the next two sections we will discuss the status of experimental information on the reactions involved in both of these cycles, followed by a discussion of the resulting elemental abundances.

6.3.1. The NeNa Cycle

Information on reaction rates for the proton-induced reactions involving the neon isotopes has become available following the development of rather sophisticated gas targets (chap. 5). The first reaction in this cycle, ^{20}Ne$(p, \gamma)^{21}$Na, was studied over a wide range of beam energies (Rol75). Preliminary analysis of the results indicates that the stellar rates are dominated by direct capture (DC) into bound states rather than the tails of the resonances at higher energies (Fig. 6.28). However, it was recognized that the 2425 keV state, 7 keV below the proton threshold, could contribute significantly to the overall stellar rate. Available data indicate that its width is in the range $\Gamma = \Gamma_\gamma = 0.7\text{--}3.0$ eV, and the DC studies reveal a reduced proton width of $\theta_p^2(l = 0) = 0.9$. With this information the high-energy tail of this subthreshold resonance, assuming a single-level Breit-Wigner shape (chap 4), can be calculated. Such calculations show that the high-energy tail of the 2425 keV state may dominate the stellar rate and may be experimentally detectable by observing a transition to the ^{21}Na ground state from the 2425 keV state, at $E_p \geq 0.5$ MeV, since the $S(E)$ factors are not far below those observed for the DC $\to$ 332 keV transition (Fig. 6.28). The unique identifying feature should be the predicted energy dependence of the γ-ray intensity, that is, an $S(E)$ factor decreasing with increasing beam energy. Experimental results confirmed these predictions, as shown in Figure 6.28, where the fit (the dash-dot curve) was obtained for a γ width at resonance of $\Gamma_\gamma = 0.31$ eV (Rol74). These results illustrate the importance of even subthreshold resonances to overall stellar reaction rates (chap. 4).

The possible importance of bound states close to threshold for stellar-burning rates was noted long ago. For example, in the case of ^{20}Ne$(p, \gamma)^{21}$Na, in 1957 Marion and Fowler (Mar57) estimated the contribution of the 2425 keV state to stellar rates. However, at that time there was little accurate information available on the nuclear properties of this state, and the calculated rates were, therefore, quite uncertain. It should also be pointed out that the laboratory observation of the high-energy tail of the 2425 keV bound state is possible because of the nearly maximum partial widths and because the state in question can be made by s-wave protons. The exact position of the state relative to the threshold is very important for the stellar-burning rate, but not

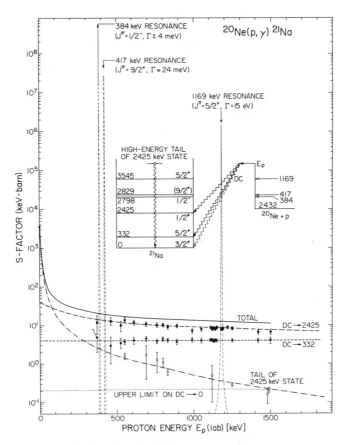

FIGURE 6.28. S-factors for the ^{20}Ne$(p, \gamma)^{21}$Na reaction. The dashed curves through the DC → 332 keV and DC → 2425 keV data points are direct-capture predictions. The dash-dot-dash curve is the prediction using the high-energy tail of the 2425 keV bound state, fitted to the observed data with the free parameter Γ_γ. A possible DC → 0 (ground-state) process would have almost the same energy dependence as the DC → 332 keV transition and is indicated as a dotted curve. Also shown for comparison are the S-factor curves for the observed resonances (Rol75).

for its laboratory observation (Fig. 6.28). As far as proton angular momentum is concerned, if d-wave protons were required to form the state from ^{20}Ne, the yield of the high-energy tail of the state would be 2 orders of magnitude less. In this case, the subthreshold state would not significantly affect the stellar rate, and also it would be very difficult to detect the high-energy tail experimentally.

The studies of the ^{21}Ne$(p, \gamma)^{22}$Na and ^{22}Ne$(p, \gamma)^{23}$Na reactions cover a wide range of beam energies, and numerous resonances as well as direct captures were found to contribute to the capture mechanisms (Rol75, Hie75, Ber77, Smi79, Goe82, Goe83). The total reaction rate depends almost entirely on the contribution of the resonances, particularly those at low energy.

In the case of ^{22}Ne$(p, \gamma)^{23}$Na, nonresonant captures to final states usually reveal (Goe83) an energy dependence characteristic of the direct-capture mechanism. The exception is the capture transition into the ^{23}Na ground state, the $S(E)$ factor for which, while generally increasing with decreasing

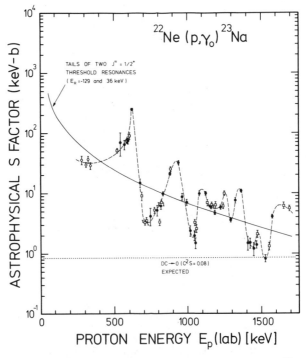

FIGURE 6.29. Astrophysical $S(E)$ factor for the ground-state transition in ^{22}Ne$(p, \gamma)^{23}$Na. The solid line represents the summed high-energy tails of the two $J^{\pi} = \frac{1}{2}^{+}$ threshold resonances at $E_{R} = -129$ and 36 keV, where interference effects are also taken into account (Goe83). The dotted line represents the expected yield for the direct-capture process DC $\rightarrow$ 0. The observed structures oscillating around the solid line are not understood.

beam energy, has an unusual oscillatory structure (Fig. 6.29). From the known level structure of ^{23}Na (End78) two s-wave resonances are expected near the proton threshold: one, a subthreshold resonance at $E_p = -129$ keV, and the other at $E_p = 36$ keV. The situation is similar to that for the ^{20}Ne$(p, \gamma)^{21}$Na reaction discussed above, except that in this instance there are two resonances involved. From the known properties of both resonances, the high-energy tails of both, including interference effects, can be calculated. The result, shown as a solid line in Figure 6.2, is consistent with the general trend of the data. However, the structure superposed on the smooth curve cannot be explained with this analysis. The discovery of this structure emphasizes the importance of detailed measurements of excitation functions. Previous studies of this reaction over a similar energy range but with larger energy steps failed to detect this unexplained structure.

Available information reveals that the stellar rate for the ^{23}Na$(p, \alpha)^{20}$Ne reaction is large enough compared with the competing reaction ^{23}Na$(p, \gamma)^{24}$Mg to guarantee cycling (Zys81). The reaction rates for these two reactions, as well as for the (p, γ) reactions on ^{21}Ne and ^{22}Ne, are well determined for certain stellar temperature ranges. For others, substantial uncertainties still exist. The reduction of these uncertainties clearly requires additional experimental effort.

6.3.2. The MgAl Cycle

Considerable information on the stellar rates for most of the reactions involved in the MgAl cycle (Fig. 6.27) was obtained from studies carried out in the early years of nuclear astrophysics. However, these studies were conducted using relatively high-energy protons, corresponding to the higher stellar temperatures, with very little data available for the lower range. With the discovery of an excess concentration of ^{26}Mg in certain inclusions of the Allende meteorite, suggesting the incorporation of radioactive ^{26}Al $(T_{1/2} = 7.2 \times 10^5$ y) in the material of the solar cloud just before the beginning of its condensation phase (chap. 10), there was considerable renewed interest in the MgAl cycle, since it was thought to be the site for the synthesis of ^{26}Al. Consequently, the need for more information on stellar rates for the MgAl reactions, especially at lower stellar temperatures, was recognized, and additional experimental efforts were stimulated. In the course of these experiments, many new low-lying resonances were found (Tra75, Eli79, Buc80). These experiments provided more information on the stellar rates for the lower stellar temperatures. An exception is the ^{27}Al$(p, \alpha)^{24}$Mg reaction, which provides the return flow in the MgAl cycle (Fig. 6.27) and for which resonances have been reported only above $E_p = 505$ keV (End78). Because numerous resonances at still lower beam energies are expected, experimental investigation of this reaction at low beam energies could be very important [for recent experimental work, see Tim87]. Another exception is the ^{26}Al$(p, \gamma)^{27}$Si reaction for which, owing to the absence of natural ^{26}Al, no data have been available until recently.

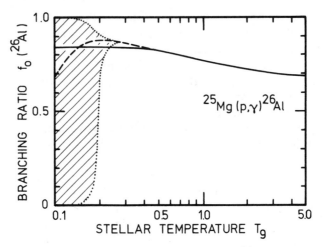

FIGURE 6.30. Temperature dependence of the branching ratio $f_0(^{26}\text{Al})$ for the formation of the ground state of ^{26}Al from the reaction $^{25}\text{Mg}(p, \gamma)^{26}\text{Al}$. The solid curve is from Ward and Fowler (War80). The dashed curve incorporates recent experimental data (Eli79). The dotted curves illustrate the present uncertainty due to expected low-energy resonances (see also Cha83a, End87).

In the MgAl cycle, the isotope ^{26}Al is produced by the $^{25}\text{Mg}(p, \gamma)^{26}\text{Al}$ reaction (Fig. 6.27). This reaction can produce ^{26}Al either in its ground state $(^{26}\text{Al}^0)$ or in its isomeric state $(^{26}\text{Al}^m)$. The 0^+ isomeric state $(^{26}\text{Al}^m)$ at $E_x = 228$ keV has a β-decay half-life of $T_{1/2} = 6.3$ s, which is many orders of magnitude shorter than that of the 5^+ ground state $(^{26}\text{Al}^0)$. Because of the short half-life of $^{26}\text{Al}^m$, its production in the low-temperature operation of the cycle, although of importance for ^{26}Mg nucleosynthesis, is of no relevance as regards the presence of ^{26}Al in the early solar system. Therefore, only the fraction of the radiative capture yield that eventually decays to the ground state must be known. Note that the existence of two separate nuclear species of ^{26}Al produced in the same reaction $^{25}\text{Mg}(p, \gamma)^{26}\text{Al}$ is very unusual. Using experimental data (Nei74, Eli79), the branching ratio $f_0(^{26}\text{Al})$ to the ground state of ^{26}Al has been calculated (War80). The results (Fig. 6.30) show that the branching ratio lies in the range of 0.68–0.85 over a wide range of stellar temperatures, with a considerable uncertainty at temperatures below $T_9 = 0.2$.

In the temperature range $T_9 \leq 0.4$, the half-lives of all of the radioactive species in the MgAl cycle (Fig. 6.27) are short compared with the nuclear-burning times, except for the ground state of ^{26}Al. As a consequence, the $^{26}\text{Al}^0$ nuclei produced via the $^{25}\text{Mg}(p, \gamma)^{26}\text{Al}$ reaction may be destroyed via the $^{26}\text{Al}^0(p, \gamma)^{27}\text{Si}$ reaction. Therefore, this reaction is critical in determining the amount of ^{26}Al present at the completion of the MgAl cycle. Clearly, an experimental determination of the $^{26}\text{Al}(p, \gamma)^{27}\text{Si}$ rate would be an important link in locating the site for ^{26}Al nucleosynthesis. Such an experimental program would require, as a first step, fabrication of an ^{26}Al target, for example, via the ^{26}Mg$(p, n)^{26}$Al reaction, followed by chemical separation. These targets have been fabricated (Buc84a) and used in the search for resonances in the ^{26}Al$(p, \gamma)^{27}$Si reaction over the energy range $E_p = 0.17$–1.58 MeV (Buc84). Because of the unusually high spin of 5^+ for the ^{26}Al target nuclei, and since the proton capture at energies far below the Coulomb barrier involves predominantly s-waves, formation of mostly high-spin resonance states in ^{27}Si $[J^\pi = (9/2)^+, (11/2)^+]$ is expected. In the work of Buchmann and coworkers (Buc84), this expectation was borne out for all observed resonances. Calculations show that only two of these resonances contribute significantly to the stellar reaction rates over a wide range of stellar temperatures. In such a situation, the validity of calculations of stellar reaction rates based on a statistical model (i.e., of the Hauser-Feshbach type; chap. 8) would be questionable. For the ^{26}Al$(p, \gamma)^{27}$Si reaction, such model calculations underestimate the rates by a factor of about 2 (Fig. 6.31).

6.3.3. Elemental Abundances

As previously discussed (sec. 6.1.5), the time dependence of the abundances of the elements involved in a cycle are described by a set of coupled differential equations. Under conditions of equilibrium, the abundance ratios for any pair

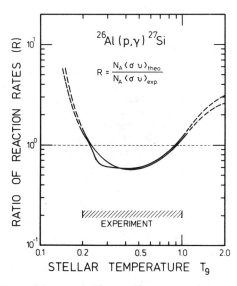

FIGURE 6.31. The stellar reaction rate of ^{26}Al$(p, \gamma)^{27}$Si calculated with a Hauser-Feshbach type statistical model (Woo78, Arn80, War80) is compared with experimental data (Buc84). For a better comparison only the ratio of the rates is plotted. Also indicated is the temperature range for which the rates are determined experimentally.

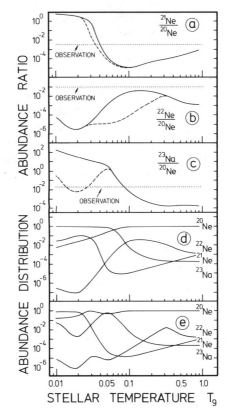

FIGURE 6.32. Abundance ratios and abundance distribution of elements involved in the NeNa cycle shown as functions of stellar temperature. The calculations (Goe83) assume the equilibrium condition, burning times much longer than the β-decay lifetime of ^{22}Na, and no loss of catalytic material from the cycle. The solid lines in (a)–(c) are the results obtained for an uncertainty factor (Fow67, Fow75, Goe83) of $f = 0$, and the dashed lines are those for $f = 1$, which are included to accommodate the lack of experimental data. The curves in (d) and (e) are obtained for $f = 0$ and $f = 1$, respectively. The dotted lines in (a), (b), and (c) represent observed abundance ratios.

of elements may be derived from the ratios of relevant reaction rates. For the NeNa cycle, the abundance ratios and the distribution of abundances are shown in Figure 6.32 as a function of stellar temperature for the equilibrium condition. The figure shows that the isotope ^{21}Ne is extremely low in abundance. Based on this result, the neutron flux resulting from the ^{21}Ne(α, n)^{24}Mg reaction ($Q = 2.56$ MeV) cannot play a significant role in the synthesis of the heavier elements via the s-process (chap. 9).

For the MgAl reactions the available information on stellar rates has been used to calculate the abundance distribution of the elements involved in the

MgAl cycle for the equilibrium condition. The results (Fig. 6.33) indicate that the radioactive isotope ^{26}Al is among the most abundant nuclei remaining at the completion of the cycle. The significance of this result with regard to the ^{26}Mg anomaly (Lee79, Was82) is discussed in chapter 10.

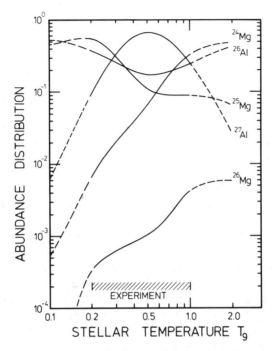

FIGURE 6.33. Abundance distribution of the elements involved in the low-temperature operation of the MgAl cycle shown as a function of stellar temperature. The calculations (Buc84) assume an equilibrium condition in the cycle and no loss of catalytic material. The temperature region, for which the rate of the ^{26}Al$(p, \gamma)^{27}$Si reaction is determined by experiment, is also indicated.

7

Helium Burning

After the hydrogen burning in the interior of an evolved star has consumed its fuel, the stellar core consists mainly of ^{4}He, the ashes of the hydrogen burning. This helium core contracts slowly, with a resulting increase in density and in temperature as the gravitational energy is converted into internal energy. The contraction also heats the hydrogen in a shell around the core, and in this relatively thin shell (Fig. 7.1) hydrogen burning continues. This shell gradually expands as the hydrogen in the shell burns and the helium core becomes more massive, more dense, and hotter as a result of the continuing gravitational contraction. The increasing temperature in the stellar interior, or, equivalently, the increase in thermal pressure, causes an expansion of the outer regions of the star, sometimes by more than a factor of 50 times its original radius. Indeed, the envelope of the star expands so much to accommodate the increasing energy production that the surface temperature actually falls. As the surface cools, the radiation shifts to longer wavelengths, resulting in a shift in color to the red. Because of the red color and the huge expansion, these stars are called super–red giants (Sch62, Dei65, Ibe67).

The evolution of stars to the red giant stage is illustrated for the stars of the globular cluster M3 in the H-R diagram shown in Figure 7.2. Some stars, probably with low mass, remain on the main sequence, with hydrogen burning still taking place in the core. Other stars that have evolved beyond this stage are in transition from the main sequence to the red giant region. After the main-sequence phase, stars evolve comparatively quickly (in a few hundred million years) along the red giant branch, up and to the right. When the helium core becomes hot and dense enough, nuclear reactions among the helium nuclei can occur with an energy release of about 8 MeV per reaction. This phase of stellar evolution is called helium burning. As a result of the onset of this process, a star may become unstable and helium detonations referred to as helium flashes may occur (chap. 2). When the thermal pressure in the helium core exceeds the electron-degeneracy pressure, the core starts to expand. This expansion reduces the energy production of the core, resulting in a cooling and a concomitant reduction of energy production due to hydrogen burning in the surrounding shell (Fig. 7.1), and the outer regions (the envelope)

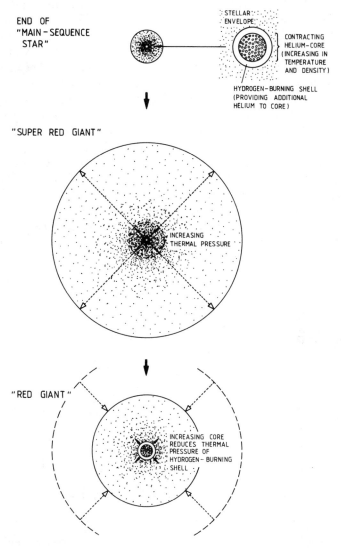

FIGURE 7.1. Cross-sectional view of a main-sequence star after the hydrogen fuel in the core is depleted. At this stage, hydrogen burning continues in a concentric shell around the contracting helium core, providing additional helium to the core. The star evolves into what is called a super-red giant. Finally, the core expands with a contracting envelope, and the star enters a stabilized period of quiescent helium burning called the red giant stage.

of the star contract. A period of quiescent burning in the stellar interior follows, and the star is now called a red giant. This period of stellar evolution probably corresponds to the horizontal branch of the H-R diagram (Fig. 7.2), wherein the helium-burning core reaches a temperature $T_9 = 100\text{–}200$ and a density of $\rho = 10^2\text{–}10^5$ g cm^{-3}. Since the temperature and density are limited by the original stellar mass (chap. 2), the evolution described above occurs only for massive stars with $M \geq 0.5\ M_\odot$. Because of this limitation, some stars in the globular cluster (Fig. 7.2) do not progress beyond the main-sequence branch but rather move directly to the left and become helium white dwarfs. Mass loss reduces the mass of $M \leq 10\ M_\odot$ stars to much smaller masses, and they eventually become white dwarfs (Ibe83 and references therein).

Helium burning begins with reactions among helium nuclei forming ^{12}C in what is commonly called the triple-α process. In turn, ^{12}C nuclei can radiatively capture another helium nucleus to form ^{16}O. In principle, this α-particle capture process can continue producing ^{20}Ne, ^{24}Mg, ^{28}Si, and so on. However, since the α-capture reaction ^{16}O(α, γ)^{20}Ne is a very slow process (sec. 7.3), the helium is mainly converted into ^{12}C and ^{16}O in porportions that depend on the relative cross sections for the triple-α and ^{12}C(α, γ)^{16}O reactions (secs. 7.1 and 7.2) and on stellar temperature and density. Although these two reactions dominate the helium burning in red giants, other reactions do take place and are of importance for the nucleosynthesis of some heavier elements.

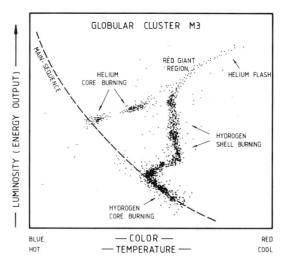

FIGURE 7.2. Shown schematically is the H-R diagram for stars of the globular cluster M3. The locus for main-sequence stars is also shown for comparison. This group of stars was presumably formed about the same time. Therefore, their H-R diagram provides a "snapshot" of the state of evolution of stars of different mass. For this old and highly evolved cluster only its least massive stars are still on the main sequence. The majority of the stars are off the main sequence and have evolved to the subgiant or giant stages.

In second-generation stars, helium-burning reactions can also take place with other nuclei, such as ^{14}N (sec. 7.3), which are present in the material that condensed to form the star. In some of these reactions, such as ^{13}C$(\alpha, n)^{16}$O and ^{22}Ne$(\alpha, n)^{25}$Mg, neutrons are produced. These reactions are thought to be the source of the neutrons necessary for the synthesis of the heavier elements via the s-process (chap. 9). In conclusion, for massive stars evolving to and beyond the red giant stage, helium burning plays an important role in both energy production and nucleosynthesis of the elements.

7.1. The Detour around the Mass Stability Gaps, and the Creation of ^{12}C

In first-generation stars the ash resulting from hydrogen burning via the p-p chain (chap. 6) is entirely ^{4}He, the creation of heavier elements having been blocked by the instabilities at $A = 5$ and $A = 8$. These are referred to as the mass gaps. Since ^{12}C, the fourth most abundant nuclear species observed in the universe, could not be synthesized in its observed abundances in the early universe (chap 2), the site for its creation has to be in stars. Thus, a major question in the early studies of nucleosynthesis was how the stability gaps were bridged to create ^{12}C using only helium. While the simultaneous inter- action of three α-particles to form ^{12}C is energetically possible, the probability for this direct process is much too small to account for the observed ^{12}C abundances.

The solution to this problem was provided in principle by Salpeter and Öpik (Sal52, Sal53, Sal57, Öpi51), who proposed that ^{12}C was created via a two-step process. In the first step, two α-particles combine to form ^{8}Be in its ground state (Fig. 7.3). The ground state of ^{8}Be is known to be unstable ($Q = -92.1$ keV) against decay into two α-particles with a lifetime of 1×10^{-16} s (Ben66, Ben68, and references therein), which is the reason for the mass-8 stability gap. However, as Salpeter pointed out, this lifetime is long compared with the 10^{-19} s transit time of two α-particles with kinetic energies corresponding to Q. As a result, a small concentration of ^{8}Be nuclei builds up in equilibrium with the decay products, two α-particles:

$$\alpha + \alpha \rightleftarrows {}^8\text{Be} .$$

The actual equilibrium concentration of ^{8}Be in a helium environment can be calculated using the Saha equation described in chapter 4. For a temperature of $T_6 = 100$ and a density of $\rho = 10^5$ g cm^{-3}, the Saha equation leads to $N(^8\text{Be})/N(^4\text{He}) = 5.2 \times 10^{-10}$. Thus there is one ^{8}Be nucleus for every 10^9 ^{4}He nuclei. In the second step, Salpeter suggested that, since the first step provides an appreciable concentration of ^{8}Be nuclei, these ^{8}Be nuclei capture an addi- tional α-particle, thus completing the ^{12}C creation process:

$$^8\text{Be}(\alpha, \gamma)^{12}\text{C} .$$

THE TRIPLE-ALPHA PROCESS

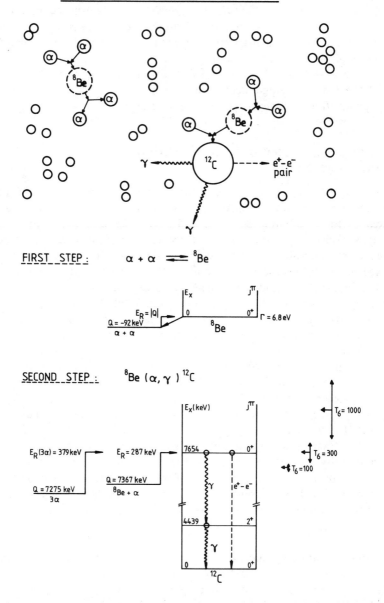

FIGURE 7.3. Schematical representation of the process by which ^{12}C can be synthesized using only ^4He nuclei, commonly called the triple-α process or "Salpeter-process." In the first step of this process, a small abundance of ^8Be nuclei is built up to equilibrium with its α-particle decay products. An additional α-particle is captured by the ^8Be nuclei, thus completing the ^{12}C creation process. This capture reaction proceeds via an s-wave resonance, which is located close to the Gamow energy region indicated for several temperatures.

Since the combined effect of these two steps is to transform three α-particles into a ^{12}C nucleus, this set of reactions is referred to as the triple-α process:

$$3\alpha \rightarrow {}^{12}C \ .$$

Assuming the triple-α process to be the correct mechanism for the synthesis of ^{12}C, Hoyle showed (Hoy53) that the amount of ^{12}C produced in this way (calculated at temperatures for the top of the red giant branch) is insufficient to explain the observed abundance. To surmount this difficulty, Hoyle (Hoy53, Hoy54) suggested that sufficient ^{12}C nuclei could be synthesized if the ^{8}Be$(\alpha, \gamma)^{12}$C reaction took place through an s-wave resonance near the ^{8}Be $+ \alpha$ threshold ($E_x \simeq 7.68$ MeV; Hoy53), since the existence of such a resonance would greatly accelerate the rate of the triple-α process. The suggestion stimulated considerable experimental effort to locate this resonance state in ^{12}C and to determine its properties. The result of this concentrated effort (Hoy53, Coo57) was a verification of Hoyle's prediction with regard to both the energetic location and the $J^\pi = 0^+$ assignment of the state. This is an impressive example of a prediction of the properties of a nucleus based on purely astrophysical grounds and subsequent verification of the prediction by experiments.

The reaction ^{8}Be$(\alpha, \gamma)^{12}$C proceeds predominantly through the $J^\pi = 0^+$ resonance state in ^{12}C; consequently, its reaction rate can be described by the formalism for resonant reactions (chap. 4). In this formalism the reaction rate (eq. [4.56]) depends on the properties of the resonant state, according to the relationship

$$\langle \sigma v \rangle \propto \omega\gamma \exp\left(-E_R/hT\right) ,$$

where E_R is the energy of the resonance with respect to the ^{8}Be $+ \alpha$ threshold and $\omega\gamma$ is the resonance strength, which is in this case related to the partial and total widths by

$$\omega\gamma = \Gamma_\alpha \Gamma_{\rm rad}/\Gamma \ .$$

In the above expression Γ_α is the α-particle width and $\Gamma_{\rm rad}$ is the sum of the electromagnetic decay widths to the ^{12}C ground state (Fig. 7.3) via γ-ray emission (Γ_γ) or via electron-positron pair emission ($\Gamma_{\rm pair}$):

$$\Gamma_{\rm rad} = \Gamma_\gamma + \Gamma_{\rm pair} ,$$

and the total width Γ is the sum of all partial widths:

$$\Gamma = \Gamma_\alpha + \Gamma_\gamma + \Gamma_{\rm pair} \ .$$

The rate of the triple-α process is given by (chap. 3)

$$r_{3\alpha} = N_{{}^8{\rm Be}} N_\alpha \langle \sigma v \rangle_{{}^8{\rm Be} + \alpha} ,$$

where $N_{{}^8{\rm Be}}$ and N_α are the number densities of the interacting ^{8}Be and ^{4}He nuclei.

The overall process can also be looked upon as an equilibrium between three α-particles and the excited carbon nuclei ^{12}C**:

$$\alpha + \alpha \rightleftarrows {}^8\text{Be} ,$$

$$\alpha + {}^8\text{Be} \rightleftarrows {}^{12}\text{C**}(7.65 \text{ MeV}) ,$$

or in a combined notation

$$3\alpha \rightleftarrows {}^{12}\text{C**}(7.65) \rightarrow {}^{12}\text{C}(0.0) + \gamma\gamma \text{ (or } e^+\text{-}e^- \text{ pair) .}$$

As indicated in the last notation, the overall process is not a true equilibrium between three α-particles and ^{12}C**(7.65 MeV), since there is an occasional irreversible leakage out of the equilibrium due to electromagnetic decay to the ground state of ^{12}C (Fig. 7.3). This leakage is very small (1 out of 2500 cases), and there is very little departure from equilibrium in the reaction $\alpha + \alpha \rightleftarrows {}^8$Be. Therefore, it is safe to use the number density $N(^8\text{Be})$ of ^{8}Be nuclei in equilibrium with the ^{4}He bath as described by the Saha equation. Combining this equation with the reaction rate $\langle \sigma v \rangle_{^8\text{Be}+\alpha}$ for resonance reactions (eq [4.56]), one arrives at

$$r_{3\alpha} = \frac{N_\alpha^3}{2} 3^{3/2} \left(\frac{2\pi\hbar^2}{M_\alpha kT} \right)^3 \frac{\omega\gamma}{\hbar} \exp\left(-\frac{Q}{kT} \right), \tag{7.1}$$

where the quantity Q is the sum of $E_R(^8\text{Be} + \alpha)$ and $E_R(\alpha + \alpha)$. The sum is equivalent to the difference in mass between ^{12}C in the 7.654 MeV excited state and three α-particles:

$$Q = (M_{^{12}\text{C**}} - 3M_\alpha)c^2 ,$$

Usually, the relevant quantities $\omega\gamma$ and E_R or Q can be measured directly in the laboratory. However, since the lifetime of ^{8}Be is extremely short, a ^{8}Be target to study the ^{8}Be(α, γ)^{12}C reaction directly cannot be fabricated. Furthermore, a study of the inverse reaction, ^{12}C(γ, α)^{8}Be, is not possible, since the relevant excited state in ^{12}C (Fig. 7.3) cannot be excited by a single photon absorption (a $J^\pi = 0^+$ to $J^\pi = 0^+$ γ-ray transition is forbidden). The exception is the monopole excitation via inelastic electron scattering (see below). As a consequence, the quantities $\omega\gamma$ and E_R have to be obtained indirectly. The experimental determination of these quantities involves a number of different and novel techniques. It provides a fascinating example of the way in which nuclear astrophysics must draw upon many facets of nuclear knowledge.

From equation (7.1), one can see that one quantity needed for the calculation of the triple-α rate is the value of Q. Several high-precision determinations have been made (Nol76, Bar82) by measuring the excitation energy of the second excited state of ^{12}C and then using the mass-table value for the mass difference between the ground state of ^{12}C and three ^{4}He atoms. The weighted mean of the measurements shown in Table 7.1 gives 7654.07 ± 0.19 keV for the excitation energy of the ^{12}C** state and

TABLE 7.1 Measurements of $Q(^{12}C** - 3\ ^4He)$

Reaction	Excitation Energy (keV)	Reference
$^{12}C(p, p')^{12}C**$	7656.2 ± 2.1	Aus71
$^{12}C(p, p')^{12}C**$	7655.9 ± 2.5	Sto71
$^{15}N(p, \alpha)^{12}C**$	7654.2 ± 1.6	McC73
$^{12}C(p, p')^{12}C**$	7655.2 ± 1.1	Jol74
$^{12}C(p, p')^{12}C**$	7654.00 ± 0.20	Nol76
Weighted average[a]	7654.07 ± 0.19	...
Derived Q-value[a]	379.38 ± 0.20	...

[a] Nol76.

379.38 ± 0.20 keV for the value of Q. A direct determination of the Q-value that does not depend on the assumed accuracy of mass tables has also been made (Bar73) by implanting ^{12}B nuclei into a solid-state counter. Boron 12 decays by β-decay to various states in ^{12}C (Ajz80). About 1.5% of the time, the decays of ^{12}B go to the $^{12}C**$ second excited state, which then mainly decays by the emission of three α-particles (Fig. 7.3). By measuring the total energy of the three (coincident) α-particles, Q is obtained directly. The usefulness of this experiment lies not only in the determination of Q but also in showing directly that the excited state in $^{12}C**$ does indeed break up into three α-particles. Thus, via the reciprocity theorem, the state can also be formed out of three α-particles (Coo57). The resulting value of $Q = 379.6 \pm 2.0$ keV agrees well with the weighted mean of the other values, although it has a lower assigned precision. The error in Q is now so small that it produces an uncertainty in the triple-α rate of only about 1% at $T_8 = 2$, even though Q appears in the exponent of the rate equation (7.1).

The pair-production partial width Γ_{pair} was determined by using inelastic scattering of high-energy electrons produced by a linear accelerator (chap. 5) to measure the E0 monopole matrix element connecting the ground state and the 7.654 MeV excited state in ^{12}C. The result was found to be $\Gamma_{pair} = 60.5 \pm 3.9$ μeV. Note that this is an absolute measurement and is used to determine the values of the other partial widths (Table 7.2).

The branching ratio Γ_{rad}/Γ has been the subject of several experimental studies. Note that an error was discovered (Cha74) in the earlier work (See63) and was subsequently verified (Table 7.2). The basic experimental techniques used in these measurements are the associated-particle technique and the particle-γ-γ triple-coincidence technique (Fig. 7.4). The 7.654 MeV state in ^{12}C is populated in both methods through a particle channel of a selected nuclear reaction. In the associated-particle technique for the reaction $^{13}C(^3He, \alpha)^{12}C$ (Mak75), illustrated in Figure 7.4, the emitted α-particles of the reaction are detected at 85° to the beam direction with a Si surface-barrier detector. The yield of the α_2-particle group $(= N_{\alpha_2})$ measures the number of times the

TABLE 7.2 Measurements of Partial Widths for the 7.654 MeV State in ^{12}C

Quantity	Value	Reference
	Ratios	
Γ_{rad}/Γ	$(3.3 \pm 0.9) \times 10^{-4}$	Alb61
Γ_{rad}/Γ	$(2.82 \pm 0.29) \times 10^{-4}$	See63
Γ_{rad}/Γ	$(3.5 \pm 1.2) \times 10^{-4}$	Hal64
Γ_{rad}/Γ	$(4.20 \pm 0.22) \times 10^{-4}$	Cha74
Γ_{rad}/Γ	$(4.30 \pm 0.22) \times 10^{-4}$	Dav75
Γ_{rad}/Γ	$(4.15 \pm 0.34) \times 10^{-4}$	Mak75
Γ_{rad}/Γ	$(3.87 \pm 0.25) \times 10^{-4}$	Mar76
Γ_{rad}/Γ	$(4.12 \pm 0.11) \times 10^{-4}$	Weighted average[a]
Γ_{γ}/Γ	$(4.02 \pm 0.28) \times 10^{-4}$ [b]	Obs76
Γ_{pair}/Γ	$(6.8 \pm 0.7) \times 10^{-6}$	Obs72, Alb77, Rob77
	Absolute Values	
Γ_{pair}	$60.5 \pm 3.9 \ \mu$eV	Cra67, Str68, Cra64, Gud65
$\Gamma \simeq \Gamma_{\alpha}$	8.90 ± 1.08 eV	
Γ_{rad}	3.67 ± 0.46 meV	Derived values
Γ_{γ}	3.58 ± 0.50 meV	

[a] Mar76.

[b] When corrected for pair decay, yields $\Gamma_{rad}/\Gamma = (4.09 \pm 0.29) \times 10^{-4}$ (Obs76).

7.654 MeV excited state in ^{12}C is formed in the reactions. Kinematical constraints demand that the associated carbon recoil ions ^{12}C**(7.65 MeV) be emitted with an energy of 3.60 MeV at an angle of 63° to the beam direction on the side opposite to the α-detector (Fig. 7.4). Usually, these recoil ions ^{12}C**(7.65 MeV) decay by α-particle emission to the ground state of ^{8}Be, which in turn breaks up quickly into two more α-particles. However, in rare instances (1 in 2500 cases) the ^{12}C**(7.65 MeV) ions decay to the stable ^{12}C ground state ^{12}C(0.0) by electromagnetic transitions (γ-ray emission and less likely e^+-e^- pair emission). The kinematics of the ^{12}C(0.0) recoil ions are nearly identical with those for ^{12}C**(7.65 MeV) because the radiative decay has very little influence on the process. The ^{12}C(0.0) recoil ions are observed with the ^{12}C(0.0) detector at 63° (Fig. 7.4). This detector is also sensitive to many other reaction products and contamination products which, because of their high intensity, hamper the detection of the relatively rare ^{12}C(0.0)-induced events. It is, therefore, necessary to find the ^{12}C(0.0) events by means of the kinematic coincidences with the α_2-group of the α-detector as well as the time-of-flight differences. The radiative decay of the ^{12}C**(7.65 MeV) recoil ions is thus identified by detecting coincidences between the α_2-particles (populating the 7.65 MeV ^{12}C** state) and the associated ^{12}C(0.0) ions remaining after γ-ray or e^+-e^- pair emission. A two-dimensional display of the observed coincidence events (Mak75) is shown schematically in Figure 7.4, where the coordinates are the time-of-flight differences and the energy of the

^{12}C(0.0) recoil ions. The observed events in the area contained within the dashed line correspond to the rare events of interest. The branching ratio Γ_{rad}/Γ for the radiative decay of ^{12}C**(7.65 MeV) is then given (Fig. 7.4) by the total number of ^{12}C$_2$(0.0)-α_2 coincidence events ($=N_{coinc}$) divided by the total number of single events ($=N_{\alpha_2}$), suitably corrected for counting losses

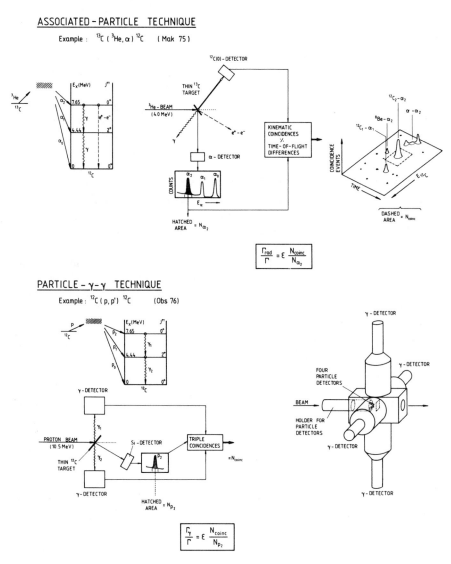

FIGURE 7.4. Shown schematically are two experimental techniques used to investigate the radiative decay of the 7.65 MeV excited state in ^{12}C, which serves as a resonance in the ^{8}Be(α, γ)^{12}C reaction.

and for the detection efficiency of the system. The resulting values for Γ_{rad}/Γ are summarized in Table 7.2. Note that the peak labeled ^{8}Be-α_2 and the intense ridge labeled α-α_2 (suppressed in scale by over an order of magnitude in Fig. 7.4) are due to the predominant decay mode of the ^{12}C**(7.65 MeV) recoil ions, which is into three α-particles. Thus, the usefulness of this experiment lies not only in the determination of Γ_{rad}/Γ but also in showing directly that the excited state in ^{12}C**(7.65 MeV) can indeed decay to the stable ^{12}C(0.0) ground state as well as break up into ^{8}Be and an α-particle or, as is more likely, into three α-particles.

In the particle-γ-γ technique, illustrated in Figure 7.4 for the example ^{12}C$(p, p')^{12}$C (Obs76), the radiative decay of the ^{12}C**(7.65 MeV) nuclei is observed through their γ-γ cascade via the 4.44 MeV state. The cascade transitions are observed by detecting the γ-rays with two γ-ray detectors and the inelastically scattered protons with a Si particle detector. In the latter detector the proton events of the p_2 group (N_{p_2}) measure the number of cases where ^{12}C**(7.65 MeV) nuclei are produced. The number of triple-coincidence events among all three detectors, when compared with the number of single events N_{p_2}, provides the ratio Γ_γ/Γ. In the actual experiment (Obs76) a high-efficiency detection system, consisting of four 4×4 inch NaI(Tl) crystals and four Si counters (Fig. 7.4), was used. The result (Table 7.2) indicates that the partial width for the decay via pair production, Γ_{pair}, is very small compared to that for the decay by γ-ray emission Γ_γ ($\Gamma_{pair} \ll \Gamma_\gamma$).

The ratio Γ_{pair}/Γ was also determined with the use of a pair spectrometer (Alb77) and using a method similar to the second technique discussed above (Rob77). The results of these most difficult measurements are in excellent agreement and yield a weighted average of $\Gamma_{pair}/\Gamma = (6.8 \pm 0.7) \times 10^{-6}$.

On the basis of these studies (Table 7.2), it is clear that the decay of the 7.654 MeV state proceeds predominantly via the α-particle channel; that is, $\Gamma \simeq \Gamma_\alpha$. Using the absolute value of the partial width for pair production, Γ_{pair}, and the measured ratios summarized in Table 7.2, the values for the other partial widths are found to be $\Gamma_\gamma = 3.58 \pm 0.50$ meV, $\Gamma \simeq \Gamma_\alpha = 8.90 \pm 1.08$ eV and $\Gamma_{rad} = 3.67 \pm 0.46$ meV. With the finding of $\Gamma \simeq \Gamma_\alpha$, the resonance strength $\omega\gamma$ is completely determined by the Γ_{rad} partial width according to the relationship

$$\omega\gamma = \Gamma_\alpha \Gamma_{rad}/\Gamma = \Gamma_{rad} .$$

This value is known (Table 7.2) to a precision of 13%. It is remarkable that the rate for the triple-α process, which could not possibly be measured in the laboratory, is believed to be known with a total precision of better than 15%, at least for α-particle energies where equation (7.1) holds for the reaction rate (for a discussion of the burning properties at high stellar temperatures see Bar71).

The triple-α process is an excellent example of a reaction involving an extremely short-lived unstable nucleus having very important astrophysical consequences. In this instance, measurement of the properties of the relevant nuclear reaction rate involving the radioactive nucleus ^{8}Be can only be done indirectly. As discussed in chapter 6, the astrophysically important ^{7}Be$(p, \gamma)^8$B and ^{26}Al$(p, \gamma)^{27}$Si reactions also involve unstable target nuclei. However, in these cases the half-lives are long enough that artificially produced targets of ^{7}Be and ^{26}Al can be used for direct measurement of the relevant cross sections. Clearly, there are many other cases for which imaginative experimental procedures need to be developed (chap. 5).

In the discussion of energy production via hydrogen burning (chap. 6), the stellar rate of energy production was shown to be very sensitive to temperature. For the triple-α process one finds for $T_6 = 100$ a dependence of

$$\epsilon_{3\alpha}(T) \propto T^{41} \; .$$

Because of this extremely strong temperature dependence, energy production via the triple-α process occurs mainly in the highest temperature regions of the star. Indeed, it is the triple-α process that is predominantly responsible for the luminosity of red giants.

Since the triple-α process bypasses the stable elements between $A = 6$ and $A = 11$, these nuclei are essentially not produced in stars, which is consistent with their observed extremely low abundances. These nuclei must, therefore, result from other nucleosynthetic processes, such as those that take place in the big bang and spallation reactions (chaps. 2 and 10; Ree70a).

7.2. The Survival of ^{12}C in Red Giants

In the previous section we saw how the stability gaps at masses $A = 5$ and $A = 8$ can be bridged by combining three α-particles to form ^{12}C. Carbon is, after hydrogen, helium, and oxygen, the most abundant element in the universe (Cam82). The carbon-oxygen ratio is C/O = 0.6. Under the assumption that nuclidic material is synthesized mostly during the major quiescent burning phases of stellar evolution, one would expect the bulk of the carbon abundance in the universe to be a direct product of the triple-α process, and oxygen to be the ash of the subsequent ^{12}C$(\alpha, \gamma)^{16}$O reaction. From this simple consideration, it appears that helium burning of ^{12}C must proceed at a moderate rate in order for sufficient carbon to remain after the helium fuel is exhausted to give the above ratio.

7.2.1. Expected Properties of the ^{12}C$(\alpha, \gamma)^{16}$O Reaction

If, as in the triple-α process, there is a resonance in the ^{12}C$(\alpha, \gamma)^{16}$O reaction near the Gamow energy window corresponding to helium-burning temperatures $(T_9 \simeq 0.1\text{--}0.2)$, this reaction must proceed at a very rapid rate, quickly destroying the available carbon nuclei. However, the energy-level diagram of ^{16}O (Fig. 7.5) indicates that for temperatures up to $T_9 \simeq 2$ there is no level available to foster such resonant behavior. One might naively expect

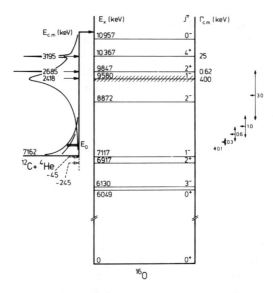

FIGURE 7.5. Level scheme of ^{16}O near and above the α-particle threshold (Ajz82, Ket82). Shown also are the astrophysically interesting energy regions for different stellar temperatures T_9. In the most effective stellar-energy region near $E_0 = 0.3$ MeV, the ^{12}C$(\alpha, \gamma)^{16}$O reaction rate (shown schematically in the form of the $S(E)$ factor) appears to be influenced by the low-energy tail of the 2.42 MeV resonance, corresponding to the 1^-, 9580 keV state, as well as the high-energy tails of the two subthreshold resonances at $E_R = -45$ and -245 keV.

that reactions do not occur and oxygen is not synthesized. However, since oxygen is among the most abundant elements in the universe and can only be produced in stars (chap. 2), one must seek other nuclear mechanisms that enable the reaction ^{12}C$(\alpha, \gamma)^{16}$O to proceed at a rate consistent with the known carbon-oxygen ratio. As discussed in chapter 4, two such mechanisms are available. One is the nonresonant direct-capture process, and the other a nonresonant type of capture into the tails of nearby resonances.

A look at the observed resonances above the α-particle threshold (Fig. 7.5) discloses that only one resonance, the one at $E_{c.m.} = 2.42$ MeV with a width of 400 keV, is sufficiently broad to influence the ^{12}C$(\alpha, \gamma)^{16}$O reaction through its low-energy tail (sketched in Fig. 7.5) in the relevant stellar energy region near $E_0 = 0.3$ MeV. In addition to this level, there are two levels, one at 45 keV and one at 245 keV below the α-particle threshold, with spin-parity assignments of $J^\pi = 1^-$ and 2^+, respectively, that can, by means of their high-energy tails, enhance the stellar burning (sketched also in Fig. 7.5). The effects of subthreshold resonances were discussed in chapter 4.

For the $E_{c.m.} = 2.42$ MeV resonance ($J^\pi = 1^-$), the relevant nuclear parameters are given in Table 7.3. With these parameters and a knowledge of the energy dependence of the partial and thus total widths (chap. 4), the energy dependence of the electric dipole (E1) capture cross section into the ^{16}O ground state can be calculated using equation (4.59). Recalling the relationship (eq. [4.16])

$$\sigma(E) = S(E) \frac{1}{E} \exp\left(-2\pi\eta\right),$$

the energy dependence of the corresponding $S(E)$ factor can be determined. The results are plotted as a dotted curve in Figure 7.6a and lead to a value of $S(E_0 = 0.3\ \text{MeV}) \simeq 1.5 \times 10^{-3}\ \text{MeV b}$.

Similarly, the energy dependence of the high-energy tails of the two sub-threshold resonances can be calculated using the available nuclear parameters listed in Table 7.3. The results of these calculations are also plotted as dotted curves in Figures 7.6a and 7.6b. At a stellar energy $E_0 = 0.3\ \text{MeV}$, one finds values of $S_{1-}(E_0) \simeq 0.1\ \text{MeV b}$ and $S_{2+}(E_0) \simeq 0.2\ \text{MeV b}$.

In addition to capture into the tails of resonances, direct-capture process into the ground state of ^{16}O must also be considered (chap. 4). In this case, the capture process is an electric quadrupole (E2) transition from an initial d-partial wave into an s-orbital state. The $S(E)$ factor for this process is essentially energy-independent, with an absolute value proportional to the reduced α-particle width θ_α^2 of the ^{16}O ground state. Using the value arrived at from α-transfer reactions (Table 7.3), one finds $S_{DC}(E_0) \simeq 5 \times 10^{-3}\ \text{MeV b}$. From Figure 7.6 it is clear that the $S(E_0)$ factors resulting from both the direct-capture process and the low-energy tail of the 2.42 MeV resonance are rela-

TABLE 7.3 Level Parameters of ^{16}O States Involved in the ^{12}C$(\alpha, \gamma_0)^{16}$O Stellar Reaction[a]

Level E_x (keV)	J^π	Resonance Energy E_R (keV)	Gamma Width Γ_{γ_0} (meV)	Reduced α-Particle Width θ_α^2			
				α Elastic Scattering	α Transfer[b]	(α, γ_0) Data[c]	(α, γ_0) Data[d]
0	0^+	...	...	...	0.19	0.04 ± 0.01[e]	$0.25^{+0.25}_{-0.08}$
6917	2^+	-245 ± 1	100 ± 6	...	1.0	...	1.0 ± 0.2
7117	1^-	-45 ± 1	57 ± 5	...	0.40[f]	0.10[g]	$0.19^{+0.14}_{-0.10}$
9580[h]	1^-	2418 ± 12[h]	23 ± 3[i]	0.55 ± 0.02[j]	$\equiv 0.61$	...	0.67 ± 0.03[k]

[a] Deduced from the compilation (Ajz82) except where quoted.
[b] Bec80; relative to $\theta_\alpha^2(9.580\ \text{MeV}) = 0.61$.
[c] Fit to the E1 capture data (Dye74). See also Tom82.
[d] Fit to the E1 + E2 capture data (Ket82).
[e] See also Tom82.
[f] Values of $\theta_\alpha^2(7.117\ \text{MeV}) = 0.10 \pm 0.04$ (Loe67), 0.03 ± 0.02 (Pue70), 0.15 ± 0.09 (Bar71a), 0.06 ± 0.05 (Wer71), and 0.25 ± 0.15 (Bec78) have been reported.
[g] Relative to $\theta_\alpha^2(9.580\ \text{MeV}) = 0.55$ (see also Tom82).
[h] From α elastic-scattering data (Ket82).
[i] The analysis of the data shown in Fig. 7.10b leads to 25 ± 4 meV (Ket82).
[j] Deduced from the total width $\Gamma_{c.m.} = 400 \pm 10$ keV (Ajz82). The recent elastic-scattering data (Ket82) lead to $\theta_\alpha^2 = 0.67 \pm 0.03$. The average value is $\theta_\alpha^2 = 0.61$.
[k] Deduced from elastic-scattering data obtained concurrently (Ket82).

tively insignificant compared with those of the two subthreshold resonances. Thus it appears that the $^{12}C(\alpha, \gamma)^{16}O$ burning rate is determined almost entirely by the effects of these subthreshold resonances. Summing the values for these two resonances, one arrives at a value of $S(E_0 = 0.3 \text{ MeV}) \simeq 0.3 \text{ MeV b}$ (see also Lan85).

This S-value translates into a value for the capture cross section of $\sigma(E_0 = 0.3 \text{ MeV}) \simeq 5 \times 10^{-17}$ b. A direct measurement of this cross section is clearly beyond present technical capability (Chap. 5). Attempts were made to predict the $S(E_0)$ factor based on astrophysical information and current stellar models (Arn72, Arn73). Since such estimates could be susceptible to fundamental systematic errors, improved experimental information was required. For example, in calculating the S-factor curves in Figure 7.6, the reduced α-particle width θ_α^2 is an important parameter. Its values (Table 7.3) are obtained from α-transfer reactions and are generally in fair agreement with the corresponding values

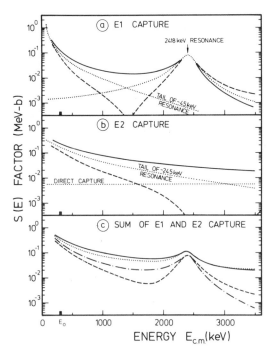

FIGURE 7.6. The capture reaction $^{12}C(\alpha, \gamma_0)^{16}O$ can have contributions from both E1 and E2 amplitudes. (a) In the case of E1 capture, the resonances at $E_R = 2418$ and -45 keV can both contribute to the yield (*dotted lines*), where constructive or destructive interference between the two E1 sources may occur (*solid or dashed lines*). (b) Similarly, the two E2 sources arising from the $E_R = -245$ keV subthreshold resonance and the direct-capture process (*dotted lines*) lead to interference effects (*solid and dashed lines*). (c) The total capture cross section is the incoherent sum of the E1 and E2 yields, leading to four possible curves depending on the sign of the interference effects.

deduced from resonance data. The analysis of α-transfer reactions is very complex (Bet70, Bec80), and again the possibility of systematic errors limiting the reliability of the resulting θ_α^2 values (Bar71, Tom82) cannot be ruled out.

Clearly, direct measurement of the cross sections for the ^{12}C$(\alpha, \gamma)^{16}$O reaction over a wide range of energies, including as low an energy as is technically possible, would provide the best information, since it would allow for the most accurate extrapolation to the lowest energies of interest. Data at the higher energies are also important, because, from the energy dependence of these data, information on the reaction mechanisms involved in the capture process can be extracted.

As previously pointed out, the electric dipole (E1) capture transition into the ^{16}O ground state can proceed through both the $J^\pi = 1^-$ subthreshold resonance ($E_R = -45$ keV) and the $J^\pi = 1^-$ resonance at $E_R = 2.42$ MeV (Fig. 7.6a). Consequently, the total cross section is influenced by interference effects between the two resonances. These effects will be maximum at those energies where the two amplitudes are of comparable magnitude, which according to Figure 7.6a is in the neighborhood of $E_{\text{c.m.}} \simeq 1.4$ MeV. Of course, the interference may be either constructive or destructive, leading to either a doubling or a vanishing of the effective cross section in this energy region (solid and dashed curves in Fig. 7.6a). If direct measurements of the ^{12}C$(\alpha, \gamma)^{16}$O reaction could be extended into this energy region, the data should provide information on the sign of the interference effect (constructive or destructive interference). The size and the energy dependence of the cross section in this energy region should also provide information from which the shape and magnitude of the high-energy tail of the $J^\pi = 1^-$ subthreshold resonance ($E_R = -45$ keV) could be deduced.

The electric quadrupole (E2) capture transition into the ground state can also arise from two sources, namely, the direct-capture process and the $J^\pi = 2^+$, $E_R = -245$ keV subthreshold resonance. From Figure 7.6b one would expect that, for the E2 capture transition, maximum interference effects should occur in the energy region near 3.0 MeV, causing either a doubling or a vanishing of the effective E2 capture cross section (solid and dashed curves in Fig. 7.6b). If there were no E1 contributions to the capture cross section, direct measurement of the ^{12}C$(\alpha, \gamma)^{16}$O reaction in this energy region would determine the sign of the interference effect as well as the magnitude of the sum of the two E2 contributions.

Unfortunately, an E1 process is present, and the total capture cross sections as a function of the beam energy are affected by both E1 and E2 processes. Because of the different multipolarities involved in both capture processes, there are no interference effects between them in the total capture cross sections. The observed energy dependence of the total capture cross section can in principle be used to determine the effects of the individual E1 and E2 processes because they have different energy dependences (Fig. 7.6c).

Even though there are no interference effects between the E1 and E2 parts for the total cross sections, the differential cross sections should show interference effects because of their differing individual γ-ray angular distribution patterns (Fig. 7.7) and because of the differing parities involved in the two multipole radiations (E1 and E2). These interference effects show up as an asymmetry in the γ-ray angular distributions around $\theta_\gamma = 90°$ (Fig. 7.7d). Analysis of these asymmetric patterns provides information on the relative magnitude of the E1 and E2 contributions. While the angular distribution patterns for E1 and E2 capture separately (Figs. 7.7a and 7.7b) do not change with beam energy, the pattern arising from the E1 and E2 interferences (Fig. 7.7d) varies with energy because of the differing energy dependences of the individual total capture cross sections $\sigma_{E1}(E)$ and $\sigma_{E2}(E)$ (Fig. 7.6). Therefore, analysis of these patterns provides the ratio of E1 and E2 total capture cross sections as a function of energy and clearly improves the precision with which data can be extrapolated to stellar energies.

From the above discussions, the $^{12}C(\alpha, \gamma)^{16}O$ burning rate appears to be controlled almost completely by the tails of the two subthreshold resonances.

GAMMA ANGULAR DISTRIBUTIONS FOR $^{12}C(\alpha,\gamma)^{16}O$

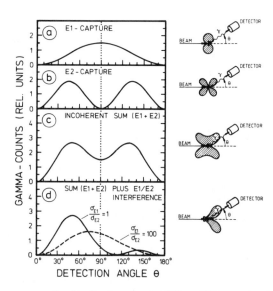

FIGURE 7.7. Gamma-ray angular distributions for the $^{12}C(\alpha, \gamma_0)^{16}O$ reaction. In the case of pure E1 capture or of pure E2 capture, the angular distributions exhibit the familiar patterns. The pattern shown in (c) is not a realistic situation, since the two multipole radiations interfere, leading to patterns asymmetric around 90°. One of the patterns shown in (d) assumes equal cross sections for the two multipole radiations. Even if the yields differ, for example, by 2 orders of magnitude, the presence of the smaller component is largely amplified through the interference term (dashed line in panel d). Note that γ-ray yield measurements carried out at 90° are only sensitive to E1 capture.

We have seen the effect of subthreshold resonances earlier, for example, in the reactions ^{20}Ne$(p, \gamma)^{21}$Na and ^{22}Ne$(p, \gamma)^{23}$Na (chap. 6). In these cases, observation of the high-energy tails is possible because of the combination of s-wave formation and maximum partial widths. In the ^{12}C$(\alpha, \gamma)^{16}$O case, the subthreshold resonances involve p- and d-wave formation; thus experimental observation of the effects of their high-energy tails appears to be impossible. Fortunately, since the interference effects discussed above lead to a considerable enhancement of these tails, experimental observations appear to be possible. In summary, these experiments should consist of measurements of both the total cross sections and the γ-ray angular distributions over a wide range of beam energies and to as low an energy as feasible.

7.2.2. Measurements of the ^{12}C$(\alpha, \gamma)^{16}$O Reaction

As discussed in the previous section, direct measurement of the ^{12}C$(\alpha, \gamma)^{16}$O reaction should be extended to energies as far below the 2.42 MeV resonance as is technically feasible. Since the capture cross section at the peak of this resonance is only 40–50 nb (Fig. 7.9), observation of interference effects in the low-energy tail of this resonance requires measurement of cross sections considerably smaller than 1 nb. Formidable problems are encountered in these measurements. They arise from the combination of a low γ-ray capture yield and a high neutron-induced γ-ray background. The ^{13}C$(\alpha, n)^{16}$O reaction is a prolific source of neutrons (Bai73); thus any target containing ^{13}C is undesirable. Deposition of any natural carbon on the target or beam-defining collimators (chap. 5) is also undesirable. As a consequence, measurements were carried out using targets of high ^{12}C isotopic purity (Jas70, Dye74). In addition, the difference between the neutron and γ-ray time of flight was used (technique 1 in Fig. 7.8) to separate the γ-rays produced by neutrons and the capture γ-rays. In another experiment (Ket82), the neutron-induced γ-ray background from the ^{13}C$(\alpha, n)^{16}$O reaction was essentially eliminated by interchanging the role of projectiles and target nuclei (technique 2 in Fig. 7.8). In these experiments the ^{4}He target nuclei, contained in a windowless gas-target system, were bombarded with an intense ^{12}C beam (up to a 50 μA particle current). The results are shown in Figure 7.9. In spite of enormous experimental difficulties, the overall agreement of these measurements appears to be satisfactory. However, a closer inspection reveals some differences among the data sets that should be mentioned because even small discrepancies in the measured cross sections can lead to major differences in the extrapolations.

The data of Jaszczak and coworkers (Jas70) and Kettner and coworkers (Ket82) are in good agreement at the lower energies. The results of Dyer and Barnes (Dye74) also agree very well with the measurements mentioned above in the energy dependence of the cross section for all reported data points at $E_{c.m.} \geq 1.9$ MeV. However, the Dyer and Barnes absolute cross sections are about 30% smaller, and the cross-section values reported at $E_{c.m.} = 1.41$–1.87 MeV drop faster with decreasing beam energy. In the work of Dyer and

Barnes only the E1 capture cross sections are reported. Perhaps the discrepancy lies in the difference between the partial (E1) and the total (E1 + E2) cross sections. In the work of Kettner and coworkers the detectors were placed in close geometry to the target (Fig. 7.8), resulting in angle-integrated total capture cross sections.

The E1 capture data of Dyer and Barnes are plotted in S-factor form in Figure 7.10a. The dashed curve in this figure represents the S-factor curve resulting from the 2.42 MeV resonance alone. The data clearly show an enhancement at the lower energies, indicating constructive interference with the -45 keV subthreshold resonance (sec. 7.2.1). The data have been analyzed using various models (Dye74, Wei74, Koo74, Hum76). A discussion of these analyses (Tom82) reveals that $S_{E1}(E_0 = 0.3 \text{ MeV}) \simeq 0.1\text{–}0.2$ MeV b for the E1

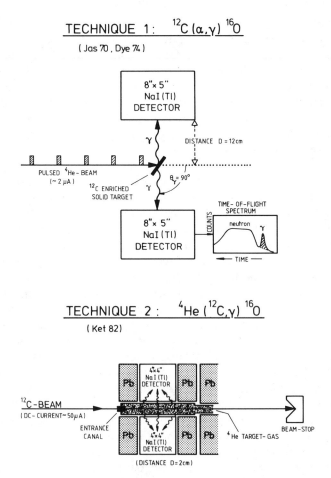

FIGURE 7.8. Schematic representation of two different experimental techniques that have been used in the measurements of the $^{12}\text{C}(\alpha, \gamma)^{16}\text{O}$ reaction.

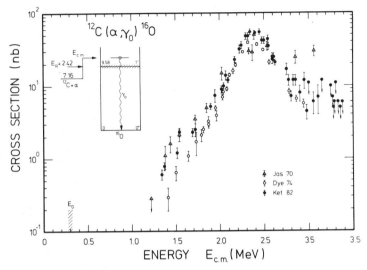

FIGURE 7.9. Cross-section data for the ground state transition in ^{12}C$(\alpha, \gamma)^{16}$O as reported by several investigators. Note that the cross section at the maximum of the $E_R = 2.42$ MeV resonance is only 40–50 nb and that the lowest data points are in the picobarn range. Two of the data sets (Jas70, Dye74) have been obtained by technique 1 (Fig. 7.8), while the more recent data (Ket82) involved technique 2 (Fig. 7.8). Note also that one data set (Dye74) represents only E1 capture.

part of the capture mechanism only (see also Lan85). An analysis of the data of Jaszczak and coworkers (Jas70) yields (Bar71) an $S(E_0)$ factor of about 0.6 MeV b ($\pm 200\%$).

Total cross-section data from the recent investigation (Ket82) are shown in Figure 7.10b. Using the Breit-Wigner formalism and incorporating data from elastic scattering, which was obtained concurrently (Ket82; see also Jon62, Cla68a), the cross sections resulting from the E1 transitions were calculated. They are shown as a dotted line in Figure 7.10b. It is clear from this figure that E1 capture mechanisms alone cannot account for the total capture cross section on either side of the 2.42 MeV resonance. By including E2 capture mechanisms (sec. 7.2.1) in the analysis, the solid line in Figure 7.10b is obtained. Extrapolation to $E_0 = 0.3$ MeV leads to a value of S_{E1+E2} ($E_0 = 0.3$ MeV) $\simeq 0.4$ MeV b (see also Lan85).

The extrapolated S-factors arrived at by analysis of the two sets of data (Fig. 7.10) differ by a factor of about 2–4. This difference might be expected if, in one case, the effects of the E2 capture mechanisms were not taken into account. The presence of the E2 part can also be inferred using the γ-ray angular distribution data obtained by Dyer and Barnes (Dye74) over a rather limited energy range by taking the ratios of the E2 and E1 contributions. These results are consistent with the analysis shown in Figure 7.10b. Clearly, having such measurements over a much wider range of energies is desirable.

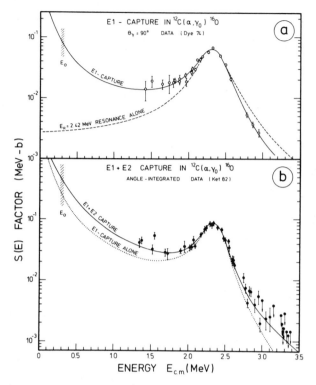

FIGURE 7.10. (a) The E1 capture yield in $^{12}C(\alpha, \gamma_0)^{16}O$ is shown in $S(E)$ factor form together with a theoretical analysis (Koo74). The data cannot be explained by the $E_R = 2.42$ MeV resonance alone. They require an additional contribution from the $E_R = -45$ keV subthreshold resonance, where the two resonances interfere constructively at energies between them. (b) The observed angle-integrated γ_0-ray yields indicate the presence of E1 and E2 capture mechanisms.

For recent experimental and theoretical work see Red85 and Pla87 and references therein.

In spite of enormous experimental efforts in studying this reaction, considerable uncertainties in its stellar reaction rate still exist. Consequently, at present, neither the relative amounts of ^{12}C and ^{16}O produced by red giant stars nor their subsequent evolution can be predicted with great confidence.

7.2.3. Elemental Abundances

For type II supernovae, model calculations indicate (Arn72, Arn78, Wea80, Woo82) that the distribution of chemical elements resulting from hydrostatic burning in the shells of supernovae is only mildly affected by the processing that occurs during the explosion that ejects the envelope. Note, however, that the isotopic ratios of the elements are strongly affected. Thus, such hydrostatic calculations should also give the elemental abundance pattern of the ejected

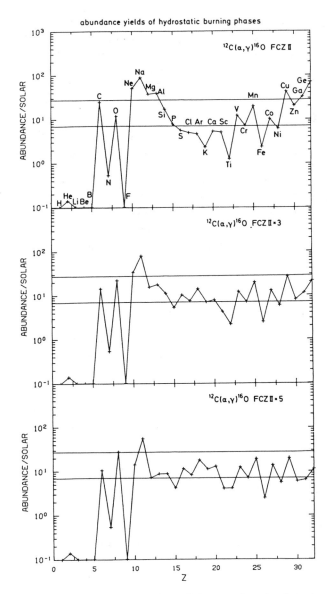

FIGURE 7.11. Calculated elemental abundances (relative to solar abundances) resulting from hydrostatic stellar-burning phases for a 25 $M_\odot$ model star (Thi85). In the top panel the rate of the $^{12}C(\alpha, \gamma)^{16}O$ reaction is based on the 1975 compilation (Fow75) and labeled *FCZII*, while for the middle and lower panels it is increased by factors of 3 and 5, respectively. A considerable improvement in the scattering of the abundances around the average is noted for the higher rates. Exceptions are hydrogen and helium, which are produced mainly in the big bang, and nitrogen, which is known to be copiously produced by stars of lower mass. The elements Li, Be, and B are low in abundance because they are produced predominantly by spallation (chap. 10). The origin of fluorine is not yet known.

matter. The results of such calculations for a 25 $M_\odot$ star, which represents a typical stellar mass (Thi85) for galactic nucleosynthesis considerations, are shown in the top diagram in Figure 7.11, where the $^{12}C(\alpha, \gamma)^{16}O$ rate of Fowler, Caughlan and Zimmerman (Fow75) has been used. The figure shows that, compared with oxygen, carbon is overabundant. As a consequence, the products of carbon burning, such as Ne, Na, and Mg (chap. 8), are overabundant, while the products of oxygen and silicon burning, including elements from Al up to the iron region (chap. 8), are underabundant. An increased $^{12}C(\alpha, \gamma)^{16}O$ reaction rate increases the oxygen and its subsequent reaction products and lowers the carbon and its subsequent reaction products, resulting in an abundance distribution more consistent with solar abundances. In Figure 7.11 are shown the results of such calculations when the $^{12}C(\alpha, \gamma)^{16}O$ reaction cross section is increased by factors of 3 and 5 (sec. 7.2.2); indeed, considerable improvement in the comparison with observed solar abundances can be noted. Recent calculations show (Woo85) that the reaction rate for $^{12}C(\alpha, \gamma)^{16}O$ is also critical in determining the lower mass limit for black hole formation.

7.3. The Blocking of Quiescent Helium Burning

In principle, the α-capture process can continue after the $^{12}C(\alpha, \gamma)^{16}O$ stage, producing ^{20}Ne, ^{24}Mg, ^{28}Si, and so on. However, this does not happen,

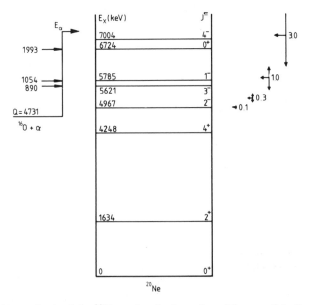

FIGURE 7.12. Energy levels of the ^{20}Ne nucleus in the region of the α-particle threshold (Ajz83). Shown also are the most effective energy regions for temperatures from $T_9 = 0.1$ to $T_9 = 3.0$. Note that the $J^\pi = 2^-$ and $J^\pi = 4^-$ levels do not produce resonances in $^{16}O(\alpha, \gamma)^{20}Ne$, because of their unnatural parity.

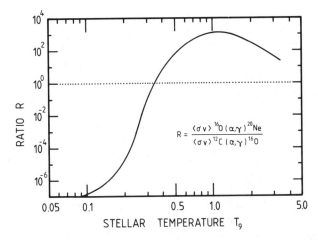

FIGURE 7.13. The ratio of the stellar reaction rates for $^{16}O(\alpha, \gamma)^{20}Ne$ and $^{12}C(\alpha, \gamma)^{16}O$ (Fow75) is shown as a function of temperature. Note that this ratio is very small at helium-burning temperatures of $T_9 = 0.1$–0.2, thus essentially blocking the nucleosynthesis beyond ^{16}O.

because of the increase in the Coulomb barriers and the properties of the resonances in the critical energy region for the first relevant reaction, $^{16}O(\alpha, \gamma)^{20}Ne$. Because the Q-value for this reaction is relatively low, the level density in ^{20}Ne is also low in the astrophysically important energy region (Fig. 7.12). There is an excited state in ^{20}Ne at $E_x = 4.97$ MeV that could give rise to a resonant behavior at helium-burning temperatures. However, this level has unnatural parity ($J^\pi = 2^-$) and, therefore, cannot be formed in this reaction (chap. 4). The observed resonances at high energies are too weak and too narrow (Ajz83) to contribute to the cross section at ordinary helium-burning temperatures. The $J^\pi = 4^+$ subthreshold state (Fig. 7.12) lies too far below the threshold, and, in addition, its g-wave formation reduces its contribution to the stellar rate. Therefore, the only mechanism left appears to be the direct-capture process, which in general for (α, γ) reactions leads to cross sections in the range of nanobarns and below. As a consequence, the $^{16}O(\alpha, \gamma)^{20}Ne$ reaction proceeds at a rather low speed at helium-burning temperatures. Stellar model calculations involving this reaction rely on theoretically derived direct-capture cross sections, which to date have not been verified experimentally (Toe71, Toe71a). With increasing temperatures the observed resonances at higher energies begin to play a role by enhancing the $^{16}O(\alpha, \gamma)^{20}Ne$ reaction rate. Indeed, at temperatures near and above $T_9 = 0.3$, this rate exceeds that for the $^{12}C(\alpha, \gamma)^{16}O$ reaction (Fig. 7.13). This situation can occur at the very high temperatures found in very massive stars. It is also clear from this figure that, at ordinary helium-burning temperatures, the $^{16}O(\alpha, \gamma)^{20}Ne$ reaction rate is slower by several orders of magnitude compared with the $^{12}C(\alpha, \gamma)^{16}O$ case, thus essentially blocking nucleosynthesis through the helium-burning process beyond ^{16}O.

Although not produced to any large extent in quiescent helium burning, ^{20}Ne is produced (Cou71a) in shell burning, advanced burning stages, and explosive burning (chap. 8). It has been pointed out (Woo79) that near $T_9 = 0.8$ the ^{20}Ne(α, γ)^{24}Mg reaction is chiefly responsible for the destruction of ^{20}Ne produced by carbon burning (chap. 8) and that current massive star models result in too much ^{20}Ne by a factor of 2. This discrepancy could be removed if the ^{20}Ne(α, γ)^{24}Mg reaction proceeded near $T_9 = 0.8$ at a rather high rate. Experimental studies of this reaction (Sch83) have shed some light on the situation. The stellar reaction rate is now known for all temperatures of practical interest ($T_9 \geq 0.2$). However, these improved data have not resolved the above problem. It should be noted that an enhanced rate of the ^{12}C(α, γ)^{16}O reaction in helium burning also leads to less ^{12}C fuel in ^{12}C burning, which reduces this problem partially (Fig. 7.11). A comparison of the reaction rates for ^{16}O(α, γ)^{20}Ne (Fow75) and ^{20}Ne(α, γ)^{24}Mg (Sch83) shows that the latter reaction is the faster one at helium-burning temperatures of $T_9 \geq 0.18$ despite the fact that the Coulomb barrier of the ^{20}Ne + α pair is higher. Thus helium burning appears not to be a promising way to produce ^{20}Ne (chap. 8).

7.4. Other Helium-Burning Reactions

Nitrogen 14 is the major residue resulting from the operation of the CNO cycles during hydrogen burning in second-generation stars (chap. 6). Typically, in addition to helium ash, there will be on the order of 1%–2% by mass of nitrogen in the stellar core. Therefore, its fate under helium-burning conditions must be considered. In fact, it had been suggested that helium flashes might be triggered by the ^{14}N(α, γ)^{18}F reaction (Ree66, Ibe67, Egg68, Asa68). This would be the case if this reaction generated enough energy below $T_9 = 0.1$, the critical energy for the triple-α process. However, present knowledge of the nuclear physics involved (Ajz83) indicates that the energy production is insufficient for this to occur (Cou71).

Although the reaction ^{14}N(α, γ)^{18}F is not fast enough to trigger the helium flashes, it is still important because of its role in synthesizing ^{18}O through the positron decay of ^{18}F (How71, Mol71). The fate of ^{18}O itself depends on the relative reaction cross sections for the ^{18}O(α, γ)^{22}Ne and ^{18}O(α, n)^{21}Ne reactions. The relevant Q-values are $Q_{\alpha\gamma} = 9.67$ MeV and $Q_{\alpha n} = -0.70$ MeV. Thus for low temperatures the capture reaction predominates, while at higher temperatures the neutron-producing reaction prevails. The available experimental information for these two processes is shown in Figure 7.14. The data indicate that below $T_9 = 0.6$ mainly ^{22}Ne is produced, while above this temperature the major ash is ^{21}Ne. Below $T_9 = 0.3$ the ^{18}O(α, γ)^{22}Ne rate might be enhanced if additional resonances were detected at low energy (Tra78).

One of the earliest suggestions for the principal source of neutrons for s-process nucleosynthesis in stars (chap. 9) was (Cam54, Gre54) the reaction

$^{13}\text{C}(\alpha, n)^{16}\text{O}$ ($Q = 2.21$) MeV), since it might be expected to follow immediately after the CNO hydrogen-burning phase. The abundant ^{14}N, the
dominant product of hydrogen burning in the CNO cycles, poses serious
problems, since it has a large cross section for absorbing neutrons via the
reaction $^{14}\text{N}(n, p)^{14}\text{C}$ ($Q = 0.63$ MeV). The proton produced in the latter reaction could, in principle, be used to produce another ^{13}C nucleus via the reactions $^{12}\text{C}(p, \gamma)^{13}\text{N}(e^+ v)^{13}\text{C}$, but it could also, with comparable probability,
produce another ^{14}N through the $^{13}\text{C}(p, \gamma)^{14}\text{N}$ reaction. Further, the ^{14}C produced in the (n, p) reaction would β-decay in due course to regenerate the ^{14}N
neutron poison. It is, of course, required that any process postulated to
produce neutrons for the s-process must be really a net neutron producer and
not a net consumer of neutrons (Sch67, San67, Bar71, Bar82).

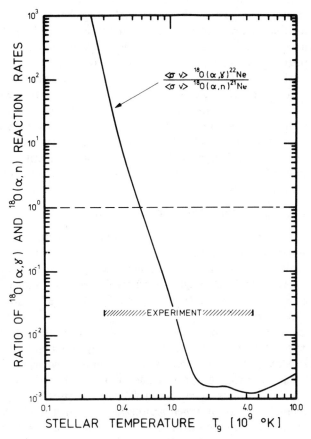

FIGURE 7.14. The ratio of the stellar reaction rates for the reaction $^{18}\text{O}(\alpha, \gamma)^{22}\text{Ne}$ (Tra78) and
$^{18}\text{O}(\alpha, n)^{21}\text{Ne}$ (Fow75) is plotted as a function of stellar temperature. Also indicated is the temperature region for which the capture reaction rates are known experimentally.

At present the most attractive s-process neutron source among the theorists who construct stellar models is the reaction ^{22}Ne$(\alpha, n)^{25}$Mg (Cam60, Ibe75, Sca77; however, see also Cam85). This reaction has a threshold at 0.48 MeV, low enough for the reaction to proceed in the late helium-burning stages. The seed nuclei ^{22}Ne for the neutron source are thought to be produced by the reactions ^{14}N$(\alpha, \gamma)^{18}$F$(e^{+}v)^{18}$O$(\alpha, \gamma)^{22}$Ne (Cam60). There are several other helium-induced reactions that might, in some instances, be significant neutron sources, for example, ^{17}O$(\alpha, n)^{20}$Ne $(Q = 0.59$ MeV), ^{18}O$(\alpha, n)^{21}$Ne $(Q = -0.70$ MeV), ^{21}Ne$(\alpha, n)^{24}$Mg $(Q = 2.56$ MeV), ^{25}Mg$(\alpha, n)^{28}$Si $(Q = 2.65$ MeV), and ^{26}Mg$(\alpha, n)^{29}$Si $(Q = 0.03$ MeV).

The rates of some of these (α, n) reactions seem to be known well enough experimentally (Tan65, Ash69, Dav68, Bai73, Haa73, Haa74, Mak74, Dav75a, Ram77). There is, however, need for further experimental information. In particular, in the mass region of the s–d shell nuclei ($A \simeq 24$) there are likely to be a few states close to the particle thresholds. Although there are hardly enough resonances for a statistical average treatment to be valid (chap 8), there may still be enough so that the astrophysical yield is dominated by one or two resonances near threshold. Each case has to be treated as a major, separate problem as in the ^{12}C$(\alpha, \gamma)^{16}$O case and other such cases (chap. 6). In addition to the desire for improved reaction rate data for the (α, n) reactions, more information about the competing (α, γ) capture reactions is needed in order to determine more precisely the rate of neutron production as a function of stellar temperature (see Fig. 7.15). There is very little information available for the ^{22}Ne$(\alpha, \gamma)^{26}$Mg reaction $(Q = 10.61$ MeV), the competitor of the very likely neutron-producing source ^{22}Ne$(\alpha, n)^{25}$Mg $(Q = -0.48$ MeV) (End78).

7.5. Perspectives on Helium-Burning Reactions

Figure 7.15 summarizes and puts in perspective the main nuclear reactions involved in quiescent helium burning in the cores of red giant stars. The ^{12}C nuclei are built with sufficient abundance, owing to the small difference between the masses of a ^{8}Be nucleus and two α-particles and a fortuitously located state in ^{12}C that provides a "thermal" resonance to enhance greatly the radiative capture of α-particles by ^{8}Be. The resulting ^{12}C nuclei survive further bombardment with particles from the α bath because of the lack of a resonant state in ^{16}O near the most effective energy window E_0. However, the 7.12 and 6.92 MeV subthreshold states provide enough yield at E_0 to let the ^{12}C$(\alpha, \gamma)^{16}$O reaction proceed at a rate such that carbon and oxygen are produced roughly in amounts such that C/O $\simeq 0.1$ This reaction provides a classic example of the way in which simple facts of nuclear physics can have profound consequences for the evolution of the universe as we know it. For example, if the E1 γ-decay of the 7.12 MeV state was not inhibited by isospin selection rules, ^{12}C would not have survived helium burning. Furthermore, the ^{16}O nuclei are not subsequently consumed because the 4.97 MeV state in

^{20}Ne, although located exactly in the most effective burning region E_0, cannot be formed via the $^{16}O(\alpha, \gamma)^{20}$Ne reaction because of parity conservation. Since the nuclear properties of the 4.25 MeV state prevent it from acting as a sub-threshold resonance, the $^{16}O(\alpha, \gamma)^{20}$Ne reaction proceeds at an extremely low rate, blocking nucleosynthesis via helium burning beyond ^{16}O.

We have seen that the major ashes of helium burning in red giants are carbon and oxygen, and it is generally believed that the ^{12}C and ^{16}O in galactic matter had their origin in these red giants. Both elements are also essential for the evolution of life, and it is only through some fortuitous nuclear properties and selection rules that both elements were produced so plentifully and survived the red giant phase of stellar evolution. It is perhaps instructive to speculate how our life and the universe as a whole might have looked if the

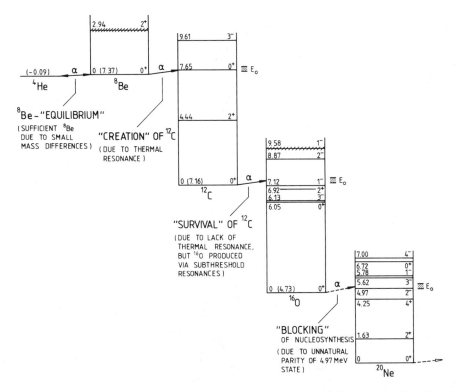

FIGURE 7.15. Shown in perspective are the level schemes of the nuclei involved in the helium-burning reactions in red giants. It is only through a number of fortuitous nuclear properties as well as parity and isospin conservation laws that carbon and oxygen, the essential elements for life as we know it, are the major products of nucleosynthesis in red giants.

mass of ^{8}Be had not been close to the mass of two α-particles, if there were no enhancing resonant state in ^{12}C, or if there were no parity and isospin conservation laws.

Einstein is quoted as saying, "God does not throw dice." This has not been verified one way or the other; but if He (or "She") does, She (of He) is incredibly lucky.

8

Advanced and
Explosive Burning

As discussed in the previous two chapters, thermonuclear reactions in the interior of a star start with hydrogen burning and gradually transform the core of the star into increasingly heavier and more tightly bound nuclei. The nuclear energy released in the reactions prevents the collapse of the overlying matter under its own gravitational force. When a given nuclear fuel in the core is exhausted, the thermal pressure is unbalanced against the gravitational force. In most cases this causes the core to contract until the resulting compressional heating is sufficient to ignite either the ashes of the previous burning stage or a shell of unburned fuel farther out in the star. For example, the ashes produced in helium burning in a massive star become fuel for further nuclear-burning processes, leading to the synthesis of most of the elements with $A \geq 20$. Immediately following helium burning come the carbon-, neon-, and oxygen-burning phases followed by silicon burning, and, as the final stage, explosive burning can occur. Since it is the balance between the gravitational force tending to collapse the star and the thermal pressure produced by the various nuclear reaction processes that determines the temperature of the core, whether or not a star will experience all of these burning phases depends critically on its mass. The time scales involved in all these heavy-element–burning phases are considerably shorter than for the hydrogen and helium burning (Table 8.1), mainly because of the large losses from neutrino emission that occur at the temperatures and densities necessary to burn the heavier elements. For explosive burning the time is of the order of the free-fall time—a few seconds. In the explosive phase, not only are the ultimate relative abundances of all the elements determined (Arn78, Woo82), but also a mechanism is provided for the propulsion of the synthesized elements into interstellar space via the star's final collapse and explosion—the supernova. The freshly synthesized elements dispersed in these explosions mix with the ashes of other stars and with primordial hydrogen and helium, gradually increasing the

TABLE 8.1 Evolutionary Stages of a 25 $M_\odot$ Star[a]

Stage	Time Scale	Temperature (T_9)	Density (g cm^{-3})
Hydrogen burning	7×10^6 y	0.06	5
Helium burning	5×10^5 y	0.23	7×10^2
Carbon burning	600 y	0.93	2×10^5
Neon burning	1 y	1.7	4×10^6
Oxygen burning	6 months	2.3	1×10^7
Silicon burning	1 d	4.1	3×10^7
Core collapse	seconds	8.1	3×10^9
Core bounce	milliseconds	34.8	$\simeq 3 \times 10^{14}$
Explosive burning	0.1–10 s	1.2–7.0	Varies

[a] From Wea80a.

heavy-element content of the interstellar medium, from which new stars are formed. Details of these nuclear-burning processes as well as of the supernova phenomenon (core collapse and core bounce) are discussed below.

8.1. Quiescent Heavy-Ion Burning

As helium burning progresses, a core develops composed primarily of carbon and oxygen. Surrounding this carbon-oxygen core are (Fig. 8.1) a helium-burning shell, a helium-rich region, and farther out a hydrogen-burning shell and a hydrogen-rich envelope. For sufficiently massive stars gravitational collapse causes sufficient increases in temperature and density in the core to ignite the carbon and oxygen ashes (Ree59, Fow64). Since the Coulomb barrier is lowest for the carbon nuclei, these will be the first to interact, resulting in the formation of neon, sodium and magnesium as follows:

$$^{12}C + {}^{12}C \rightarrow {}^{20}Ne + \alpha \quad (Q = 4.62 \text{ MeV}),$$

$$\rightarrow {}^{23}Na + p \quad (Q = 2.24 \text{ MeV}),$$

$$\rightarrow {}^{23}Mg + n \quad (Q = -2.62 \text{ MeV}).$$

This set of reactions is referred to as carbon burning.

It is important to note that the manner in which carbon is burned depends critically on the mass of the star. In stars not much more massive than our sun, the central temperature does not become high enough to initiate carbon burning, and burning stops. Since no further energy release by thermonuclear reactions can occur, the core begins to contract, increasing its density and radiating away its thermal energy. When most of the thermal energy of the carbon core has been radiated away, electron degeneracy provides sufficient pressure to balance gravity indefinitely. At this point there is no further collapse of the star, and no further nuclear burning occurs. In fact, for low-mass stars this effect can even stop helium burning in the core after hydrogen burning is completed. Thus, any single star with a mass below the Chandrasekhar limit becomes a stable object; no further evolution occurs. Such "exhausted" stars are called white dwarfs (chap. 2). They eventually cool and die as black dwarfs.

If stars with masses above the Chandrasekhar limit shed enough mass, they may also get below this limit and become white dwarfs. They may be associated with planetary nebulae, such as the beautiful ring nebula in the constellation Lyra (Fig. 1.14), which are thought to be the result of such mass losses. A mass-loss rate $dM/dt \geq 10^{-5} M_{\odot} y^{-1}$ is required in this case, a rate much bigger than normal red giant winds. Observation of white dwarfs in some clusters of stars of known age indicates that for stars with initial masses less than approximately 8–10 $M_{\odot}$ electron degeneracy pressure becomes dominant prior to the onset of carbon burning and these stars are likely to become carbon-oxygen white dwarfs. It has been suggested that in such stars electron degeneracy could possibly result in explosive carbon burning, leading to carbon-detonation supernovae (chap. 2 and Arn70, Arn73). However, such supernovae have not yet been observed.

If the mass of a star is greater than about 8–10 $M_{\odot}$, the contracting carbon-oxygen core remains in a nondegenerate state and, for temperatures of about 5×10^{8} K and densities of about 3×10^{6} g cm^{-3}, carbon begins to burn. The burning generates sufficient energy to offset losses (primarily by neutrinos), and contraction halts. Thus the core can be stabilized, and quies-

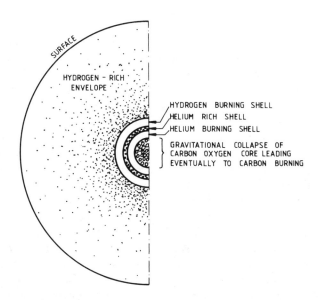

FIGURE 8.1. Schematic showing the cross-sectional view of a massive star ($M \geq 8 M_{\odot}$) after the helium fuel in the core is depleted. At this stage, helium and hydrogen burning continue in concentric shells around the contracting carbon-oxygen core. Contraction of the core eventually triggers the burning of carbon, and the gravitational collapse stops.

cent carbon burning takes place at a relatively constant temperature. The critical stellar mass separating these two different paths of stellar evolution is determined by the rate of the $^{12}C + ^{12}C$ reaction at energies as low as $E_{c.m.} = 1$ MeV. The rate of this reaction is, therefore, one of the basic quantities needed to understand the nature of stellar evolution and nucleosynthesis. Note that, at this burning stage, neutrino losses directly from the star's core have replaced electromagnetic radiation loss from its surface as the dominant energy-loss mechanism. This causes the core to evolve so rapidly that the envelope and thus the outward appearance of such a massive star (red supergiant) do not have time to change significantly before the rest of the core's nuclear fuel has been burned and the entire star is destroyed in a supernova explosion.

The carbon-burning phase results in the synthesis of neon, sodium, and magnesium. Smaller amounts of aluminum and silicon are also produced in this phase by the capture of α-particles, protons, and neutrons released in the carbon-burning process (Arn69). Near the end of the carbon-burning phase some reactions such as $^{12}C + ^{16}O$ and $^{12}C + ^{20}Ne$ may occur as well. These reactions are not expected to play a major role because of their greatly increased Coulomb barriers relative to that of the $^{12}C + ^{12}C$ reaction.

When the carbon fuel is exhausted, the core experiences a rise in temperature and density until photons from the high-energy tail of the Planck radiation distribution begin to photodisintegrate ^{20}Ne. Because of the relatively low α-particle separation energy of 4.73 MeV in ^{20}Ne, this process can occur at moderate temperatures (sec. 8.2). The resulting α-particles react with the undissociated ^{20}Ne to build ^{24}Mg. At the conclusion of this relatively brief neon-burning episode ($T_9 \simeq 1$), the temperature and density in the core, now consisting primarily of ^{16}O with some ^{24}Mg, again rise until the oxygen reacts with itself, giving rise to the oxygen-burning phase ($T_9 \simeq 2$). In this phase, mainly silicon and sulfur are synthesized. Nuclei such as argon and calcium are also produced, as well as smaller amounts of chlorine, potassium, and other nuclei up to the neighborhood of scandium in a complicated network of nuclear reactions (Woo72, Woo78). This phase is followed by silicon burning at still higher temperatures ($T_9 \simeq 3$). Since the temperature and density are rising throughout this sequence of processes, the time scale of each successive process becomes increasingly shorter. The phases may have no distinct separation; rather, one may merge immediately into the next.

The late phases in the evolution of the core of massive stars are characterized by a complex interaction of neutrino losses, degeneracy pressure, and nuclear reactions, whereby the core loses its battle against gravitational collapse at an ever more rapid rate. Above a density of about 10^6 g cm^{-3} the electrons of the star's core become degenerate and resist further compression. A core of less than the Chandrasekhar mass can cool by neutrino emission and temporarily lie dormant by resting on its degeneracy pressure following the exhaustion of its own current nuclear fuel. However, this can occur only if a nuclear-burning shell remains active somewhere within the inner Chandra-sekhar mass to maintain the thermal pressure support of the overlying regions. Under these circumstances continued nuclear burning gradually increases the mass of the fuel-exhausted core and typically leads to an inverted temperature structure, in which the outlying nuclear-burning shell is substantially hotter than the degenerate core. When the current nuclear fuel has been exhausted in the entire inner Chandrasekhar mass, gravity overwhelms the core's residual degeneracy pressure, and the core contracts and heats until the next fuel is ignited.

From the above discussions it follows that the heavy-ion reactions $^{12}C + ^{12}C$, $^{12}C + ^{16}O$, $^{12}C + ^{20}Ne$, and $^{16}O + ^{16}O$ play an important role in

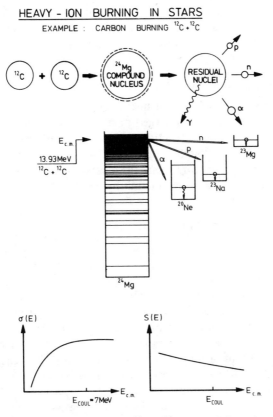

FIGURE 8.2. The heavy-ion reactions such as $^{12}C + ^{12}C$ proceed through compound nuclei at high excitation energies, where the level density is high and the states overlap appreciably. As a result the cross section and the S-factor should vary smoothly with energy. The reaction can be studied either by detecting the light particles evaporated from the compound nucleus or the γ-rays emitted from the residual nuclei. Direct detection of the recoil nuclei is difficult and has to date not been performed at low interaction energies (chap. 5).

the behavior of highly developed stars. The properties of these reactions govern the way stars evolve, and produce the many heavier elements observed in the universe. They also play a role in the explosion of massive stars, and in determining whether remnants are black holes, neutron stars, or white dwarfs (chap. 2).

To extrapolate the heavy-ion reaction yields obtained at higher beam energies down to the relevant stellar energies, the reaction mechanisms involved must be very well understood. For a well-behaved heavy-ion interaction, the dominant feature of the cross sections should be given by Coulomb barrier effects, while the behavior of the exit channels should be given primarily by the statistical averages of an evaporation model (Fig. 8.2). The S-factor is expected, therefore, to be rather featureless and to vary in a systematic way as the heavy-ion species is changed. However, studies of the heavy-ion reactions involved in astrophysics, at energies near and far below the Coulomb barrier, have revealed richer and more puzzling structures, triggering a great deal of experimental and theoretical research in the areas of nuclear structure and heavy-ion nuclear physics (Bar85).

8.1.1. Absorption under the Barrier

The fusion cross sections of the four heavy-ion reactions of astrophysical interest have been studied by either charged-particle or γ-ray spectroscopy (Fig. 8.2 and chap. 5). The $^{12}C + {}^{12}C$ reaction was investigated over a wide range of energies (Alm60, Alm64, Pat69, Maz73, Spi74, Ket77, Ket80, Bec81, Hig77, Erb80). The resulting S-factors from the charged-particle measurements are shown in Figure 8.3a. Note that in this figure the S-factor is plotted in the modified form (Pat69)

$$\tilde{S}(E) = S(E) \exp\left(-gE\right),$$

where $S(E)$ is the usual S-factor (chap. 4). The value of the constant g has been taken as 0.46 MeV^{-1} (Pat69). The fusion cross sections for the reactions $^{12}C + {}^{16}O$, $^{12}C + {}^{20}Ne$, and $^{16}O + {}^{16}O$ also have been determined over wide ranges of energies (Pat71, Spi74, Cuj76, Swi76, Chr77, Wu78, Hul80). The ordinary S-factors for these reactions are illustrated in Figure 8.4.

The $^{12}C + {}^{12}C$ reaction reveals pronounced resonance structure in the subbarrier region, as does to a lesser extent the $^{12}C + {}^{16}O$ reaction. However, such structure is absent in the $^{12}C + {}^{20}Ne$ and $^{16}O + {}^{16}O$ reactions. In addition to the observed narrow resonances ($\Gamma \simeq 100$ keV), the $\tilde{S}$-factor for $^{12}C + {}^{12}C$ shows an anomalously steep rise with decreasing beam energy at $E_{c.m.} \leq 3.5$ MeV (Fig. 8.3a), in contrast to the relatively smooth behavior in the other reactions. The theoretical interpretation of this steep rise at energies far below the Coulomb barrier has led to two descriptive models that differ in their predictions of the extrapolated stellar rates by more than 2 orders of magnitude. The essential difference between these models relates to the question of whether the fusion of two nuclei can be initiated far out in the tenuous tail of the nuclear-mass distribution (i.e., several times the radii of the nuclei) in what might be called the "nuclear stratosphere." In the standard model of Coulomb barrier penetration (Fow75), extensive use is made of the conventional optical model to fit the gross features of the available data (i.e., averages over the observed resonance structures) by varying the parameters specifying the real and imaginary potentials. Since the steep rise in the $\tilde{S}$-factor at low

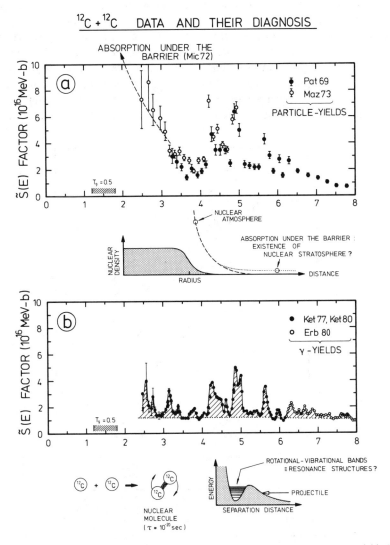

FIGURE 8.3. Fusion rates of $^{12}C + {}^{12}C$ as derived from particle yields (a) and γ-ray yields (b) are shown in the form of the modified $\tilde{S}(E)$ factor. The abscissae are $E_{c.m.}$ in units of MeV. Also shown are features used in the diagnosis of the observed data.

energies cannot be explained with this model, it has been interpreted as the high-energy wing of a broad resonance that may again decrease as a function of energy around stellar energies.

From a detailed study of the optical-model potentials, Michaud and Vogt (Mic69, Mic73) focused attention on an interesting new physical concept known as "absorption under the barrier." If the absorptive imaginary part of the optical potential has a surface thickness large enough (i.e., if the nuclear density extends to large radii), the nuclear fusion process at energies far below the Coulomb barrier will take place at large internuclear separations before a significant penetration of the Coulomb barrier occurs. In this case the reaction cross section is enhanced, since the effect of the Coulomb barrier is partially compensated for by shifting the maximum of the integrand in the integral for the reaction to larger distances. Consequently, a significant rise in the S-factor at very low energies can be expected. Since the shape of the imaginary potential far out in the tail is not known and some reaction rates at stellar energies

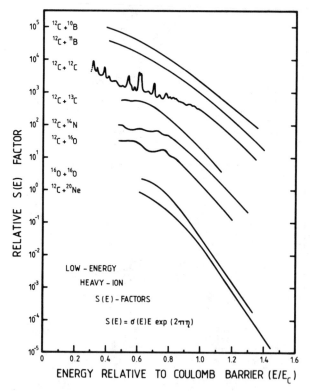

FIGURE 8.4. Total reaction cross-section $S(E)$ factors for several heavy-ion nuclear reactions at sub-Coulomb energies are shown in relative units (Sto76) as a fraction of Coulomb barrier height. The large differences observed in the shape of such curves are still not understood. Note that for the $^{16}O + ^{16}O$ system there is also some disagreement in the absolute cross section (Bar82).

may very well be dominated by the tail of this imaginary part, it is argued that the standard optical model may in such cases lose its utility in extrapolating high-energy data down to stellar energies. The steep rise in the $\tilde{S}$-factor for $^{12}C + {}^{12}C$ fusion at low energies has been cited by Michaud (Mic73) as evidence for this new phenomenon of absorption under the barrier. In this case, the rise in the $\tilde{S}$-factor will continue, resulting in a very different extrapolated rate than that obtained from the standard model (Fow75). Because of the widely differing results from extrapolation using the two models, and since experimental studies of reactions of ions in the mass range $A = 9\text{--}16$ provided no additional evidence for this phenomenon (Fig. 8.4), a more precise remeasurement of the $^{12}C + {}^{12}C$ reaction rate far below the Coulomb barrier was needed.

This measurement was done via γ-ray spectroscopy (Ket77, Hig77, Ket80). The ever-present hydrogen contamination in the targets seriously hampered the measurements at low energies, owing to an intense γ-ray background created by interactions of the ^{12}C beam with the hydrogen and deuterium contaminating nuclei. Use of targets with low hydrogen content was, therefore, crucial for this experiment. For reliable results at low beam energies, extreme care must be taken in determining the target thickness and the exact absolute beam energy (chap. 5). The resulting yields are shown in Fig. 8.3b. The steep increase observed earlier for the rates at low energies has not been confirmed, and the differences in the data could largely be removed if the energy of the ^{12}C beam in the previous data (Maz73) were actually higher by 100 keV than it was thought to be. The new data also reveal that the pronounced resonance structure continues down to the lowest energies (Fig. 8.3b). These results obviate the need for absorption under the barrier down to at least 2.4 MeV and suggest the use of the standard model for extrapolations. The absolute cross sections of the $^{12}C + {}^{12}C$ fusion reactions (Pat69) have been confirmed by subsequent measurements (Bec81).

8.1.2. Intermediate Structure in the Continuum

Since the structure in the $^{12}C + {}^{12}C$ cross sections can continue into the stellar energy region and since such pronounced structures can also exist at stellar energies for the other reactions, a significant uncertainty remains in the extrapolation procedure for this reaction. This uncertainty has stimulated experimental studies of other systems in the $A = 9\text{--}16$ mass range (Voi72, Han74, Day76, Swi77) to try to improve the extrapolation through a better understanding of the nature of these structures. However, in the course of this work (Fig. 8.4), such structures were not found in the energy regions studied. The physical explanation of the presence or absence of such structures remains as one of the long-standing questions in heavy-ion physics. Many speculations and contrary opinions (Fes76, Cin78, Bro78a) exist.

The narrow width of the structures (Fig. 8.3) does not allow their identification as compound nuclear resonances, and an explanation in terms of statistical fluctuations fails because of the strong correlations observed among all exit channels. It was realized, therefore, that some new type of intermediate structure was present in the continuum. The earliest data were interpreted as resonances in a ^{12}C + ^{12}C single-particle shallow potential well and were termed "quasimolecular" resonances (Vog60, Vog64, Dav60, Ima68). These molecular states are considered (Cin78) to be an intermediate step (doorway) in the reaction mechanism lying between entrance channel and compound nucleus formation, which may decay either into the compound nucleus or back into the entrance channel. The resonance-like creation of the doorway state then results in cross-section resonances for the particles evaporated from the compound nucleus. Thus, the resonances appear as "entrance-channel" effects. This hypothesis is supported by experimental data, since other heavy-ion reactions leading to the same compound nucleus do not show evidence for intermediate structure. In the model of Imanishi (Ima68) the incident channel is coupled to a doorway state where one of the carbon nuclei is in its first excited state. Although this model and its extensions generate resonance structures that can be linked with observation at low energies, they do not provide an explanation for the absence of structure in other systems. It has also been suggested that the density of levels of the compound nucleus as well as the number of open channels may enter in an important way into the presence or absence of structures in a given system, since both quantities are small for ^{12}C + ^{12}C and have increased significantly for ^{16}O + ^{16}O. However, there are other systems, such as ^{9}Be + ^{12}C and ^{12}C + ^{13}C, for which the density of levels and the number of open channels are both small, and neither of these systems exhibits resonance structures. The above considerations are, therefore, inconclusive. As a possible solution to this problem, Michaud and Vogt suggested a mechanism in which the resonances correspond to intermediate α-cluster states. This model considers the excitation of one carbon nucleus to its $J^\pi = 0^+$, 7.65 MeV state, which has a good three–α-particle structure (chap. 7); this doorway state is trapped in the effective potential of both colliding nuclei. This model requires the presence of resonance structures in the ^{12}C + ^{20}Ne heavy-ion reaction (Mic69, Mic73). However, such structures have not been observed (Fig. 8.4).

Although some of these models have successfully provided partial explanation of the experimental features, none is able to give an entirely satisfactory answer. Obviously, a great deal of experimental and theoretical work remains to be done.

8.1.3. Gross Energy Dependence

Cross-section measurements with ions in the mass region $A = 9$–16 do not reveal any systematic trends at subbarrier energies (Fig. 8.4). The gross energy dependence varies markedly from one system to the next, even though differences of only one or two nucleons are involved. This observation sharply contrasts with the situation usually encountered at energies above the barrier and in systems of heavier ions. The observed features for different ion pairs at subbarrier energies cannot be reproduced with a consistent set of optical-model parameters, suggesting that nuclear structure interactions on a microscopic level may determine the reaction mechanism between heavy ions in the $A = 9$–16 mass region. These features may also reflect interesting variations in the ion-ion interaction potential. All these data provide a rich source of information on the macroscopic and microscopic physics involved in heavy-ion reactions. The explanation of these data is an important and interesting subject for further experimental and theoretical work.

Once the mechanisms involved in these features, as well as in the intermediate resonance structures, are completely clarified, we may also understand how two deformed nuclei interact (Won73, Sto78) when they are in close contact and under what conditions nuclear-molecule systems actually exist. For example, some studies (Arn76) indicate that the intrinsic deformation of ^{12}C nuclei and their relative orientation during the collision $^{12}C + ^{12}C$ may play a critical role in determining low-energy fusion cross sections (see also Hul80).

8.2. Silicon Burning

Following oxygen burning, the remaining core consists mainly of ^{28}Si. As the core subsequently contracts and the temperature and density again increase, one might expect, by analogy to the previously discussed heavy-ion–burning phases, that ^{28}Si would begin to react with itself, resulting in a silicon-burning phase with the accompanying synthesis of elements up into the iron region. In fact, direct Si burning ($^{28}Si + ^{28}Si$) does not occur in stars because of the extremely high stellar temperatures, on the order of $T_9 > 3.5$, required to provide sufficient penetration of the Coulomb barrier. Before these high temperatures are reached, the silicon-destroying photodisintegration process sets in. Since the number of photons increases as the fourth power of temperature, the ^{28}Si nuclei are essentially swimming in an intense sea of high-energy γ-rays. At this stage and beyond in the evolutionary process, photodisintegration plays a dominant role in nucleosynthesis. Here, α-particles, protons, and neutrons are ejected by photodisintegration from the most abundant nuclei present and are made available for building nuclei up to and slightly beyond the iron peak in the abundance curve (Fig. 8.7), a process called silicon burning.

Beyond the iron peak, more energy is consumed by the photodisintegration process than is gained by adding nucleons to nuclei, since at iron the binding energy per nucleon is maximum. For this reason and because of the very large Coulomb barriers encountered as the nuclear charge increases, charged-particle–induced nucleosynthesis must, for the most part, terminate with silicon burning, and elements heavier than iron must be produced mainly by neutron-capture reactions (chap. 9).

8.2.1. The Photodisintegration Era

The electromagnetic force is involved in all phases of stellar evolution. In the phases involving thermonuclear reactions of charged particles, the Coulomb repulsion, which can be thought of as resulting from the exchange of virtual photons, determines the reaction rates and also gives rise to the distinctly separate burning phases. Additionally, by providing radiative energy transport, a major mechanism by which energy is carried from the stellar interior to the surface of the star, the electromagnetic force in the form of real photons makes possible the hydrostatic stability required for quiescent burning. As a star evolves and its temperature continues to rise, the number of photons capable of causing photodisintegration becomes greater and greater and photons begin to play a direct role in stellar evolution and in the nucleosynthesis of elements.

In chapter 3 the mean binding energy per nucleon was shown to be about 8 MeV. Since the photoejection of two nucleons requires approximately twice the photon energy required for the ejection of one, it does not occur in stars. Therefore, the first photodisintegration processes involve the ejection of a single nucleon, provided that the photon energy is above 8 MeV. The exception to this general rule is the α-particle–producing process, which, because of the high binding energy released in the formation of the α-particle, can occur at approximately the same photon energy as the single-particle reactions. Thus the major reactions involved in stellar photodisintegration are the (γ, p), (γ, n), and (γ, α) reactions.

Direct measurement of the cross section for these photodisintegration processes as a function of photon energy, that is, measurement of excitation functions, is difficult because of the lack of intense and monochromatic γ-ray beams of variable energy. A tunable γ-ray laser would be perfect for these measurements. However, the photodisintegration cross sections for the reaction $Y(\gamma, a)X$ to the ground state of X can be determined from measurement of the inverse capture reaction $X(a, \gamma)Y$, using the principle of detailed balance (chap. 3). The corresponding reaction rates $\langle \sigma v \rangle$, and thus the decay rates λ and lifetimes $\tau = 1/\lambda$, are determined through the relationship (chap. 3)

$$\frac{\lambda(\gamma, a)}{\lambda(a, \gamma)} = \frac{1}{N_a} \frac{(2\pi\mu kT)^{3/2}}{h^3} \frac{(2j_a + 1)(2j_x + 1)}{(2j_y + 1)} \exp\left(-\frac{Q}{kT}\right). \tag{8.1}$$

Note that this ratio is independent of detailed nuclear properties, except for the spins of the nuclei involved and through the Q-value, which is a measure of the mass differences. Since in this equation the Q-value and the stellar temperature T appear in the exponent, the ratio of the rates is very sensitive to these quantities.

To illustrate this sensitivity as well as the lifetime for photodisintegration, consider the capture reaction $^{12}C(p, \gamma)^{13}N$ and the inverse reaction $^{13}N(\gamma, p)^{12}C$. In this case the photodisintegration process is induced on a radioactive nucleus of mean lifetime $\tau_\beta(^{13}\text{N}) = 840$ s, and therefore the photo-disintegration cross section would be extremely difficult to measure even with the hypothetical γ-ray laser. Using the known nuclear properties of the nuclei involved ($Q = 1.94$ MeV, $j_a = \frac{1}{2}$, $j_x = 0$, $j_y = \frac{1}{2}$) and using stellar conditions $\rho = 10^6$ g cm^{-3} for the density of a stellar plasma composed predominantly of hydrogen nuclei ($X_\text{H} = 1$), the ratio of decay rates is given by the equation

$$\frac{\lambda(\gamma, p)}{\lambda(p, \gamma)} = 8.84 \times 10^3 T_9^{3/2} \exp\left(-\frac{22.55}{T_9}\right).$$

The evaluation of this ratio as well as the associated photodisintegration lifetime are shown for several temperatures in Table 8.2. Also shown in Table 8.2 are the results assuming only that the Q-value is at a more normal value, namely, 8.0 MeV. As indicated, the calculations verify the extreme sensitivity of the ratio of decay constants and the photodisintegration lifetime to the Q-value of the reaction as well as to the temperature. On the basis of Table 8.2, one can see that for normal Q-values the photodisintegration process becomes important only at temperatures above $T_9 = 2$.

The strong Q-value dependence of the photodisintegration process has an interesting and important consequence with regard to the relative abundance of neighboring even-even and even-odd nuclei. For example, in the case of ^{12}C the Q-value for the photoejection of a proton or a neutron is 16.0 or 18.7 MeV, while for ^{13}C and ^{13}N the Q-values for neutron and proton photodisintegration are 4.9 and 1.9 MeV, respectively. Clearly, very early in the photodisintegration episode of stars, the ^{13}C and ^{13}N nuclei with their lower nuclear stabilities are converted into the more stable even-even nuclei ^{12}C, thus increasing the relative abundance of even-even nuclei over that of even-

TABLE 8.2 Features of the $^{13}\text{N}(\gamma, p)^{12}\text{C}$ and $^{12}\text{C}(p, \gamma)^{13}\text{N}$ Reactions for Two Reaction Q-Values

| | $Q = 1.94$ MeV | | $Q = 8.00$ MeV | |
| | Ratio of Decay Rates $\dfrac{\lambda(\gamma, p)}{\lambda(p, \gamma)}$ | Lifetime $\tau(\gamma, p)^a$ (s) | Ratio of Decay Rates $\dfrac{\lambda(\gamma, p)}{\lambda(p, \gamma)}$ | Lifetime $\tau(\gamma, p)^a$ (s) |
T_9				
0.3	3.0×10^{-30}	1.7×10^{24}	6.3×10^{-132}	8.9×10^{125}
0.5	8.1×10^{-17}	6.5×10^8	7.8×10^{-78}	6.7×10^{69}
0.7	5.3×10^{-11}	1.0×10^2	1.4×10^{-54}	3.8×10^{45}
1.0	1.4×10^{-6}	8.2×10^{-4}	4.3×10^{-37}	2.7×10^{27}
2.0	3.2×10^{-1}	8.2×10^{-11}	1.8×10^{-16}	1.5×10^5
3.0	2.5×10^1	9.1×10^{-12}	1.7×10^{-9}	1.3×10^{-1}
5.0	1.1×10^3	2.2×10^{-13}	8.6×10^{-4}	2.8×10^{-7}

^{a} The photodisintegration process will take place in stars if $\tau(\gamma, p) \leq \tau_\beta(^{13}\text{N}) = 840$ s. The lifetime is calculated using the reaction rate of $^{12}\text{C}(p, \gamma)^{13}\text{N}$ as quoted by Fow75.

PHOTODISINTEGRATION   Y (γ, a) X

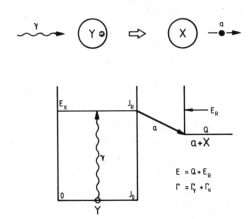

FIGURE 8.5. Shown schematically is the photodisintegration process Y(γ, a)X proceeding via a resonant state at E_x. The partial widths Γ_γ and Γ_a of the state describe the formation and decay via the γ and particle channel, respectively, with the total width given by $\Gamma = \Gamma_\gamma + \Gamma_a$. This diagram depicts photodisintegration from the ground state. Photodisintegration from excited states is also possible and may be important at high temperatures.

odd nuclei. In other words, the net effect of photodisintegration in stars is to reduce nuclei to their most stable form.

So far we have considered the ratio of rates for inverse reactions. In the case of photodisintegration through a resonance state (Fig. 8.5), one can arrive directly at an expression for the photodisintegration decay rate $\lambda(\gamma, a)$. As in any nuclear system, the population of an excited state E_x relative to the ground state of the nucleus is described by the Boltzmann probability function:

$$P(E_x) = \exp\left(-\frac{E_x}{kT}\right) \frac{2J_R + 1}{2J_0 + 1},$$

where J_0 and J_R are the spins of the ground state and excited resonant state, respectively. It is assumed here that photodisintegration occurs only from the ground state. For an unbound state, this expression has to be modified by the factor Γ_γ/Γ to take into account the fact that a fraction of the nuclei in this excited state can decay by the emission of particle a (Fig. 8.5):

$$P(E_x) = \exp\left(-\frac{E_x}{kT}\right) \frac{2J_R + 1}{2J_0 + 1} \frac{\Gamma_\gamma}{\Gamma}.$$

The probability for photodisintegration is then simply given by the product of the decay probability of the excited state, which is $\lambda_a = 1/\tau_a = \Gamma_a/\hbar$, times the probability of finding a nucleus in this resonant state:

$$\lambda(\gamma, a) = \exp\left(-\frac{E_x}{kT}\right) \frac{2J_R + 1}{2J_0 + 1} \frac{\Gamma_\gamma}{\Gamma} \frac{\Gamma_a}{\hbar} .$$

Since $E_x = Q + E_R$ (Fig. 8.5), one arrives at (see also Bar71)

$$\lambda(\gamma, a) = \frac{\exp(-Q/kT)}{\hbar(2J_0 + 1)} (2J_R + 1)\left(\frac{\Gamma_a \Gamma_\gamma}{\Gamma}\right) \exp\left(-\frac{E_R}{kT}\right) .$$

When more than one resonance is involved in the photodisintegration, their effect must be taken into account by summing over all the participating resonances:

$$\lambda(\gamma, a) = \frac{\exp(-Q/kT)}{\hbar(2J_0 + 1)} \sum_i (2J_{R_i} + 1)\left(\frac{\Gamma_a \Gamma_\gamma}{\Gamma}\right)_i \exp\left(-\frac{E_{R_i}}{kT}\right) . \qquad (8.2)$$

This expression can also be arrived at from equation (8.1) simply by inserting for $\lambda(a, \gamma)$ its value for narrow resonances (chap. 4). In this equation, the energy of the resonance state E_R and the temperature enter exponentially. Thus the photodisintegration lifetime is highly sensitive to these parameters, as in the case discussed previously and shown in Table 8.2.

When there are a number of resonances, the question arises as to which one to use in equation (8.2). Owing to the exponential nature of the Boltzmann factor, neutron photodisintegration will involve mainly resonances located near the threshold. In the case of charged-particle photodisintegration, the ejected proton or α-particle also must penetrate through the Coulomb barrier of the a + X reaction channel. As in the inverse capture reaction (chap. 4), the combined effects of the Boltzmann factor and the Coulomb barrier penetration lead to the result that the most effective resonances will be those lying energetically in the neighborhood of the Gamow energy window, E_0; that is,

$$E_R \simeq E_0 = 0.122(Z_1^2 Z_2^2 \mu T_9^2)^{1/3} \text{ MeV} ,$$

and the photodisintegration rate becomes

$$\lambda(\gamma, a) \propto \exp\left[-(Q + E_0)/kT\right] .$$

For example, consider the photodisintegration of ^{32}S nuclei at a stellar temperature $T_9 = 2.5$. From the Q-values alone (Table 8.3) one would expect the magnitude of the relative reaction rates to be related as

$$\lambda_{\gamma a} \gg \lambda_{\gamma p} \gg \lambda_{\gamma n} .$$

When the effect of the Gamow energy is taken into account (Table 8.3), the expected relationships are

$$\lambda_{\gamma a} \simeq \lambda_{\gamma p} \gg \lambda_{\gamma n} .$$

The sum in equation (8.2) (i.e., sum over the resonances in the Gamow window) is determined from experimental data (Fow75, Woo78, End78) and leads to the lifetimes for photodisintegration given in Table 8.3. The results

TABLE 8.3 Photodisintegration of ^{32}S near $T_9 = 2.5$

Reaction	Q (MeV)	E_0 (MeV)	$Q + E_0$ (MeV)	$\tau(\gamma, a)$ (s)
^{32}S$(\gamma, \alpha)^{28}$Si	6.948	3.146	10.094	$\simeq 10^4$
^{32}S$(\gamma, p)^{31}$P	8.864	1.352	10.216	$\simeq 10^3$
^{32}S$(\gamma, n)^{31}$S	15.092	0	15.092	$\simeq 10^{12}$

confirm the above expectation and show that ^{32}S nuclei, which are among the ashes of oxygen burning, are destroyed near $T_9 = 2.5$ predominantly through the (γ, p) channel, leading to the production of ^{31}P nuclei. Since the proton-binding energy of ^{31}P is only 7.29 MeV, these nuclei are destroyed immediately by photoinduced proton ejection to form ^{30}Si nuclei. The latter nuclei themselves are mainly converted into ^{29}Si nuclei by the (γ, n) process ($Q_{\gamma, n} = 10.62$ MeV, $Q_{\gamma, p} = 13.51$ MeV, $Q_{\gamma, \alpha} = 10.65$ MeV), which nuclei, again by (γ, n) photodisintegration, are made into ^{28}Si nuclei ($Q_{\gamma, n} = 8.48$ MeV, $Q_{\gamma, p} = 12.33$ MeV, $Q_{\gamma, \alpha} = 11.13$ MeV). The ^{28}Si nuclei are not usually destroyed by photodisintegration ($Q_{\gamma, \alpha} = 9.98$ MeV, $Q_{\gamma, p} = 11.58$ MeV, $Q_{\gamma, n} = 17.18$ MeV). From the Q-values and the effects of the Coulomb barrier, one expects the relationship

$$\lambda_{\gamma p} > \lambda_{\gamma \alpha} \gg \lambda_{\gamma n} .$$

The corresponding lifetimes, calculated on the basis of experimental data (Fow75) for the inverse capture reactions ^{27}Al$(p, \gamma)^{28}$Si and ^{24}Mg$(\alpha, \gamma)^{28}$Si, are shown in Figure 8.6 as a function of stellar temperature. From these results it is clear that less stable nuclei such as ^{32}S are reduced at $T_9 > 1$ to the more stable form of ^{28}Si and that the more stable ^{28}Si nuclei will not photodisintegrate until the stellar core has completely contracted and the temperature increased to $T_9 = 3-4$. This is the start of the silicon-burning phase.

In the above discussion we have neglected equilibrium between photodisintegration and the inverse process, capture reactions. If equilibrium is taken into account (Woo73), the abundances of elements from magnesium to nickel produced in silicon burning are found to depend solely on reaction Q-values, stellar temperature, and density but are independent of nuclear reaction rates (sec. 8.2.2).

8.2.2. Photodisintegration in Silicon Burning

At the end of oxygen burning ($T_9 \simeq 2$) the stellar core contracts gravitationally and the temperature and density increase continuously. At temperatures $T_9 > 3$ photodisintegration starts to destroy the complex nuclei already present from previous burning phases, releasing α-particles, protons, and neutrons. At relatively low temperatures the nuclei with the smallest binding energies will be destroyed first. The released light particles can then combine with undissociated nuclei to build heavier nuclei. For those nuclear reactions that occur

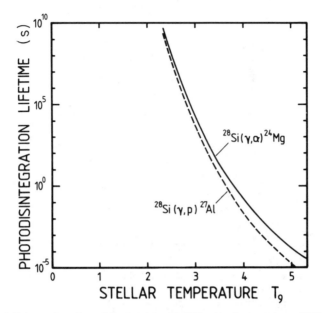

FIGURE 8.6. Lifetimes for photodisintegration of ^{28}Si via the reactions ^{28}Si$(\gamma, p)^{27}$Al and ^{28}Si$(\gamma, \alpha)^{24}$Mg are shown. Note the high sensitivity of the lifetimes to stellar temperature.

most rapidly, a quasi-equilibrium (Hoy46, Fow64) exists between competing nuclear-capture and photodisintegration reactions. Many of the resulting light particles will be captured by nuclei, forming new nuclei in which they have a higher nuclear binding energy than those from which they were photoejected. The result, then, is that loosely bound nuclei are transformed into nuclei of higher stability. This is a type of rearrangement process of stellar material into nuclei with the highest nuclear stability, which, as we know, is in the mass region of iron. The observed abundance peak near iron reflects this property and is due mainly to the rearrangement of nuclear matter by photo-disintegration (Fig. 8.7).

The ashes of prior burning phases are transformed early in the photo-disintegration process into the more stable nuclei ^{28}Si, which themselves are photodisintegrated at much higher temperatures. Since not all ^{28}Si nuclei are destroyed, the light particles are captured by the undissociated ^{28}Si nuclei, thereby forming heavier nuclei:

$$^{28}\text{Si} + \alpha \rightleftarrows {}^{32}\text{S} + \gamma ,$$

$$^{32}\text{S} + \alpha \rightleftarrows {}^{36}\text{Ar} + \gamma ,$$

$$\cdot$$
$$\cdot$$
$$\cdot$$

$$^{52}\text{Fe} + \alpha \rightleftarrows {}^{56}\text{Ni} + \gamma .$$

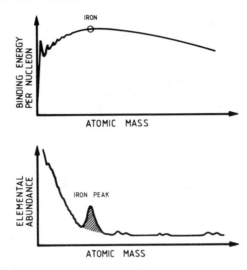

FIGURE 8.7. The observed iron peak in the elemental abundance distribution is to a large extent the result of silicon burning, reflecting the maximum binding energy per nucleon in this mass region. The steep drop in abundances going from hydrogen to iron reflects the charged particle's increasing difficulty in overcoming the higher Coulomb barriers. Elements above iron are synthesized mainly by neutron-capture reactions, which are independent of the Coulomb barrier and lead to a relatively flat distribution of elements above iron (chap. 9).

In addition to the capture reactions (α, γ), (p, γ), and (n, γ) and the inverse photodisintegration processes, reactions such as (α, n), (α, p), (p, n), and their inverse processes are also involved. In each pair of reactions the system tries to reach a state of equilibrium. The heavier nuclei can be made in this network of nuclear reactions because of their increasing nuclear stability. In this way, the ^{28}Si nuclei are slowly transformed in a system of changing equilibrium into the heavier nuclei, up to and including the nuclei of the abundance peak at iron. Because of the high stability of the ^{28}Si nuclei, these nuclei photodisintegrate very slowly (Fig. 8.6) compared with the time scales of the processes that build up the heavier nuclei. Consequently, all other processes are essentially in equilibrium with ^{28}Si, where the time scale is controlled by its photodisintegration rate. Because of this feature, there is an awesome array of nuclear reactions, numbering many hundreds, involved in this stellar phase, the total of which is called silicon burning. This phase is also referred to as silicon melting because of the slow rearrangement ("melting") of Si nuclei into heavier nuclei.

In the rearrangement of lighter nuclei into nuclei up to the iron region, nuclear energy is liberated. However, weak-interaction processes such as positron decay and electron capture also play an important role (i.e., electron capture of ^{56}Ni, $T_{1/2} + 6.1$ d). As a consequence, in silicon burning a high flux of neutrinos will be created that will escape the star, carrying away a significant fraction of the nuclear energy produced. As a result, a star in this stage of evolution exhibits a high neutrino luminosity. This neutrino luminosity provides a mechanism for cooling the stellar core, which therefore must burn its nuclear fuel faster in order to balance the gravitational pressure. The lifetime of a star in this stage of stellar evolution is rather short, and the probability of observing it in this stage is very small.

To the extent that a quasi-equilibrium between forward and backward reactions exists, the nucleosynthesis calculations of the elements produced in silicon burning do not require a detailed knowledge of nuclear reaction rates. The elemental abundances that would result in this case depend only on the number densities of neutrons, protons, and α-particles and on the masses and partition functions (statistical weights; chap. 4) of the nuclei at the prevailing temperatures. While this situation might be achieved for rapidly burning nuclear reactions, slower nuclear reactions may not achieve equilibrium because of the rapid changes in temperature and density occurring during the silicon-burning phase. At freeze-out of silicon burning, individual nuclear reaction rates (sec. 8.2.3) become especially important in determining the resulting nuclidic abundances. The growing capacity of modern computers now makes it possible to include huge networks of nuclear reactions as well as the evolutionary dynamics of the star in calculating expected nuclear abundances (Woo73, Wea78). Figure 7.11 shows the calculated nuclidic abundances resulting from presupernova nucleosynthesis, including silicon burning. These results reproduce fairly well the trend of the observed abundances, supporting the hypothesis that silicon burning is the dominant mechanism for element synthesis in the mass range $A = 28$–65.

8.2.3. The Nuclear Physics of Silicon Burning

Silicon burning involves many nuclear reactions, wherein α-particles, protons, and neutrons interact with all the nuclei in the mass range $A = 28$–65. In this mass range, the levels in the compound nucleus are dense and generally overlap. Furthermore, it seems that many of the reactions go so fast at the high temperatures involved ($T_9 = 3$–5) that a quasi-equilibrium between forward and backward reactions is established; thus a detailed knowledge of nuclear reaction rates is not necessary. Nuclear masses (Q-values), excitation energies, and spin and parity (J^π) values, which provide the statistical weights, are all that is required. However, as the temperature drops as a result of energy loss by neutrino radiation and reduction of available nuclear fuel due to its consumption, various nuclear reactions can no longer occur. Because of this freeze-out of some reactions, a detailed knowledge of their rates will be necessary to calculate the nuclidic abundances. These reactions frequently involve nuclei far from the valley of stability.

Because many hundreds of nuclear reactions may be involved in silicon burning, it is essential to develop a nuclear reaction model around which to describe and systematize laboratory reaction-rate information. Such a model is also essential because many of the nuclear reactions in silicon burning involve short-lived radioactive nuclei or nuclei thermally excited to very short-lived excited states. It is sometimes possible to bypass the problem of a radioactive target by inverting the nuclear reaction and then using the principle of time-reversal invariance. Of course, this technique will not work if both target and residual nucleus have short lifetimes (chap. 5), or if excited states are involved in the reaction processes.

Numerous experimental studies show that direct nuclear reactions are rarely significant in the range of nuclei and at the energies that are relevant for astrophysics, and that compound nucleus reactions dominate the processes. The Hauser-Feshbach statistical reaction model (Hau52, Vog68) is therefore usually adequate for a general description of these compound nucleus reactions.

The statistical theory of nuclear reactions (Bla62) is based on the compound nucleus picture of nuclear reactions. The closely spaced narrow resonances, which are characteristic of compound nucleus processes, are individually of interest only at low energies and for the lightest nuclear systems. At higher energies and for the heavier nuclei, the number of resonances increases drastically; the physical interest is in nuclear cross sections averaged over these resonances, which also is usually what is measured. The statistical theory of nuclear reactions deals with such energy-averaged cross sections.

To a large extent, the treatment of average cross sections employs an evaporation model with only a partial basis in nuclear reaction theory. The model depicts the nuclear reaction as proceeding in two stages. In the first stage a compound nucleus is formed by the collision of the projectile with the target nucleus. In the second stage, the compound nucleus decays into one of many possible pairs of reaction products. The evaporation model assumes that the compound nucleus, in its decay, loses all memory of the way in which it was formed. Thus, the decay treats all possible decay products in the democratic fashion envisaged in evaporation of a classical liquid drop. In this evaporation model a nuclear reaction of the type

$$A(x, y)B$$

is assumed to have an energy-averaged cross section $\bar{\sigma}(E)$, which may be factored as

$$\bar{\sigma}(E) = \sigma_{\text{form}}(x + A) \frac{\Gamma(y + B)}{\Gamma_{\text{tot}}}, \tag{8.3}$$

where the average of the cross section is over all the resonances involved. The first factor in the above equation, which depends only on the initial pair of particles in the entrance channel, is called a cross section for the formation of the compound nucleus. The second term represents a branching ratio that describes the probability of decay into the channel of interest, namely, $y + B$, compared with the total decay probability, which is the decay into all pairs of reaction products (exit channels) available to the compound nucleus. This branching ratio is proportional to the ratio of partial width $\Gamma(y + B)$ and total width Γ_{tot}. Clearly the total decay probability is a property of the compound nucleus only and does not depend on the particular choice of the entrance and exit channels. The factorization assumed in equation (8.3) is an expression of the independence of formation and decay of the compound nucleus.

The cross section for the formation of the compound nucleus is assumed to be a maximum; that is (chap. 4),

$$\sigma_{form}(x + A) = \sigma_{max} = \pi \lambda^2 \omega \ ,$$

where the quantity ω is the statistical factor and where for neutrons $\pi \lambda_n^2 = \pi R^2$ (R is the nuclear radius of the colliding pair). In the case of charged particles, the penetration through the Coulomb barrier (chap. 4) must be taken into account by multiplying by the penetration factor $P_l(E, R)$:

$$\sigma_{form}(x + A) = \pi \lambda^2 \omega P_l(E, R) \ .$$

For both neutrons and charged particles, it is assumed that particles reaching the nuclear domain are fully absorbed, i.e., that the reduced particle widths (chap. 4) are unity. This assumption is also applied for the exit channels through the reciprocity theorem. Hence all partial widths $\Gamma(y + B)$ as well as the total width $\Gamma_{tot} = \sum_i \Gamma_i$ can be replaced by their respective penetration factors. These factors, often called transmission functions, are given by $T_l(E, R)$. The basic Hauser-Feshbach expression for the energy-averaged cross section for any orbital angular momentum l is then

$$\bar{\sigma}(E) = \pi \lambda^2 \omega \frac{T_x(l) T_y(l)}{\sum_i T_i(l)} \ .$$

If J and π are the angular momentum and parity of the compound nucleus resonance states, the reaction can proceed through all such states, and one has to sum over these quantities:

$$\bar{\sigma}(E) = \frac{\pi \lambda^2}{(2j_p + 1)(2j_t + 1)} \sum_{J, \pi} (2J + 1) \frac{T_x(J, \pi) T_y(J, \pi)}{\sum_i T_i(J, \pi)} \ . \tag{8.4}$$

Here j_t and j_p are the spins of the target nucleus and projectile, and the quantities $T_x(J, \pi)$ and $T_y(J, \pi)$ are the transmission functions for formation and dissociation of the compound nucleus state (J, π) via the A + x entrance channel and the B + y exit channel, respectively.

The sum in the denominator of equation (8.4) includes all states in the residual nuclei that are energetically accessible at the interaction energy E of the entrance channel. Each transmission function contains an implicit sum over orbital angular momentum and channel spin, if applicable. When the accessible excitation energy in the residual nucleus is higher than the energy for which excitation energies, spins, and parities of the excited states are known, the sums will include integrals over the regions where the states are not known. This aspect requires the development of analytic expressions for

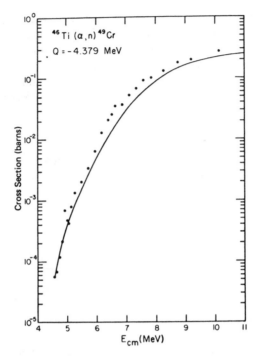

FIGURE 8.8. Comparison of Hauser-Feshbach cross-section calculations with measurements of the ^{46}Ti$(\alpha, n)^{49}$Cr reaction (How74). The energy dependence of this reaction cross section is mainly governed by the α-particle transmission functions, except very close to the neutron threshold.

the density of excited states as a function of E, J, and π. The other essential ingredients in the cross-section formula are the transmission functions, which are commonly obtained with the use of optical-model potentials.

Many authors have contributed to the development and improvement of statistical nuclear reaction models (Mic70, Tru66, Tru72, Man75, Hol76, Woo78). This theoretical effort is still going on, and it is possible to hope that a model will be produced that will predict reaction rates to better than a factor of 2. In any case, it is imperative to check, wherever possible, these theoretically predicted reaction rates—in part as an incentive for further improvements in theory. For this reason, there has been considerable experimental effort to measure absolute cross sections, as a function of bombarding energy (so-called excitation functions), for a wide range of silicon-burning nuclear reactions (How74, Rio75). That the Hauser-Feshbach model gives an acceptable description of excitation functions, even with global parameters in the potentials, was quickly verified. Figure 8.8 shows an example. Usually, such excitation functions are rather featureless, and they usually test the energy dependence of only a few of the partial widths, often only that of the entrance channel. Many of the silicon-burning reactions have not been experimentally measured and represent an open field for future activity in the nuclear laboratory.

Charged-particle–induced cross sections have been calculated for most silicon-burning reactions, with the reaction model as then developed (Woo78). These calculations predict cusplike behavior at neutron thresholds for many reactions. These cusps are seen as relatively sharp drops in the cross section for a given reaction as a second reaction channel opens up. If the incoming flux is constant, there is a reduction in the number of reactions of the first kind. This situation is analogous to the decrease in flow from a water faucet as another faucet is turned on.

The first of these competition cusps, as they are now called (Bar82), was found (Man75a) in the reaction ^{64}Ni$(p, \gamma)^{65}$Cu in the energy range where thresholds for several low angular momentum states appear in the competing reaction ^{64}Ni$(p, n)^{64}$Cu (Fig. 8.9). The observed competition cusp is about a factor of 3 smaller than predicted. It was immediately recognized that competition cusps would provide a more sensitive test of the nuclear reaction model than reactions without such cusps because all of the significant partial widths must be described correctly to reproduce the observed structure. A program to study such cusps was then undertaken in several laboratories. The resulting improvements on the statistical reaction model are continuing (Bar82).

The current global Hauser-Feshbach models already give acceptable predictions of reaction cross sections for nuclei near the valley of stability. Thiele-

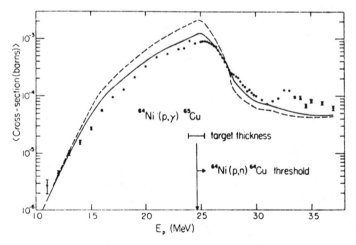

FIGURE 8.9. Competition cusp in ^{64}Ni$(p, \gamma)^{65}$Cu compared with the Hauser-Feshbach calculations. The solid (dashed) line is a calculation with (without) width fluctuation corrections (Man75a). Note that, as the successive neutron thresholds open, the (p, γ) yield shows a large drop to about one-tenth of the maximum value reached just below the first neutron threshold.

mann (Thi80) has suggested alternative potentials and level-density expressions, which are claimed to improve the capability of the theory to predict cross sections involving nuclei far from the valley of stable nuclei. At this time, there is no satisfactory way to test experimentally the extrapolation of nuclear reaction models far from the valley of stability, since these nuclei are instable (chap. 5).

8.3. The Final Bursts of Nucleosynthesis in Massive Stars

We have seen that the silicon core built up in the sequence of quiescent burning phases will eventually contract. The resulting increase in temperature will result in some of the silicon nuclei being broken down into α-particles. These α-particles, in turn, are captured by the rest of the silicon and other nuclei such as sulfur, argon, and calcium, resulting in the formation of iron and nickel. Because of prodigious neutrino losses the energy output of silicon burning is not very large and conversion of the material of the inner region of the star to iron and nickel is rapid. Iron and nickel, the most tightly bound of all nuclei, cannot participate in any reactions providing nuclear energy needed to support the hydrostatic pressure which stabilizes the iron-nickel core and thus prevents gravitational collapse. We have now arrived at a crucial moment in the history of a massive star. We will pause here and examine the existing inner structure of the star (the supernova progenitor) before going on to the inevitable gravitational collapse and the final catastrophic evolutionary event, a giant explosion called a supernova.

8.3.1. The Inner Structure of a Presupernova Star

By the time the iron-nickel core of a massive star forms and the core is just beginning to collapse, the star has gone through many burning stages, and the mantle of the star has evolved into a complex structure, an onion-shell–like composition. The core is surrounded by many distinct layers, each consisting primarily of a specific element. As shown in Figure 8.10, the layers, assuming a star considerably more massive than the sun, and starting at the core, consists of silicon, oxygen, neon, carbon, helium, and hydrogen. The envelope of the star consists of matter similar to that in our sun, i.e., mostly hydrogen. Since temperature and density increase as we go toward the center of the star (Fig. 8.10), the interior boundary of each of these layers is a nuclear-burning zone contributing newly formed heavier matter to the next interior layer. For example, between the hydrogen layer and the helium layer, there is a narrow zone where the temperature is high enough to fuse hydrogen nuclei into helium nuclei. The shells are typically convective but are separated by density gradients sufficiently large to prevent convective mixing between them. From the surface toward the core, such a presupernova structure recapitulates the various evolutionary stages of the star during its quiescent hydrostatic evolution. If these " onion " shells could be ejected into interstellar space, they would furnish about the observed abundances of the elements between carbon and iron. The question, however, is what the mechanism is by which the star blows these shells into space (sec. 8.3.2).

After helium burning, the most important way for energy to escape from the central core is through production of neutrinos. These unusual particles have no mass (or only very small mass) and no charge and react very weakly with matter. As a result, they escape from a star with relative ease, creating little interaction between the inner and outer regions. These neutrinos carry off energy, cooling the system, just as evaporation of moisture from our skin cools us.

Because the core of a massive star is quite dense, the electron-degeneracy effect remains important during each stage of evolution. This effect will prevent the core from contracting significantly at each burning phase until

INNER STRUCTURE OF A PRESUPERNOVA STAR

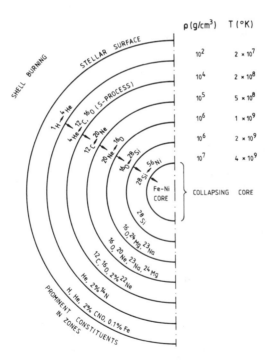

FIGURE 8.10. Schematic diagram of the inner structure and chemical composition of a massive star prior to its explosive disruption. The concentric "onion" shell structure indicated, where in each zone certain elements dominate, is highly idealized. The approximate pre-explosion temperature and density of each shell-burning zone are also given. The s-process (chap. 9) is most likely occurring in the helium-burning zone of second-generation stars. At the borders of the various zones, complex mixing processes might occur.

enough mass has accumulated to overcome the electron-gas pressure. The last stage of hydrostatic burning in a presupernova star is silicon burning. It occurs over a relatively long time scale (long compared to the subsequent collapse and bounce of the core) and at high density such that protons in the nuclei of the iron-peak elements produced in such burning can be transformed into neutrons by capturing electrons. This process of neutronization results in "iron" cores slightly exceeding the Chandrasekhar mass. The composition is dominated by very neutron-rich iron-peak species such as ^{48}Ca, ^{66}Ni, ^{54}Cr, and ^{50}Ti near the center of the core. From the center outward it gradually shifts to progressively less neutron-rich species dominated successively by ^{58}Fe, ^{56}Fe, and ^{54}Fe. Thus, for stars with $M \geq 25 \ M_\odot$ the final configuration prior to collapse would have an iron core with a mass greater than about $2 \ M_\odot$ surrounded by the various burning shells. Since the actual mass of the iron core depends critically on the ^{12}C$(\alpha, \gamma)^{16}$O rate (chap. 7), this rate is also important in determining which stars ultimately become black holes. The basic configuration just discussed would hold qualitatively for all stars with masses greater than about 8 $M_\odot$ and less than about 100 $M_\odot$. For even more massive stars, the evolution of the star takes a different course (Ibe63, Hoy65, Fow66, Bar71, Fri73, Fri80, Woo84).

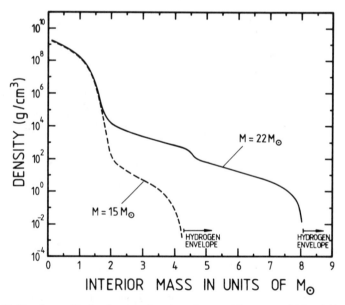

FIGURE 8.11. Density profile for the central region of stars with masses of 15 and 22 $M_\odot$. Note that the abscissa measures the distance from the center of the star in terms of the mass interior to a given layer. The ordinate shows the density at a given layer. Thus, since the interior layers consist of heavier elements and are more compressed, they can be seen to have much greater density.

It is instructive to look at the relative chemical composition in the shells of stars of different masses. Since stars of different masses have different temperatures, and hence different elemental profiles, the configuration of their layers will be different as well. This feature is illustrated in Figure 8.11, where the pattern of density changes is displayed for the interior region of two stars, one 15 and the other 22 times the mass of the sun. Notice that both stars have a dense interior core, measuring about 1.4 $M_\odot$, with the same density. Outside this region, however, the density for the lower mass star drops off much more rapidly than that for the higher mass star. This lower mass star has formed a far thinner layer of heavier elements around its iron-nickel core. As a consequence, stars of different masses contain different abundances of the various heavier elements produced. Since these massive stars may explode (sec. 8.3.2), the material outside the 1.4 $M_\odot$ core may be hurled into space to join the gas and dust that make up the interstellar medium. It is from this enriched mixture of elements in the interstellar medium that new stars will eventually form. Our sun, for example, was formed from material already enriched by heavy elements ejected by earlier supernova explosions.

Thus one would like to see the abundances of the heavier elements observed in the solar system and elsewhere (chap. 1) consistent with the abundances of these elements calculated using theories of element production in massive stars. Since stars of different mass contain different amounts of the various elements, one must average over the element production in stars of different mass (8–100 $M_\odot$). This, in turn, must be weighted by the observed number of stars of that mass as well as by the amount of mass that the stars of differing mass actually eject, leading finally to a quantity termed the yield function. When all statistics are done, the typical star undergoing supernova explosions induced by iron-core collapse (supernovae of Type II) and contributing to the heavy-element content of the interstellar medium turns out to have a mass of about 20–30 $M_\odot$. It is interesting to note that the relative abundances from carbon to iron outside the core of such a typical star appear to be quite similar to the relative abundances of these elements in the solar system (Fig. 7.11). Thus, the solar system reflects the heavy-element composition of the interstellar gas as it was 4.6 billion years ago.

8.3.2. Theories of Supernovae of Type II

Supernovae of Type II (chap. 2), because of their enormous energy output, their ability to create heavy elements and to form neutron stars and black holes, and their association with the death of massive stars, have long played a central role in the models of the cosmos. Yet, after decades of study, the detailed sequence of events within the dying star is still not known with any certainty. The difficulties stem, in part, from the incomplete nature of the observational sample (Tam77). Models of supernovae of Type II (SNII) are also quite diverse, and, probably, more than one mechanism will be required to explain all events. Currently fashionable models include (Arn80a, Woo81,

Whe81, Hil82, Tri82, Bow82, Bro82, Bet82) iron-core collapse accompanied by hydrodynamical bounce at nuclear density and various magnetohydro-dynamic phenomena. We follow here, to a large extent, a recent discussion by Bethe and Brown (Bet85).

As long ago as the 1930s Zwicky suggested that the source of the huge energy in supernovae is the gravitational collapse of a massive star, and that supernovae collapse into neutron stars. Since these early suggestions there have been many ideas about the general mechanisms for supernova explosions, mostly initiated by Hoyle and Fowler (Hoy 46, Hoy60, Fow64, Col66, Fow69, B[2]FH). One scenario is that at the end of the life of massive stars, the center of these stars collapses owing to gravitation, and the enormous energy set free by this collapse expels most of the mass of the star and gives rise to the large emission of light. There is observational evidence that only singular stars with masses $M \geq 8\ M_\odot$ can lead to supernovae of Type II (supernovae of Type I have been discussed in chap. 2). Such massive stars live only about 10 million years and then die very fast (Table 8.1). The end of the star comes with a bang, implosion of the central part into a dense matter core taking a time of the order of one second.

The relatively stable hydrostatic existence of a massive star ends when it has used up its central supply of nuclear fuel and is left with a core consisting of iron-peak elements. At the surface of the core there is still silicon, which constantly is burned to iron (Fig. 8.10), gradually adding more iron to the core. As long as the mass of the iron core is small, the situation is similar to that of a white dwarf and the core is stabilized by electron-degeneracy pressure. But when it reaches the Chandrasekhar mass, the core can no longer support itself against gravity and it begins to collapse. The exact value of this mass depends on many conditions, such as the degree to which electrons have been captured by nuclei (before the collapse starts) and the entropy of the core. When such effects are taken into account, the masses of typical collapsing cores turn out to be in the range 1–2 $M_\odot$. Note that in a typical supernova progenitor more than 80% of the stellar mass lies outside the collapsing core.

The collapse of the core is accelerated by the combination of two major mechanisms. As the core contracts and heats up, the resulting electromagnetic radiation begins to decompose a small proportion of iron-peak elements into α-particles. This iron photodisintegration costs energy ($^{56}\mathrm{Fe} \rightarrow 13\alpha + 4n$, $Q = -124.4$ MeV), which comes from the electrons decreasing their thermal pressure. Because of the high temperature and density in the core, a second mechanism occurs: some electrons are captured, converting protons into neutrons within the nuclei (called neutronization of the material) with the emission of neutrinos. The neutrinos easily escape from the star, since their mean free path greatly exceeds the stellar radius. This effect also decreases the thermal pressure of the electrons. Both effects absorb so much energy from the star's core that the core contracts and compressionally heats at an ever-increasing rate. The times involved in the collapse of the iron core are fractions of a second. Indeed, the most important part of the collapse (see below) takes place in milliseconds. These times are very short compared with the times involved in the burning of the layers outside the iron core (Table 8.1). Thus, the collapse of the iron core can be considered, to a good approximation, to be independent of what happens in the outer parts of the star, the more massive mantle and envelope. The iron core cannot pull away completely from the rest of the star, but it makes a pretty good try at doing so, leaving a region of low density between it and the rest of the star.

At this point an important feature of the collapsing core is its entropy. For the presupernova cores one finds that the entropy per nucleon (in units of Boltzmann's constant k) is slightly less than unity. This very low entropy at the end of stellar evolution seems paradoxical. One would have expected that entropy increases with time. But this expectation is fulfilled only for isolated systems. A star is not an isolated system, since it emits energy through most of its life as ordinary light and toward the end as neutrinos. The entropy decreases from a value of about 15 in the original hydrogen star to less than unity in the iron core of a star. The 56 nucleons in ^{56}Fe are forced to move together as the iron nucleus moves. Consequently, there are few particle states, and thus there is much more order (low entropy) in iron material than in the original hydrogen star. The burning of a massive star can therefore be characterized as a steady progress toward greater order. The electrons are already highly degenerate in the core, so that their entropy is also low. The low entropy of the presupernova core is roughly evenly distributed between nuclei and electrons. Although the times involved in the collapse of the core are short, of the order of milliseconds, typical nuclear interaction times are much shorter, from 10^{-15} to 10^{-23} s, bringing the system to equilibrium at each stage of the collapse. In equilibrium the entropy is constant; no entropy is created during the entire collapse, and thus the collapse is orderly. At first sight, one might have expected that the collapse is quite chaotic. Calculations show that entropy is the key thermodynamic variable of the collapsing core.

The dynamical collapse of the core starts at a central density of about 10^9 g cm^{-3} (Fig. 8.11) and continues with ever-increasing density. Given the low entropy in the collapse, the nucleons of the collapsing core have no choice but to remain inside nuclei. Nuclei therefore are preserved right up to the point where they begin to touch, i.e., up to nearly nuclear matter density ($\rho_0 = 2.7 \times 10^{14}$ g cm^{-3}). Calculations show that the collapse cannot be stopped at any density below nuclear density. Once the latter density is reached, the nuclei merge completely into "one giant macroscopic nucleus," which might be considered as nuclear matter. The situation then changes dramatically. Inside this matter, the nucleons move independently and exert pressure. These nucleons may now be considered as a degenerate Fermi gas with interactions (chap. 2). The interactions are such that at nuclear matter density the energy is a minimum and hence the pressure is zero. As the implosion carries the matter to densities somewhat greater than nuclear matter density, the pressure increases rapidly because the kinetic (Fermi) energy of the nucleons dominates over the attractive interaction between the nucleons. The effect is almost as if the material were running into a "brick wall." Because of the extreme stiffness of nuclear matter as it is compressed above ρ_0, the nucleons rapidly take over from the degenerate electrons (and neutrinos) in determining the pressure, and the density cannot be increased much further. After this density is reached, the collapse is halted quickly and the core rebounds, or "bounces." Such a hydrodynamical bounce occurs because of the extremely stiff equation of state that applies when the density exceeds nuclear matter density, i.e., nuclear matter is nearly incompressible. The core bounce is analogous to that of a very stiff spring. The bounce of the stellar material is very hard, and results in a velocity reversal and the production of an outward-moving compression wave reflected from the center of the star. This wave steepens into a shock wave as it advances into regions of decreasing density. The shock wave forms at the interface of the rebounding core and the infalling matter of the envelope and then propagates outward. After bounce the inner regions oscillate a few times, but the oscillations rapidly damp out. Calculations show that the higher the density at bounce, the higher the energy put into the shock wave. A dense inner part of the star is thought to remain and to evolve quickly into a neutron star or even a black hole. If the shock wave is strong enough, the surrounding material of the star acquires a substantial outward velocity after the shock wave originating from the "brick wall" reaches it. In fact, for a large portion of this material, the velocity exceeds the escape velocity, leading to ejection of most of the mass of the star and thus to the actual supernova phenomenon. It has also been argued (Fre74) that outward-moving neutrinos might provide additional pressure to blow off the mantle and envelope of the star.

Prior to collapse the gravitational binding energy of the core (chap. 2), whose structure resembles that of a white dwarf, is about 10^{50} ergs, and its binding energy after collapse (when the core is a neutron star) is about 10^{53} ergs. Observations of supernova explosions in nearby galaxies and supernova remnants in our Galaxy indicate that the total energy in optical radiation plus the kinetic energy of the ejected debris is about 10^{51} ergs. For the production of a supernova, therefore, it is only necessary that about 1% of the enormous gravitational energy released by stellar core collapse be converted into light and mass motion. The fundamental problem is to understand how the gravitational energy released in the collapse to nuclear matter density is transferred to the outer layers of the star. The matter within the collapsed core is so strongly bound gravitationally that it cannot be ejected by the subsequent evolution of the supernova, and thus it evolves in a neutron star or a black hole.

A complete mathematical description of a supernova is extremely complex and requires a detailed treatment incorporating many physical processes, in addition to the conditions in massive presupernova stars just before the start of the collapse. Mass motion at about 10% the speed of light is described by hydrodynamics, which includes general relativistic corrections to Newtonian gravitation, homology of the core during collapse, shock heating, energy transport, and neutrino momentum exchange with matter. A wealth of neutrino physics is involved, such as neutrino emission and absorption (by electron and positron capture on nucleons and by thermal pair production, including degeneracy effects), Compton scattering of neutrinos on electrons (redistributing the neutrino energies and exchanging energy with matter), and neutrino diffusion at various density regimes (including coherent scattering off heavy nuclei). Additional input physics involves convection (transporting the state variables of the matter), thermonuclear burning rates (occurring in zones which contain nuclear fuels), and electron-capture rates by the more loosely bound protons in heavy nuclei (transforming material on infall and changing the average number of electrons per nucleon, i.e., the charge per baryon). All these processes determine the state variables of the material such as radii and velocities, density and temperature, damping rate of the kinetic energy of the collapsing core, charge per baryon, atomic charges, and mass fractions of unburned nuclei. These state variables are interconnected by the equation of state of the material. Clearly, a simple gas-law equation of state in the hydrodynamic calculation of stellar collapse and bounce does not adequately describe the matter near nuclear density. A source of uncertainty in the equation of state is connected with the question of heat capacity of a nucleus and the bulk viscosity heating during collapse. Heavy nuclei are known to have a large number of excited states, i.e., a large number of internal degrees of freedom. In a gas of nuclei this feature is thermodynamically equivalent to a large heat capacity. The primary uncertainty is how the distribution in energy of the excited states changes at densities prior to bounce. If much energy can be stored in nuclei, nuclear matter is more compressible, which makes the core rather like a rubber ball: after being strongly compressed by the initial collapse of the star, it bounces back elastically beyond its equilibrium size. As a consequence, the rebound is stronger, enhancing the strength of the outward shock wave.

The equation of state determines quantities, such as pressure and energy of the core as well as the conditions for thermonuclear burning rates and neutrino cross sections, which in turn are input parameters for the complex physical processes just discussed. Thus, the input physics, the state variables, and the equation of state represent a coupled system. A considerable simplification in the calculations occurs when, after a certain stage in the collapse, the density of the core exceeds about 10^{11} g cm^{-3}. Then the neutrino mean free path becomes small relative to the core radius, and thus neutrinos are trapped and no energy can get out of the system that way. Neutrino trapping is due to the "neutral weak currents" (allowing neutrinos to be involved in reactions where no charge is exchanged), which cause substantial scattering of neutrinos by nucleons. As this happens, the neutrino density increases and the electron-capture rate thus decreases. This effect does not reduce the collapse rate, which is adiabatic after this time. The equation of state of the central region of the star during the gravitational collapse is obviously one of the most critical ingredients in determining the characteristics and outcome of the implosion. Another critical consideration is the degradation of the shock wave (see below) as it passes through the outer layers, in particular the neutrino photosphere, where an adequate model of complicated neutrino interactions is not available.

A comprehensive model of the evolution of massive stars over their entire life has not yet been developed. Such a model should start from the observable initial configuration of such stars and continue through their various hydrostatic burning stages, their gravitational collapse and the bounce of their cores, and their ensuing death in supernova explosions. Instead, the results of collapsing presupernova models have been used as input to detailed supernova-core calculations. Essentially all present calculations are in general agreement in the general features of the collapse and the general mechanism of shock-wave formation. However, what happens beyond this point is not yet established. In the simplest scenario, the shock wave rushes outward, reaching the edge of the iron core in a fraction of a second, and continuing on through the successive onion-shell–like layers in the star, erupting after some days from the outer layers of the star in a supernova explosion. Beyond a certain radius, all matter is blown off, spreading various elements into interstellar space. The remaining core relaxes into a neutron star or a black hole. Although there appears to be no fundamental obstacle in the way of such a scenario, computer calculations have not produced shocks of sufficient strength to eject the mantle and envelope of the stars, at least not until recently. Some investigators have been able to simulate such an explosion on the computer, while others have not, depending on which assumptions are made about the complex physics issues taking place in the core.

There are many processes by which the shock wave can lose energy as it moves away from the inner core. For example, the shock wave has to penetrate through the overlying iron and loses a lot of energy by dissociating some of the iron nuclei into nucleons, eventually getting stuck in these quagmires. Calculations indicate that the shock does not fade away but is stalled at some radius. Originally it was felt that all the material of the star might fall back into the core, and with the accumulation of so much material the entire star would collapse into a black hole. However, recent calculations of Wilson (Wil83) show that the revival of the shock can occur as a result of heating by neutrinos. According to calculations, nearly one-half the energy set free by gravitational contraction to a neutron star is emitted by the core in the form of neutrinos. Outside the neutrino photosphere (defined as the radius at which the average neutrino undergoes its last interaction with the stellar matter) the neutrinos undergo very few collisions, but when they do, they give away most of their energy. The energy given up serves primarily to dissociate nuclei into nucleons, the process which stalled the original shock. Since the absorbed energy increases the temperature as well as the number of particles, additional pressure is generated, forcing the matter outward. Following a relatively long period of shock stagnation (of the order of 0.2–0.5 s), the revitalized shock appears to produce a delayed supernova explosion. Improved calculations of this effect require an elaborate neutrino transport scheme.

As a summary, the situation in understanding supernovae of Type II is at present encouraging, whereas for several years it appeared as if no mechanism could produce a supernova explosion. Many ideas about the mechanisms of supernovae have been around for some time. Working them out in detail—the hydrodynamics, the energy transport, and the wealth of complex physics issues characterizing core calculations—awaited the availability of large computers. Some of the underlying physics is still poorly understood, and future work will include continued efforts to improve the input physics as well as to build a comprehensive evolutionary model. We seem to be nearing a general understanding of the supernova phenomenon, but we have hardly begun to reproduce the many observations in detail.

As noted above, most of the gravitational energy released by the core collapse is carried off by neutrinos. The detection of neutrinos from supernova explosions and from the cooling of the neutron star may tell us in some detail what goes on. The neutrinos originate in the core of the star and travel unhindered through the mantle and envelope. In contrast, the core is opaque to electromagnetic radiation, and light from the explosion arrives only some days after the core collapses. Neutrino detectors have been set up (chap. 10) in anticipation of the next supernova, and one just has to wait for a supernova to occur in our Galaxy, which, unfortunately, occurs only once in about 50 years.

Coincidentally on 23 Feburary 1987, as the manuscript for this book was being edited, a supernova occurred in the Large Magellanic Cloud, our companion galaxy. It has been designated "Supernova 1987A (Shelton)" and is the brightest since the one seen by Kepler in 1604. It has been observed optically as well as, for the first time, with neutrino detectors. At the time of this writing, a wealth of observational and experimental information was being collected and critical comparison with theoretical expectations was in progress. Plans for a continuing observational program of this supernova also were being made.

8.3.3. Explosive Nucleosynthesis

Clearly, supernovae of Type II are one means by which large quantities of heavy elements produced in the hydrostatic burning phases of presupernova stars (Fig. 8.10) are expelled and mixed with interstellar material, enriching it with heavy ("metallic") elements. However, the abundances of these elements are expected to change as a result of the supernova explosion itself. The production of elements and isotopes during the explosion is called explosive nucleosynthesis (Tru69, Arn 71, How71, Woo73). To calculate the changes of elemental abundances during the supernova explosion, a model is needed to describe the dynamics of the explosion; that is, data are needed on the changes in temperature and density as a function of time. In principle, this requires a comprehensive evolutionary model for massive stars as discussed in section 8.3.2. However, because of the widely differing hydrodynamic response times of the core and the rest of the star (Table 8.1), a decoupling can be made in a simplifying model by replacing the core with a time-dependent inner boundary condition (including an outward neutrino flux). This inner boundary first falls in, then accelerates sharply outward, driving a shock wave into the mantle. Calculations show that the processes occurring outside the star's core follow- ing the supernova explosion are relatively insensitive to the exact behavior of the core, provided that after collapse it generates a sufficiently strong outgoing shock wave to eject the mantle and envelope. It turns out that the most impor- tant parameter describing the supernova is the explosion energy (the sum of kinetic and electromagnetic energies, but not neutrino energies). Thus, super- nova models have been considered with different explosion energies to lie in the range of observation [(1–4) $\times$ 10^{51} ergs].

The shock wave generated by the bounce of the core propagates out through the rest of the star, temporarily heating the material to temperatures sufficient to cause substantial nuclear processing in the burning zones of a presupernova star (Fig. 8.12). The extent of this nuclear processing is limited

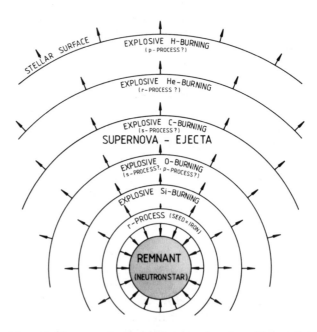

FIGURE 8.12. Schematic diagram of an exploding massive star with a collapsing core (the remnant) and various explosive burning shells in the supernova ejecta.

by the rapid expansion of the shock-heated material, since the momentum initially deposited by the shock accelerates the matter beyond its escape velocity. For explosion energies considered, the shock is not particularly strong with respect to the energy density of the supernova mantle (Wea80a, Woo82). This circumstance implies that most of the energy in the material that the shock has just passed resides as internal rather than kinetic energy. It is found further (Wea80a, Woo82) that the density and temperature in the mantle at any given time are roughly constant behind the shock. Because of these characteristics, it appears reasonably accurate to think of the shock as an expanding bubble of radiation containing most of the energy of the core explosion. In this simple radiation-dominated model, the temperature T_s that a given zone in the mantle reaches is approximately determined by the presupernova radius r_0 and the explosion energy E_0 through the relation (Stefan's law)

$$T_s(r_0) \simeq (3E_0/4\pi a r_0^3)^{1/4} ,$$

where a is the radiation density constant (chap. 2). For typical explosion energies the silicon layer reaches a peak temperature above 500 keV, the oxygen-neon layer above 100 keV, and the hydrogen layer above 10 keV. These temperatures are sufficient to cause a substantial amount of neon, oxygen, and silicon to be burned explosively, while the cooler outer layers of carbon, helium, and hydrogen (Figs. 8.10 and 8.12) are ejected without substantial explosive nuclear burning. During the few tenths of a second for which these temperatures typically persist, the material of the silicon layer will be processed into iron-peak elements (predominantly ^{56}Ni, which decays radioactively first to ^{56}Co and then to ^{56}Fe) so rapidly that their neutronization via the weak interaction (electron capture) cannot take place.

The final composition profile of the ejected ashes for a 25 $M_\odot$ stellar model is in remarkably good agreement with observed solar abundances (Wea80a, Woo82, Woo85). It shows that the elements between silicon and iron are modified or formed primarily by explosive processing, while those from oxygen to magnesium are formed hydrostatically in the presupernova star and have abundances largely unmodified by the explosive ejection. Using the above model, the observed light curves of several supernovae of Type II have also been reproduced remarkably well (Wea80a), strongly supporting the occurrence of explosive nucleosynthesis.

Another exciting possibility is that regions of a supernova may be blasted into space more or less intact, that is, without much mixing with outer regions. If this occurs, these inhomogeneous regions of matter, or remnants of supernovae, should carry with them a record of their individual nucleosynthetic history. Confirmation of the essential correctness of this scenario has come from studies of certain regions of the 300 year old supernova remnant Cas A. These regions are strongly overabundant in sulfur, argon, and calcium relative to the lighter elements (Che78, Che79, Kir80), which are the principal products to be expected in a portion of a star that has gone through oxygen burning (sec. 8.1). Similarly, the X-ray spectra of Cas A and of the remnant of Tycho's supernova (1575) show strong enhancements of silicon, sulfur, argon, and calcium (Hol79, Bec80a).

8.3.4. Nuclear Physics Aspects of Explosive Burning

For a given nuclear reaction the problems encountered in extrapolation of measured cross sections to astrophysically relevant energies, so difficult for ordinary hydrostatic-burning episodes, are greatly eased in explosive burning. Owing to the much higher temperatures encountered in explosive situations, the Gamow energy window lies much higher, often near the height of the associated Coulomb barrier. As a consequence, the energy regions in which reaction rates are needed for explosive nucleosynthesis are often the same or nearly the same as those in which they are directly measurable, and thus no extrapolation procedures are needed. Because of the high explosive temperatures, reactions induced on short-lived radioactive nuclides, as well as on excited states of nuclei, may also play an important role in explosive nucleosynthesis. The Hauser-Feshbach statistical model (sec. 8.2.3) should eventually result in acceptable predictions for the rates of these reactions. However, this statistical formalism breaks down for light nuclei ($A \leq 30$), since the rates may be governed by a single state. The experimental investigation of reactions induced on light radioactive nuclei represents an enormous challenge to the experimentalists (chap. 5). Such experiments need to be carried out only near and above the associated Coulomb barrier, where the cross sections are usually appreciable.

At the temperatures involved in explosive nucleosynthesis many reactions come into equilibrium with their inverse reactions. In these cases, the abundances depend more upon nuclear properties, such as binding energies, β-decay times, and statistical factors, than upon reaction rates. The knowledge of these properties for nuclei far from the valley of β-stability (in particular those involved in r-process nucleosynthesis; chap. 9) clearly represent important ingredients in the explosive synthesis, and their experimental determinations represent an active area of research in many laboratories.

9

Nucleosynthesis
beyond Iron

9.1. The Quest for the Origin of the Trans-Iron Elements

As discussed in chapter 2, the ashes of the big bang consist predominantly of nuclei of hydrogen and helium. The ashes themselves are subsequently burned in the nuclear fires of the stellar interiors with the metallic elements as residue. The increasing densities and temperatures in stellar interiors provide the necessary environment in which all elements up to the region of the iron peak can be synthesized by charged-particle–induced nuclear reactions (chaps. 6, 7, and 8). The sharp drop in the elemental abundances from $A = 1$ to $A = 50$ (Fig. 1.23) reflects the fact that the Coulomb barrier, rising as the nuclear charge increases, increasingly impedes the reactions required to synthesize these elements. An exception to this general trend of the abundance curve is the iron peak ($A \simeq 56$), because iron has the greatest stability of the nuclei involved in the rearrangement processes occurring during silicon burning.

If the heavier elements above the iron peak ($A \geq 70$) had been synthesized by silicon burning or other charged-particle–induced fusion reactions, their abundances would drop very steeply with increasing mass, continuing the trend seen in the $A = 1$ to $A = 50$ mass region (Fig. 1.23). Actually, the observed abundance curve of the heavy elements exhibits a much slower decrease with mass number (Figs. 1.23 and 9.1), with absolute abundances much higher than expected from charged-particle–induced reactions. We have seen that the rapidly rising Coulomb barrier for charged particles greatly inhibits the synthesis of the heavy elements; where, then, did these heavy elements originate? The generally accepted answer was originally suggested by the abundance data of Suess and Urey (Sue56). These data have been periodically updated (Cam82, And82, and chap. 1); a plot of the data shows double peaks labeled r and s in Figure 9.1. Obviously these peaks are associated with neutron shell filling at the magic neutron numbers $N = 50$, 82, and 126. Suess and Urey had made the breakthrough that led to the extension of nucleo-synthesis in stars by neutron capture, unhindered by the Coulomb barriers; this continued all the way up to ^{238}U.

The details of nucleosynthesis in stars by neutrons were subsequently pre-sented in the classical paper by Burbidge, Burbidge, Fowler, and Hoyle in 1957 (B^{2}FH) and independently by Cameron (Cam57). It was hypothesized by B^{2}FH and Cameron that, with iron and other nuclei of intermediate mass as seeds, processes similar to those suggested by Gamow, involving neutron capture and β-decay for primordial nucleosynthesis (chap. 2), operate within stars to synthesize the heavier elements. The synthesis proceeds in steps of 1

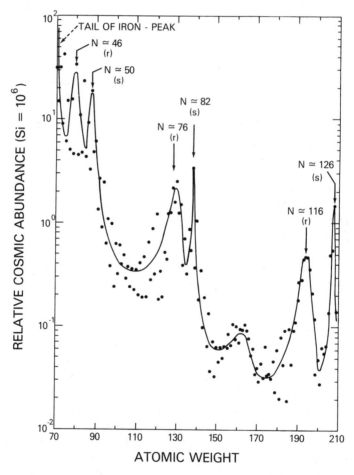

FIGURE 9.1. Cosmic abundances of the heavy elements are shown as a function of atomic weight (Cam82). The line through the data points is to guide the eye. Note the narrow peaks for nuclei with closed neutron shells ($N = 50$, 82, and 126) and the broader peaks for nuclei 4–10 mass units below the closed neutron shells.

mass unit and occurs either at a slow rate (*s*-process) or at a rapid rate (*r*-process). That these two types of neutron-capture processes are required is dictated predominantly by the abundance data (Fig. 9.1). Hence, Gamow's early idea of elemental nucleosynthesis is operative in nucleosynthesis within stars; but now the chain which plays a dominant role in the formation of heavy elements starts well beyond the "missing links" at $A = 5$ and $A = 8$ (chap. 2).

The above hypothesis is supported by the following features: (1) Only about 3% of the iron-peak elements are needed to synthesize all heavy elements ($A \geq 70$), i.e., there is enough seed material available. (2) At certain stages of a star's lifetime, large neutron fluxes are produced in the stellar interior. (3) The neutron-capture cross sections of the heavy elements are very large compared with those of the light elements (Fig. 9.6), i.e., the heavy elements eagerly absorb neutrons. (4) As pointed out above, the abundance curve (Fig. 9.1) has structure, which can be explained only by neutron-capture reactions. (5) Finally, the discovery of technetium lines in the atmosphere of S-type stars (red giants) by Merrill in 1952 (Mer52) demonstrates conclusively that the formation of heavy elements in stars by neutron capture occurs and is a continuing process (Cos84). Technetium, which no longer exists at a detectable level on earth, is an unstable element with a half-life of less than a million years.

To establish the quantitative details of these processes, accurate energy-averaged neutron-capture cross sections at neutron energies of about 30 keV (sec. 9.2) are needed. Such data provide information on the mechanisms of the neutron-capture processes and time scales, as well as temperatures involved in the processes. The data should also shed light on neutron sources, required neutron fluxes, and possible sites of the processes.

9.2. Neutron-Capture Cross Sections

The neutrons produced in stellar interiors are quickly thermalized through elastic scattering (in about 10^{-11} s; All71), after which their velocities are represented by a Maxwell-Boltzmann distribution (chap. 3). Since the expected energy dependence of the neutron-capture cross section (chap. 4) has the form

$$\sigma_{n\gamma} \propto 1/v \propto 1/E^{1/2} ,$$

the most probable energy for the process to occur is near $E_0 = kT$ (Fig. 9.2), or, equivalently, the most probable thermal velocity is $v_T = (2kT/m)^{1/2}$, with m as the reduced mass. The neutrons involved in the *s*-process (sec. 9.4) are most likely produced during the helium-burning phase in red giants via (α, n) reactions, in which temperatures are in the range of $T_9 = 0.1$–0.6; hence $E_0 \simeq 30$ keV.

For the above energy dependence of the capture cross section, one obtains

$$\sigma v = \text{constant} = \sigma_T v_T .$$

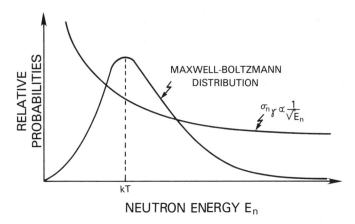

FIGURE 9.2. Schematic diagram of the Maxwell-Boltzmann energy distribution and the expected energy dependence of the neutron-capture cross section (chap. 4). The most probable energy for the capture process in stars is near $E_n = kT$.

Therefore, the reaction rate per particle pair is constant:

$$\langle \sigma v \rangle = \text{constant}$$
$$= \langle \sigma \rangle v_T \, .$$

In the latter equation an averaged cross section $\langle \sigma \rangle$ has been defined such that the product of $\langle \sigma \rangle$ with v_T leads to $\langle \sigma v \rangle$. Frequently, $\langle \sigma \rangle$ is nearly equal to σ_T, that is, the cross section measured at $v = v_T$. For the above dependence this is precisely true ($\langle \sigma \rangle = \sigma_T$). For slightly different dependences ($\sigma \propto \text{constant}$ or $\sigma \propto 1/v^2$), it is very nearly fulfilled ($\langle \sigma \rangle = 2\sigma_T/\sqrt{\pi} = 1.13\sigma_T$). Thus, in these most common cases, a measurement of σ near v_T provides a good value for $\langle \sigma \rangle$. Note that, while there is some uncertainty as to the actual stellar temperature (sec. 9.4), Maxwellian-averaged capture cross sections $\langle \sigma \rangle$ are relatively independent of temperature for most nuclides between 10 and 100 keV. From the point of view of data compilation, it is then sufficient to choose the most convenient and common energy. That is found to be 30 keV. Data obtained at nearby energies are extrapolated to 30 keV using the above relations, thereby introducing a small uncertainty.

In the rare case where the capture cross section is derived from a few narrow resonances or where a wider temperature range is involved in the capture process (e.g., r-process), the cross section must be measured over a wider range of energies and folded numerically with the Maxwell energy distribution:

$$\langle \sigma \rangle = \frac{\langle \sigma v \rangle}{v_T} = \frac{2}{\sqrt{\pi}} \frac{1}{(kT)^2} \int_0^\infty \sigma(E) E \exp\left(-\frac{E}{kT}\right) dE \, .$$

In the neutron energy range of interest to s-process nucleosynthesis ($E_n \simeq$ 1–300 keV), capture cross sections can be measured using several different techniques and a variety of neutron sources. Neutrons are produced most efficiently using accelerators (chap. 5). Electron linear accelerators (linac) and Van de Graaff accelerators (VdG) are most frequently used for this type of work, each having specific advantages.

An electron linac, such as ORELA at Oak Ridge, provides a very powerful neutron source. Intense neutron bursts with broad energy distributions are produced by pulsed high-power electron beams via (γ, n) reactions on heavy-metal targets (Fig. 9.3). Repetition rates are typically 1 kHz with pulse widths of a few nanoseconds. With this time structure, capture cross section measurements can be carried out with excellent neutron energy resolution using time-of-flight (TOF) techniques (chap. 5) in combination with flight paths of about 50 m. A problem with the linac is the intense bremsstrahlung from the neutron target, which requires heavy shielding of the target area (Fig. 9.3). This kind of neutron source has been used extensively for capture cross section measurements, in particular in the work of Macklin and Gibbons at Oak Ridge National Laboratory (Mac65, Mac71, All71, Bro81, Bee82).

In contrast to a linac, the VdG accelerator produces neutron fluxes smaller by 3–4 orders of magnitude. However, because the ion beams used for neutron

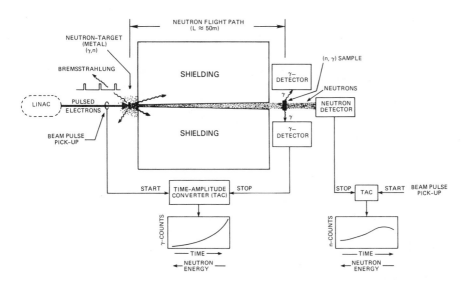

FIGURE 9.3. Schematic diagram of the experimental arrangement for capture cross section measurements. The pulsed electron beam from a linac produces neutrons via (γ, n) reactions on a metal target. The bursts of neutrons produced are collimated and impinge on a sample of interest. The resulting prompt γ-rays from the (n, γ) capture reaction are observed with γ-detectors. Using the time-of-flight technique, information on the yield of capture γ-ray transitions as a function of neutron energy is obtained.

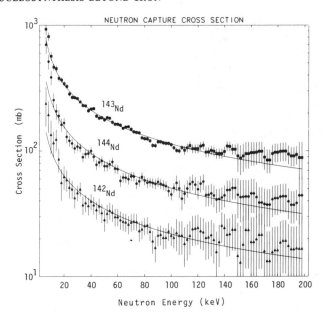

FIGURE 9.4. Neutron-capture cross sections from 6 to 200 keV are shown for Nd isotopes (Mat84). The lines through the data are fits with $\sigma(E) \propto 1/E^b$ ($b \simeq 0.66$–0.69). The data lead to Maxwellian-averaged cross sections at 30 keV of $\langle \sigma \rangle = 49$, 259, and 119 mb for the ^{142}Nd, ^{143}Nd, and ^{144}Nd isotopes, respectively. These values are quite close to the respective cross sections $\sigma_T = 49$, 257, and 117 mb measured at 30 keV.

production do not generate bremsstrahlung radiation, the need for target shielding is greatly reduced. As a consequence, significantly shorter neutron flight paths (≤ 60 cm) can be used (Mac63, Wis78). This, in turn, allows the use of larger solid angles that compensate to a large extent for the lower neutron source strength. The (p, n) reactions on ^{7}Li or ^{3}H are used most commonly for neutron production. The proton energy is adjusted slightly above the reaction threshold, so that the center-of-mass velocity of the system exceeds the velocity of the emitted neutrons. In this situation the neutrons are kinematically collimated in a forward cone.

In most experiments, capture reactions are detected (All71) via the prompt γ-cascades by which the newly formed nucleus deexcites (Fig. 9.3). These so-called direct detection methods, which are supplemented by the activation technique (see below), are discussed in a review by Chrien (Chr75). Examples of such measurements are shown in Figure 9.4 for three Nd isotopes.

If neutron capture leads to an unstable nucleus with a half-life of less than 0.5 y, measurements of the activity of these nuclei can be used to determine the capture rate. In this case the sample is placed close to the neutron target (Fig. 9.5) and irradiated over a specific time interval (adjusted to the half-life of the unstable nucleus produced). The neutron source is shut off and the induced activity counted with calibrated high-resolution detectors. Measurement of the activity as a function of time is used to verify that the correct assignment was made of the (n, γ) reaction being studied. The basic advantage of activation measurements is their inherent sensitivity. However, this can be fully exploited only if certain criteria are fulfilled (Bee80, Käp82). This technique fails, of course, for neutron captures that result in stable (or very long-lived) isotopes. In such cases the prompt-capture γ-rays must be detected (Fig. 9.3). Since the activation technique provides an average cross section over the spectrum of the neutrons used for activation (Fig. 9.5), the neutron spectrum must be accurately known. Many investigators have tried to circumvent this difficulty by using monoenergetic neutrons. Beer and Käppeler (Bee80) solved the problem by tailoring a spectrum that almost perfectly matches the Maxwell spectrum for $kT = 25$ keV. This was possible because of the experimentally determined properties of the $^7Li(p, n)^7Be$ reaction when using protons 25 keV above threshold. Integration over the emission angles of the kinematically collimated neutron beam (Fig. 9.5) gives a spectrum, which is—as shown in the histogram in the lower part of Figure 9.5—in 95% agreement with the Maxwell distribution for $kT = 25$ keV (*solid line*). Thus activation measurements done with

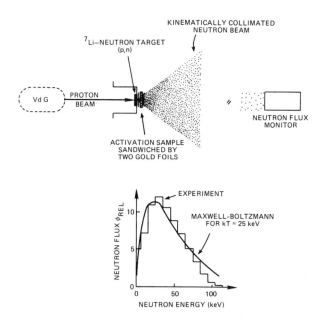

FIGURE 9.5. Schematic setup for the irradiation of samples using a kinematically collimated neutron beam (Bee80). All neutrons are emitted in a forward cone because of the reaction kinematics (for protons about 25 keV above threshold). The time dependence of the neutron flux and of its energy at zero degrees is monitored with a ^6Li glass detector (Mac71). The lower figure shows the approximation to the Maxwell-Boltzmann energy distribution at $kT = 25$ keV (*solid line*) of the neutron spectrum produced via the $^7Li(p, n)^7Be$ reaction during activation (histogram).

such a neutron spectrum already provide the proper Maxwellian average for the cross section:

$$\langle\sigma\rangle = \frac{\langle\sigma v\rangle}{v_T} = \frac{2}{\sqrt{\pi}} \int_0^\infty \sigma(E)E \exp\left(-\frac{E}{kT}\right)dE \bigg/ \int_0^\infty E$$

$$\times \exp\left(-\frac{E}{kT}\right)dE \simeq \frac{2}{\sqrt{\pi}} \sigma_{exp}\ .$$

In the actual measurements (Bee80) the sample to be investigated is sandwiched between two gold foils (Fig. 9.5). Since the neutron-capture cross section for gold is accurately known and it can be activated as well, gold serves as an in situ standard.

Neutron cross sections of astrophysical interest are also being measured using the 24 keV neutron beam from the Brookhaven High Flux Beam

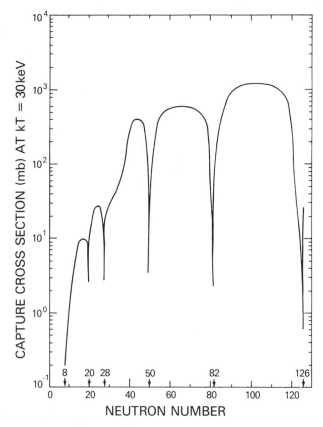

FIGURE 9.6. The measured average neutron-capture cross sections $\langle\sigma\rangle$ at 30 keV are shown schematically as a function of neutron number N of the nuclei (All71, Ulr82, Käp82, Alm83). Note the large dips near the neutron closed shells at $N = 8, 20, 28, 50, 82,$ and 126.

Reactor (Bra79). Thus, the measurement of keV neutron-capture cross sections remains a flourishing enterprise, motivated in part by the general worldwide interest in neutron cross sections for reactor purposes.

The experimental efforts of many research groups contributed data on the average capture cross section $\langle \sigma \rangle$ at 30 keV for a wide range of nuclei (Ulr82, Käp82, Alm83). These data are shown schematically in Figure 9.6 as a function of neutron number N. Note that the light nuclei have small capture cross sections compared with the heavier ones ($N > 28$, $A \geq 60$). This feature is due in part to the low energy-level density in light nuclei, where often just a few resonances contribute to the capture cross section. Note also the large dips that appear near the neutron closed shells with $N = 8, 20, 28, 50, 82$, and 126. These dips are caused by the drastically reduced level density of nuclei with closed neutron shells as compared with neighboring nuclei. The data verify that the heavier elements are particularly likely to absorb neutrons. The data suggest a relatively flat abundance curve of the trans-iron nuclei, in agreement with observation. The exceptions are the cross-section dips at $N = 50, 82$, and 126, which, as will be seen below, lead to the abundance peaks in Figure 9.1 for these nuclei with closed neutron shells (peaks labeled s). The other peaks, labeled r and lying 4–10 mass units below the peaks labeled s (Fig. 9.1), are also due to effects of closed neutron shells at these magic neutron numbers (sec. 9.5). Note that the existence of the iron peak in the abundance curve (Fig. 9.1) cannot be similarly explained, since the iron-peak nuclei exhibit no anomalously small capture cross sections. Recall that these iron-peak nuclei are not synthesized by neutrons but rather in silicon burning (chap. 8). Finally, it should be pointed out that the cross sections and thus the stellar reaction rates $\langle \sigma v \rangle$ for neutron-induced reactions can be measured directly at relevant stellar energies, in contrast to charged-particle–induced reactions, where the reaction rates at stellar energies must be obtained by extrapolation from data obtained at higher energies (chap. 4).

9.3. Basic Mechanisms for Nucleosynthesis beyond Iron

As a result of each (n, γ) capture reaction, a nucleus (Z, A) is transformed into the heavier isotope $(Z, A + 1)$. If this isotope is stable, an additional neutron capture leads to the isotope $(Z, A + 2)$, and so on. Thus (n, γ) reactions increase the mass number by one unit at a time, providing the mechanism for synthesizing the elements beyond iron. If, in this chain of capture reactions, the final isotope produced is unstable, subsequent processes depend on the intensity of the neutron flux impinging on the nuclei as well as on the lifetimes of the unstable nuclei against β-decay.

For unstable nuclei, when the time between successive neutron captures $\tau_{n\gamma}$ is much larger than the β-decay lifetimes τ_β ($\tau_{n\gamma} \gg \tau_\beta$), the network of processes involved is called the s-process. Inspection of a chart of the nuclides reveals that the s-process closely follows the valley of β-stability. A section of the nuclear chart is shown in Figure 9.7. Starting, for example, with ^{127}I, neutron capture leads to the unstable isotope ^{128}I, which decays ($T_{1/2} = 25$ minutes) to the stable isotope ^{128}Xe. Subsequent neutron captures then produce the heavier Xe isotopes until the unstable ^{133}Xe isotope ($T_{1/2} = 5.3$ d) is reached. After β-decay to ^{133}Cs and neutron capture to ^{134}Cs ($T_{1/2} = 2.3$ y), which β-decays to ^{134}Ba, the stable Ba isotopes are synthesized by successive neutron capture, and so on (Fig. 9.7). Typical lifetimes τ_β are in the range of seconds to years, and hence the above condition requires $\tau_{n\gamma}$ to be much greater than these lifetimes. If one takes an average capture cross section (Fig. 9.6) of $\sigma \simeq 0.1$ b at 30 keV ($v_T = 3 \times 10^8$ cm s^{-1}), the reaction rate per particle pair is $\langle \sigma v \rangle \simeq 3 \times 10^{-17}$ cm^3 s^{-1}. The product of neutron-capture lifetime $\tau_{n\gamma}$ times the neutron density N_n is (chap. 3) $\tau_{n\gamma} N_n = 1/\langle \sigma v \rangle = 3 \times 10^{16}$ s neutrons cm^{-3}. For a capture lifetime of, say, $\tau_{n\gamma} = 10$ y, the required neutron density for the s-process is then $N_n \simeq 10^8$ neutrons cm^{-3}.

As seen from Figure 9.7, no stable nuclei exist at the neutron number $N = 61$. Among the unstable nuclei at this neutron number is the isotope ^{107}Pd, which has a rather long lifetime of $T_{1/2} = 7 \times 10^6$ y. Hence, if $\tau_{n\gamma} \lesssim T_{1/2}(^{107}$Pd), or equivalently if $N_n \geq 10^2$ neutrons cm^{-3}, the s-process track continues to the stable isotope ^{108}Pd (Fig. 9.7). There are similar complications at neutron numbers $N = 35, 39, 45, 89, 115,$ and 123 (Fig. 9.13), which cause branchings in the s-process path (sec. 9.4). These branchings, together with other features, led to the concept (sec. 9.4) that several neutron exposures with differing fluxes must be involved in the synthesis of the s-process elements (Cla61, See65, Ibe83).

If neutron capture proceeds on a rapid time scale compared with β-decay lifetimes (the other extreme, with $\tau_{n\gamma} \ll \tau_\beta$), the network of reactions involved is called the r-process. As with the s-process, the r-process involves the addition of neutrons to seed nuclei, which are primarily ^{56}Fe, with minor contributions from other iron-peak species. In this r-process (sec. 9.5) neutrons are added to ^{56}Fe with capture times of the order of 10^{-3} s, i.e., much shorter than the β-decay lifetimes for nuclei near the valley of stability. Only when very neutron-rich nuclei are reached, for which the neutron binding energy is close to zero, can β-decay again compete with capture reactions to increase the nuclear charge by one unit. Subsequent neutron captures and β-decays produce heavier and heavier nuclei with increasing charge. The r-process follows a path (Fig. 9.13) at the extreme neutron-rich side of the valley of stability with neutron binding energies close to 1.2–2.0 MeV (the so-called neutron drip line). Since the β-decay lifetimes of nuclei far from the valley of stability are very short ($\tau_\beta \ll 1$ s), the neutron-capture times $\tau_{n\gamma}$ must then be shorter (of the order of 10^{-3} to 10^{-4} s). For $\tau_{n\gamma} = 10^{-4}$ s, a neutron density of $N_n \simeq 3 \times 10^{20}$ neutrons cm^{-3} is required, which is 12 orders of magnitude higher than that for the s-process. Once the neutron flux ceases, the neutron-rich matter produced evolves quickly by several β-decays (keeping a given mass A constant) to the region of stability (Fig. 9.7). Thus, the r-process syn-

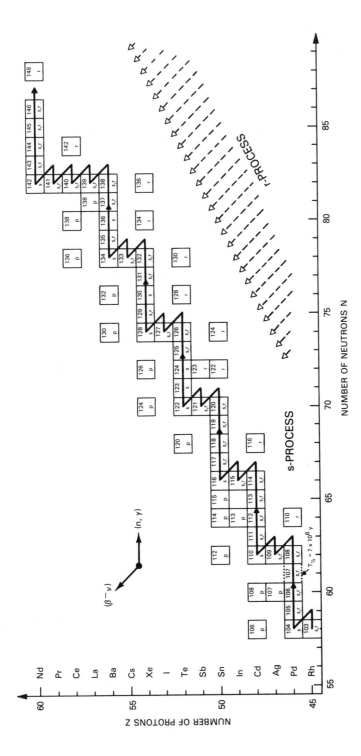

FIGURE 9.7. A section of the chart of nuclides and the s-process path through the elements in this mass region. Note that the path bypasses the p- and r-process nuclei. The r-process nuclei are the end products of an isobaric β-decay chain, the flow of which is indicated by the inclined arrows, from neutron-rich progenitors produced in an intense neutron flux. When this flux ceases, the progenitors "rain" down to the valley of stability via β-decay.

thesizes neutron-rich isotopes not reached by the *s*-process. Of course, the *r*-process also produces many *s*-process isotopes labeled *s, r* in Figure 9.7. There are, however, a few nuclei (28 in number) along the *s*-process path, which are shielded by stable *r*-only nuclei and which consequently are called *s*-only isotopes (Fig. 9.7).

Now we see how the two types of evolutionary mechanisms, the *s*-process and the *r*-process, synthesize heavier elements with the iron-peak nuclei as seeds. Both processes bypass the proton-rich nuclei along the valley of stability (labeled *p* in Fig. 9.7), which are on the average a factor of 10^2–10^3 less abundant than the *s*- and *r*-process isotopes (Fig. 9.8). These proton-rich isotopes must be produced by a third mechanism, called the *p*-process. The *p*-process involves positron production and capture, proton capture, and (γ, n) or (p, n) reactions starting from the *s*- and *r*-isotopes as seed nuclei (Woo78a, Aud80, Tru84). Since the *p*-process is not as well understood as the other two processes, we will, in what follows, discuss predominantly the *s*- and *r*-processes.

One concludes, then, that the trans-iron elements are produced in three essentially different processes (*s*, *r*, and *p*). From the required time scales it is evident that the *r*-process must occur in an explosive environment (i.e., supernova), while the *s*-process is most likely to occur inside stars (Lam77,

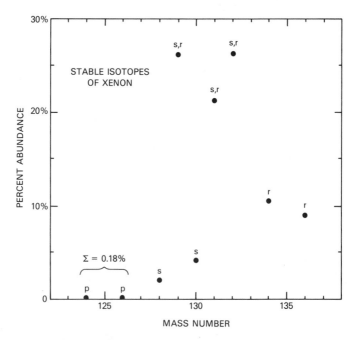

FIGURE 9.8. Evidence for the operation of three separate processes (*p*, *s*, and *r*) in the formation of the stable isotopes of the element xenon is given by the isotopic abundances (Cam82). The first two isotopes can be produced only in the relatively rare *p*-process, and their abundances are seen to be quite small (together only 0.18%). The most abundant isotopes are produced by neutron capture at a slow rate (*s*-process) and/or at a rapid rate (*r*-process).

Tru77, Ibe83, Tru84). These basic conclusions have been among the most durable of those reached in early pioneering studies (B²FH, Cam57).

9.4. The s-Process

As stated above, the nuclei of the s-process follow a track that wanders along the valley of β-stability (Figs. 9.7 and 9.13), i.e., it stays close to the most bound isotopes of a given atomic weight (or mass number) A. Neutron capture occurs all the way from ^{56}Fe (the seed nuclei) on up to ^{209}Bi (the last stable nucleus). The capture path is finally terminated by rapid α-decay above bismuth (Cla67). The time dependence of the abundance N_A of an s-only isotope A is then given by the differential equation (Cla61 and chaps. 3 and 6):

$$\frac{dN_A(t)}{dt} = N_n(t)N_{A-1}(t)\langle\sigma v\rangle_{A-1} - N_n(t)N_A(t)\langle\sigma v\rangle_A - \lambda_\beta(t)N_A(t) , \qquad (9.1)$$

where $N_n(t)$ is the neutron density at time t and $\lambda_\beta(t) = 1/\tau_\beta(t) = (\ln 2)/T_{1/2}(t)$ gives the β-decay rate at time t if the isotope A is radioactive. It is important to note that under stellar conditions the β-decay lifetimes are time-dependent through their dependence on the stellar temperature (chap. 6 and below). The first term in the above equation describes the production of isotope A by neutron capture of its lighter neighbor $A - 1$; the second term, its destruction by neutron capture; and the last term, its destruction by β-decay if applicable. All components in the terms are time-dependent because of possible changes in stellar temperature or stellar evolution. Note that the abundance N_A is labeled here only by the mass number A and not additionally by the nuclear charge Z, since the latter quantity is (in most cases) uniquely defined in the s-process path. Equation (9.1) defines a set of coupled differential equations, which unfortunately cannot be solved for the most general case. For an analytic approach one must, therefore, make the following simplifying assumptions. (1) With $\lambda_n = 1/\tau_{n\gamma} = N_n\langle\sigma v\rangle_A$ (chap. 3), the destructive terms in equation (9.1) can be combined into $N_n(t)[\lambda_n(t) + \lambda_\beta(t)]$. It is then assumed that one of the inequalities, either $\lambda_\beta \gg \lambda_n$ or $\lambda_\beta \ll \lambda_n$, is valid. In the case $\lambda_\beta \gg \lambda_n$ (or equivalently $\tau_\beta \ll \tau_{n\gamma}$) the radioactive nuclei decay quickly to their adjacent isobar of higher Z, and their own abundances can be completely neglected. In the other case ($\tau_\beta \gg \tau_{n\gamma}$) the radioactive nuclei are treated as stable nuclei. Although in general this treatment is justified, there are cases, in particular at the so-called s-process branchings (see below), where neither extreme condition prevails and a more detailed treatment is called for. (2) The stellar temperature is constant during the s-process. One deals then with well-defined capture cross sections. With the definition $\langle\sigma v\rangle = \langle\sigma\rangle v_T = \sigma_A v_T$, where σ_A is the Maxwellian-averaged neutron-capture cross section for isotope A (sec. 9.2), equation (9.1) can be rewritten as

$$\frac{dN_A(t)}{dt} = v_T N_n(t)(\sigma_{A-1}N_{A-1} - \sigma_A N_A) . \qquad (9.2)$$

In this equation the thermal velocity v_T is nearly the same for all processes, since the reduced mass m is nearly constant. It should be pointed out that the s-process can occur at different stellar temperatures, depending on the neutron sources in the stellar interior. However, during a given neutron irradiation, the stellar temperature changes very little with time, since the relevant stellar-mass zone (helium burning) burns with nearly constant temperature until the neutron-producing fuel is consumed.

With the neutron flux $\phi(t)$ given by

$$\phi(t) = v_T N_n(t) \text{ neutrons cm}^{-2} \text{ s}^{-1} ,$$

a new quantity, the time-integrated neutron flux τ, is introduced:

$$\tau = \int_0^t \phi(t)dt = v_T \int_0^t N_n(t)dt \text{ neutrons cm}^{-2} . \tag{9.3}$$

It is a measure of the total neutron irradiation per unit area and is given commonly in units of neutrons per millibarn $= 10^{27}$ neutrons cm^{-2}. With the dynamic variable time t being replaced by this neutron exposure variable τ, equation (9.2) transforms to

$$\frac{dN_A}{d\tau} = \sigma_{A-1}N_{A-1} - \sigma_A N_A , \tag{9.4}$$

where the rate of change of the abundance N_A is now with respect to neutron exposure τ. As in the situations discussed in chapter 6, the processes described by these coupled equations have a tendency to be self-regulating. That is, the effect is to minimize the difference between the products $\sigma_A N_A$ and $\sigma_{A-1}N_{A-1}$ and to reach a state of equilibrium where $dN_A/d\tau = 0$. Note that in the mass region between magic neutron numbers the capture cross sections are very large (Fig. 9.6), and the equilibrium condition

$$\sigma_A N_A = \sigma_{A-1}N_{A-1} = \text{constant} , \tag{9.5}$$

called the local equilibrium approximation, should be reached first. It means that the abundance of an s-only element builds up in the s-process until its rate of destruction is equal to its rate of production ($dN/d\tau = 0$). In first-order approximation $\sigma_A N_A = \text{constant}$ implies that a nucleus with a small (large) neutron-capture cross section must have a large (small) abundance to maintain continuity in the s-process path. Quantitative tests of the $\sigma_A N_A = \text{constant}$ prediction of the s-process theory are permitted at several mass regions, notably for the element tellurium. This element is unique in that it has three s-only isotopes (^{122}Te, ^{123}Te, and ^{124}Te) shielded from the r-process by the heavy tin isotopes and by antimony (Fig. 9.7). The ratio of products $\sigma N(A)$ is found (Käp82) to be $\sigma N(122):\sigma N(123):\sigma N(124) = 1.0:0.99:0.97$. This result, as well as the plateaus seen in the $\sigma_A N_A$ correlation (Fig. 9.9) in the mass regions between closed neutron shells ($A = 90$–140 and $A = 140$–206), provide excellent confirmation of the local equilibrium approximation for the s-process.

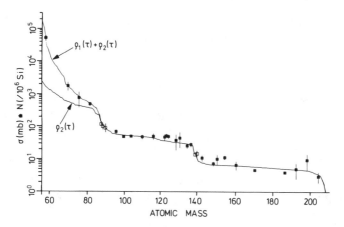

FIGURE 9.9. The product of neutron-capture cross section (at 30 keV in mb) times solar system abundances (relative to Si = 10^6) is shown as a function of atomic mass for nuclei produced only in the s-process (Alm83). The solid lines represent theoretical calculations for a single exponential distribution (neutron exposure τ) $\rho_2(\tau)$ and for the sum of two such distributions, $\rho_1(\tau) + \rho_2(\tau)$.

Nuclear shell structure introduces the characteristic ledge-precipice breaks shown in Figure 9.9 at $A \simeq 84$, 138, and 208, which correspond to the s-process abundance peaks in Figure 9.1. At these values of the mass number A, the neutron numbers are magic ($N = 50, 82, 126$) and the cross sections for neutron capture into new neutron shells are very small (Fig. 9.6). With a finite supply of neutrons available, the σN product must drop to a new plateau, just as observed (Fig. 9.9). The appearance of these breaks in the σN correlation clearly identifies the synthesizing mechanism. Quantitative explanations of the precipice structures are discussed below.

For a true equilibrium condition in the s-process path ($dN/d\tau = 0$), the σN values should be equal over the entire mass region from ^{56}Fe up to ^{209}Bi. Although this situation nearly prevails at the plateaus, the precipices at closed neutron shells, as well as the anomalous behavior below mass $A = 80$ (Fig. 9.9), clearly indicate that equilibrium is not reached over the entire path of the s-process. It is necessary then to investigate the products σN as a function of time or, equivalently, as a function of neutron exposure τ. The coupled differential equations (9.4) (with $dN_{56}/d\tau = -\sigma_{56} N_{56}$ for the seed nuclei ^{56}Fe) are usually solved with the initial boundary condition $N_A(0) = N_{56}(0)$ for $A = 56$, and $N_A(0) = 0$ for $A > 56$. For a given value τ of the total neutron irradiation, the distribution of the $\sigma N(\tau)$ products can be calculated using coupled differential equations. The number of neutrons n_c absorbed by an Fe isotope is then determined by means of the equation (Cla61)

$$n_c(\tau) = \sum_{A=56}^{209} (A - 56)N_A(\tau)/N_{56} . \qquad (9.6)$$

The first quantitative calculations (Cla61) assumed a single irradiation of the iron-group elements. However, for a uniform exposure of ^{56}Fe to a single neutron flux, the observed distribution of σN products could not be generated. Near a closed shell (Fig. 9.10), depending on the exposure level, σN decreased sharply, had a peak at the closed shell, or increased sharply. That is to say, σN depends on whether the seed nuclei typically capture enough neutrons to produce nuclei lighter than, equal to, or heavier than the closed-shell nuclei. It was concluded that the observed σN curve (Fig. 9.9) is the result not of the irradiation of a unique target but rather of quite a few irradiated targets with different neutron exposures τ. The data called for a distribution of neutron exposures. In other words, the observed σN curve (Fig. 9.9) can be thought of as a superposition of the results of different amounts of iron nuclei being exposed to different neutron fluxes, i.e., a superposition of various curves shown in Figure 9.10 (a type of Fourier transformation in the parameter n_c or τ). It was also argued that this concept of successive neutron irradiations is not surprising in view of the hypothesis (chap. 2 and Fig. 2.10) that the observed elemental abundances are the result of a mixing in interstellar space of the ashes of several former generations of stars. The concept also assumes that, for the s-process production of a given species A, only the total number of neutron irradiations is relevant. Thus, the final result is not influenced by interruptions in time or differences in location (i.e., production in several stars). For example, there could be a short period of irradiation of the iron-peak elements synthesizing the neighboring elements, followed by a long irradiation period during which the heavier elements up to bismuth are synthesized. From this work the panorama of the origin of the s-process elements began to unfold.

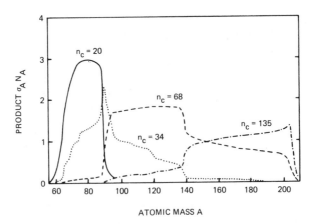

ATOMIC MASS A

FIGURE 9.10. Calculated distributions $\sigma_A N_A$ for differing levels of neutron irradiation (Cla61). Each curve is labeled by the parameter n_c, which is the average number of neutrons captured per initial iron seed nucleus. Nuclei with closed neutron shells are at masses $A \simeq 84$, $A \simeq 138$, and $A \simeq 208$.

If the function $\rho(\tau)$ represents the continuous distribution of multiple neutron exposures, the number of ^{56}Fe seed nuclei that had an exposure in the range of τ and $\tau + d\tau$ is given by $\rho(\tau)d\tau$. Each product $\sigma N(\tau)$ then has to be weighted by the function $\rho(\tau)$ and integrated over all values of τ, hence

$$\sigma_A N_A = \int_0^\infty \rho(\tau)\sigma N(\tau)d\tau \ . \tag{9.7}$$

Thus, the function $\sigma_A N_A$ is a measure of the integrated neutron irradiation to which seed nuclei have been exposed. The function $\rho(\tau)$ contains information on the history of the s-process synthesis. This function $\rho(\tau)$ is the Laplace transform[1] of the $\sigma_A N_A$ distribution and has been found (Cla61, See65, Cla74) to have the form

$$\rho(\tau) = \frac{fN_{56}^\odot}{\tau_0} \exp\left(-\frac{\tau}{\tau_0}\right), \tag{9.8}$$

where the parameter τ_0 is the mean neutron exposure and $N_{56}^\odot$ is the total number of initial seed nuclei (solar abundance $= 8.25 \times 10^5$ per 10^6 Si). The parameter f represents the fraction of the iron seed nuclei that have been subjected to the exponential distribution of neutron flux exposures; thus the total number of irradiated seeds is

$$N_{56} = fN_{56}^\odot = \int_0^\infty \rho(\tau)d\tau \ .$$

For the above function $\rho(\tau)$ the differential equation (9.4) has the simple solution (Cla68, Ulr82)

$$\sigma_A N_A = \frac{\sigma_{A-1}N_{A-1}}{1 + 1/\tau_0\sigma_A} \ . \tag{9.9}$$

In the mass region between closed neutron shells, the capture cross sections are large, and hence the product $\tau_0\sigma_A$ is very large. Therefore, the previously discussed local approximation $\sigma_A N_A \simeq \sigma_{A-1}N_{A-1}$ holds. However, near closed shells the product $\tau_0\sigma_A$ is relatively small; consequently, the denominator in equation (9.9) becomes large, producing the stepwise decreasing structure in the σN curve at closed neutron shells (Fig. 9.9). Since the parameter τ_0 causes the precipices in the σN curve, its value is essentially determined by these structures (their slopes and step heights). The best fit to the data (Fig. 9.9) was actually obtained (War78, Käp82) for two such exponential distributions, $\rho_1(\tau)$ and $\rho_2(\tau)$, with $\tau_{01} = 0.06$ neutrons mb^{-1} and $\tau_{02} = 0.24$ neutrons mb^{-1}, and with $f_1 = 2.7\%$ and $f_2 = 0.09\%$. The corresponding average numbers of neutrons captured per initial ^{56}Fe seed nucleus are $n_{c1} = 1.1$ and $n_{c2} = 8.2$ (Käp82, Mat84, Bee84a). These numbers n_c are smaller than one might naively think necessary for making, for example, ^{209}Bi from ^{56}Fe. They

---

[1] Laplace transforms are used to solve differential equations.

have small values because n_c is averaged over a range of neutron exposures required to produce the s-process distribution, whereas only the largest exposures contribute to the heaviest elements. The history of s-process nucleosynthesis is described by two exponential forms of neutron exposures: practically all s-process abundances of elements with $A > 90$ were synthesized by a strong component $\rho_2(\tau)$ in the neutron fluence distribution, while the weaker component $\rho_1(\tau)$ accounts for the steep decrease of the σN curve from the ^{56}Fe seed to the nuclei with closed shells around $A = 90$. Also, at most, 2.8% of the iron seed nuclei have been irradiated by neutrons, and only 0.09% of them have received large irradiations with $n_c = 8$–14 (Käp82, Mat84).

As discussed above, for the conditions of either $\lambda_\beta \gg \lambda_n$ or $\lambda_\beta \ll \lambda_n$ the s-process path is uniquely defined. For the intermediate condition $\lambda_\beta \simeq \lambda_n$ at species A, there is competition between neutron capture and β-decay that leads to different paths in the s-process near this mass region A, called s-process branchings (Cam59, War76, War77, War80, Ulr82). The species A can produce by neutron capture its heavier isotopic neighbor $(A + 1)$ or by β-decay its isobaric neighbor ($Z + 1$ for β^- and $Z - 1$ for β^+ or electron capture). The branching ratio R for the two s-process paths starting at species A is given by the ratio of the two rates:

$$R = \lambda_\beta/\lambda_n = 1/\tau_\beta\, N_n\langle\sigma v\rangle_A . \tag{9.10}$$

This ratio R can be deduced directly from the observed σN values of the neighboring nuclei involved in the s-process branching (for example, $R = (\sigma N)_{Z+1}/(\sigma N)_{A+1}$ or $R = (\sigma N)_{A-1}/(\sigma N)_{A+1} - 1$). If the reaction rate $\langle\sigma v\rangle_A$ is known and if the β-decay lifetime is temperature-independent, the observed ratio R leads to a value for the s-process neutron density N_n. Analysis of various branchings in the s-process path tends to give values around $N_n \simeq 10^8$ neutrons cm^{-3} (War76, Ulr82, Käp82).

In the analysis of such s-process branchings, the β half-lives for processes taking place in the stellar interior must be used with caution. In principle there are three effects that alter half-lives (Cam59, War76, War80, Ulr82): (1) β^-- or β^+-decays are hindered in the presence of electron or positron degeneracy; (2) electron-capture rates are affected by the temperature and density through the population of the K electronic shell or, in the case of complete ionization, by the density of free electrons in the vicinity of the nucleus (chap. 6); (3) β^-- or β^+-decays may occur from excited isomeric states that are maintained in equilibrium with the ground state by radiative transitions called (γ, β^-) decays (Ulr82, Tak83, Yok85). For a few cases, in the stellar environment relevant to the s-process nucleosynthesis, the last two effects play a role and must be included.

The above discussions on s-process branching and on alteration of β-lifetimes are illustrated using the example ^{176}Lu (Fig. 9.11). The s-process path leads to the isotope ^{175}Lu, which captures a neutron to produce ^{176}Lu. The latter production proceeds (Bee80) to the long-lived ground state ^{176}Lu^0

$(T^0_{1/2} = 3.6 \times 10^{10}$ y) 36% of the time and to the isomeric state $^{176}\text{Lu}^m$ $(T^m_{1/2} = 3.7$ h) 64% of the time. For a neutron density of $N_n \simeq 10^8$ neutrons cm^{-3}, the lifetime for neutron capture by both the $^{176}\text{Lu}^0$ and the $^{176}\text{Lu}^m$ state (Fig. 9.11) is of the order of $\tau_{n\gamma} \simeq 1$ y. Hence, the ground state $^{176}\text{Lu}^0$ can be regarded as a stable nucleus with regard to further neutron capture, while the isomeric state $^{176}\text{Lu}^m$ decays quickly to the stable isobar ^{176}Hf before further neutron capture can occur. We expect, then, to have an s-process

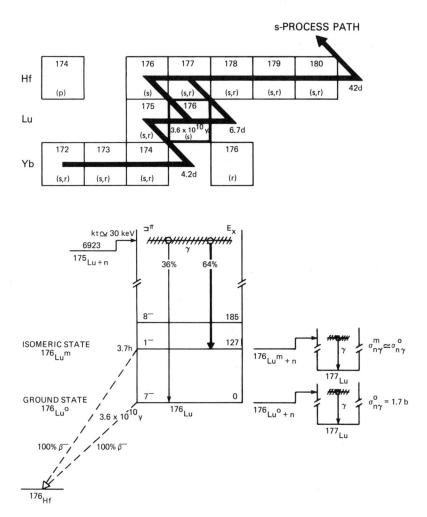

FIGURE 9.11. The s-process branching at the isotope ^{176}Lu is due to the isomeric state in ^{176}Lu at 127 keV excitation energy, which is significantly populated by neutron capture on ^{175}Lu.

branching at ^{176}Lu with a 36% branch to ^{177}Lu and a 64% branch to ^{176}Hf (Fig. 9.11). The importance of direct isomeric branching was first discussed in connection with this nuclide by Audouze and coworkers (Aud72) and Arnould (Arn73a). The large abundance of ^{176}Hf relative to p-only ^{174}Hf (^{176}Hf/^{174}Hf = 29; Cam82) rather strongly demanded a significant amount of s-process branching at ^{176}Lu to account for it. However, it was stressed (War76, War77, Bee81) that the branchings quoted above might be altered by the thermal effects in the hot stellar environment (Boltzmann distribution of excited-state populations, chap. 4). If the ground state and isomeric state could be rapidly linked internally by collisions with surrounding energetic photons and/or charged particles in the synthesizing stellar plasma, then the resulting thermal distribution of level populations would yield a dramatically shorter overall half-life against β-decay,

$$T_{1/2}(^{176}\text{Lu}) = 18.5 \exp(14.7/T_8) \text{ hours} \qquad \text{for } T_8 > 1 \,,$$

enhancing the ^{176}Lu to ^{176}Hf branch. Because of the exponential dependence on temperature, the s-process branching at ^{176}Lu constitutes a sensitive s-process "thermometer." From the observed abundances in this mass region, Beer and coworkers (Bee81) concluded that the temperature was in the range $kT = 18$–42 keV (see also Alm83, Bee84).

The s-process branching at ^{176}Lu has also been used as a "cosmic clock" (because of the rather long half-life of ^{176}Lu^0) to determine the average time scale of s-process nucleosynthesis. While first analyses (Bee80, Bee81) led to an upper limit for the mean s-process age of 12 billion years, more detailed investigations (Bee84a) showed that the clock was probably "reset" by the influence of temperature on the decay rate. Since s-process branchings involving electron capture depend critically on density, they can serve as s-process "barometers" (Ulr82, Bee84a). Information derived from the s-process abundances can help to determine details of the environment in which this process took place.

Finally, one must ask, where is the actual site of the s-process and what is the source of the neutrons? Many locations and sources have been suggested (Cam55, Cam60, Ulr82). A very convincing case was made by Iben (Ibe75), who argues that the site is in the helium-burning shell of a pulsating red giant and that the neutrons come from the ^{22}Ne(α, n)^{25}Mg reaction. That this is not unreasonable can be shown by the following discussion. Consider that only 2.7% of the ^{56}Fe available was required to provide the seed for all the heavier elements (see above). With the solar abundance of ^{56}Fe equal to 8.25×10^5 (Si = 10^6) we find that 2.2×10^4 neutrons are needed to transmute the ^{56}Fe into ^{57}Fe, and 3.4×10^6 neutrons to transmute all the ^{56}Fe into ^{209}Bi (representing an upper limit). Now, the ^{22}Ne for the process ^{22}Ne(α, n)^{25}Mg originates in the helium burning of the elements CNO left after completion of hydrogen burning (chaps. 6 and 7). The sum of the abundances of these elements is 3.2×10^7 (Si = 10^6), which is also the number of ^{22}Ne nuclei available for the $^{22}Ne(\alpha, n)^{25}Mg$ process and thus represents the maximum number of neutrons available. Thus, there are more neutrons available than needed. The problem of neutron poisoning (chap. 7) has been studied in detail (Alm83). From this work the number of neutrons appears to be sufficient even in the light of loss by neutron poisoning. The Iben model, which produces periodic neutron irradiations, is closest to reproducing the required exponential neutron irradiations as well as other features of the s-process (Tru77, Ulr82). In subsequent work Iben (Ibe82, Ibe84b) also found that in the He-shell flashes (pulses) mixing of protons leads to the reactions $^{12}C(p, \gamma)^{13}N(e^+v)^{13}C$, and thus $^{13}C(\alpha, n)^{16}O$ could be another important neutron source. The quantity of ^{13}C produced is, however, more difficult to predict. As a summary, it is fair to say that the s-process has the clearest phenomenological basis of all processes of nucleosynthesis.

9.5. The r-Process

Among the trans-iron elements ($A \geq 70$), there are 27 nuclei that are produced only by the r-process (labeled r in Fig. 9.7). However, the majority of all isotopes have their origin in both the s-process and the r-process (labeled s, r in Fig. 9.7). By demanding that the s-process contributions N_s to these elements be such that the product σN_s falls on the smooth calculated curve $f(A)$ of Figure 9.9, the difference between solar abundances $N_\odot$ (Cam82) and s-process abundances yields a good approximation to the r-process abundance contributions N_r:

$$N_r \simeq N_\odot - N_s = N_\odot - f(A)/\sigma(A) .$$

In performing such a subtraction, good values of the neutron-capture cross sections $\sigma(A)$ are clearly needed. The resulting N_r abundances from a recent analysis (Käp82) are shown in Figure 9.12. It has been pointed out that, because of numerical uncertainties, the above subtraction procedure will be of little value when N_r turns out to be a small difference between two large numbers. For example, a change of about 25% in the capture cross sections is sufficient to resolve the discrepancy found for the Zr isotopes and bring them up to the generally smooth distribution curve. However, the N_r abundance curve clearly exhibits pronounced maxima around $A = 80$, $A = 130$, and $A = 195$. It was the existence of these maxima (displaced by several mass units below the s-process peaks), as well as the existence of the naturally radioactive elements ^{232}Th, ^{235}U, and ^{238}U (not produced in the s-process, since it terminates at ^{209}Bi), that led to the recognition of the need for an additional and different nucleosynthesis mechanism known as the r-process (B^{2}FH and Cam57). It is generally believed that many of the trans-iron elements, including all of those heavier than ^{209}Bi, were synthesized by this rapid neutron-capture process.

The classical r-process mechanism is based on a flow concept similar to that of the s-process, except that very high densities of free neutrons are assumed, so that the neutron-capture rates are much faster than those for β-decay ($\tau_{n\gamma} \ll \tau_{\beta}$). A nucleus (Z, A) in such an environment transmutes quickly to its neutron-rich isotopic neighbors $(Z, A + i)$ by a chain of (n, γ) capture reactions, whose rates are fast compared with the β-decay rates of the resulting unstable isotopes $(Z, A + i)$. Since the β-decay lifetimes of the neutron-rich isotopes are short (τ_{β} = milliseconds to seconds), the required neutron den-

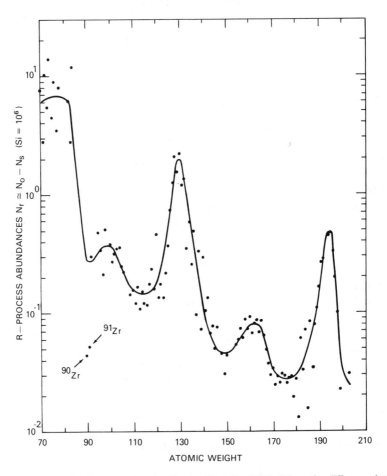

FIGURE 9.12. Approximate r-process abundances N_r derived (Käp82) as the difference between solar abundances $N_{\odot}$ and calculated s-process abundances N_s. The line through the data points is to guide the eye. The Pb and Bi isotopes are omitted because of their more complex s-process history (Cla67). The isotopes ^{90}Zr and ^{91}Zr appear to be produced by the s-process to 99% and 97%, respectively, hence their N_r determinations are critically dependent on neutron-capture cross sections. The most important characteristics are the three main abundance peaks at $A \simeq 80$, $A \simeq 130$, and $A \simeq 195$, which led to the identification of the r-process.

sities (see above) for, say, $\tau_{n\gamma} \leq 1$ ms are $N_n \geq 10^{19}$ neutrons cm^{-3}. As the neutron number increases, the neutron binding energy Q_n decreases, and the rapid addition of neutrons stops when Q_n approaches zero energy (the so-called neutron drip line). This will occur, for a given neutron density N_n and a stellar temperature T, when the rate of the photodisintegration process (γ, n) is equal to that of the (n, γ) capture process (chap. 8), i.e.,

$$\lambda_{\gamma n} \propto \frac{T^{3/2}}{N_n} \exp\left(-\frac{Q_n}{kT}\right)\lambda_{n\gamma}. \tag{9.11}$$

For example, with $N_n = 10^{24}$ neutrons cm^{-3} and $T_9 = 1$, both rates become equal at a neutron binding energy of $Q_n \simeq 2$ MeV. At the point at which the (n, γ) reaction and the (γ, n) inverse reaction reach equilibrium, the chain of rapid neutron captures stops, and the nuclei wait until β-decay transforms neutrons inside the nuclei into protons:

$$(Z, A + i) \rightarrow (Z + 1, A + i) + \beta^- + \bar{\nu}.$$

The isobaric neighbor $(Z + 1, A + i)$ then rapidly absorbs neutrons until the balance between neutron capture and photodisintegration is again reached, this time for its isotopic neighbor $(Z + 1, A + i + k)$. This recurring sequence of events leads to a "waiting point" (B^{2}FH) for each nuclear charge Z, at which point β-decay must occur before any further neutrons can be added (see, however, Cam83). For these conditions, the abundances are not characterized by the atomic mass (N_A) as in the s-process but rather by the nuclear charge (N_Z). The time dependence of the abundances is determined by the set of coupled differential equations

$$\frac{dN_Z(t)}{dt} = \lambda_{Z-1}(t)N_{Z-1}(t) - \lambda_Z(t)N_Z(t), \tag{9.12}$$

with the boundary conditions $N_Z(0) = N_{26}(0)$ for ^{56}Fe as seed and $N_Z(0) = 0$ for $Z > 26$. The first and second terms describe the production via the waiting point $Z - 1$ and the destruction via the waiting point Z, respectively. Note that this set of equations has a tendency to reach a state of equilibrium $[dN_Z(t)/dt = 0]$ with $N_Z \propto 1/\lambda_Z = \tau_\beta(Z)$. As a consequence, the N_Z abundances in the r-process correlate with the β-decay lifetimes at the waiting points of charge Z, while in the s-process the abundances are correlated inversely with the capture cross sections $(N_A \propto 1/\lambda_A \propto 1/\sigma_A)$.

The r-process drives the nuclear matter far to the neutron-rich side of the line of β-stability (Fig. 9.13), typically 10–20 neutrons in excess of their stable isobaric neighbors. If the path reaches nuclei with magic neutron numbers $(N_m = 50, 82,$ and $126)$, the next heavier isotopes with $N_m + 1$ neutrons have very small neutron binding energies Q_n and relatively long half-lives, and thus these nuclei with closed neutron shells represent a special set of waiting points in the r-process path. After neutron capture and β-decay the (Z, N_m) nucleus is transmuted to $(Z + 1, N_m)$, which is again a nucleus with the same closed

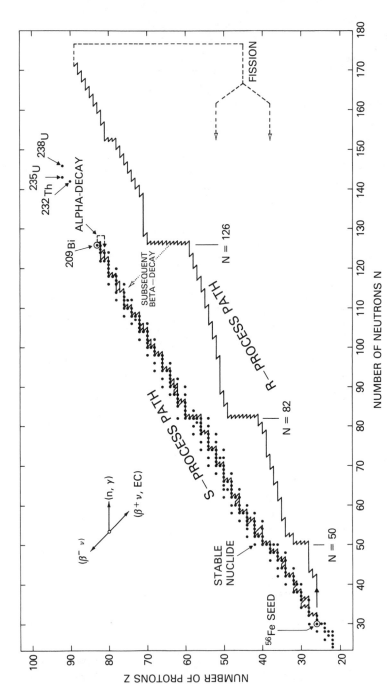

FIGURE 9.13. Neutron-capture paths for the s-process and the r-process are shown in the (N, Z)-plane. Both paths start with the iron-peak nuclei as seeds (mainly ^{56}Fe). The s-process follows a path along the stability line and terminates finally above ^{209}Bi via α-decay (Cla67). The r-process drives the nuclear matter far to the neutron-rich side of the stability line, and the neutron capture flows upward in the (N, Z)-plane until β-delayed fission and neutron-induced fission occur (Thi83). The r-process path shown was computed (See65) for the conditions $T_9 = 1.0$ and $N_n = 10^{24}$ neutrons cm^{-3}.

neutron shell N_m. One expects, therefore, a sequence of waiting points at the same magic neutron number N_m, increasing the nuclear charge Z slowly, by one unit at a time. Finally, after several such "neutron capture plus β-decay" processes, the resulting nuclei are sufficiently close to the stability line that the neutron binding energy Q_n becomes large enough to break through the neutron magic bottleneck at N_m and to resume the normal sequence of neutron-capture events (Fig. 9.13). The effect of this sequence of waiting points at nuclei with magic neutron numbers is demonstrated in Figure 9.13.

At the cessation of the r-process, the neutron-rich nuclei β-decay to their stable isobars (keeping A constant). In first-order approximation $[dN_A(t)/dt = 0]$, this means that the abundance of an r-process nucleus divided by the β-decay lifetime of its neutron-rich r-process isobaric progenitor is roughly constant (Bec59, Fow64a). Since the waiting points at the magic neutron numbers involve nuclei with significantly longer than average β-decay lifetimes, these nuclei build up to relatively large abundances in the r-process track. Accordingly, the abundances of the progenitors with $N_m = 50$, 82, and 126 are large. Since their associated numbers of protons will be less than in the corresponding s-process nuclei with the same magic number of neutrons, it follows that the stable isobars of the r-process progenitors will have smaller mass numbers (Fig. 9.13). For example, $^{124}_{50}\mathrm{Sn}_{74}$ is not neutron magic, but $^{124}_{42}\mathrm{Mo}_{82}$, the progenitor in the r-process, is. Similarly, we see in Figure 9.12 the relatively sharp r-process abundance peaks occurring near the atomic weights $A = 80$, $A = 130$, and $A = 195$, each of which is 8–12 mass units below the corresponding s-process peak (Fig. 9.1). Thus, within the r-process hypothesis, there is no obvious correlation between the nuclear properties of the progenitors and those of the stable isotopes reached via the chain of β-decays. For example, the product of the abundances of the stable r-process nuclei and their neutron-capture cross sections, σN, should not exhibit a smooth curve of σN versus A (as in the case for the s-process, Fig. 9.9), and, indeed, they do not (B^{2}FH).

The overall r-process path bypasses nuclei with natural α-radioactivity (which stops the s-process path) and terminates (Fig. 9.13) only by neutron-induced fission or, more important, by β-delayed fission near or somewhat beyond $A = 270$ (See65, Thi83, Kla83). There is considerable uncertainty as to what is the largest mass ($A_{\max}$) produced in the r-process; this is the result of uncertainties of mass formulae and of fission barrier predictions far from stability (Thi83, Kla83; Fig. 9.14). The fission reactions feed nuclear matter back into the process roughly at $A \simeq A_{\max}/2$. Thus cycling and a true steady flow can occur in the mass range $A_{\max}/2$ to $A_{\max}$, depending on the time scales available to the r-process (See65). If this cycling occurs, the nuclear species in this mass range will grow in number as long as the neutron supply lasts, provided that there are enough seed nuclei to start the process off. Following the synthesizing episode, the very heavy nuclei near and below $A_{\max}$ begin their succession of β-decays, thereby increasing their nuclear charge. During the decay phase, some nuclei will fission spontaneously if their fission parameter Z^2/A (Cla68) becomes too large. For the heaviest nuclei produced in the r-process, this fate seems inevitable as the decay carries them toward the valley of stability. Only for values of $A \leq 256$ does it appear possible that the isobaric decay chain reaches β-stability without fissioning and that α-decay proceeds more rapidly than spontaneous fission. In summary, the r-process is singularly responsible for the production of the trans-bismuth elements—in particular, the long-lived nuclei such as ^{232}Th, ^{235}U, and ^{238}U (Fig. 9.13), which serve as chronometers in estimating the age of the elements (sec. 9.6).

It follows from equation (9.11) that the exact position of the waiting points in the (N, Z)-plane, i.e., the path of the r-process, depends on neutron density N_n, stellar temperature T, and neutron binding energy Q_n. For example, larger values for N_n shift the waiting points to more neutron-rich matter, and higher temperatures displace them to neutron-poorer nuclei. Since Q_n appears in the exponent of equation (9.11), the positions of the waiting points are very sensitive to the neutron binding energy Q_n. A detailed treatment of the r-process (eqs. [9.11] and [9.12]) requires thorough knowledge of basic nuclear physics properties of the r-process nuclei. Unfortunately, since the nuclei in question are so far from the line of stability and have such short lifetimes, they have not yet been produced in terrestrial laboratories; thus their properties cannot be obtained from direct experimental measurements (chap. 5). One must, therefore, rely on semiempirical extrapolations of the observed properties of neutron-rich nuclei currently accessible to experiments (i.e., those unstable nuclei relatively close to the stability line). However, it is not known whether extrapolation to the very neutron-rich nuclei of the r-process (see below) is sufficiently accurate. These uncertainties limit the reliability of conclusions reached in all discussions of the r-process.

To calculate the neutron binding energy Q_n, one needs to know the mass of all the neutron-rich nuclei. To date, the nuclear mass estimates used have been based on the liquid-drop model of atomic nuclei (Mye66) and (more recently) on the droplet model (Mye77, Hil76), including closed-shell corrections and effects of deformed nuclei. It is expected (Sch82) that the Q_n values along the r-process path can be estimated to about 1 MeV. The absolute β-decay lifetimes, particularly for nuclei with magic neutron numbers, also enter sensitively into calculations of the path of the r-process and thus into the location and width of the r-process abundance peaks (Fig. 9.12). Furthermore, these lifetimes play an important role in any nonstatic r-process calculations, i.e., the length of time required for the r-process. Since the major waiting points along the r-process path are at the nuclei with magic neutron numbers (B^{2}FH), a lower limit for this time may be obtained simply by adding up the time spent at each waiting point. The β-decay rate λ_β (or, equivalently, the lifetime $\tau_\beta = 1/\lambda_\beta$) is given by the relation (chap. 6; Ful80, Sch82) $\lambda_\beta \propto W_\beta^5 M^2$, where W_β is the effective β-decay energy and M^2 represents the squared matrix element involved in the decay. However, since W_β depends on Q_n, and since there exist uncertainties of the order of 1 MeV for Q_n, one expects large uncertainties in these rates. Additional questions arise (Sch82) in regard to (1) level densities available for β-decay to excited states in the daughter nucleus, (2) β-strength functions (including the effects of both allowed and first-forbidden transitions) averaged over the final states in the daughter nucleus (the so-called gross theory of β-decay; Kod75, Tak73) and (3) β-strength functions taking into account the effects of nuclear structure (Kla83). The predictions of the various models suggest that the total time spent at the *r*-process waiting points is in the range of 0.1–30 s (Sch82). The actual *r*-process should last longer than this because of the time required for neutron captures and for fission cycling if indeed it occurs. At present it is not clear which of these model calculations is most nearly correct for nuclei far from the stability line.

As pointed out above, the termination of the *r*-process is due to fission because eventually, for high-*A* nuclei, neutron-induced or β-delayed fission will be faster than β-decay. In addition, during the decay back to the stability line, spontaneous fission can be faster than β-decay, thus preventing some nuclei from making it back to the valley of stability. Thus the details of the *r*-process path depend on the characteristics of fission barriers for *r*-process nuclei (Sch71a, Sch73, Haw74a, Sch82, Thi83). The size of these fission barriers also determines the possibility of producing the theoretically predicted superheavy elements (nuclei with closed neutron and proton shells at $N = 184$ and $Z = 114$) in the *r*-process (Sch82). The *r*-process thus provides a good demonstration of how much knowledge of nuclear physics is required for astrophysical studies.

The first *r*-process calculations based on the waiting-point method were carried out (See65) with constant temperature and neutron density (Fig. 9.13). However, it was recognized that the extremely high neutron densities and temperatures needed were probably attainable only in dynamic events such as supernovae. At the present time, it is generally believed that the waiting-point approximation should be replaced (Cam83) by dynamic *r*-process flow calculations, taking into account (n, γ), (γ, n), and β-decay rates, as well as time-varying temperature and neutron density (for recent experimental work, see Kra86). Schramm (Sch82) has discussed such calculations in some detail, emphasizing that nonequilibrium effects are particularly important during the freezeout at the end of the *r*-process when the temperature drops and the neutron flux falls to zero. These nonequilibrium calculations include dynamic reaction networks and thus are quite sensitive to another nuclear physics input, namely, the size of the neutron-capture cross sections for nuclei far from the stability line (Hol76, Woo78, Thi80, Mat83).

Although many recent dynamical *r*-process calculations (Sch82, Thi83, Tru83, Tru84, Cam85, Fow84) have met with impressive success in reproducing the *r*-process abundances (Fig. 9.14), the problem is still far from being fully understood. If all the nuclear physics input parameters were known to high accuracy, the observed *r*-process abundance distribution would, like the

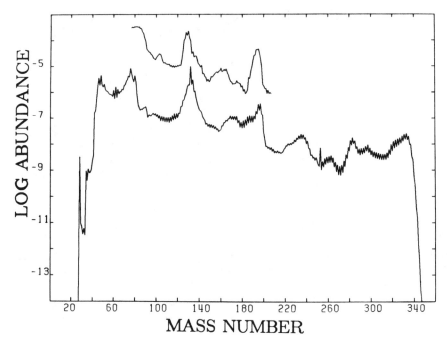

FIGURE 9.14. Plot of *r*-process abundances produced in the thermal runaway model (*lower curve*) of Cameron, Cowan, and Truran (Cam85) compared with the solar system *r*-process abundances (*upper curve*, Fig. 9.12). The calculations do not include β-delayed fission (Thi83, Kla83) and thus lead to the production of nuclei up to $A = 340$. The vertical normalization is arbitrary.

s-process, reflect the history of the *r*-process nucleosynthesis, i.e., its dynamics and its site (Laplace transform, sec. 9.4). Many suggestions have been made for possible sites of the *r*-process, almost all in supernova explosions or other catastrophic events where the basic requirement of a large neutron flux of short duration is met (Sch82, Tru83, Tru84).

In summary, in spite of the problems associated with it, the *r*-process continues to be an important nucleosynthetic process. It is directly applicable in both long- and short-lived nucleochronometric dating (sec. 9.6) as well as in explanations of observed isotopic anomalies in meteorites (chap. 10). The general nuclear physics of a neutron-rich path of nucleosynthesis seems to be well established. However, details of the properties of nuclei along the way are exceedingly uncertain. This will require greater attention in the future. Similarly, the astrophysical site remains a mystery. Its detailed determination must await improvement in our knowledge of nuclei far from stability. A major mystery of the synthesizing source is that the *r*-process path appears to be very narrow, implying a restricted range of source conditions, and yet all the proposed sites seem to have rather violent and wide-ranging characteristics. Another mystery is that the *s*- and *r*-process abundance peaks have about the same magnitude even though the processes appear to be quite different. Is this a coincidence, or an important clue to the astrophysical site? The whole field is clearly in need of further study.

9.6. Nucleocosmochronology—the Age of the Chemical Elements

Study of the very distant past has itself had quite a long history. Dating specific events such as the formation of the earth and the sun has moved from the realm of religious revelation to quantitative scientific inference based largely on physical measurements. Dating of various events in the history of the universe is referred to as cosmochronology. There are about four major chronological milestones: the age of the universe, the age of our Galaxy (the Milky Way), the age of the chemical elements, and the age of the solar system. It seems trivial, but note that the universe is older than our Galaxy and that our Galaxy is older than the solar system.

The determination of the age of the universe is based on redshift and luminosity observations of the brightest galaxies in clusters of galaxies (chap. 1). These observations led to the Hubble time or reciprocal of Hubble's constant, which is related to the age of the universe (Hod81, San82). A recent value for the Hubble time is 19.5 ± 2.7 billion years (San84). However, as we have seen, the Hubble time is truly equal to the age of the expanding universe only for a completely open universe. The observed visible matter in galaxies is estimated to be 10% of the critical density for closure, which because of the increase in gravitational attraction over a completely open universe reduces the age of the universe to 16.5 billion years (Fow84, Fow85). Invisible matter, neutrinos, black holes, and so on, also may add to the gravitational attraction, further decreasing the velocity of expansion perhaps even to that corresponding to the critical density, for which the age would be 13.0 billion years. Other investigators obtain results for the Hubble time equal to about one-half that quoted above (deV79). Thus, the time back to the beginning, the age of the universe, appears to fall between 9 and 23 billion years (Hod81, Fow85).

The age of our Galaxy—and to a good approximation of other galaxies—is derived from the age of its oldest members, the stars in globular clusters (Fig. 1.7). The key assumption is that all of the stars in the cluster formed at roughly the same time. Since the ages of the stars can be determined from their relative evolutionary characteristics (Hertzsprung-Russell diagram, Fig. 7.2), the oldest systems in the Galaxy can be dated (Ibe74, Fow77). The results of such analyses (San82, Van83, Ibe82, Jam83, Fow85) indicate that the age of the globular clusters and thus the age of the Galaxy is between 12 and 18 billion years. The time for the formation of large galaxies is estimated to be about 1 billion years (Pee84, Fow85), which must be added to the age of galaxies to obtain the age of the observable universe—between 13 and 19 billion years.

Since the discoveries of natural radioactivity by Becquerel and the Curies, the stepwise decays of uranium and thorium to lead isotopes have been studied extensively. In 1929 Rutherford investigated a sample of pitchblende. He determined the amount of uranium and radium in the rock, calculated the annual output of α-particles, measured the amount of helium retained in the rock, and by simple division found the period during which the rock had existed in compact form to be 700 million years. The pioneering work of Rutherford (Rut29) on natural radioactivity led to the science of nuclear geo-chronology and nuclear cosmochronology. Rutherford's age for the piece of pitchblende was soon exceeded by many other minerals, and today there is abundant evidence (Pat56, Ost63, Kir78) pointing to major solidification of the earth and the meteorites $t_m = 4.55 \pm 0.07$ billion years ago (Fig. 9.16). The formation of the sun and its planets occurred only slightly earlier (up to 0.2 billion years; Ost63), thus a value of $t_s = 4.6 \pm 0.1$ billion years is commonly adopted (Fow77) as the age of the solar system, i.e., the age of its isolation and condensation from the interstellar medium (Fig. 9.16). These results are fully consistent with the ages reported for lunar samples (Moo70).

Since most of the chemical elements and their isotopes found on the earth, moon, and meteorites are, as far as one can measure, stable, one might think that they have existed forever. However, several unstable nuclei still exist at measurable quantities in the solar system owing to their long lifetimes. Their existence argues clearly against their having existed forever, and thus one has to conclude that the elements themselves have a finite history. The key ques-tions in relation to the science of nucleosynthesis are then: When were the elements formed, and what kind of cataclysmic events in the distant past were required to produce these unstable elements as a side effect? To what extent is it possible to use the abundances of the radioactive nuclei and their daughters, as they existed at the time of the formation of the solar system, to learn some-thing about the prior history of nucleosynthesis? It must be kept in mind that the heavy elements found in the sun and its planetary system were synthesized elsewhere. Nuclear processes presently occurring in the sun only transform hydrogen into helium (chap. 6). Therefore, the origin of the solar system ele-ments must date back to a time prior to formation of the solar system—a time beyond 4.6 billion years ago. It is through the use of radioactive nuclei that it is possible to gain insight into the history of the elements. It is similar to using the radioactive isotope ^{14}C to date archeological samples. This is the science of nucleocosmochronology, which was started by Rutherford (Rut29) and revised and extended with the papers of Burbidge, Burbidge, Fowler, and Hoyle (B^2FH) and Fowler and Hoyle (Fow60). Their seminal work provided the basis upon which modern nucleocosmochronology is built. Nucleo-cosmochronology is the use of the observed or derived relative abundances of radioactive nuclei to determine time scales for the nucleosynthesis of these nuclei. These time scales in turn determine or constrain the time scales for the duration of nucleosynthesis of all the elements, the age of the Galaxy, and the age of the universe itself (Sym81, Sch82, Fow84, Fow85).

Since the history of nucleosynthesis of the elements involves time scales beyond 4.6 billion years, the radioactive nuclei with lifetimes of that order, such as ^{232}Th, ^{238}U, and ^{235}U, are particularly useful as chronological clocks (Fow60, Cla68, Fow72a). These radioactive nuclei and their daughters are analogous to hourglasses (Fig. 9.15) and are called often aeon glasses (one aeon = 10^9 y).

Think of the sand in the top of the hourglass as the radioactive parent ^{232}Th, ^{238}U, and ^{235}U. The sand in the bottom of the glass represents the daughter, which, after many α- and β-transformations of short lifetime com-

THE RADIOACTIVE AEON GLASSES

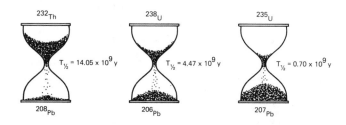

COMPLICATIONS IN NUCLEOCOSMOCHRONOLOGY

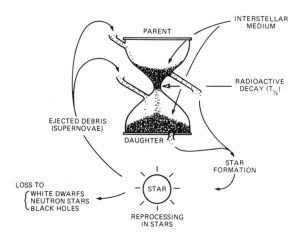

FIGURE 9.15. Radioactive aeon glasses ^{232}Th, ^{238}U, and ^{235}U. The sand (nuclei) in the glasses represents the abundance of the radioactive parents (*top*) and of the stable great-great-granddaughters (*bottom*) as they are in the interstellar medium. Both can be lost to star formation, while some is regained, for example, via supernova events (Fow77). This leakage and return of the material of the aeon glasses from and to the interstellar medium makes the application of these nuclei as chronometers more complicated.

pared to that of the parents, is ^{208}Pb, ^{206}Pb, and ^{207}Pb. Table 9.1 shows additional radioactive aeon glasses with lifetimes covering a wide range of values (Led78). As in hourglasses, the amount of sand in the top and in the bottom of the aeon glass, together with the radioactive decay rate of the parents, yields the age of the parents. Unfortunately, the situation of these nuclear chronometers is much more complicated than the analogous hourglass. There could be changes in the ratio of daughter to parent from causes other than radioactive decay—for example, by chemical or physical fractionation processes. Since the daughter is not the same chemical element as the parent, the daughter and its compounds have chemical and physical properties different from those of the parent and its compounds. Fractionation may occur at any time and reset the clock.

TABLE 9.1 Selection of Potential Cosmochronometers

Radioactive Nucleus	Half-Life[a] (10^9 y)	Radioactive Nucleus	Half-Life[a] (10^9 y)
^{187}Re	50	^{244}Pu	0.0826
^{87}Rb	47	^{129}I	0.0157
^{232}Th	14.05	^{107}Pd	0.0065
^{238}U	4.47	^{26}Al	0.000716
^{235}U	0.704		

[a] Source: Kernforschungszentrum Karlsruhe, *Chart of Nuclides* (4th ed.; Karlsruhe, 1974).

Another complication arises if the aeon glass is leaky (Fig. 9.15). The primary interest is in the history of the Galaxy's interstellar medium consisting of gas and dust. In particular, there is great interest in the buildup of the abundance of the heavy elements in the interstellar medium from essentially zero at the time of formation of the Galaxy to their abundance at the time of formation of the solar system. When the solar system condensed from the interstellar medium 4.6 billion years ago, it contained heavy-element abundances, including both radioactive parents and daughters, characteristic of the interstellar medium as it was at that time and as it was in the region of the Galaxy where the system formed. After condensation from the interstellar gas, the solar system received no further contributions from galactic nucleosynthesis. It is a closed chemical and physical system with no elemental or isotopic abundance changes except for radioactive decay. Note that it is not necessary to assume that the Galaxy is completely mixed or homogeneous but only that the entire Galaxy formed at nearly the same time, so that the abundances of the heavy elements in our sample do in fact yield an age that is characteristic of the Galaxy as a whole. The main leak from the interstellar medium is star formation and the eventual loss of material to white dwarfs, neutron stars, and black holes. Meanwhile, other stars return material to the interstellar medium. This material will in general be enhanced in heavy-element abundance over that in the original material of the star, owing to nuclear processes which generate the stellar energy and transform lighter elements into heavier ones (stellar nucleosynthesis). The material is returned via several mechanisms such as stellar winds, mass loss during the red giant stage, the formation of planetary nebulae, and nova events, and most spectacularly via supernova explosions.

The standard point of view in this age game (Fig. 9.15) is that shortly after the big bang (historical time t_u), the Galaxy and other galaxies formed with their first-generation stars (time t_g). The stars evolved while burning their nuclear fuels and began at some time (t_n) to produce the heavier elements beyond hydrogen and helium, which were the only major nuclear species in the debris of the big bang (chap. 2). The heavy elements produced in stars include the radioactive elements that serve as nuclear chronometers or aeon glasses. Since times are generally taken to be increasing from the present ($t = 0$), we find that $t_u > t_g > t_n$.

The trans-bismuth radioactive parents ^{232}Th, ^{238}U, and ^{235}U are all produced in the r-process (sec. 9.5). A knowledge of the absolute abundances of the parent and daughter in each of the aeon glasses (Fig. 9.15) would reveal their history. However, absolute abundances still cannot be precisely measured or calculated.

Fortunately, ratios of abundances, such as ^{235}U/^{238}U and ^{232}Th/^{238}U, which can be precisely determined, also can be used in the age game, since these nuclei are produced via the same stellar process. Let us set the conceptual stage for such investigations starting with the abundance ratio ^{235}U/^{238}U, which has a measured value of $(7.25 \pm 0.01) \times 10^{-3}$ (Str58, Led78, Tat76, Che80). Since both are isotopes of the same element, this ratio is not subject to errors of measurement or to changes by chemical fractionation. Using their known lifetimes (Table 9.1), the ratio was calculated backward in time to the closure of the solar system ($t_s = 4.6 \times 10^9$ y), at which time it was increased to $(^{235}$U/^{238}U$)_s = 0.33$ (Fig. 9.17) owing to the shorter lifetime of ^{235}U. Beyond this time, r-process production plus decay during the period of galactic nucleosynthesis influence this ratio, and thus extrapolation into this epoch requires a chronological (galactic) model (Fig. 9.16). Since both nuclei are produced in the same stellar environment, the ratio of their production rates is independent of the actual site as well as of the time of their stellar synthesis. Armed with a knowledge of this primary production ratio $\lambda_{235}/\lambda_{238}$, and with a given galactic model, the abundance ratio at the time the solar system became a closed system can be calculated and compared with observation. This procedure, though model-dependent, enables one to determine the duration of r-process nucleosynthesis from its beginning in the first stars in the Galaxy up to the last events before the formation of the solar system.

The important physics input in the age game is, therefore, the r-process production rates for the radioactive trans-bismuth nuclei, which have been calculated (B^{2}FH) using the methods outlined in section 9.5. However, the calculations of these production rates involved a special twist. For example, the abundance of ^{235}U is derived not only from the β-decay of its r-process progenitor at $A = 235$ but also from r-process material at $A = 239, 243, 247, 251,$ and 255 because, although β-stable, the nuclei at these atomic weights rapidly α-decay to ^{235}U (Fow77). At $A = 259$ and beyond, very short-lived spontaneous fission is the fate of the β-stable forms. Similarly, ^{238}U has progenitors at $A = 238, 242, 246,$ and a small contribution (10% α-activity) from $A = 250,$ but not at all from $A = 254,$ which decays 100% by spontaneous fission (Fow77). Thus, ^{235}U has 6 progenitors that decay rapidly to it after an r-process event, and ^{238}U has $3 + 0.1 = 3.1$ progenitors. If the production rate curve of the r-process were uniform in this mass region, one would find from the number of progenitors an a priori ratio of $\lambda_{235}/\lambda_{238} = 6/3.1 = 1.93$. Detailed r-process calculations come out very close to this value and, because of the slightly humped curve in this mass region (Fig. 9.14) caused in part by the nuclear physics properties (Kla83, Thi83) of the r-process path, one finds values in the range of $\lambda_{235}/\lambda_{238} = 1.24$–$1.42$ (Fow77, Thi83).

As pointed out above, the analogy to hourglasses in the age game (Fig. 9.15) fails because the sand in the aeon glass is being added or removed,

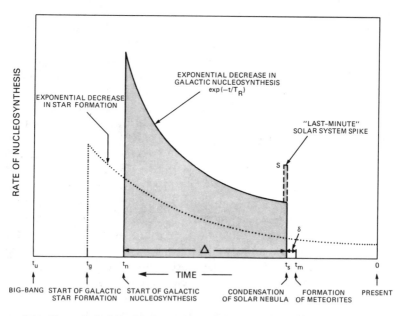

FIGURE 9.16. Chronological model for nucleosynthesis within the Galaxy, which is used to extract the time interval Δ of solar system element production and hence the age t_n of these elements. The time is measured backward from the present epoch. Note that radioactive elements produced in galactic nucleosynthesis are continuously decaying. A last-minute solar system spike, caused, for example, by a nearby supernova, is indicated as occurring shortly before the material of the solar system stopped receiving further contamination from stellar activity in the Galaxy.

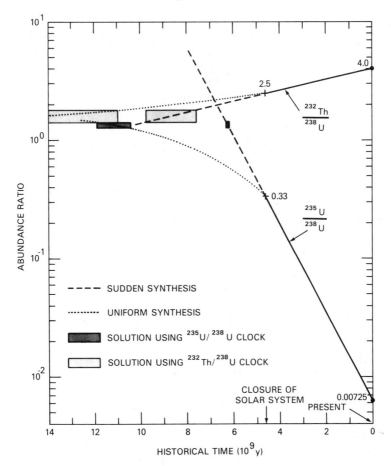

FIGURE 9.17. Abundance ratios $^{235}U/^{238}U$ and $^{232}Th/^{238}U$ are shown as a function of time. Using the present ratios and the known half-lives, the ratios were calculated by extrapolation backward in time to an epoch when the solar system became a closed system. Beyond this time the curves are calculated for the assumptions of sudden and uniform synthesis. The intersection of the curves with the production ratios calculated for the r-process then yields the time of onset of nucleosynthesis in the Galaxy.

top and bottom, by nucleosynthesis (production in stars) and astration (destruction in stars). Properly expressed, differential equations can compensate for this failure. By measuring time backward from the present epoch (Figs. 9.15 and 9.17), the amount N of a long-lived radioactive nucleus whose progenitors in the synthesis mechanism are all relatively short-lived is given by the solution of the differential equation (Fow60, Sch70)

$$\frac{dN(t)}{dt} = -\frac{N(t)}{\tau} + f(t, t_n, \lambda, T_R) \tag{9.13}$$

for the time interval $t_s < t < t_n$ (Fig. 9.16). The first term describes the normal radioactive decay of species N, and the second term its production via galactic nucleosynthesis. This latter term can be a complicated function of parameters such as time t, onset of nucleosynthesis t_n, primary production rate λ per unit time for the radioactive nucleus under consideration, and characteristic time T_R for element synthesis. Thus this function f describes the chemical evolution of the Galaxy. At the condensation of the solar nebula (t_s), the solar system material became a closed system with $f(t, t_n, \lambda, T_R) \equiv 0$ for $0 < t < t_s$, and thus the species N change only by normal radioactive decay (free decay):

$$N(t) = N(0) \exp(t/\tau) , \qquad (9.14)$$

where $N(0)$ represents the number of radioactive nuclei (the parents) observed today and τ its mean lifetime ($\tau = T_{1/2}/\ln 2$). In this time interval the ratio of uranium abundances is given by the relation

$$\left(\frac{{}^{235}U}{{}^{238}U}\right)_t = \left(\frac{{}^{235}U}{{}^{238}U}\right)_0 \frac{\exp(t/\tau_{235})}{\exp(t/\tau_{238})} . \qquad (9.15)$$

Equation (9.15) was used to calculate the abundance ratio back in time, arriving at $({}^{235}U/{}^{238}U)_s = 0.33$ at $t_s = 4.6 \times 10^9$ y (Fig. 9.17). If we assume a single r-process event responsible for the primary ratio $\lambda_{235}/\lambda_{238} = 1.24$–$1.42$, the production function f is described by a δ-function and equation (9.13) becomes

$$\frac{dN(t)}{dt} = -\frac{N(t)}{\tau} + \lambda\delta(t_n - t) ,$$

with the solution [for the boundary condition $N(t_n) = \lambda$]

$$\frac{\lambda_{235}}{\lambda_{238}} = \left(\frac{{}^{235}U}{{}^{238}U}\right)_s \frac{\exp(\Delta/\tau_{235})}{\exp(\Delta/\tau_{238})} ,$$

where $t = t_s + \Delta$. As expected, the time dependence of the abundance ratio is identical with that for normal radioactive decay in this time interval (dashed line in Fig. 9.17). From the intersection of the dashed line with the calculated production ratio (black square box in Fig. 9.17) one finds then that the single assumed r-process event occurred about $t_n = 6.2 \times 10^9$ years ago. Since such nucleosynthesis is produced by a single, narrow spike in time, it is called sudden synthesis. This model represents clearly one extreme of the more general production function f. On the other hand, there might have been numerous such r-process events that have contributed to the solar system material over long periods of time before the condensation of the solar nebula. Thus, if we assume, as the other extreme, a uniform element production over the time interval $t_s < t < t_n = t_s + \Delta$, the solution to the differential equation

$$\frac{dN(t)}{dt} = -\frac{N(t)}{\tau} + \lambda$$

becomes [for the boundary condition $N(t_n) = 0$]

$$\lambda = \frac{N_s}{\tau} \frac{1}{1 - \exp(-\Delta/\tau)},$$

and the uranium abundance ratio is described by

$$\frac{\lambda_{235}}{\lambda_{238}} = \left(\frac{^{235}U}{^{238}U}\right)_s \frac{\tau_{238}}{\tau_{235}} \frac{1 - \exp(-\Delta/\tau_{238})}{1 - \exp(-\Delta/\tau_{235})}.$$

The calculated ratio $^{235}U/^{238}U$ as a function of time is shown in Figure 9.17 as a dotted curve. Since production as well as decay of the aeon glass material occurs over the period of galactic nucleosynthesis, the ratios are no longer represented by straight lines. The curve intersects the primary production ratio at a time $t_n \simeq (10.4{-}11.8) \times 10^9$ y (black square box in Fig. 9.17). On the basis of this model, called uniform synthesis, the start of the r-process nuclei production goes back to about 10–12 billion years ago. Since both sudden synthesis and uniform synthesis are extreme models, one might expect that a more realistic model would provide an age in between the extreme values quoted above, that is, an age for solar r-process material between 6 and 12 billion years.

Additional information can be obtained from the other long-lived chronometric pair, $^{232}Th/^{238}U$. Again, in the age game one needs as input parameters the present abundance ratio deduced from meteoritic samples ($^{232}Th/^{238}U = 4.0 \pm 0.2$; Fow77) and the calculated primary r-process abundance ratio ($\lambda_{232}/\lambda_{238} = 1.39{-}1.80$; Fow77, Thi83). Owing to the longer lifetime of ^{232}Th compared with that of ^{238}U (Table 9.1), the abundance ratio drops to a value of $(^{232}Th/^{238}U)_s = 2.50$ (or 2.30 for $^{232}Th/^{238}U = 3.75$; And82) at the closure time t_s of the solar nebula (Fig. 9.17). The ratio decreases further beyond this time, with a slope depending on the model chosen, either sudden or uniform synthesis. The intersection of the curves with the primary abundance ratio (dotted square boxes in Fig. 9.17) leads to an age of $t_n = (7.5{-}9.7) \times 10^9$ y for sudden synthesis and $t_n \geq 10.9 \times 10^9$ y for uniform synthesis. Since the nuclei of both chronometric pairs are produced in the same process, there ought to be concordant results. (This requirement must, of course, be applied to all aeon glasses.) Clearly, the sudden-synthesis model provides no concordant value for the two pairs of aeon glasses, while the results from the uniform synthesis model only overlap at $t_n \simeq (11{-}12) \times 10^9$ y. It is expected, then, from both chronometric pairs that a more realistic model might provide an age value in the concordant range of $t_n \simeq (7{-}12) \times 10^9$ y. In any case, one sees that stellar activity and nucleosynthesis existed in our Galaxy long before the solar system formed. We have seen earlier (chap. 2) that the solar system is not central in space; we now see that it is not unique in time, either. Also, one sees a certain self-consistency in that there was plenty of time for the early stars to evolve and produce the heavy elements present before our solar system condensed. Thus, natural radioactive nuclei are the most direct evidence for the existence of nucleosynthesis, just as the existence of stars is the most convincing evidence that interstellar gas and dust may somehow collapse into a star.

The primordial matter of the big bang (mainly hydrogen and helium) condenses into the galaxies, and within each galaxy, such as our Milky Way, stars form at time t_g (Fig. 9.16). After a relatively short period of stellar evolution, the stellar interiors are hot enough to begin the nucleosynthesis of the chemical elements, including the r-process nuclei, such as those of the aeon glasses Th and U. The stars glow for a period of time determined primarily by their masses (chap. 2) and finally burn out or explode as supernovae. The heaviest stars produce the highest concentrations of gas and dust. Since about 1.5 $M_\odot$ of a burned-out star is left behind as a glowing cinder after the explosion, the gas and dust in a galaxy available for forming new stars is progressively reduced in comparison with that available to stars that formed earlier. On the average, then, the rates of supernova explosions and heavy-element synthesis are expected to have been high soon after the formation of a galaxy and to have decreased progressively thereafter. For star formation in a galaxy proportional to the total mass of gas in the galaxy at any time, Salpeter (Sal59) has shown that the time variation of that mass is exponential (Fig. 9.16) if one neglects the gas returned to the interstellar medium by evolving stars. However, even when gas return is considered, he finds the time variation to be roughly exponential during the early stage before the total mass of gas has fallen below 10% of its initial value (for observational evidence of this time dependence see Sal59, Sch59, Sch63, Mil79, Pal83). In the more realistic galactic models, it is then implied, that, since both star formation and evolution have decreased exponentially with time (at least for the largest part of them), galactic nucleosynthesis must also have varied exponentially with time (Fig. 9.16). To summarize, stellar nucleosynthesis started sometime (t_n) after the origin of the Galaxy and the beginning of galactic star formation and then decreased exponentially with a time constant T_R. The heavy elements, including the r-process aeon glass elements, were produced in supernovae and mixed into the interstellar material. For the solar system, inclusion of this material terminated on the contraction of solar material at time t_s. Thus, the synthesis of the solar system material occurred during the time interval $\Delta = t_n - t_s$ (Fig. 9.16). Although most of the material in the solar system resulted from the general process of heavy-element production outlined above, the observation of small isotopic anomalies (a few percent and less) caused by extinct (relatively short-lived) radioactive progenitors (minute glasses) indicates that the solar system also has a unique history of its own (chap. 10).

If the production function f for galactic nucleosynthesis decreases exponentially with time, the more general differential equation for the time dependence of aeon glasses is

$$dN(t)/dt = -N(t)/\tau + \lambda \exp\left[-(t - t_n)/T_R\right] .$$

For the boundary condition $N(t_n) = 0$, the analytic solution is

$$\lambda = N_s(T_R - \tau)/\{T_R \tau[\exp(-\Delta/T_R) - \exp(-\Delta/\tau)]\},$$

which can be used to calculate abundance ratios as a function of the parameters Δ and T_R. For the limit $T_R = 0$ (or $T_R = \infty$) the model reduces to the extreme model of sudden (or uniform) synthesis. In this popular model of exponential synthesis with two free parameters, at least two chronometric pairs are needed for a solution. (Also note that, when matter is lost from interstellar gas to white dwarfs, neutron stars, and black holes, modification of the above equation is required; see Tin80, Thi83.)

The actual history of galactic nucleosynthesis could involve a superposition of various combinations of exponential, uniform, and sudden synthesis. Each synthetic process produces a fraction S of the total heavy elements. The magnitude of the fraction S is determined from analyses of radioactive nuclear chronometers. In Fowler's model (Fow77) the superposition of exponential synthesis and sudden synthesis (due to a last-minute final synthesis occurring before solar system formation) was investigated with the four free parameters T_R, Δ, S, and δ (Fig. 9.16). The quantity δ represents the time between the collapse of the solar nebula (end of galactic nucleosynthesis for solar system material) and the time when the parent bodies of the meteorites became closed systems. Using the four chronometric pairs ${}^{235}U/{}^{238}U$, ${}^{232}Th/{}^{238}U$, ${}^{234}U/{}^{238}U$, and ${}^{129}I/{}^{127}I$ developed at that time (Fow60, Fow72a, Fow77), concordance was found for values of the free parameters near $T_R = 8.9 \times 10^9$ y, $\Delta = 6.0 \times 10^9$ y, $S = 2.7\%$, and $\delta = 0.16 \times 10^9$ y. In this model, the age of the r-process elements ($t_n = 4.0 \times 10^9 + \delta + \Delta$), and thus probably of all chemical elements in our Galaxy, is about 11 billion years, which is a lower limit on the age of the Galaxy itself. It should be noted that the long-living chronometers (the aeon glasses), such as ${}^{232}Th$, ${}^{238}U$, and ${}^{235}U$, probe most sensitively the time interval Δ ($T_{1/2} \simeq \Delta$), since they should have received substantial contributions over the entire period of nucleosynthesis and consequently their abundances are practically independent of last-minute production spikes (S, δ). In contrast, the shorter-lived chronometers (the minute glasses), such as ${}^{129}I$ and ${}^{244}Pu$ ($T_{1/2} \ll \Delta$), are not very sensitive to the entire duration Δ but only to events within the range of their lifetimes (Table 9.1). They are probes, then, of last-minute local synthetic events (last-minute "salting" of solar system material), hence they determine predominantly the parameters S and δ of the presolar spike and its time occurrence before the meteorites were formed. The above results show that r-process nucleosynthesis has occurred throughout galactic history and extends up to the time of solar system formation.

Recent analyses of these four actinide r-process chronometers (Thi83) have been carried out using revised inputs of primary production ratios (due to effects of β-delayed fission) and of meteoritic final abundances. They indicate a value of $\Delta = (13 \pm 4) \times 10^9$ y and thus suggest that r-process nucleosynthesis in the Galaxy started 18 ± 4 billion years ago (Thi83a). It has been pointed out (Thi83a), however, that another model, perhaps including an initial spike in galactic nucleosynthesis, might reduce the age (Fow85). With the additional assumptions that the r-process started soon (less than a billion years) after the formation of the Galaxy and that the Galaxy formed soon (less than a billion years) after the big bang, the expanding universe appears, on the basis of nucleocosmochronology, to be about 20 ± 4 billion years old. This value is completely independent of the redshift-distance correlation observations on distant galaxies done by astronomers. While this number appears to agree within uncertainties with that obtained from globular clusters (16 ± 3 billion years), some astrophysicists believe there is no concordance between them. In such a case (Bro81), either a gravitational repulsion (acceleration) is operating at large distances or the big-bang cosmology is itself not valid. In view of these cosmological implications and in order to draw really safe conclusions, the uncertainties in the age values as derived from nucleocosmochronology, the evolution of globular clusters, and the redshift-distance observations (Hubble time) in astronomy must be reduced significantly. Achieving this goal will require much work on all fronts.

In order to improve the results of nucleocosmochronology using the classical actinide clocks, uncertainties in the meteoritic and calculated abundance ratios have to be reduced. The actinide chronometer production ratios are affected by the nuclear physics of the r-process, that is, by the characteristics of nuclear masses, fission barrier heights, and β-strength functions for nuclei far of the stability line (sec. 9.5). Extension of present experimental knowledge toward more neutron-rich (short-lived) nuclei, although an enormous experimental challenge (chap. 5), will provide information to improve the extrapolations toward the waiting points. Besides the nuclear and meteoritic uncertainties, the galactic model used also influences the results (Sch82), and thus more theoretical work is also needed. Several spikes could be superposed with various contributions to the exponential model. The use of several spikes, however, adds more free parameters; more chronometric pairs are therefore needed to cover a wide range of half-lives required to determine sensitively the duration as well as the history of galactic nucleosynthesis.

Several new cosmochronometers have been suggested (Sch82), including the very short-lived chronometric pairs ^{26}Al/^{27}Al and ^{107}Pd/^{110}Pd (Table 9.1) as well as ^{205}Pb/^{205}Tl (Yok85a), associated with minor isotopically anomalous components of the solar system (Was82). These clocks may indicate the injection of elements with mass numbers $A < 110$ on the order of a few million years before the solar system became closed. Such a last-minute "salting" of the solar system material (chap. 10) is certainly of key importance in understanding the history of the early solar system (Arn81), but it is not expected to reset the classical clocks, except perhaps the shortest-lived ones (Ree79). The ^{187}Re/^{187}Os r-process chronometer, proposed by Clayton (Cla64), is of special interest because of the very long half-life of ^{187}Re (Table 9.1). It is generally regarded as one of the best candidates for providing information on the duration Δ of r-process nucleosynthesis (see, however, Arn83, Yok83). Other chronometers for the s- and p-processes have also been proposed (Sym81, Sch82), among which the ^{176}Lu/^{176}Hf chronometric pair (Fig. 9.11) appears to be the best candidate to investigate the duration of s-process nucleosynthesis in our Galaxy. However, these cosmochronometers, particularly the long-lived radiogenic ^{187}Re/^{187}Os and ^{176}Lu/^{176}Hf pairs, have problems (Bee80, Sch82, Bee84, Bee84a, Fow84, Fow85). The problems are associated with unknown nuclear physics parameters such as neutron-capture cross sections of ground states and of thermally populated excited states, the dependence of half-lives on temperature and density of stellar environments, the galactic environment, and, finally, the precision of terrestrial half-life measurements.

In summary, in spite of recent tremendous efforts (Sch82, Fow84), much exciting experimental and theoretical work remains to be done on these new chronometers, as well as on the classical ones, before they can be used with full confidence as minute or aeon glasses to trace the history of our Galaxy. To attain this goal, sophisticated models of the chemical evolution of our Galaxy (Tin75, Sym81, Sch82, Yok83) also will be needed. Those who wrestle with problems in the science of cosmochronology are grateful for the small peculiarities of nuclear properties that have provided these marvelous clues to history (Cla68).

10

Miscellaneous Topics

10.1. The Case of the Missing Solar Neutrinos

10.1.1. The Quest

The probable role of thermonuclear reactions in stellar interiors was recognized as early as the 1920s by astrophysicists such as Eddington (Edd26). This recognition was based not on any direct observation but rather on calculations showing that energy produced in nuclear reactions deep inside the stars could readily support the observed luminosities over the long stellar lifetimes. Also, it was recognized that the natural progression of these reactions could be the basis of stellar evolution (chap. 2). Yet firsthand knowledge of stellar structure is limited, consisting largely of observations of surface phenomena. How certain, then, is our understanding of the processes governing energy production and synthesis of the elements taking place deep within the fiery stellar cores?

The sun, a typical star and—because of its proximity to earth—the best-known star, provides unmatched opportunities for testing theories of stellar structure and processes. The generally accepted point of view is that the fusion of hydrogen into helium through the *p-p* chain and the CNO cycles (chap. 6) powers the sun. The resulting nuclear radiations travel only a short distance before being changed into internal thermal energy. After perhaps as long as several million years, this energy diffuses to the solar surface, where it is converted into earth-warming electromagnetic radiation containing essentially no detailed information on the parent nuclear reactions. The operation of the thermonuclear furnace deep in the sun's interior is hidden by an enormous mass of cooler material.

Of the many particles resulting from the proposed thermonuclear reactions in the solar interior, only the neutrinos can penetrate the 1 million kilometers of solar material and escape into space. Neutrinos are produced in the solar furnace by a number of weak interactions involving electron captures and β-decays. These massless (or nearly massless) particles travel at the speed of light. They react so weakly that only one in every 100 billion is intercepted on its flight to the sun's surface. If the fusion of hydrogen into helium $(4p \rightarrow {}^4He + 2e^+ + 2v; \ Q = 26.73 \ MeV)$ in the interior of the sun does indeed power the solar luminosity $(L_\odot = 2.4 \times 10^{39} \ MeV \ s^{-1})$, the total number of neutrinos leaving the sun every second (the neutrino flux) is

$$N_v(sun) = 2(L_\odot/Q) = 1.8 \times 10^{38} \ neutrinos \ s^{-1} \ .$$

Taking into account the solid angle subtended by the earth at a distance of 150 million kilometers, we find that every second 64 billion neutrinos fall on every square centimeter of the earth and pass on without notice. It takes these neutrinos only about 2 seconds to get to the surface of the sun and then another 8 minutes and 20 seconds to travel to the earth. Because they escape the sun essentially unhindered, they provide the possibility of seeing directly into the solar interior. If detected, they could tell us what was going on in the interior of the sun just over 8 minutes ago and provide the possibility of directly determining the role thermonuclear reactions play in the solar interior, thus helping to clarify the general picture of stellar structure and evolution (Cam58, Fow58). Unfortunately, the very fact that neutrinos escape so easily from the sun makes it enormously difficult to capture them on earth, and great ingenuity is needed to do so.

Before deciding on the type of neutrino detector, it is necessary to know the energy spectrum of the solar neutrinos. While, except for the assumption of thermonuclear fusion as the solar energy source, the total neutrino flux is practically model-independent, the energy distribution of the neutrinos depends sensitively on the solar model, in particular on its central temperature.

10.1.2. The Standard Solar Model

Formed some 4.6 billion years ago, the sun has now progressed through half of the main sequence of its evolution, a phase in which all of its energy is derived from hydrogen burning (p-p chain and CNO cycles). Stellar theory depicts the main sequence as a relatively simple, steady-state period in a star's evolution. The actual calculations of the solar interior in such a solar model involve the following principles (chap. 2).

1. The sun is regarded as a spherically symmetric object.
2. The sun is in a stable situation, in spite of the enormous gravitational forces tending to collapse it. At every point predominantly gas pressure balances the gravitational forces (hydrostatic equilibrium):

$$dP(r)/dr = -G[M(r)\rho(r)]/r^2 \ ,$$

$$where \ M(r) = \int_0^r 4\pi r^2 \rho(r)dr \ .$$

3. The pressure in the sun is proportional to the temperature according to the equation of state of an ideal gas:

$$P(r) = (k/m)\rho(r)T(r) .$$

4. The sun was chemically homogeneous at its birth, and the chemical composition of the present solar surface reflects the abundances of the elements ($X = 0.73$, $Y = 0.25$, $Z = 0.02$) present in the prestellar cloud. Because of the operation of the hydrogen-burning reactions, the composition in the solar core has changed (today, $X = 0.42$, $Y = 0.56$, $Z = 0.02$), where X, Y, and Z are the mass fractions of hydrogen, helium, and the metallic elements, respectively.

5. The present luminosity of the sun reflects the rate ϵ_{nuc} of nuclear energy production in the solar interior:

$$L_\odot(\text{present}) = \int_{sun} \varepsilon_{nuc}\, \rho(r)4\pi r^2 dr .$$

The quantity ε_{nuc} depends on the cross sections σ (stellar reaction rates) for the hydrogen-burning reactions in the p-p chain and/or CNO cycles, the density, the temperature, and the chemical composition X_i in the solar interior:

$$\epsilon_{nuc}(r) = f(\sigma, \rho, T, X_i) .$$

6. Thermal equilibrium prevails throughout the star:

$$dL(r)/dr = \epsilon_{nuc}(r)\rho(r)4\pi r^2 dr ,$$

where the energy is transported by radiation and convection. These transport mechanisms depend on the opacity of the solar matter, which in turn depends on the density, temperature and chemical composition.

7. The sun has evolved because it has been burning its nuclear fuel for 4.6 billion years.

The above system of differential equations can be solved uniquely by applying boundary conditions. For example, for the present age of the sun, the model must reproduce the observed macroscopic features such as surface temperature, luminosity, and radius (for a constant solar mass).

The above set of principles defines the standard solar model (Bah82). Using it with input data from nuclear and atomic physics, the temperature at the solar core is found to be $T_6 = 15.5$ and the central density to be 156 g cm^{-3} (Bah82). According to this model, 98.5% of the energy in the sun comes from the p-p chain reactions and 1.5% from the reactions in the CNO cycles. The energy spectrum of the solar neutrinos produced is shown in Figure 10.1. It extends from a very luminous region of low-energy neutrinos (mainly from the $p + p$ reaction) up to 14.1 MeV, where there is a comparatively low flux of neutrinos from the β-decay of ^{8}B. Because of the strong temperature dependence of the charged-particle reactions producing ^{8}B, the expected flux of

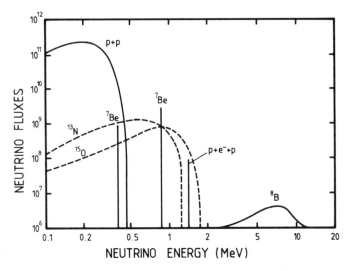

FIGURE 10.1. The energy spectrum of solar neutrinos as predicted by the standard model (Bah82). The solid (broken) lines are neutrinos from the p-p chain (CNO cycles). The neutrino fluxes are in units of $cm^{-2} s^{-1} MeV^{-1}$ for continuum sources and in units of $cm^{-2} s^{-1}$ for line sources.

these high-energy neutrinos is very dependent on the central temperature of the sun $[N_\nu(^8B) \propto T^{13}]$, and thus a measure of their intensity could serve as a solar "thermometer." Clearly, the neutrinos retain a detailed record in their energy spectrum and corresponding fluxes of the thermonuclear reactions that created them (Table 10.1). Thus a successful program of solar neutrino spectroscopy could test the principal assumption of the standard model.

10.1.3. Detection of Solar Neutrinos

Precisely because neutrinos react so weakly with matter, terrestrial experiments to detect them are enormously difficult. Since neutrinos are sensitive

TABLE 10.1 Solar Neutrino Sources, Energies, and Fluxes

Source Reaction	Energy (MeV)	Flux at the Earth[a] $(cm^{-2} s^{-1})$
$p + p \rightarrow d + e^+ + \nu$	≤ 0.42	6.1×10^{10}
$p + e^- + p \rightarrow d + \nu$	1.44	1.5×10^8
$^7Be + e^- \rightarrow {}^7Li + \nu$	0.86 (90%)	3.9×10^9
	0.38 (10%)	4.0×10^8
$^8B \rightarrow {}^8Be + e^+ + \nu$	≤ 14.1	5.6×10^6
$^{13}N \rightarrow {}^{13}C + e^+ + \nu$	≤ 1.20	5.0×10^8
$^{15}O \rightarrow {}^{15}N + e^+ + \nu$	≤ 1.73	4.0×10^8

[a] From Bah82.

only to the weak-interaction force, their detection must involve reactions such as the inverse β-decay, which are governed by the weak force:

$$v + A \rightarrow B + e^- \ .$$

Since a large amount of target material A is needed, it must be relatively inexpensive and readily available. Furthermore, since in such neutrino detectors only a few inverse β-decays occur, detection of the few residual nuclei B must be definitive and relatively easy.

The possibility of using ^{37}Cl material as a neutrino trap was suggested in the 1940s by Pontecorvo (Pon46), a suggestion subsequently discussed in more detail by Alvarez (Alv49). (For a complete history of the development of the ^{37}Cl solar neutrino experiment see Bah82a.) The threshold for this reaction is $E_v = 0.81$ MeV:

$$v + {}^{37}\text{Cl} \rightarrow {}^{37}\text{Ar} + e^- \ .$$

The cross section for the inverse β-decay is of the order of 10^{-44} cm². Since it increases steeply with neutrino energy, the ^{37}Cl cross section is sensitive predominantly to the high-energy neutrino flux from the ^{8}B decay ($N_v \simeq 10^7$ neutrinos cm^{-2} s^{-1}; Table 10.1), which has a mean absorption cross section of $\sigma_v \simeq 10^{-42}$ cm². The number of captured neutrinos per unit of time is then given by the relation (chap. 5)

$$N_{\text{cap}} = N(^{37}\text{Ar}) = N_v N(^{37}\text{Cl})\sigma_v \ .$$

In order to capture at least one neutrino per day, 70 tons of ^{37}Cl must be used, constituting a giant neutrino trap. Fortunately, the natural abundance of ^{37}Cl is 24% (the other stable chlorine isotope is ^{35}Cl) and large amounts of a chlorine compound are readily available in the form of tetrachlorethylene (C_2Cl_4), an ordinary inexpensive cleaning fluid. To achieve a capture rate of one neutrino per day, about 330 tons are needed. Due to the extraordinary small neutrino-capture rates, a special unit for the detection rate of solar neutrinos has been introduced (Bah69b, Bah82a):

1 SNU = 1 solar neutrino unit

$$= 10^{-36} \text{ neutrino-captures per second per target atom} \ .$$

Thus, an atom of ^{37}Cl has to wait 10^{36} s before capturing a neutrino. Assuming a capture rate of 1 SNU, a large tank with about 570 tons (2×10^{30} atoms of ^{37}Cl) would see a single capture event about every 6 days.

Such a giant neutrino trap with about 610 tons of C_2Cl_4 was built between 1964 and 1968 by R. Davis, Jr., and his collaborators (Bah82a). It is located 1500 m underground in the Homestake gold mine at Lead, South Dakota. The depth of overburden is necessary to shield the detector against cosmic rays, which can also initiate reactions leading to ^{37}Ar. Clearly, it is necessary to go deep enough in the mine to reduce the cosmic-ray effects below those expected from the solar neutrinos. To shield against neutrons, the entire tank is surrounded by water.

The rare gas atom ^{37}Ar is ejected from the C_2Cl_4 molecule and soon forms a neutral atom. It is unstable and reverts with a half-life of 35 days to ^{37}Cl by capturing one of its orbital electrons:

$$^{37}\text{Ar} + e^- \rightarrow {}^{37}\text{Cl}^* + \nu \; .$$

In the deexcitation process the residual atom ^{37}Cl* ejects one of its orbital electrons (2.8 keV Auger electrons), which can be detected in a proportional counter (chap. 5). In order to have a reasonable amount of argon gas to work with in this radiochemical experiment, a nonradioactive form of argon (^{36}Ar or ^{38}Ar) is introduced into the tank as a carrier gas. The number of radioactive ^{37}Ar atoms is then allowed to build up in the tank for about a month. The tank is subsequently flushed for a day with large quantities of helium gas (about 0.3 m^3 per minute), whereby the helium bubbles collect the argon in the liquid. The argon-laden helium is passed over a cold trap, which collects the argon, separating it from the helium. Heating the cold trap releases the collected argon into the proportional counter, where it is assayed for the ^{37}Ar radioactivity over a period of several months. The efficiency of ^{37}Ar recovery from the tank has been demonstrated by a number of clever tests (Bah69, Bah82a). It is well to remember that the big tank consists of about 10^{30} atoms, and the experimental task is to find, extract, and identify the few atoms of ^{37}Ar produced by the capture of solar neutrinos. This makes looking for a needle in a haystack seem easy.

The measured neutrino-capture rate (average over the runs of the years 1969–1983) is 2.1 ± 0.3 SNU (Bah85), which is equivalent to the production of one ^{37}Ar atom in the entire tank about every 3 days. This rate is in disagreement with the current theoretical predictions of 5.8 ± 2.2 SNU (Bah85) arrived at by using standard solar models with the best values for the various input parameters. This disagreement between observation and standard theory (about a factor of 3) constitutes what has come to be known as "the solar neutrino problem." In referring to this as a "problem," the tremendous and successful effort of Davis and his group is somehow lost sight of. Since the ^{37}Cl neutrino detector has no directional sensitivity, it is assumed that the observed neutrinos are from the sun. It is argued (Bah69b) that the flux of neutrinos from the rest of the universe bears roughly the same relation to the solar neutrino flux as starlight does to sunlight, and hence can be ignored. However, if the detected neutrino flux does arise from some other cosmic sources such as supernovae, then the disagreement between experiment and theory is even greater, and the case of the missing solar neutrinos is even more puzzling.

The solar neutrino problem may be a symptom of some fundamental difficulty with our understanding of the structure and evolution of stars, some input parameters, or even some physical laws. Until this problem is solved, we are forced to conclude that after 50 years we still have only circumstantial and indirect evidence for thermonuclear reactions in the interior of the sun and other stars (i.e., chap. 9).

10.1.4. Suggested Solutions

About 80% of the predicted solar neutrino rate in the ^{37}Cl experiment comes from the high-energy neutrinos emitted in the ^{8}B β-decay (Fig. 10.1 and Table 10.1). The predicted rate of these rare neutrinos is very dependent on the calculated temperature of the center of the sun and is thus very model-dependent. The ^{37}Cl experiment seems to indicate that the standard model is faulty. Many of the proposed variants of the standard model suggest that the central temperature of the sun is lower than 15 million K, since that would decrease the predicted flux of the temperature-sensitive ^{8}B neutrinos. The lowering of temperature necessary to provide the observed rate is relatively small but is still about a million K. The variant model must then somehow compensate for this decrease in central temperature and the consequent decrease in pressure; otherwise the sun would collapse, which it is obviously not doing. Thus, all of these variant models must somehow maintain the sun's stability against gravitational collapse in spite of a lower central temperature and pressure. One model assumes rapid internal rotation, so that centrifugal force due to rotation keeps the sun from collapsing against gravity. Another model postulates a high internal magnetic field, which can withstand compression and the forces of gravity. It has also been suggested that the inner parts of the sun have somehow been mixed with the outer parts, thus reducing composition differences attributable to thermonuclear reactions in the solar interior. These and many other models (Kuc76, Bah82a) cannot easily be dismissed a priori, but all of them raise more questions than they answer (Bah82).

The predictions of the ^{37}Cl experiment are particularly sensitive to the reaction rates of the p-p chain leading to ^{8}B. The dependence of the predicted neutrino flux on these rates can be parameterized as follows (Par85):

$$N_\nu(^{37}\text{Cl}) \propto S_{11}^{-2.5} S_{33}^{-0.37} S_{34}^{0.8} (1 + 3.47 S_{17} \tau_{e7}) .$$

The subscripts i, j on the S-factors are usually taken to be the mass numbers in the entrance channel (i.e., S_{34} for ^{3}He$(\alpha, \gamma)^7$Be). An obvious exception is τ_{e7} for the electron-capture lifetime of ^{7}Be. Because of changes in the values for S_{11}, S_{34}, and S_{17} resulting from laboratory efforts since 1979 (chap. 6 and Par85), the present value, 5.8 SNU, is nearly 30% smaller than previous calculations (Bah82, Bah82a). However, there is still a discrepancy of more than 3 standard deviations between the predicted results and those observed in the ^{37}Cl neutrino detector. A systematic program to unravel nuclear physics aspects of the current solar neutrino problem is currently underway in several laboratories.

The most imaginative idea has come from Pontecorvo (Pon67, Bil78) concerning the properties of the neutrinos themselves (Bah82). We know that there are neutrino species other than the electron neutrinos—the mu and tau neutrinos. Pontecorvo suggested that, in the 8 minutes it takes to get from the sun to the earth, electron neutrinos might transform into mu neutrinos, and this suggested transformation can be extended today to tau neutrinos. According to some grand unified theories (chap. 2), when this transformation of neutrinos does occur (referred to as neutrino oscillations), it is possible, starting with one kind of neutrinos, to wind up with an equal number of all three. In this case, only one-third of the electron neutrinos born in the sun would survive the trip to the earth. Since the ^{37}Cl experiment can only detect electron neutrinos, this would be consistent with what Davis and collaborators observe. It is an intriguing possibility but is viewed with great skepticism by many elementary-particle theorists. It should be noted that theories predicting neutrino oscillation also predict a finite rest mass for the neutrinos. At present there are many laboratories involved in experiments to measure neutrino masses as well as neutrino oscillations. At present, no evidence for neutrino oscillations has been found (Boe84).

A next step in improving the available information about the solar neutrino problem would seem to be the use of detectors with different neutrino energy thresholds, with which one could cover the entire spectrum of solar neutrinos (Fig. 10.1). The intense flux of low-energy neutrinos from the fundamental reaction of the p-p chain, $p + p \rightarrow d + e^+ + v$, is effectively fixed by the observed solar luminosity and by the assumption that hydrogen burning is the source of solar energy. If the flux of these neutrinos also proves to be strongly suppressed, one could conclude that some failure in our understanding of neutrino propagation, rather than solar physics, is responsible for the solar neutrino puzzle.

At present the best hope for measuring these low-energy neutrinos is an experiment based on the reaction

$$v + {}^{71}\text{Ga} \rightarrow {}^{71}\text{Ge} + e^- \qquad (Q = -233 \text{ keV}) ,$$

where 63% of the ^{71}Ge produced would arise from the $p + p$ neutrinos (Bah82). As in the ^{37}Cl detector, the ^{71}Ge residual atoms are radioactive (half-live = 12 days) and their decay ejects an atomic orbital electron which can be detected in a counter. It is a costly experiment. Present estimates suggest that a meaningful experiment would require 30 tons of ^{71}Ga with a cost in excess of 25 million dollars. This experiment has been shown to be technically feasible in small tests (Bah82). Other neutrino detectors such as ^{7}Li, ^{79}Br, ^{81}Br, ^{97}Mo, ^{98}Mo, and ^{115}In have also been suggested, each of which would be sensitive to a different part of the solar neutrino spectrum (Bah82). It has also been suggested that suitable shielded ores containing ^{41}Ca, ^{81}Br, ^{98}Mo, or ^{205}Tl could be used (because of the long half-lives of the resulting radionuclides) to study the time-integrated solar neutrino flux over periods of several million years (Bah82, Cow82, Par85). A meaningful interpretation of any such geochemical solar neutrino measurements will clearly require not only a thorough understanding of the nuclear physics involved in the neutrino-capture reactions but also a reliable knowledge of the geologic history of the ore sample and its shielding form cosmic rays.

Future solar neutrino experiments must more clearly indicate what is missing from our present understanding or even whether the primary problem is in physics or in astrophysics. Once the nature of thermonuclear reactions and the resulting solar neutrinos is well known, the neutrinos might be used as a probe to obtain new information about a wide variety of cosmic phenomena such as supernovae, X-ray double stars, active galactic nuclei, and quasars (chaps. 1 and 2). The neutrinos interact so rarely with other particles that they can cross the entire universe, even traversing substantial aggregations of matter, without being significantly absorbed or even deflected. The surviving neutrinos carry with them information about the site of their production, possibly in regions of space from which light and other kinds of electromagnetic radiation are blocked by matter. Cosmic neutrinos may also provide answers to basic questions about the overall structure and history of the universe. One question is whether all galaxies are made of ordinary matter, or roughly half of them are made of antimatter. Photon astronomy cannot directly resolve the issue because the photon is its own antiparticle and hence carries no information about whether its source is matter or antimatter. It is different with neutrinos; a star composed of matter (antimatter) radiates mainly neutrinos (antineutrinos). Thus, detection of neutrinos from a particular galaxy should reveal whether it consists of matter or antimatter.

The detection of the solar neutrinos is clearly the first chapter of what might be called neutrino astronomy. The detection of cosmic neutrinos will extend this branch of astronomy into the vast universe, and it is likely that neutrino astronomy will bring its own share of surprises (Lan79). Proposals for building giant neutrino telescopes have already been made (Lea81).

10.2. Isotopic Anomalies and the Early History of the Solar System

10.2.1. The Homogeneous Isotopic Composition

It is in general true that the isotopes of an element are chemically identical, and thus chemical processes do not distinguish between them. In reality, the isotopes do exhibit small differences in their chemical properties resulting strictly from atomic mass differences. The small mass difference has a slight influence on the rate and the equilibrium point of chemical reactions, which in some cases can be significant.

Because the chemical segregation of isotopes is always dependent on atomic mass, it follows a simple pattern. Oxygen, for example, has three stable isotopes ^{16}O, ^{17}O, and ^{18}O with terrestrial abundances of 99.756%, 0.039%, and 0.205%, respectively. If the ratio of $^{17}O/^{16}O$ is slightly increased by a chemical process, the ratio of $^{18}O/^{16}O$ must be increased by twice as much, since the difference in mass is twice as large (Fig. 10.2). Using this relation, it is possible to compensate for the effects of chemical fractionation and retrieve the "true" isotopic abundances (Cla78, Pod78, Lee79, Beg80, Was82). When these minor chemical effects are taken into account, it is found that the relative abundances of isotopes are constant to high accuracy. For example, the same isotopic ratios can be observed in oxygen from the atmosphere, from sea water, and from sedimentary or igneous rocks, even though the amount of oxygen in these substances varies widely.

Isotopic ratios for oxygen and for other elements have been measured in numerous samples from the earth, the moon, and different classes of meteorites, and have been found (after adjustments for chemical effects) to be essentially the same. This high degree of isotopic homogeneity is most remarkable, since the nucleosynthesis of different isotopes of the same element often requires drastically different sources and sites (chaps. 6–9). This homogeneity implies great efficiency of the processes which mixed different nucleosynthetic components before the formation of planetary bodies. The uniform isotopic composition is referred to as "normal." Coupled with the elemental composition derived from carbonaceous meteorities and solar atmosphere, a "cosmic" composition can be derived (chap. 1). It is thought to represent the average of many nucleosynthetic events which contributed to the solar system matter but whose individual records have been erased by the mixing (Fig. 2.10). This com-

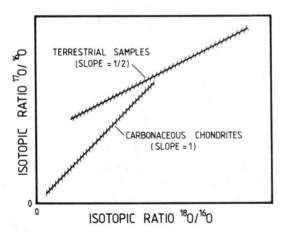

FIGURE 10.2. Owing to chemical fractionation, the isotopic composition of the earth's oxygen is not quite uniform, but the variations follow a simple pattern: if the ratio $^{17}O/^{16}O$ is increased by one unit, then the ratio $^{18}O/^{16}O$ is increased by two units. A graph of the two ratios is a straight line with a slope of $\frac{1}{2}$, shown here schematically. In the meteorites called carbonaceous chondrites (including the Allende meteorite) a different relation with a slope of nearly unity has been discovered (Cla73, Cla81). Note that such relations always require that at least three stable isotopes of a given element exist (Was82).

position is widely used as an initial or boundary condition for the evolution of planets, the solar system, stars, the interstellar medium, and galaxies.

As just discussed, there are well-understood variations of isotopic abundances resulting from the modification of the normal composition by solar system processes. These processes are mass-dependent fractionation (diffusion, chemical reactions, and so on), decay of radionuclides, and nuclear reactions with energetic particles (cosmic rays; sec. 10.3). "Isotopic anomalies" can be defined operationally as variations which cannot be explained by these processes and, for lack of a better explanation, are attributed to the incomplete mixing of nucleosynthetic products. Before about 1973, almost all isotopic variations could be explained by these solar system processes. The exceptions were rare gases, especially Ne and Xe, but they were largely ignored because of their low abundance in solid bodies.

The solar system is thought to have had its origin in the gravitational collapse of a diffuse and swirling cloud of gas and dust, the "solar nebula" or "solar cloud." The lightest and most abundant elements in the nebula, hydrogen and helium, were primordial materials from the early universe (chap. 2), but the metallic elements ($Z \geq 6$) were mainly the debris from stellar nucleosynthesis (novae and supernovae), the so-called stardust. In these scenarios the debris incorporated into the solar nebula need not have been identical in isotopic composition or even in elemental composition. In fact, it is more likely that each scenario involved a distinctive mix of isotopes and elements. However, once these diverse mixtures joined the solar cloud, they were thoroughly blended. By the time the solar system condensed from the nebula the isotopic composition was homogeneous, reflecting only the average isotopic content of all the stellar debris and the primordial gases. The solar system is thus a 4.6 billion year old sample of the metallic elements in the interstellar medium of our Galaxy (chap. 2).

The isotopic composition of various chemical elements in the sun, planets, and meteorites is one of the most important sources of information on the physical and chemical processes that shaped the solar system (Lee79, Was82, Boc83). The time scales related to these processes can be established using long-lived radioactive isotopes (aeon glasses; chap. 9). Isotopes, which show low natural abundances and are mainly produced by the decay of radioactive nuclides, have been used for dating events such as the solidification of minerals (4.6 billion years ago). Isotopes of noble gases have always been especially suited for such investigations, since they are present in low concentrations in most natural samples. The most popular case is the ratio $^{40}K/^{40}Ar$: by far the largest part of ^{40}Ar present in the solar system is due to the decay of ^{40}K ($T_{1/2} = 1.3$ billion years). Isotopes produced by the decay of short-lived, extinct isotopes (minute glasses) such as ^{129}I ($T_{1/2} = 15.7$ million years) have been used to compute the time δ between the last nucleosynthetic source contributing material to the solar system and the formation of the solid bodies (Fig. 9.16). The varying content of ^{129}Xe in meteorites (Rey60), generated from the decay of ^{129}I, indicates that the formation of the solar system, although it did not occur in one instantaneous event, must have begun soon after the last nucleosynthetic event producing ^{129}I (about 160 million years ago).

During the 4.6 billion years between the formation of the solar system and the present time, small bodies such as parent bodies of meteorites stayed relatively undisturbed, while material in large and active planetary bodies received further thermal and chemical processing. For example, the oxygen in the atmosphere, the water, the rocks, and the living organisms passes continually through an interlocking web of chemical cycles. Thus, the traces of isotopically distinctive oxygen present when the planet first condensed are today mixed with all the rest and diluted to an undiscernible level. Because of these blending processes, the possible presence of isotopic anomalies in the original solar material cannot now be recognized by analyzing material from the large planetary bodies. Such anomalies can be expected to persist today only in material that has not been subjected to much chemical processing. One likely place to search for distinctive isotopical anomalies would be in comets, which probably have been altered very little by chemical processes. Comets, however, are quite inaccessible. In 1986 a space probe was sent to Halley's comet to investigate its isotopic composition. Several probes were launched, and data obtained are now beginning to appear in the literature. The best readily available sources of primitive solar system materials are meteorites, which, because of their size, never experienced geochemical cyling. Among the meteorites the most primitive seem to be thermally fragile carbonaceous chondrites. They are distinguished by the presence of carbon and by small, round inclusions called chondrules, which show evidence of having once been melted. It is thought that the carbonaceous chondrites solidified early in the history of the solar system and that little has happened to alter them since. Examinations of meteoritic material before 1973 failed, however, to detect isotopic anomalies predicted on theoretical grounds (Lee79, Was82).

10.2.2. The Discovery of Isotopic Anomalies

The underlying assumption for all dating methods was that the material under investigation once had uniform isotopic composition and that all deviations from this uniform composition were caused by processes (chemical fractionation, spallation reactions, radioactive decay) which happened after the separation of the solar system material from the interstellar medium. For an amazingly large number of purposes this assumption can still be used as a working hypothesis. However, since about 1973 it has become clear that the original picture of an isotopically well-mixed early solar system no longer holds and that the history of the early solar system was much more complicated than had been expected. There are two major causes of this change of view.

First, as a consequence of the Apollo mission to the moon and the resulting availability of lunar samples for investigation, several laboratories have developed improved mass spectrometers as well as efficient procedures for cleanly separating minerals from meteoritic samples on a microscopic scale. This improved capability for measuring isotopic compositions with high efficiency (Was82) has been essential to the success of isotopic anomaly searches in meteorites.

Second, in 1969 a giant meteorite (about 2 tons recovered) fell near the village of Pueblito de Allende in northern Mexico. Now known simply as the Allende meteorite, it is a carbonaceous meteorite providing for the first time large specimens of this important type of meteor for study. The most significant Allende samples are large with peculiar inclusions (sometimes as large as 1 cm across) containing minerals rich in refractory elements. The presence of such inclusions is expected for samples which have condensed from a gas of solar composition at temperatures above 1500 K and at low pressure, and then cooled further. It was shown that the inclusions were the oldest samples ever dated in the solar system (Was82). Studies of these minerals looked like a promising way to gain information about the birth of the solar system. Investigating these grains of stardust corresponds to "sampling pieces of time."

Since 1973 isotopic anomalies have been seen for many elements and in various classes of meteorites, giving evidence for incomplete mixing of solar material. Isotopic anomalies have been found for the elements O, Ne, Mg, Si, Ca, Ti, Kr, Sr, Te, Xe, Ba, Nd, and Sm and are now expected to be found in elements across the periodic table. An account of most of these discoveries over the last several years is given in several review articles (Cla78, Pod78, Lee79, Beg80, Was82). However, there still is not an understanding of the early chemical history of the solar system because it is difficult to reconcile the many pieces of evidence now available. Certainly, these discoveries have opened up a real Pandora's box. In the following subsections we will discuss a few outstanding cases.

10.2.3. Oxygen Isotopic Anomalies

Oxygen is a rock-forming element; its content in solid solar system materials is of the order of 40% by weight (Boc83). The isotopic composition of oxygen is extremely insensitive to any external influence such as irradiation of neighboring target elements. Also, there are no long-lived radioactive nuclei decaying into the oxygen isotopes ^{16}O, ^{17}O, and ^{18}O. The only possible process affecting the isotopic composition of oxygen in a closed system is isotope fractionation by chemical and physical processes. These processes will always influence the $^{18}O/^{16}O$ ratio twice as strongly as the $^{17}O/^{16}O$ ratio. Thus, in a plot of $^{17}O/^{16}O$ versus $^{18}O/^{16}O$ ratios, isotopic fractionation will be represented by points wandering along a straight line of slope $\frac{1}{2}$ (Fig. 10.2). Terrestrial and lunar samples as well as samples from some classes of meteorites follow this line.

If solar system material had been separated from one reservoir and had only been affected by fractionation, all solar system samples should lie close to this line. Clayton, Grossman, and Mayeda (Cla73) discovered in 1973 that the anhydrous mineral assemblages in the Allende meteorite, and also samples from other carbonaceous meteorites, did not fall on this fractionation line but rather fell on a straight line with a slope of nearly unity (Fig. 10.2). The most extreme samples showed a -5% deviation in the $^{18}O/^{16}O$ ratio from the fractionation line. This observation could not be explained as a result of fractionation of (what had previously been thought to be) solar system oxygen, and a new explanation was required.

Since the $^{17}O/^{16}O$ ratio was unperturbed, it was suggested (Cla73) that the oxygen in the carbonaceous meteorites is a mixture of two components—one having the normal isotopic composition of the solar system oxygen and the other being made up of pure ^{16}O which had been added in various proportions ranging up to about 5%. This implies that the solar nebula was not entirely homogeneous, and that some components of it had a different isotopic composition of a major chemical element. If the ^{16}O-rich component had been in the form of a gas, it would have mixed with the rest of the solar cloud and would have been diluted very rapidly (Cla73). It is more likely that the anomalous oxygen entered the cloud in chemically combined form, possibly as interstellar dust and solid grains, while the second component was gaseous.

Two possible origins for such grains have been proposed (Cla73, Cla78, Sch78, Cla81). They might have been primordial grains; indeed, all interstellar grains might have an isotopic composition different from that of interstellar gas (see, however, Lee79, Was82). That difference would not be detected in other bodies of the solar system because elements from the solid phase and the gas phase have been thoroughly blended. Alternatively, most gas and dust grains in the cloud might have been identical in isotopic composition, with only a few anomalous grains that were added too late for complete mixing to occur. These grains could have been preserved in some meteorites such as Allende, whereas they would have been chemically broken down and recombined in the earth and the other planets.

It must be emphasized that the oxygen isotopes carry no information about the time of formation of the grains. All three oxygen isotopes are stable, so an unusual isotopic ratio (such as 100% ^{16}O) could have been frozen into some dust grains hundreds of millions of years before those grains were incorporated into the solar system. The only requirement for explaining the observations is that the grains must have had no chance to mix completely with more typical solar system materials. At present there is no simple, generally accepted theory of the nucleosynthetic origin of the observed oxygen anomalies. One suggested source for the pure ^{16}O component is a nearby supernova (Cam77, Sch78) that exploded too late for the ejected debris to become thoroughly mixed with the rest of the protosolar cloud. Even without a theory for the creation of isotopic anomalies, however, they can produce useful information. For example, the oxygen isotopic data provide increasing evidence for generic relationships between different classes of meteorites and other planetary bodies. For instance, it appears that the moon, earth, and meteorites are closely related. Such a relationship clearly excludes certain models of the origin of the moon (Boc83).

10.2.4. Magnesium and the Discovery of Extinct ^{26}Al

The discovery of oxygen isotopic anomalies in Allende and other carbonaceous meteorites strongly indicated that the material from each distinct nucleosynthetic event was not completely mixed and that searches for isotopic anomalies in other elements should be renewed. Magnesium also has three stable isotopes, ^{24}Mg, ^{25}Mg, and ^{26}Mg, with normal isotopic abundances of 78.99%, 10.00%, and 11.01%, respectively. Lee, Papanastassiou, and Wasserburg (Lee77, Lee79, Was82) investigated the early condensates of Allende for isotopic anomalies in magnesium. During this search they found instead evidence for the presence of active ^{26}Al in the early solar system.

Aluminum has only one stable isotope, ^{27}Al. The radioactive isotope ^{26}Al has a half-life of 0.72 million years and transforms by β-decay into the stable magnesium isotope ^{26}Mg. Suppose that a quantity of ^{26}Al had been introduced into the protosolar cloud just before the carbonaceous chondrites condensed. It would have essentially the same chemical properties as ^{27}Al and so would appear in a constant ratio with ^{27}Al in all minerals that include aluminum. After a few million years, however, most of the ^{26}Al would have decayed in situ, and after the 4.6 billion years that have passed since the formation of the solar system, essentially all of it would have been converted to ^{26}Mg. If the rock was never melted or subjected to other processes that could separate the elements, then the amount of ^{26}Al originally present could be determined simply by measuring the amount of the daughter products ^{26}Mg in ^{27}Al-rich phases. In practice the measurement is not that simple because virtually all minerals include magnesium (and thus ^{26}Mg). Indeed, magnesium is much more abundant than aluminum. As a consequence, any ^{26}Mg created by the decay of ^{26}Al could be detected only as an excess over the normal abundance of that isotope.

The isotope ^{26}Al is an extinct nuclide, i.e., a radioactive nuclide that may have been present in the early solar system but has now decayed. However, its previous presence might be inferred from isotopic variations in the daughter element ^{26}Mg. The isotopic variations due to extinct nuclides are not isotopic anomalies from nucleosynthetic heterogeneities as was the case for the oxygen isotopes. Rather, they are caused by radioactive decay in objects with different parent-to-daughter chemical ratios. In ideal cases, these variations are expected to show a characteristic correlation which provides definite evidence for the presence of an extinct nuclide. The first clear evidence for the presence of extinct nuclides was reported by Reynolds (Rey60), where large ^{129}Xe excesses were identified as the decay products of the extinct nuclides ^{129}I ($T_{1/2} = 15.7$ million years).

If an isotopically homogeneous object formed early when, for example, ^{26}Al was still active, and if this object remained undisturbed until now, then the isotopic ratio $(^{26}Mg/^{24}Mg)_p$ of its present (p) measured phases are given by the relation (Lee79)

$$(^{26}Mg/^{24}Mg)_p = (^{26}Mg/^{24}Mg)_0 + (^{26}Al/^{27}Al)_0(^{27}Al/^{24}Mg)_p \, ,$$

where $(^{26}Mg/^{24}Mg)_0$ and $(^{26}Al/^{27}Al)_0$ are the initial isotopic ratios at the time of formation and $(^{27}Al/^{24}Mg)_p$ depends on the chemical composition of the phases. Thus, the two measurable quantities $(^{26}Mg/^{24}Mg)_p$ and $(^{27}Al/^{24}Mg)_p$ for distinct phases of Al/Mg should strictly correlate with each other linearly. Such a linear array is called an "isochron." The initial Al and Mg isotopic ratios can be obtained from the intercept and the slope of the isochron, respectively.

Enhancements of ^{26}Mg were indeed found in the Allende material (Lee77, Lee79, Was82), and in some specimens the percentage of ^{26}Mg was increased from its normal level of 11.01 to about 11.5. Anomalies that large were observed only in minerals, such as anorthite, that are rich in aluminum and poor in magnesium. Most of the anomalies were much smaller. These ^{26}Mg enhancements could not have been caused by chemical fractionation; if they had been, there would have been similar but smaller enhancements in ^{25}Mg, which were not observed. Furthermore, a clear correlation of the $^{26}Mg/^{24}Mg$ ratio with that of $^{27}Al/^{24}Mg$ was found: the greater the Al/Mg ratio, the larger the excess of ^{26}Mg. A typical example for such a correlation is shown in Figure 10.3, which was obtained (Ste81) from an analysis of different minerals

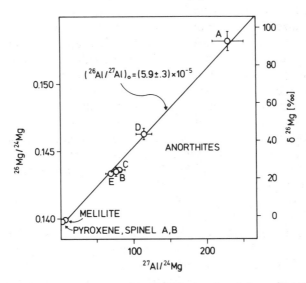

FIGURE 10.3. Magnesium isotopic composition in different minerals from a Ca-Al–rich inclusion in the Leoville carbonaceous meteorite (Ste81). The straight line connects samples of equal $^{26}Al/$ ^{27}Al isotopic ratios during formation of the samples.

in an inclusion of the Leoville carbonaceous meteorite. Clearly, that is what one would expect if active ^{26}Al nuclei were included when the samples formed. It should be noted, however, that here, as in the case of oxygen, the straight line in this three-isotope plot could also be interpreted as a mixing line. This line would connect two end members of different magnesium composition, one of them with normal magnesium and the other enriched in "fossil" ^{26}Al (from the decay of the parent in preserved presolar grains). Meanwhile, additional data have given strong evidence that the two-component explanation can be excluded. Thus, there is almost no doubt left today that ^{26}Al was indeed present in at least certain parts of the early solar system, and that the ^{26}Mg enhancements are due to ^{26}Al decay in the primitive inclusions of carbonaceous meteorites.

The correlation of the aluminum content and ^{26}Mg excess in many minerals makes possible an estimate of how much ^{26}Al relative to ^{27}Al was initially present. Most inclusions studied so far lead to a "typical" value of $(^{26}Al/^{27}Al)_0 \simeq 5 \times 10^{-5}$ (within a factor of 2). However, substantially lower ratios as well as larger ratios (up to 10^{-3}) have been observed (Lee79, Was82). In any case, it seems that ^{26}Al was a minor contaminant in the early solar system material.

The discovery of extinct ^{26}Al was particularly important because it supplies a date for the creation of the anomalous material. Since ^{26}Al is a radioactive isotope with a cosmologically short half-life of 0.72 million years, it had to be incorporated into the minerals no more than a few half-lives or a few million years after it was created in some extrasolar event (nova or supernova). If there had been a longer delay, all the ^{26}Al would have decayed to ^{26}Mg, which would have mixed freely with other magnesium, thereby removing any correlation between aluminum and magnesium. The ^{26}Al results fix a maximum time interval of about 3 million years—and not 160 million years as inferred previously from ^{129}I—for the duration between the last injection of freshly synthesized material into the solar nebula and the formation of the closed solar system. The results prove that nucleosynthetic activities were going on until about 3 million years before the solar system formed. Thus, the time gap δ inferred before does not exist: essentially, nucleosynthesis was going on right up to the closure of the solar system. The discovery of ^{107}Ag excesses attributable to extinct ^{107}Pd, with $T_{1/2} = 6.5$ million years, further strengthens these conclusions (Lee79, Was82).

10.2.5. Neon and Extinct ^{22}Na

The three stable neon isotopes, ^{20}Ne, ^{21}Ne, and ^{22}Ne, have terrestrial abundances of 90.5%, 0.3%, and 9.2%, respectively. In 1972 Black (Bla72) discovered neon that was remarkably enriched in ^{22}Ne with $^{20}Ne/^{22}Ne \leq 1.5$ (the terrestrial ratio is 9.8) in the Orgueuil meteorite. He called it "Ne-E" (E for extraordinary). Black hypothesized that Ne-E was of extrasolar origin produced under nucleosynthetic conditions different from the normal solar system material, the anomalies being due to subsequent incomplete mixing of solar material. Subsequently, refined measurements of grains of this meteorite led (Ebe81) to extreme ratios of $^{20}Ne/^{22}Ne \leq 0.15$ (and $^{21}Ne/^{22}Ne \leq 0.0022$). It was thus concluded that the separation of Ne-E from other types of neon must be explained chemically. No natural physical process such as nucleosynthesis in supernovae or irradiation of a solid target with energetic particles (spallation processes; sec. 10.3) appeared to produce ^{22}Ne to such a high degree of purity. It was suggested (Ebe81), therefore, that Ne-E is essentially fossil material of extinct ^{22}Na.

This suggestion has some astrophysically interesting consequences. If Ne-E were incorporated into the solid grains in the form of ^{22}Na, this must have happened very soon after nucleosynthetic production, since the half-life of ^{22}Na is only 2.6 years. Decay of ^{22}Na into ^{22}Ne must have taken place to a large extent after the grain carriers had released ^{21}Ne and ^{20}Ne, which would originally have been incorporated in the grain as well. These and other considerations produced rather severe constraints on any astrophysical scenario which might explain the origin of the carriers. The carriers must not only survive the time between their formation and the final incorporation into larger parts of meteorites, but they also have to be kept at sufficiently low temperatures soon after separation from the nucleosynthetically active regions until now, in order to retain the loosely bound ^{22}Ne.

10.2.6. Conclusions

The studies of isotopic abundances of several elements over the last years have revealed several new clues to the formation and history of our solar system. The discovery of the extinct nuclides ^{26}Al and ^{107}Pd (with half-lives of about 1 million years) in the early solar system implies that there were nucleosynthetic activities involving a great many elements taking place up to almost the instant of solar system formation. There is strong evidence that there is an astonishingly brief time interval of about 3 million years between the last nucleosynthetic event and the creation of the sun. The observation of isotopic abundance variations (anomalies) relative to the cosmic composition in a variety of planetary objects indicates that the solar system material had not been completely mixed prior to the formation of solid bodies. The isotopic heterogeneities caused by the incomplete mixing of distinct nucleosynthetic components permeate the entire solar system. Continuing investigations of the different processes of nucleosynthesis are in order.

The extinct nuclides and the wide spectrum of isotopic anomalies across most of the periodic table represent "timekeepers of the solar system." To explain them, a nucleosynthetic event has to be invoked in the vicinity of the protosolar cloud at the instant of solar system formation. The data have led to the speculation that a nearby supernova had injected freshly synthesized material into the early solar nebula and possibly "triggered" the collapse of the solar cloud some 4.6 billion years ago (Cam77, Sch78, Her79). It is argued that the shock wave from the supernova explosion could effectively compress the diffuse solar cloud of interstellar gas and dust to a density high enough for gravity to pull the material together into clumps of matter, out of which the sun, the planets, and the meteorites were formed. The debris of this supernova became trapped in the nascent material, and, because of its close proximity in space and time, it contributed relatively more than each of the many supernovae (and novae) which built up the metallic element abundances in the interstellar medium over the lifetime of the Galaxy.

The close link between nucleosynthesis and star birth is supported by the astronomical observation that stars tend to form in clusters. While massive stars may explode within a few million years after birth, smaller stars still in the process of formation in a cluster might be contaminated with the nucleosynthetic debris (star dust and gas) of the more massive stars. Thus isotopic anomalies in solar system material can be viewed as the result of locally very fast recycling of interstellar matter. The thought of a close link between death and birth of stars seems philosophically attractive, but one should bear in mind that such attractiveness can imply a dangerous intellectual bias.

The new results have major implications for cosmochronology and cosmochemistry, nucleosynthesis, star formation, thermal evolution of the planets, and the genetic relationship between different planetary bodies (Ure55, Lee79). Various astrophysical sites have been suggested to explain the nucleosynthesis of ^{26}Al and Ne-E (Lee79, Was82), but at present no definite decision can be made. Improved information on the reaction rates of several relevant processes (chap. 6) is also needed to provide stringent tests for the models. Clearly, there is much work to be done on all fronts before the rich information embedded in these new data can be fully exploited.

As a final remark, note that recent observations in γ-ray astronomy with the *HEAO 3* satellite led to the discovery of the 1809 keV γ-ray line (Mah82), which is known to be produced in the β-decay of ^{26}Al. From its intensity it was estimated (Cla84) that about 4.2 $M_\odot$ of ^{26}Al nuclides at present exist in the interstellar medium of our Galaxy. Thus, while the Mg isotopic variations show that ^{26}Al must have been produced not later than some 4.6 billion years ago, the observations of the 1809 keV γ-ray line provide evidence that the ^{26}Al nucleosynthesis continued at least until "recently," about 1 million years ago (the half-life of ^{26}Al), if not up to now. Clearly, any astrophysical model for ^{26}Al nucleosynthesis must be concordant with both observations (Cam84, Cla84).

10.3. The Origin of the Light Elements Li, Be, and B

10.3.1. The Problem

We have seen that nucleosynthesis in the big bang and in stellar interiors can account for nearly all the elements of the periodic table. However, the abundances of the very light, low-abundance elements, lithium, beryllium, and boron (LiBeB), sometimes collectively referred to as the *l*-elements (*l* for light), cannot be accounted for by these two production mechanisms. This is because of the mass stability gaps at $A = 5$ and $A = 8$. Because of these gaps hydrogen burning in stellar interiors via the *p-p* chain produces mainly ^{4}He (chap. 6), with the subsequent helium burning bypassing the LiBeB elements and producing ^{12}C via the triple-α process (chap. 7). Even if these *l*-elements were produced in stars, they would readily be destroyed in hydrogen burning in stellar interiors, particularly via (p, α) reactions. Thus their abundances, calculated from stellar synthesis, are many orders of magnitude less than their observed solar abundances (Fig. 10.4). Thus, their origin cannot be in stellar nucleosynthesis. Similar arguments apply to nucleosynthesis in the big bang (chap. 2), although some ^{7}Li may be produced. Clearly, the remaining isotopes

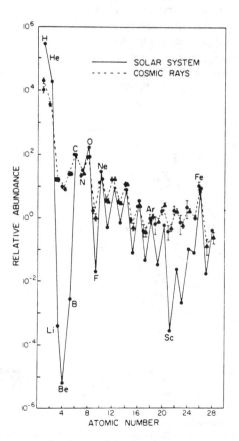

FIGURE 10.4. The solar system abundances of the elements (Cam82), sometimes referred to as the "local galactic abundances," are compared with those of the cosmic-ray matter as measured at the top of the earth's atmosphere (Mey74), normalized to carbon = 100. The most outstanding differences are in the light elements Li, Be, and B.

of these elements, ^{6}Li, ^{9}Be, ^{10}B, and ^{11}B, must have been synthesized by some other mechanism(s). In their classical paper Burbidge, Burbidge, Fowler, and Hoyle (B^{2}FH) recognized this problem and postulated an undefined l-process for their production. It was also noted that this l-process would very likely require an environment of low temperature and/or low density, so that synthesis is not immediately followed by destruction. A compelling clue to this l-process and thus to the origin of LiBeB was the discovery that relative to solar system abundances these light elements are enriched by a factor of about 10^6 in the galactic cosmic rays.

10.3.2. Properties of Galactic Cosmic Rays

In recent years advances in space technology, in particular the availability of satellite observatories, made it possible to determine the chemical and isotopic composition of samples of matter from elsewhere in our Galaxy. Although the direct sampling of matter from objects outside the solar system is still not possible, one can obtain some information by intercepting authentic starry messengers, the galactic cosmic rays. These are nuclei arriving at the earth from outside the solar system (Fig. 10.5). Their detection and study must be

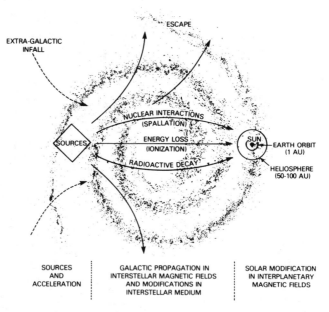

FIGURE 10.5. Shown schematically is the life history of a cosmic ray from acceleration in the source through propagation and modification in the Galaxy to observation on earth. For details of the propagation and modification of cosmic-ray matter in the Galaxy and the solar system, as well as the determination of the cosmic-ray composition at their sources, see Sim83. The solar system with its heliosphere at 50–100 astronomical units (AU) is not drawn in proper scale relative to the dimensions of our Galaxy.

accomplished in space, however, because, upon entering the earth's atmosphere, cosmic-ray nuclei are shattered by collisions with atmospheric nuclei before they can reach the earth's surface.

Early investigations conducted with instruments carried by high-altitude balloons established that the cosmic rays are composed of protons, α-particles, and heavier nuclei stripped of all their orbital electrons (Sch41, Fre48). Subsequent improved observations using balloons and satellites showed (Sim83 and references therein) that the nuclei have a continuous energy spectrum extending to energies in excess of 10^{14} MeV. The energy spectra of all nuclei are found to be quite similar and to be well described at the high energies by a power-law spectrum,

$$\Phi_P(E) \propto E^x \, ,$$

where $\Phi_P(E)$ is the differential flux of nuclear species P at kinetic energy E per nucleon and with $x = 2.5\text{--}2.7$. Most of the particles have energies between 100 and 3000 MeV. In this energy range it is found that the cosmic rays consist of approximately 87% hydrogen, 12% helium, and 1% heavier nuclei. Detailed analyses of the available data lead (Sim83) to the abundance distribution shown in Figure 10.4.

There is convincing evidence (Sim83) that the bulk of these cosmic-ray nuclei below an energy of about 10^9 MeV have their origin in our Galaxy, while the ultra–high-energy nuclei of low intensities probably are extragalactic in origin (Fig. 10.5). There is at present no model for the mechanism accelerating the nuclei to cosmic-ray energies, nor is there an identification of their astrophysical sites of origin. One scenario suggests that some cosmic rays are nuclei ejected by supernovae and may be accelerated during the explosion to nearly the velocity of light. Another scenario has nuclei in the interstellar gas being accelerated to high velocities by the shock waves from later supernovae. Collisions with matter cause the energetic atomic nuclei to lose their electrons (at least partially; chap. 5). These electrically charged cosmic rays are confined by the interstellar magnetic field (about 10^{-6} gauss) for periods on the order of 10 million years (Sim83) before they finally escape into intergalactic space. Before leaving the Galaxy, some of these cosmic rays pass close to the earth, where they can be captured and analyzed (Fig. 10.5). Since most of them eventually escape from the Galaxy, they must be continually replaced; data from meteorites reveal that the average cosmic-ray flux entering the solar system has been nearly constant over the past 4 billion years (For79). Thus on a galactic scale there appears to be a steady-state condition. Several astrophysical sources of energy sufficient to accelerate and sustain the bulk of the cosmic rays have been identified in recent years (Fow81, Sim83).

The elemental and isotopic abundances of the cosmic rays are an important independent channel of information about the galaxies and intergalactic space, since the cosmic rays represent the only accessible sample of matter of extrasolar origin. With the exception of the elements LiBeB (sec. 10.3.3), there appears to be a remarkable overall correspondence of the solar system abundances and galactic cosmic-ray abundances (Fig. 10.4). Equally remarkable and most significant are the observed differences in the two sets of elemental abundances. The cosmic-ray anomalies are in some cases more than a factor of 2, in contrast to about 1% or less for isotopic anomalies in solar system matter (sec. 10.2). These similarities and differences provide the clues and constraints for the identification of astrophysical sites, for the origin of cosmic rays (interstellar medium, stellar flares, stellar explosions, and so on), and for the nucleosynthetic processes that preceded the acceleration of the cosmic rays. For example, not only do the isotopic abundances of cosmic rays convey the imprint of their nucleosynthetic origins in the Galaxy, but also their radioactive species (for example ^{26}Al, $T_{1/2} = 7.2 \times 10^5$ y) reveal the time required (Wad77) for their acceleration and their propagation in interstellar magnetic fields (Fig. 10.5). Clearly, the cosmic rays are important for critically testing the validity of current ideas and models for nucleosynthesis of matter in the Galaxy and represent one of the exciting areas of research in astrophysics.

10.3.3. Production of Li, Be, and B via Spallation

The overall similarity in abundances of cosmic-ray and solar system matter (Fig. 10.4) reinforces the validity of the previous discussions of nucleosynthesis of the elements. Thus, the cosmic-ray nuclei at their nucleosynthetic sources should also have extremely low abundances of LiBeB elements. However, these "primary" nuclei are subsequently accelerated to high energies and travel on long journeys through the interstellar medium, producing "secondary" cosmic-ray nuclei via nuclear interactions between cosmic-ray nuclei and nuclei of interstellar matter. For example, if ^{12}C nuclei in the interstellar medium are bombarded by high-energy cosmic-ray protons, the following reactions can take place in which the ^{12}C nuclei break up (or spall) into fragments such as LiBeB:

$$p + {}^{12}C \rightarrow {}^{11}B + 2p \qquad (Q = -16.0 \text{ MeV}),$$
$$\rightarrow {}^{10}B + 2p + n \qquad (Q = -27.4 \text{ MeV}),$$
$$\rightarrow {}^{10}B + {}^{3}He \qquad (Q = -19.7 \text{ MeV}),$$
$$\rightarrow {}^{9}Be + 3p + n \qquad (Q = -34.0 \text{ MeV}),$$
$$\rightarrow {}^{9}Be + {}^{3}He + p \qquad (Q = -26.3 \text{ MeV}),$$
$$\rightarrow {}^{7}Li + 4p + 2n \qquad (Q = -52.9 \text{ MeV}),$$
$$\rightarrow {}^{7}Li + {}^{4}He + 2p \qquad (Q = -24.6 \text{ MeV}),$$
$$\rightarrow {}^{6}Li + 4p + 3n \qquad (Q = -60.2 \text{ MeV}),$$
$$\rightarrow {}^{6}Li + {}^{4}He + 2p + n \qquad (Q = -31.9 \text{ MeV}),$$
$$\rightarrow {}^{6}Li + {}^{4}He + {}^{3}He \qquad (Q = -24.2 \text{ MeV}).$$

Nuclear reactions, in which several particles are emitted in the exit channel, are called spallation reactions. The relatively high negative Q-values of these spallation reactions are easily matched by the high-energy, nonthermal cosmic-ray particles. The reactions take place in an environment which is relatively cool and has a very low density; this allows the LiBeB products to survive after their formation.

The idea of nucleosynthesis of the l-elements by the galactic cosmic rays was suggested as early as 1970 by Reeves, Fowler, and Hoyle (Ree70a). Since some of the LiBeB so created will thermalize and become part of the interstellar medium, eventually to be incorporated into stars, this process clearly contributes to the light-element abundances. One must, of course, demonstrate that the process is quantitatively effective.

In the simplest model of cosmic-ray nucleosynthesis the production yield per unit time of a nuclide i is

$$(dN_i/dt) = \sum_{P,T} N_T \int_0^\infty \Phi_P(E)\sigma_{P,T}^i(E)dE ,$$

where $\Phi_P(E)$ is the flux of the high-energy cosmic-ray projectiles P inducing the spallation reactions, N_T is the number density of the target nuclei T in the interstellar medium (spallation target), and $\sigma_{P,T}^i(E)$ is the cross section for production of nuclides i in the collision of P and T. The summation is over all possible projectiles and targets. The important reactions are those for which the product $N_T\Phi_P(E)$ is large. Inspection of the abundance distribution (Fig. 10.4) suggests that these are primarily $p + \text{CNO}$ spallation reactions. In the case of collision between a cosmic-ray proton (projectile) and CNO interstellar matter (target) the spallation products LiBeB have low energy, and thus a high probability for thermalization and incorporation into the interstellar medium. In the alternative case, the spallation products induced by CNO projectiles on a hydrogen target have velocities near the projectile velocity; thus these LiBeB nuclei are likely to be destroyed by further collisions with interstellar matter or to escape from the Galaxy. Thus, it is adequate for a rough estimate to consider only protons striking CNO elements, and among these only CO, since C and O are the most abundant targets in the interstellar medium (Fig. 10.4). Since the cross sections do not depend strongly on energy, we can remove the cross section from under the integral sign to obtain

$$(dN_i/dt) \simeq (N_C \sigma_{pC}^i + N_O \sigma_{pO}^i)\Phi_p ,$$

where Φ_p is now the total flux of cosmic-ray protons above the reaction threshold near 30 MeV. Finally, assuming that the production rate is constant from the birth of the Galaxy (t_g) to that of the sun (t_s), and dividing by the number density of hydrogen H, yields

$$(N_i/H) \simeq ((N_C/H)\sigma_{pC}^i + (N_O/H)\sigma_{pO}^i)\Phi_p(t_g - t_s) .$$

Using $t_g - t_s = 9 \times 10^9$ y, $\Phi_p(E > 30$ MeV$) = 8.3$ cm^2 s^{-1}, $N_C/H = 0.048\%$ and $N_O/H = 0.085\%$ (Aus81), and spallation cross sections as given in Table 10.2, we obtain the approximate values for N_i/H from the above equation as given in Table 10.2. Except for ^{7}Li the predictions agree within the uncertainties with the observed abundances of Li, Be, and B.

TABLE 10.2 Estimates of LiBeB Production by Galactic Cosmic Rays[a]

Nuclide	σ_{pC}^i (mb)	σ_{pO}^i (mb)	$\left(\dfrac{N_i}{H}\right)_{theo}$ ($\times 10^{-12}$)	$\left(\dfrac{N_i}{H}\right)_{obs}$ ($\times 10^{-12}$)	$\left(\dfrac{N_i}{H}\right)_{theo} \Big/ \left(\dfrac{N_i}{H}\right)_{obs}$
^{6}Li	14.8	13.9	45	70	0.64
^{7}Li	20.5	21.2	66	900	0.07
^{9}Be	6.2	4.4	16	14	1.14
^{10}B	22.7	12.7	51	30	1.70
^{11}B	57.0	26.5	118	120	0.98

[a] From Aus81.

An improved model has to include a detailed treatment of features such as the cosmic-ray transport processes; the cross sections of various types of spallation reactions such as p, α, and CNO cosmic rays on p, α, and CNO interstellar matter targets; and the actual energy spectra of the cosmic rays. A discussion of the relevant spallation cross sections and techniques for their measurement can be found in the recent review article by Austin (Aus81). Improved models (Aus81) have verified that one is able to reproduce the abundances of the LiBeB elements within their uncertainties. The exception is the isotope ^{7}Li, which is underproduced by approximately an order of magnitude, thus strengthening the arguments for its origin in the big bang.

In summary, the puzzle of the origin of the l-elements appears to be solved in a natural, straightforward, and economical fashion. The solution came from the recognition that the interaction of the galactic cosmic rays with elements present in the interstellar medium leads to an additional nucleosynthesis mechanism. Undoubtedly cosmic-ray interactions also contribute to the abundances of other elements, but these additions represent only a minor perturbation of the major element abundances. There are still many questions to be answered, but it is unlikely that their answers will significantly change the above conclusions.

Epilogue

Throughout most of recorded history, matter was thought to be composed of various combinations of four basic elements: earth, air, fire, and water. Modern science has replaced this list with a considerable longer one: the known chemical elements and their isotopes. Questions about their origin and history could be answered only after the science of nuclear physics was born at the beginning of this century. Data obtained in nuclear physics laboratories provided the empirical basis for the development of theories of nucleosynthesis. These theories revealed that most of the elements were formed during the fiery lifetimes and explosive deaths of stars in the heavens, the cauldrons in the cosmos. A few of the lighter elements were formed before the stars even existed, during the birth of the universe itself. Also, a few of the lighter elements were synthesized in the intergalactic space by cosmic rays. Thus, theories of nucleosynthesis have identified the most important sites of element formation and also the diverse nuclear processes involved in their synthesis. Indeed, over the past quarter-century, the theories of nucleosynthesis have advanced at a remarkable pace and have achieved an impressive record of success. Factors contributing to these rapid developments include progress in experimental and theoretical nuclear physics and an improved knowledge of abundances in astrophysical environments.

It is the nature of astrophysics that many of the processes and the objects we are trying to understand are physically inaccessible. Thus, it is important that those aspects that can be studied in the laboratory be rather well understood. Nuclear astrophysics is one area in which laboratory data require hard work and careful theoretical analyses to understand the data. Clearly, theories of nucleosynthesis will not be on solid ground until the nuclear physics input data are thoroughly understood. With the advance of improved technologies, it might be possible to investigate several of the difficult-to-measure processes including short-lived nuclei (ground states and isomeric states) as well as to

515

extend the knowledge of nuclear properties for nuclei far from the valley of stability. Advanced technologies also will promote neutrino astronomy, providing new clues to nuclear energy generation and nucleosynthesis in diverse astrophysical scenarios.

Improved nuclear data as well as refined astronomical data are needed to solve many of the remaining problems that challenge basic aspects of theories of nucleosynthesis. Among these problems are isotopic anomalies and formation of stars and planets, the site of the r-process and nucleocosmochronology, the origin of the p-elements, details of supernova dynamics and explosive burning of the elements, galactic chemical evolution, and the source of galactic cosmic rays and the mechanism of their acceleration. Clearly, nuclear astrophysics is not a finished subject and will remain an exciting research area for many years to come. There are many open questions that confront contemporary astrophysics in general. For example: What is the large-scale structure of the universe? How do galaxies form and evolve? What role do violent events play in the evolution of the universe? Where and in what form is the missing mass? What is the fate of the universe? Is it possible, as astrophysics pushes the frontiers of time back to the moment of cosmic creation, that the existence of the universe will be recognized as a natural consequence of the working of the fundamental forces? Is it possible to unify gravitation with the other forces of nature, and what role do the quantum properties of gravitation play in cosmic expansion? Such questions, once believed to be outside the range of science, are now valid subjects of scientific investigations. With improved technologies, such as the Hubble Space Telescope and the powerful accelerators, partial answers to some of these questions will be possible, and a different and still more wonderful view of the cosmos will almost surely be revealed to us in the years ahead.

Appendix:
Notation and Units

A.1. Notation

The notation is generally described within each chapter, but many symbols are of such general use that they are not defined repeatedly. The notation here will be used without further explanation, if there exists no ambiguity.

Signs

$=$	equals	$\neq$	is not equal to
$\simeq$	approximately equals	$\equiv$	is identical with
$\triangleq$	corresponds to	$\leq$	is less than or equal to
$\geq$	is greater than or equal to	$\ll$	is much smaller than
$\gg$	is much larger than	$\propto$	is proportional to
$\rightarrow$	goes to	$\times$	times
∞	infinity	log	logarithm to the base 10
%	percent	exp	power of e
ln	natural logarithm	$\pm \Delta y$	error of y
$\bar{y}$	mean value of y	$\sqrt{y}, y^{1/2}$	square root of y
$|y|$	absolute value of y	$d^2y/dt^2 = \ddot{y}$	second time derivative of y
$dy/dt = \dot{y}$	first time derivative of y	$\int_a^b \cdots dy$	integral over y
$\sum_i y_i$	sum of y_i		from a to b

Symbols

π	ratio of circumference to diameter ($= 3.14159$)
e	base of natural logarithms ($= 2.71828$); electron charge
ν, λ	frequency, wavelength
Ω	solid angle
ω	angular frequency ($= 2\pi\nu$)
c	velocity of light
t	time (except for half-lives, where $T_{1/2}$ is used)

dx	element of quantity x
T	temperature
m (or M)	mass of object
r (or R)	radius
R	gas constant
k	Boltzmann constant
μ	reduced mass
ρ	density
h	Planck's constant ($= 2\pi\hbar$)
N	number of objects (often per unit volume)
σ	radiation constant; cross section
$\odot$, $\oplus$	sun, earth
c.m.	center-of-mass system
lab	laboratory system
LN_2	liquid nitrogen
STP	standard temperature (0 °C) and pressure (1 atm)
FWHM	full width at half-maximum

Decimal Multiples and Submultiples

Factor	Prefix (to place before a unit)	Symbol
10^{12}	tera	T
10^{9}	giga	G
10^{6}	mega	M
10^{3}	kilo	k
10^{-2}	centi	c
10^{-3}	milli	m
10^{-6}	micro	μ
10^{-9}	nano	n
10^{-12}	pico	p
10^{-15}	femto	f
10^{-18}	atto	a

A.2. Units

The units (All73, Coh73, Wap85) are expressed in the cgs (centimeter gram second) and esu (electrostatic units) system. From time to time values are quoted in other units as described below.

Length

meter	m = 100 cm
kilometer	km = 10^5 cm
millimeter	mm = 10^{-1} cm = 10^{-3} m
micron	$\mu = \mu$m = 10^{-4} cm = 10^{-6} m
angstrom	Å = 10^{-8} cm
fermi (also femtometer)	fm = 10^{-13} cm = 10^{-15} m
inch	in = 2.54000 cm

astronomical unit (= mean sun-earth distance)	$AU = 1.49598 \times 10^{13}$ cm
light-year	$ly = 9.46053 \times 10^{17}$ cm = 63240 AU
parsec	$pc = 3.08568 \times 10^{18}$ cm = 206265 AU = 3.26163 ly
solar radius	$R_\odot = 6.9599 \times 10^{10}$ cm
earth equatorial radius	$R_\oplus = 6.3782 \times 10^{8}$ cm = 6378 km $(R_\odot/R_\oplus = 109.1)$
Bohr radius (infinite mass)	$a_0 = h^2/4\pi^2 m_e e^2 = 0.52918 \times 10^{-8}$ cm
electron Compton wavelength	$h/m_e c = 2.42631 \times 10^{-10}$ cm

Area

square meter	$m^2 = 10^4$ cm^2
barn	$b = 10^{-24}$ cm^2

Volume

cubic meter	$m^3 = 10^6$ cm^3
liter	$l = 10^3$ cm^3

Time

minute	min = 60 s
hour	h = 3600 s = 60 min
day	d = 86400 s = 24 h
sidereal year	$y = 365.256$ d $= 3.15581 \times 10^7$ s
aeon	10^9 y

Mass

kilogram	$kg = 10^3$ g
metric ton	$t = 10^6$ g
solar mass	$M_\odot = 1.989 \times 10^{33}$ g
earth mass	$M_\oplus = 5.976 \times 10^{27}$ g $(M_\odot/M_\oplus = 332945)$
atomic mass unit (^{12}C = 12 scale)	$amu = 1.66056 \times 10^{-24}$ g
electron mass	$m_e = 9.10953 \times 10^{-28}$ g $= 5.4858 \times 10^{-4}$ amu
proton mass	$m_p = 1.67265 \times 10^{-24}$ g $= 1.00728$ amu
	$(m_p/m_e = 1836.15)$
mass of ^{1}H atom	$m_H = 1.67356 \times 10^{-24}$ g $= 1.00783$ amu

Density

mean density of earth	$\bar{\rho}_\oplus = 5.5$ g cm^{-3}
solar mean density	$\bar{\rho}_\odot = 1.4$ g cm^{-3}

Energy

erg	1 g cm^2 s^{-2}
joule	$J = 10^7$ ergs
electron volt	$eV = 1.60219 \times 10^{-12}$ ergs $= 10^{-3}$ keV $= 10^{-6}$ MeV $= 10^{-9}$ GeV
mass energy of 1 amu	1.49243×10^{-3} ergs $= 931.502 \times 10^6$ eV $= 931.502$ MeV
rest mass energy of electron	$m_e c^2 = 8.18727 \times 10^{-7}$ ergs $= 511.003$ keV

Power

watt $W = 10^7$ ergs s^{-1} = J s^{-1}

solar luminosity $L_\odot = 3.826 \times 10^{26}$ watts $= 3.826 \times 10^{33}$ ergs s^{-1}
                                                    $= 2.388 \times 10^{39}$ MeV s^{-1}

Force

dyne   dyn = 1 g cm s^{-2}

newton   newton = 10^5 dyn

Velocity

kilometer per second   km s^{-1} = 10^5 cm s^{-1}

velocity of light $c = 2.997925 \times 10^{10}$ cm s^{-1}

Pressure

pascal   10 dyn cm^{-2}

bar 10^6 dyn cm^{-2}

atmosphere   atm $= 1.01325 \times 10^6$ dyn cm^{-2} = 1.013 bar
                                         = 760 torr

millimeter of mercury   mm Hg = 1333.22 dyn cm^{-2}

(= 1 torr) $= 1.315 \times 10^{-3}$ atm = 1.298 mbar

Temperature

temperature comparisons   0 °C = 273.150 K
                                       100 °C = 373.150 K

T_6 in units of 10^6 K

T_9 in units of 10^9 K

surface temperature of sun $T_s = 5800$ K

central temperature of sun $T_c = 15 \times 10^6$ K ($T_6 = 15$)

Frequency

hertz   Hz = cycles s^{-1}

Angle

degree   deg = 1° = right angle/90
                                       = 60 minutes of arc (60 arcmin)
                                       = 3600 seconds of arc (arcsec = 3600″)

radian   rad = 57°29578

Solid angle

steradian   sr = 3282.8 deg^2

Angular momentum

quantum unit $\hbar = 1.0546 \times 10^{-27}$ ergs s $= 6.5822 \times 10^{-16}$ eV s

Planck's constant $h = 2\pi\hbar = 6.6262 \times 10^{-27}$ ergs s

Electric charge

electron charge $e = 4.80325 \times 10^{-10}$ esu $= 1.60219 \times 10^{-19}$ C

coulomb   C $= 2.997925 \times 10^9$ esu
                                       $= -6.24145 \times 10^{18}$ electrons

Electric potential

volt

$$V = 3.33564 \times 10^{-3} \text{ esu}$$

Resistance

ohm

$$\Omega = 1.11265 \times 10^{-12} \text{ esu}$$

Electric current

ampere

$$A = 2.997925 \times 10^9 \text{ esu}$$
$$= -6.24145 \times 10^{18} \text{ electrons s}^{-1}$$

Magnetic field

ampere-turn per meter

gauss

tesla

$$12\pi \times 10^7 \text{ esu} = 4\pi \times 10^{-3} \text{ oersteds}$$
$$1 \text{ oersted} = 79.58 \text{ amp-turn m}^{-1}$$
$$10^4 \text{ gauss}$$

Capacitance

farad

$$F = 9 \times 10^{11} \text{ cm}$$

Inductance

henry

$$H = 1.11265 \times 10^{-12} \text{ esu}$$

Radioactivity

curie

$$Ci = 3.700 \times 10^{10} \text{ disintegrations s}^{-1}$$

Some physical constants

gravitation constant

Boltzmann constant

$$G = 6.670 \times 10^{-8} \text{ dyn cm}^2 \text{ g}^{-2}$$
$$k = 1.38062 \times 10^{-16} \text{ ergs K}^{-1}$$
$$= 8.6171 \times 10^{-5} \text{ eV K}^{-1}$$

gas constant

Avogadro's number

Loschmidt's number

fine-structure constant

radiation density constant

$$R = 8.3143 \times 10^7 \text{ ergs K}^{-1} \text{ mole}^{-1}$$
$$N_A = 6.02217 \times 10^{23} \text{ mole}^{-1}$$
$$L = 2.68684 \times 10^{19} \text{ cm}^{-3} \text{ (STP)}$$
$$\alpha = e^2/\hbar c = 1/137.036$$
$$a = 8\pi^5 k^4/15 c^3 h^3$$
$$= 7.56464 \times 10^{-15} \text{ ergs cm}^{-3} \text{ K}^{-4}$$

Stefan-Boltzmann constant

Wien displacement law constant

$$\sigma = ac/4 = 5.66956 \times 10^{-5} \text{ ergs cm}^{-2} \text{ K}^{-4} \text{ s}^{-1}$$
$$hc/4.9651k = 0.2898$$

References

Journals frequently cited are abbreviated whenever possible, as shown below:

Ann. Rev. Astr. Ap.	*Annual Review of Astronomy and Astrophysics*
Ann. Rev. Nucl. Sci.	*Annual Review of Nuclear Science*
Ann. Rev. Nucl. Part. Sci.	*Annual Review of Nuclear and Particle Science*
Ann. N.Y. Acad. Sci.	*Annals of the New York Academy of Sciences*
Ap. J.	*Astrophysical Journal*
Ap. J. (Lett.)	*Astrophysical Journal Letters*
Ap. J. Suppl.	*Astrophysical Journal Supplement Series*
Ap. Space Sci.	*Astrophysics and Space Science*
At. Nucl. Data Tables	*Atomic Data and Nuclear Data Tables*
Astr. Ap.	*Astronomy and Astrophysics*
Bull. Am. Phys. Soc.	*Bulletin of the American Physical Society*
Can. J. Phys.	*Canadian Journal of Physics*
Geoch. Cosmoch. Acta	*Geochimica et Cosmochimica Acta*
M.N.R.A.S.	*Monthly Notes of the Royal Astronomical Society*
Nucl. Phys.	*Nuclear Physics*
Nucl. Instr. Meth.	*Nuclear Instruments and Methods*
Phys. Lett.	*Physics Letters*
Phys. Rev.	*Physical Review*
Phys. Rev. Lett.	*Physical Review Letters*
Rept. Progr. Phys.	*Report on Progress in Physics*
Rev. Mod. Phys.	*Reviews of Modern Physics*
Rev. Sci. Instr.	*Review of Scientific Instruments*
Scient. Am.	*Scientific American*
Trans. Nucl. Sci.	*Transactions of Nuclear Science*
Z. Phys.	*Zeitschrift für Physik*

Abr65 Abramowitz, M., and Stegun, I. A., *Handbook of Mathematical Functions* (New York: Dover, 1965).

Ajz80 Ajzenberg-Selove, F., and Busch, C. L., *Nucl. Phys.* A336 (1980) 1.

Ajz82   Ajzenberg-Selove, F., *Nucl. Phys.* A375 (1982) 1.
Ajz83   Ajzenberg-Selove, F., *Nucl. Phys.* A392 (1983) 1.
Ajz84   Ajzenberg-Selove, F., *Nucl. Phys.* A413 (1984) 1.
Alb60   Alburger, D. E., in *Nuclear Spectroscopy*, Part A, ed. F. Ajzenberg-Selove (New York: Academic, 1960), p. 228.
Alb61   Alburger, D. E., *Phys. Rev.* 124 (1961) 193.
Alb77   Alburger, D. E., *Phys. Rev.* C16 (1977) 2394.
Alb82   Albrecht, A., and Steinhardt, P. J., *Phys. Rev. Lett.* 48 (1982) 1220.
Alb82a Albrecht, A., Steinhardt, P. J., Turner, M. S., and Wilczek, F., *Phys. Rev. Lett.* 48 (1982) 1437.
Ale84   Alexander, T. K., Ball, G. C., Lennard, W. N., Geissel, H., and Mak, H. B., *Nucl. Phys.* A427 (1984) 526.
Alk82   Alkemade, P. F. A., Alderliesten, C., de Wit, P., and van der Leun, C., *Nucl. Instr. Meth.* 197 (1982) 383.
All71   Allen, B. J., Gibbons, J. H., and Macklin, R. L., *Adv. Nucl. Phys.* 4 (1971) 205.
All74   Allen, K. W., in *Nuclear Spectroscopy and Reactions*, Part A, ed. J. Cerny (New York: Academic, 1974), p. 3.
All76   Allen, K. W., Dolan, S. P., Holmes, A. R., Symons, T. J. M., Watt, F., Zimmerman, C. H., Litherland, A. E., and Sandorfi, A. M. J., *Nucl. Instr. Meth.* 134 (1976) 1.
Alm60   Almqvist, E., Bromley, D. A., and Kuehner, J. A., *Phys. Rev. Lett.* 4 (1960) 515 and *Phys. Rev.* 130 (1963) 1140.
Alm61   Almen, O., and Bruce, G., *Nucl. Instr. Meth.* 11 (1961) 257.
Alm64   Almqvist, E., Kuehner, J. A., McPherson, D., and Vogt, E. W., *Phys. Rev.* B136 (1964) 84.
Alm83   Almeida, J., and Käppeler, F., *Ap. J.* 265 (1983) 417.
Alp48   Alpher, R. A., Bethe, H. A., and Gamow, G., *Phys. Rev.* 73 (1948) 803.
Alp48a Alpher, R. A., *Phys. Rev.* 74 (1948) 1577.
Alp49   Alpher, R. A., and Herman, R. C., *Phys. Rev.* 75 (1949) 1089.
Alp50   Alpher, R. A., and Herman, R. C., *Rev. Mod. Phys.* 22 (1950) 153 and *Ann. Rev. Nucl. Sci.* 2 (1953) 1.
Alt49   Alter, W., and Garbuny, M., *Phys. Rev.* 76 (1949) 496.
Alt81   Alton, G. D., *Nucl. Instr. Meth.* 189 (1981) 15.
Alv49   Alvarez, L. W., Univ. Cal. Rad. Lab. Rept. No. UCRL-328 (1949).
Alv51   Alvarez, L. W., *Rev. Sci. Instr.* 22 (1951) 705.
And72   Anderson, P. W., *Science* 177 (1972) 393.
And77   Andersen, H. H., and Ziegler, J. F., *Hydrogen Stopping Powers and Ranges in All Elements* (New York: Pergamon, 1977).
And82   Anders, E., and Ebihara, M., *Geoch. Cosmoch. Acta* 46 (1982) 2363.
Ard56   Ardenne, M. von, *Tabellen der Elektronenphysik und Übermikroskopie* (Berlin: Deutscher Verlag der Wissenschaften, 1956).
Arn54   Arnold, W. R., Phillips, J. A., Sawyer, G. A., Stovall, E. J., and Tuck, J. L., *Phys. Rev.* 93 (1954) 483.
Arn67   Arnison, G. T. J., *Nucl. Instr. Meth.* 53 (1967) 357.
Arn69   Arnett, W. D., and Truran, J. W., *Ap. J.* 157 (1969) 339.
Arn69a Arnett, W. D., *Ap. Space Sci.* 5 (1969) 180.
Arn70   Arnett, W. D., *Ap. J.* 162 (1970) 349.

Arn71   Arnett, W. D., Truran, J. W., and Woosley, S. E., *Ap. J.* 165 (1971) 87.
Arn72   Arnett, W. D., *Ap. J.* 176 (1972) 681.
Arn73   Arnett, W. D., *Ann. Rev. Astr. Ap.* 11 (1973) 73 and *Ap. J.* 179 (1973) 249.
Arn73a   Arnould, M., *Astr. Ap.* 22 (1973) 311.
Arn76   Arnould, M., and Howard, W. M., *Nucl. Phys.* A274 (1976) 295.
Arn78   Arnett, W. D., *Ap. J.* 219 (1978) 1008.
Arn79   Arnett, W. D., *Ap. J. (Lett.)* 230 (1979) L37.
Arn80   Arnould, M., Norgaard, H., Thielemann, F. K., and Hillebrandt, W., *Ap. J.* 237 (1980) 931.
Arn80a   Arnett, W. D., *Ann. N.Y. Acad. Sci.* 336 (1980) 366.
Arn81   Arnould, M., and Norgaard, H., *Comm. Ap. Space Phys.* 9 (1981) 145.
Arn83   Arnould, M., Takahashi, K., and Yokoi, K., *Astr. Ap.* 137 (1983) 51.
Asa68   Asano, D., and Sugimoto, D., *Ap. J.* 154 (1968) 1127.
Ash69   Ashery, D., *Nucl. Phys.* A136 (1969) 481.
Ast27   Aston, F. W., *Proc. Roy. Soc. London* A115 (1927) 487.
Atk29   Atkinson, R. d'E., and Houtermans, F. G., *Z. Phys.* 54 (1929) 656.
Atk31   Atkinson, R. d'E., *Ap. J.* 73 (1931) 250 and 308.
Aud72   Audouze, J., Fowler, W. A., and Schramm, D. N., *Nature* 238 (1972) 8.
Aud73   Audouze, J., Truran, J. W., and Zimmerman, B. A., *Ap. J.* 1984 (1973) 493.
Aud80   Audouze, J., and Vauclair, S., *An Introduction to Nuclear Astrophysics* (Dordrecht: Reidel, 1980).
Aus71   Austin, S. M., Trentlemen, G. F., and Kashy, E., *Ap. J. (Lett.)* 163 (1971) L79.
Aus81   Austin, S. M., *Progr. Part. Nucl. Phys.* 7 (1981) 1.
Baa34   Baade, W., and Zwicky, F., *Phys. Rev.* 45 (1934) 138.
Bac67   Bacher, A. D., and Tombrello, T. A., quoted T. A. Tombrello, *Nuclear Research with Low-Energy Accelerators*, ed. J. B. Marion and D. M. van Patter (New York: Academic, 1967), p. 195.
Bae10   Baeyer, O. von, and Hahn, O., *Phys. Z.* 11 (1910) 488.
Bag74   Baglin, J. E., and Ziegler, J. F., *J. Appl. Phys.* 45 (1974) 1413.
Bah62   Bahcall, J. N., *Phys. Rev.* 126 (1962) 1143.
Bah62a   Bahcall, J. N., and Moeller, C. P., *Phys. Rev.* 128 (1962) 1297.
Bah68   Bahcall, J. N., and Shaviv, G., *Ap. J.* 153 (1968) 113.
Bah69   Bahcall, J. N., and May, R. M., *Ap. J.* 155 (1969) 501.
Bah69a   Bahcall, J. N., and Moeller, C. P., *Ap. J.* 155 (1969) 511.
Bah69b   Bahcall, J. N., *Scient. Am.*, July 1969.
Bah80   Bahcall, J. N., Huebner, W. F., Lubow, S. H., Magee, N. H., Merts, A. L., Parker, P. D., Rozsnyai, B., Ulrich, R. K., and Argo, M. F., *Phys. Rev. Lett.* 45 (1980) 945.
Bah82   Bahcall, J. N., Huebner, W. F., Lubow, S. H., Parker, P. D., and Ulrich, R. K., *Rev. Mod. Phys.* 54 (1982) 767.
Bah82a   Bahcall, J. N., and Davis, R., in *Essays in Nuclear Astrophysics*, ed. C. A. Barnes, D. D. Clayton, and D. N. Schramm (Cambridge: Cambridge University Press, 1982), p. 243.
Bah85   Bahcall, J. N., Cleveland, B. T., Davis, R., and Rowley, J. K., *Ap. J. (Lett.)* 292 (1985) L79.
Bai50   Bailey, C. L., and Stratton, W. R., *Phys. Rev.* 77 (1950) 194.
Bai60   Bailey, L. E., *Rev. Sci. Instr.* 31 (1960) 1147.

Bai63   Bailey, G. M., and Hebbard, D. F., *Nucl. Phys.* 46 (1963) 529 and 49 (1963) 666.

Bai70   Bailey, G. M., Griffiths, G. M., Olivo, M. A., and Helmer, R. L., *Can. J. Phys.* 48 (1970) 3059.

Bai73   Bair, J. K., and Haas, F. X., *Phys. Rev.* C7 (1973) 1356.

Bal59   Baldinger, E., *Handbuch der Physik* 44 (1959) 1.

Bam77   Bambynek, W., Behrens, H., Chen, M. H., Crasemann, B., Fitzpatrick, M. L., Ledingham, K. W. D., Genz, H., Mutterer, M., and Intermann, R. L., *Rev. Mod. Phys.* 49 (1977) 77.

Ban66   Banford, A. P., *The Transport of Charged Particle Beams* (London: Spon, 1966).

Bar33   Barber, N. F., *Proc. Leeds Phil. Lit. Soc. Sci.* 2 (1933) 427.

Bar71   Barnes, C. A., *Adv. Nucl. Phys.* 4 (1971) 133.

Bar71a   Barker, F. C., *Australian J. Phys.* 24 (1971) 777.

Bar71b   Barghoorn, E. S., *Scient. Am.*, May 1971.

Bar73   Barbon, R., Ciatti, F., and Rosino, L., *Astr. Ap.* 29 (1973) 57.

Bar73a   Barnes, C. A., and Nichols, D. B., *Nucl. Phys.* A217 (1973) 125.

Bar80   Barker, F. C., *Australian J. Phys.* 33 (1980) 177.

Bar82   Barnes, C. A., in *Essays in Nuclear Astrophysics*, ed. C. A. Barnes, D. D. Clayton, and D. N. Schramm (Cambridge: Cambridge University Press, 1982), p. 193.

Bar85   Barnes, C. A., Trentalange, S., and Wu, S. C., in *Treatise on Heavy-Ion Science*, Vol. **6,** ed. D. A. Bromley (New York: Plenum, 1985), p. 3.

Bas57   Bashkin, S., and Carlson, R. R., *Phys. Rev.* 106 (1957) 261.

Bas59   Bashkin, S., Carlson, R. R., and Douglas, R. A., *Phys. Rev.* 114 (1959) 1543.

Bas59a   Bashkin, S., Kavanagh, R. W., and Parker, P. D., *Phys. Rev. Lett.* 3 (1959) 518.

Bau81   Baumann, H., and Bethge, K., *Nucl. Instr. Meth.* 189 (1981) 107.

B^{2}FH   Burbidge, E. M., Burbidge, G. R., Fowler, W. A., and Hoyle, F., *Rev. Mod. Phys.* 29 (1957) 547.

Bec59   Becker, R. A., and Fowler, W. A., *Phys. Rev.* 115 (1959) 1410.

Bec78   Becchetti, F. D., Flynn, E. R., Hanson, D. L., and Sunier, J. W., *Nucl. Phys.* A305 (1978) 293 and 313.

Bec80   Becchetti, F. D., Overway, W., Jaenecke, J., and Jacobs, W. W., *Nucl. Phys.* A344 (1980) 336.

Bec80a   Becker, R. H., Holt, S. S., Smith, B. W., White, N. E., Boldt, E. A., Mushotzky, R. F., and Serlemitsos, P. S., *Ap. J.* (*Lett.*) 235 (1980) L5.

Bec81   Becker, H. W., Kettner, K. U., Rolfs, C., and Trautvetter, H. P., *Z. Phys.* A303 (1981) 305.

Bec82   Becker, H. W., Buchmann, L., Goerres, J., Kettner, K. U., Kräwinkel, H., Rolfs, C., Schmalbrock, P., Trautvetter, H. P., and Vlieks, A., *Nucl. Instr. Meth.* 198 (1982) 277.

Bec82a   Becker, H. W., Kieser, W. E., Rolfs, C., Trautvetter, H. P., and Wiescher, M., Z. Phys. A305 (1982) 319.

Bec84   Becker, H. W., Ph.D. thesis, Universität Münster (1984) and Becker, H. W., Rolfs, C., and Trautvetter, H. P., *Z. Phys.* A327 (1987) 341.

Bec96   Becquerel, H., *Comptes Rendus* 122 (1896) 501 and 689.

Bee80   Beer, H., and Käppeler, F., *Phys. Rev.* C21 (1980) 534.

Bee81   Beer, H., Käppeler, F., Wisshak, K., and Ward, R. A., *Ap. J. Suppl.* 46 (1981) 295.

Bee82   Beer, H., *Ap. J.* 262 (1982) 739.

Bee84   Beer, H., Käppeler, F., Yokoi, K., and Takahashi, K., *Ap. J.* 278 (1984) 388.

Bee84a   Beer, H., Walter, G., Macklin, R. L., and Patchett, P. J., *Phys. Rev.* C30 (1984) 464.

Beg80   Begemann, F., *Rept. Progr. Phys.* 43 (1980) 1309.

Bel70   Bel, B. D., Bingham, C. R., Halbert, M. L., and van der Woude, A., *Phys. Rev. Lett.* 24 (1970) 1120.

Ben36   Bennett, W. H., and Darby, P. F., *Phys. Rev.* 49 (1936) 97, 422, and 881.

Ben53   Bennett, W. H., *Rev. Sci. Instr.* 24 (1953) 915.

Ben66   Benn, J., Dally, E. B., Müller, H. H., Pixley, R. E., Staub, H. H., and Winkler, H., *Phys. Lett.*, 20 (1966) 43.

Ben68   Benn, J., Dally, E. B., Müller, H. H., Pixley, R. E., Staub, H. H., and Winkler, H., *Nucl. Phys.* A106 (1968) 296.

Ber64   Berger, M. J., and Seltzer, S. M., *Nat. Acad. Sci. Pub.* 1133 (1964) 205.

Ber77   Berg, H. L., Hietzke, W., Rolfs, C., and Winkler, H., *Nucl. Phys.* A276 (1977) 168.

Bes80   Besenbacher, F., Andersen, J. U., and Bonderup, E., *Nucl. Instr. Meth.* 168 (1980) 1.

Bet34   Bethe, H. A., and Heitler, W., *Proc. Roy. Soc. London* A146 (1934) 83.

Bet36   Bethe, H. A., and Bacher, R. F., *Rev. Mod. Phys.* 8 (1936) 82.

Bet37   Bethe, H. A., *Rev. Mod. Phys.* 9 (1937) 69.

Bet37a   Bethe, H. A., and Placzek, G., *Phys. Rev.* 51 (1937) 462.

Bet38   Bethe, H. A., and Critchfield, C. L., *Phys. Rev.* 54 (1938) 248 and 862.

Bet39   Bethe, H. A., *Phys. Rev.* 55 (1939) 103 and 434.

Bet70   Bethge, K., *Ann. Rev. Nucl. Sci.* 20 (1970) 255.

Bet82   Bethe, H. A., in *Essays in Nuclear Astrophysics*, ed. C. A. Barnes, D. D. Clayton, and D. N. Schramm (Cambridge: Cambridge University Press, 1982), p. 439.

Bet85   Bethe, H. A., and Brown, G. E., *Scient. Am.*, April 1985.

Bia68   Bianco, W. D., Lemire, F., Levesque, R. J. A., and Poutissou, M. M., *Can. J. Phys.* 46 (1968) 1585.

Bia69   Bianco, W. D., and Poutissou, J. M., *Phys. Lett.* 29 (1969) 299.

Bie55   Bieri, R., Everling, F., and Mattauch, J., *Z. Naturforschung* 10a (1955) 659.

Bil78   Bilenky, S. M., and Pontecorvo, B., *Phys. Rept.* 41 (1978) 225.

Bis68   Bishop, G. B., *Nucl. Instr. Meth.* 62 (1968) 247.

Bjo47   Bjoestrom, K. T., Huus, T., and Tanger, R., *Phys. Rev.* 71 (1947) 661.

Bla62   Blatt, J. M., and Weisskopf, V. F., *Theoretical Nuclear Physics* (New York: Wiley, 1962).

Bla72   Black, D. C., *Geoch. Cosmoch. Acta* 36 (1972) 347.

Ble67   Blewett, M. H., *Ann. Rev. Nucl. Sci.* 17 (1967) 427.

Blo67   Bloch, R., Pixley, R. E., and Winkler, H., *Helvetica Phys. Acta* 40 (1967) 832

Boc33   Bock, C. D., *Rev. Sci. Instr.* 4 (1933) 575.

Boc83   Bochsler, P., *Europhysics News* 14 (1983) 1.

Boe70   Boerma, D. O., and Smith, P. B., *Nucl. Instr. Meth.* 86 (1970) 221.

Boe84   Boehm, F., and Vogel, P., *Ann. Rev. Nucl. Part. Sci.* 34 (1984) 99.

Boh48   Bohr, N., *Kgl. Danske Videnskab. Selskab Mat.-Fys. Medd.* 18, No. 8 (1948).

Bok74   Bok, B. J., and Bok, P. F., *The Milky Way* (Cambridge: Harvard University Press, 1974).

Bok81   Bok, B. J., *Scient. Am.*, March 1981.

Bon48   Bondi, H., and Gold, T., *M.N.R.A.S.*, 108 (1948) 252.

Bon52   Bondi, H., *Cosmology* (Cambridge: Cambridge University Press, 1952).

Bon63   Bondelid, R. O., and Butler, J. W., *Phys. Rev.* 130 (1963) 1078.

Bon78   Bondarenko, L. N., Kurguzov, V. V., Prokofev, Y. A., Rogov, E. V., and Spivak, P. E., *Soviet Phys. Lett.* 28 (1978) 303.

Boo67   Booth, R., *IEEE Trans. Nucl. Sci.* NS-14, No. 3 (1967) 943.

Bor50   Borst, L. B., *Phys. Rev.* 78 (1950) 807.

Bor74   Borchers, R. R., in *Nuclear Spectroscopy and Reactions*, Part A, ed. J. Cerny (New York: Academic, 1974), p. 483.

Bou37   Bouwers, A., and Kuntke, A., *Z. Tech. Phys.* 18 (1937) 209.

Bow82   Bowers, R. L., and Wilson, J. R., *Ap. J. Suppl.* 50 (1982) 115 and *Ap. J.* 263 (1982) 366.

Boy81   Boyd, R. N., *IEEE Trans. Nucl. Sci.* NS-30 (1983) 1387.

Bra05   Bragg, R., *Phil. Mag.* 10 (1905) 318.

Bra77   Brand, K., *Nucl. Instr. Meth.* 141 (1977) 519.

Bra79   Bradley, T., Stelts, M. L., Chrien, R. E., and Parsa, Z., *Bull. Am. Phys. Soc.* 24 (1979) 871.

Bra81   Brand, K., in *Proc. Third Conf. on Electrostatic Accelerator Technology* (IEEE-81-CH1639-4), p. 205.

Bra84   Branch, D., *Ann. N. Y. Acad. Sci.* 422 (1984) 186.

Bre76   Brecher, K., in *Frontiers of Astrophysics*, ed. E. H. Avrett (Cambridge: Harvard University Press, 1976), p. 438.

Bri82   Brix, P., Max-Planck-Institut für Kernphysik, Heidelberg, MPI-H-1982-V3 (preprint).

Bro49   Brown, H. S., *Rev. Mod. Phys.* 21 (1949) 625.

Bro62   Brown, R. E., *Phys. Rev.* 125 (1962) 347.

Bro74   Bromley, D. A., *Nucl. Instr. Meth.* 122 (1974) 1.

Bro78   Brout, R., Englert, F., and Gunzig, E., *Ann. of Phys.* 115 (1978) 78.

Bro78a   Bromley, D. A., in *Nuclear Molecule Phenomena*, ed. N. Cindro (Amsterdam North-Holland, 1978), p. 3.

Bro79   Bromley, D. A., *Nucl. Instr. Meth.* 162 (1979) 1.

Bro79a   Brout, R., Englert, F., and Spindel, P., *Phys. Rev. Lett.* 43 (1979) 417.

Bro81   Browne, J. C., and Berman, B. L., *Phys. Rev.* C23 (1981) 1434.

Bro82   Brown, G. E., Bethe, H. A., and Baym, G., *Nucl. Phys.* A375 (1982) 481.

Buc80   Buchmann, L., Becker, H. W., Kettner, K. U., Kieser, W. E., Schmalbrock, P., and Rolfs, C., *Z. Phys.* A290 (1980) 273.

Buc84   Buchmann, L., Hilgemeier, M., Krauss, A., Redder, A., Rolfs, C., and Trautvetter, H. P., *Nucl. Phys.* A415 (1984) 93.

Buc84a   Buchmann, L., Baumeister, H., and Rolfs, C., *Nucl. Instr. Meth.* B4 (1984) 132.

Buc85   Buchmann, L., and d'Auria, J. M., in *Proc. Workshop on Accelerated Radioactive Beams* (Parksville) (TRIUMF, Lab. Rept. TRI-85-1).

Bud76   Budker, G. I., Dikansky, N. S., Kudelainen, V. I., Meshkov, I. N., Parchomchuk, V. V., Pestrikov, D. V., Skrinsky, A. N., and Sukhina, A. N., *Accelerators* 7 (1976) 197.

Bur70   Burbidge, G. R., *Ann. Rev. Astr. Ap.* 8 (1970) 369.

Bus68   Buss, W., Bianco, W. D., Wäffler, H., and Ziegler, B., *Nucl. Phys.* A112 (1968) 47.

Bus71   Bussiere, J., and Robson, J. M., *Nucl. Instr. Meth.* 91 (1971) 103.

Byr80   Byrne, J., Morse, J., Smith, K. F., Shaikh, F., Green, K., and Green, G. L., *Phys. Lett.*, B92 (1980) 274.

Cal65   Callan, C., Dicke, R. H., and Peebles, R. J. E., *Am. J. Phys.* 33 (1965) 105.

Cam51   Camac, M., *Rev. Sci. Instr.* 22 (1951) 197.

Cam54   Cameron, A. G. W., *Phys. Rev.* 93 (1954) 932.

Cam55   Cameron, A. G. W., *Ap. J.* 121 (1955) 144.

Cam57   Cameron, A. G. W., Atomic Energy of Canada, Ltd., CRL-41.

Cam58   Cameron, A. G. W., *Ann. Rev. Nucl. Sci.* 8 (1958) 299.

Cam59   Cameron, A. G. W., *Ap. J.* 130 (1959) 452.

Cam60   Cameron, A. G. W., *Astr. J.* 65 (1960) 485.

Cam62   Cameron, A. G. W., *Icarus* 1 (1962) 13.

Cam76   Cameron, A. G. W., in *Frontiers of Astrophysics*, ed. E. H. Avrett (Cambridge: Harvard University Press, 1976), p. 118.

Cam77   Cameron, A. G. W., and Truran, J. W., *Icarus* 30 (1977) 447.

Cam82   Cameron, A. G. W., in *Essays in Nuclear Astrophysics*, ed. C. A. Barnes, D. D. Clayton, and D. N. Schramm (Cambridge: Cambridge University Press, 1982), p. 23.

Cam83   Cameron, A. G. W., Cowan, J. J., and Truran, J. W., *Ap. Space Sci.* 91 (1983) 235.

Cam84   Cameron, A. G. W., *Icarus* 60 (1984) 416.

Cam85   Cameron, A. G. W., Cowan, J. J., and Truran, J. W., in *Nucleosynthesis*, ed. W. D. Arnett and J. W. Truran (Chicago: University of Chicago Press, 1985), p. 190.

Car73   Cardman, L. S., Fivozinsky, B. P., O'Connell, J. S., and Penner, S., *Bull. Am. Phys. Soc.* 18 (1973) 78.

Cau62   Caughlan, G. R., and Fowler, W. A., *Ap. J.* 136 (1962) 453.

Cau65   Caughlan, G. R., *Ap. J.* 141 (1965) 688.

Cau72   Caughlan, G. R., and Fowler, W. A., *Nature* 238 (1972) 1.

Cau77   Caughlan, G. R., in *CNO Isotopes in Astrophysics*, ed. J. Audouze (Dordrecht: Reidel, 1977), p. 121.

Cau85   Caughlan, G. R., Fowler, W. A., Harris, M. J., and Zimmerman, B. A., *At. Nucl. Data Tables* 32 (1985) 197.

Cha31   Chandrasekhar, S., *M.N.R.A.S.*, 91 (1931) 456 and *Ap. J.* 74 (1931) 81.

Cha32   Chandrasekhar, S., *Z. Ap.* 5 (1932) 321.

Cha35   Chandrasekhar, S., *M.N.R.A.S.* 95 (1935) 207.

Cha39   Chandrasekhar, S., *An Introduction to the Study of Stellar Structure* (Chicago: University of Chicago Press, 1939).

Cha60   Chamberlain, O., *Ann. Rev. Nucl. Sci.* 10 (1960) 161.

Cha65   Chasman, C., Jones, K. E., and Ristinen, R. A., *Nucl. Instr. Meth.* 37 (1965) 1.

Cha74   Chamberlin, D., Bodansky, D., Jacobs, W. W., and Oberg, D. L., *Phys. Rev.* C9 (1974) 69.

Cha83   Chandrasekhar, S., *The Mathematical Theory of Black Holes* (Oxford: Clarendon, 1983).

Cha83a   Champagne, A. E., Howard, A. J., and Parker, P. D., *Nucl. Phys.* A402 (1983) 159 and 179; and *Ap. J.* 269 (1983) 686.
Cha84   Chandrasekhar, S., *Rev. Mod. Phys.* 56 (1984) 137.
Che78   Chevalier, R. A., and Kirshner, R. P., *Ap. J.* 219 (1978) 931.
Che79   Chevalier, R. A., and Kirshner, R. P., *Ap. J.* 233 (1979) 154.
Che80   Chen, J. H., and Wasserburg, G. J., *Lunar Planetary Sci.* 11 (1980) 131.
Che81   Chevalier, R. A., *Ap. J.* 246 (1981) 267.
Chi66   Chiu, H. Y., *Phys. Rev. Lett.* 17 (1966) 712.
Chr64   Christenson, J. H., Cronin, J. W., Fitch, V. L., and Turlay, R., *Phys. Rev. Lett.* 13 (1964) 138.
Chr75   Chrien, R. E., *Nuclear Cross Sections and Technology*, NBS Special Pub. 425 (1975) 139.
Chr77   Christensen, P. R., Switkowski, Z. E., and Dayras, R. A., *Nucl. Phys.* A280 (1977) 189.
Chu75   Chu, W. K., in *New Uses of Ion Accelerators*, ed. J. F. Ziegler (New York: Plenum, 1975), p. 135.
Cin78   Cindro, N., *Nuclear Molecule Phenomena* (Amsterdam: North-Holland, 1978).
Cla61   Clayton, D. D., Fowler, W. A., Hull, T. E., and Zimmerman, B. A., *Ann. of Phys.* 12 (1961) 331.
Cla64   Clayton, D. D., *Ap. J.* 139 (1964) 637.
Cla67   Clayton, D. D., and Rassbach, M. E., *Ap. J.* 148 (1967) 69.
Cla68   Clayton, D. D., *Principles of Stellar Evolution and Nucleosynthesis* (New York: McGraw-Hill, 1968).
Cla68a   Clark, G. J., Sullivan, D. J., and Treacy, P. B., *Nucl. Phys.* A110 (1968) 481.
Cla73   Clayton, R. N., Grossman, L., and Mayeda, T. K., *Science* 182 (1973) 485.
Cla74   Clayton, D. D., and Ward, R. A., *Ap. J.* 193 (1974) 397.
Cla78   Clayton, R. N., *Ann. Rev. Nucl. Sci.* 28 (1978) 501.
Cla81   Clayton, R. N., *Phil. Trans. Roy. Soc. London* A303 (1981) 339.
Cla84   Clayton, D. D., *Ap. J.* 280 (1984) 144.
Cla89   Clarke, I. W., *Bull. Phil. Soc. Washington* 11 (1889) 131.
Cle60   Cleland, M. R., and Morgenstern, K. H., *Nucleonics* 18 (1960) 52.
Cle68   Cleland, M. R., in *Third Symposium on the Structure of Low-Medium Mass Nuclei*, ed. J. P. Davidson (Lawrence: University of Kansas Press, 1968), p. 230.
Cle84   Cleff, B., Universität Münster, private communication (1984).
Cle85   Cleff, B., Schulte, W. H., Schulze, H., Terlau, W., Koudijs, R., Dubbelman, P., and Peters, H. J., *Nucl. Instr. Meth.* B6 (1985) 46.
Clo83   Close, F., *The Cosmic Onion* (New York: Heinemann Educational Books, 1983).
Coc32   Cockcroft, J. D., and Walton, E. T. C., *Nature* 129 (1932) 242 and *Proc. Phys. Soc. London* A136 (1932) 619.
Coh73   Cohen, E. R., and Taylor, B. N., *J. Phys. Chem. Ref. Data* 2 (1973) 663.
Col66   Colgate, S. A., and White, R. H., *Ap. J.* 146 (1966) 626.
Col69   Colgate, S. A., and McKee, C., *Ap. J.* 157 (1969) 623.
Col81   Cole, F. T., and Mills, F. E., *Ann. Rev. Nucl. Part. Sci.* 31 (1981) 295.
Con29   Condon, E. U., and Gurney, R. W., *Phys. Rev.* 33 (1929) 127.
Con78   Conti, P. S., *Ann. Rev. Astr. Ap.* 16 (1978) 371.

Coo57   Cook, C. W., Fowler, W. A., Lauritsen, C. C., and Lauritsen, T., *Phys. Rev.* 107 (1957) 508.

Cos64   Costello, D. G., Skofronick, J. G., Morsell, A. L., Palmer, D. W., and Herb, R. G., *Nucl. Phys.* 51 (1964) 113.

Cos84   Cosner, K. R., Despain, K. H., and Truran, J. W., *Ap. J.* 283 (1984) 313.

Cot38   Cotte, M., *Ann. de Phys.* 10 (1938) 333.

Cou48   Courant, R., and Friedrich, K. O., *Supersonic Flow and Shock Waves* (New York: Interscience, 1948).

Cou68   Courant, E. D., *Ann. Rev. Nucl. Sci.* 18 (1968) 435.

Cou71   Couch, R. G., Spinka, H., Tombrello, T. A., and Weaver, T. A., *Nucl. Phys.* A175 (1971) 300.

Cou71a   Couch, R. G., and Shane, K. C., *Ap. J.* 169 (1971) 413.

Cow82   Cowan, G. A., and Haxton, W. C., *Los Alamos Sci. Rept.*, No. 47 (1982), and *Science* 216 (1982) 51.

Cox68   Cox, J. P., and Giuli, R. T., *Principles of Stellar Structure* (New York: Gordon & Breach, 1968).

Cra64   Crannell, H. L., and Griffy, T. A., *Phys. Rev.* 136 (1964) B1580.

Cra67   Crannell, H. L., Griffy, T. A., Suelzle, L. R., and Yearian, M. R., *Nucl. Phys.* A90 (1967) 152.

Cra84   Crawford, J., and d'Auria, J. M., TRIUMF-ISOL Workshop, Mont Gabriel, Quebec (1984), TRI-84-1.

Cri42   Critchfield, C. L., *Ap. J.* 96 (1942) 1.

Cro51   Cross, W. G., *Rev. Sci. Instr.* 22 (1951) 717.

Cuj76   Cujek, B., and Barnes, C. A., *Nucl. Phys.* A266 (1976) 461.

Dah60   Dahl, P. F., Costello, D. G., and Walters, W. L., *Nucl. Phys.* 21 (1960) 106.

Dan47   Dandel, R., *Rev. Sci. Instr.* 85 (1947) 162.

Dav60   Davis, R. H., *Phys. Rev. Lett.* 4 (1960) 521.

Dav68   Davids, C. N., *Nucl. Phys.* A110 (1968) 619 and *Ap. J.* 151 (1968) 775.

Dav75   Davids, C. N., Pardo, R. C., and Obst, A. W., *Phys. Rev.* C11 (1975) 2063.

Dav75a   Davids, C. N., *Ann. Rev. Astr. Ap.* 13 (1975) 69.

Dav76   Davis, M., in *Frontiers of Astrophysics*, ed. E. H. Avrett (Cambridge: Harvard University Press, 1976), p. 472.

Day76   Dayras, R. A., Stokstad, R. A., Switkowski, Z. E., and Wieland, R. M., *Nucl. Phys.* A261 (1976) 478 and A265 (1976) 153.

DeB78   DeBievre, P., in *Adv. Mass Spectrometry*, ed. N. R. Daly (London: Heydon & Son, 1978), p. 818.

Dec78   Deconninck, G., *Introduction to Radioanalytical Physics* (Budapest: Akademiai Kiado, 1978).

Dei65   Deinzer, W., and Salpeter, E. E., *Ap. J.* 142 (1965) 813.

deV79   de Vaucouleurs, G., *Ap. J.* 233 (1979) 433 and 253 (1982) 520.

Dia71   Diamond, W. T., Alexander, T. K., and Häusser, O., *Can. J. Phys.* 49 (1971) 1589.

Dic68   Dicke, R. H., and Peebles, P. J. E., *Ap. J.* 154 (1968) 891.

Dic79   Dicke, R. H., and Peebles, P. J. E., in *General Relativity*, ed. S. W. Hawking and W. Israel (Cambridge: Cambridge University Press, 1979), p. 504.

Die73   Diener, E. M., and Chang, C. C., *Nucl. Instr. Meth.* 109 (1973) 585.

Dim78   Dimopoulos, S., and Susskind, L., *Phys. Rev.* D18 (1978) 4500.

Dol81   Dolgov, A. D., and Zel'dovich, Y. B., *Rev. Mod. Phys.* 53 (1981) 1.

Don67   Donhowe, J. M., Ferry, J. A., Monrad, W. G., and Herb, R. G., *Nucl. Phys.* A102 (1967) 383.

Don67a   Donnelly, T. W., Ph.D. thesis, University of British Columbia (1967).

Dun51   Duncan, D. B., and Perry, J. E., *Phys. Rev.* 82 (1951) 809.

Dwa71   Dwarakanath, M. R., and Winkler, H., *Phys. Rev.* C4 (1971) 1532.

Dwa74   Dwarakanath, M. R., *Phys. Rev.* C9 (1974) 805.

Dwo72   Dworkin, P. B., Ph.D. thesis, University of Toronto (1972).

Dye74   Dyer, P., and Barnes, C. A., *Nucl. Phys.* A233 (1974) 495.

Ebe81   Eberhardt, P., Jungck, M. H. A., Meier, F. O., and Niederer, F. R., *Geoch. Cosmoch. Acta* 45 (1981) 1515.

Edd20   Eddington, A. S., *Rept. Brit. Assoc. Adv. Sci.* (Cardiff) (1920), p. 34.

Edd26   Eddington, A. S., *The Internal Constitution of Stars* (Cambridge: Cambridge University Press, 1926).

Egg68   Eggleton, P. P., *Ap. J.* 152 (1968) 345.

Egg76   Eggleton, P. P., Mitton, S., and Whelan, J. J. A., *Structure and Evolution of Close Binary Systems* (Dordrecht: Reidel, 1976).

Ein16   Einstein, A., *Ann. der Phys.* 49 (1916) 469.

Eli79   Elix, K., Becker, H. W., Buchmann, L., Goerres, J., Kettner, K. U., Wiescher, M., and Rolfs, C., *Z. Phys.* A293 (1979) 261.

Ell79   Ellis, J., Gaillard, M. K., Nanopoulos, D. V., *Phys. Lett.* B80 (1979) 30.

Ell81   Ellis, J., Gaillard, M. K., Nanopoulos, D. V., and Rudaz, S., *Phys. Lett.* B99 (1981) 101.

Eme72   Emery, G. T., *Ann. Rev. Nucl. Sci.* 22 (1972) 165.

End78   Endt, P. M., and van der Leun, C., *Nucl. Phys.* A310 (1978) 1.

End87   Endt, P. M., and Rolfs, C., *Nucl. Phys.* A467 (1987) 261.

Eng58   Enge, H. A., *Rev. Sci. Instr.* 29 (1958) 885.

Eng59   Enge, H. A., *Rev. Sci. Instr.* 30 (1959) 248.

Eng66   Engelbertink, A., and Endt, P. M., *Nucl. Phys.* 88 (1966) 12.

Eng68   England, J. B. A., and Trippett, J. C., *Nucl. Instr. Meth.* 63 (1968) 353.

Eng72   England, J. B. A., *Nucl. Instr. Meth.* 98 (1972) 237.

Eng74   England, J. B. A., *Techniques in Nuclear Structure Physics* (New York: Macmillan, 1974).

Erb80   Erb, K. A., Betts, R. R., Korotky, S. K., Hindi, M. M., Sachs, M. W., Willet, S. J., and Bromley, D. A., *Phys. Rev.* C22 (1980) 507.

Eva82   Evans, H. C., Ewan, G. T., Leslie, J. R., MacArthur, J. D., Mak, H. B., McLatchie, W., Turnock, R. C., Woods, M. B., and Allen, K. W., *Nucl. Instr. Meth.* 192 (1982) 143.

Fag73   Fagg, L. W., Bendel, W. L., Ensslin, N., and Jones, E. C., *Phys. Lett.* B44 (1973) 163.

Fan46   Fano, U., *Phys. Rev.* 70 (1946) 44 and 72 (1947) 26.

Fan63   Fano, U., *Ann. Rev. Nucl. Sci.* 13 (1963) 1.

Fau68   Fauska, H., Ward, N. G., Lilly, J., and Williamson, C. F., *Nucl. Instr. Meth.* 63 (1968) 93.

Faz76   Fazio, G. G., in *Frontiers of Astrophysics*, ed. E. H. Avrett (Cambridge: Harvard University Press, 1976), p. 203.

Fen73   Feng, J. S. Y., Chu, W. K., and Nicolet, M. A., *Thin Solid Films* 19 (1973) 227.

Fer82   Ferris, T., *Galaxies* (Washington: Sierra Club, 1982).

Fes76   Feshbach, H., *J. de Phys.* 37 (1976), Suppl. C5, p. 177.

Fet72   Fetisov, V. N., and Kopysov, Y. S., *Phys. Lett.* B40 (1972) 602 and *Nucl. Phys.* A239 (1975) 511.

Fia73   Fiarman, S., and Meyerhof, W. E., *Nucl. Phys.* A206 (1973) 1.

Fil83   Filippone, B. W., Elwyn, A. J., Davids, C. N., and Koethe, D. D., *Phys. Rev.* C28 (1983) 2222.

Fin40   Finkelstein, A. T., *Rev. Sci. Instr.* 11 (1940) 94.

For79   Forman, M. A., and Schaeffer, O. A., *Rev. Geophys. Space Phys.* 17 (1979) 552.

Fow26   Fowler, R. H., *M.N.R.A.S.* 87 (1926) 114.

Fow47   Fowler, W. A., Lauritsen, C. C., and Lauritsen, T., *Rev. Sci. Instr.* 18 (1947) 818.

Fow48   Fowler, W. A., Lauritsen, C. C., and Lauritsen, T., *Rev. Mod. Phys.* 20 (1948) 236.

Fow58   Fowler, W. A., *Ap. J.* 127 (1958) 551.

Fow60   Fowler, W. A., and Hoyle, F., *Ann. of Phys.* 10 (1960) 280.

Fow64   Fowler, W. A., and Hoyle, F., *Ap. J. Suppl.* 9 (1964) 201.

Fow64a   Fowler, W. A., *Proc. Nat. Acad. Sci.* 52 (1964) 524.

Fow66   Fowler, W. A., in *High Energy Astrophysics*, ed. L. Gratton (New York: Academic, 1966), p. 313.

Fow67   Fowler, W. A., Caughlan, G. R., and Zimmerman, B. A., *Ann. Rev. Astr. Ap.* 5 (1967) 525.

Fow69   Fowler, W. A., and Hoyle, F., *Ap. J.* 157 (1969) 339.

Fow72   Fowler, W. A., *Nature* 238 (1972) 24.

Fow72a   Fowler, W. A., in *Cosmology, Fusion and Other Matters*, ed. F. Reines (Boulder: University of Colorado Press, 1972), p. 67.

Fow75   Fowler, W. A., Caughlan, G. R., and Zimmerman, B. A., *Ann. Rev. Astr. Ap.* 13 (1975) 69.

Fow77   Fowler, W. A., *Proc. Welch Foundation Conf. on Chemical Research*, ed. W. D. Milligan (Houston: Houston University Press, 1977), p. 61.

Fow81   Fowler, P. H., Masheder, M. R. W., Moses, R. T., Walker, R. N. F., Worley, A., in *Origin of Cosmic Rays*, ed. G. Setti, G. Spada and A. W. Wolfendale (Dordrecht: Reidel, 1981), p. 77.

Fow84   Fowler, W. A., *Rev. Mod. Phys.* 56 (1984) 149.

Fow85   Fowler, W. A., and Meisl, C. C., in *Proc. Symposium on Cosmogonical Processes* (Boulder: University of Colorado Press, 1985).

Fow85a   Fowler, W. A., private communication (1985).

Fra74   Frauenfelder, H., and Henley, E. M., *Subatomic Physics* (Englewood Cliffs: Prentice-Hall, 1974).

Fre48   Freier, P. S., Lofgren, E. J., Oppenheimer, F., Bradt, H. L., and Peters, B., *Phys. Rev.* 74 (1948) 213.

Fre70   Freer, C. M., *Nucl. Instr. Meth.* 86 (1970) 311.

Fre74   Freedman, D. Z., *Phys. Rev.* D9 (1974) 1389.

Fre77   Freye, T., Lorenz-Wirzba, H., Cleff, B., Trautvetter, H. P., and Rolfs, C., *Z. Phys.* A281 (1977) 211.

Fri22   Friedmann, A., *Z. Phys.* 10 (1922) 377.

Fri51   Frieman, E., and Motz, L., *Phys. Rev.* 83 (1951) 202.

Fri73   Fricke, K. J., *Ap. J.* 183 (1973) 941 and 189 (1974) 535.

Fri77   Friedjung, M., *Novae and Related Stars* (Dordrecht: Reidel, 1977).

Fri80   Fricke, K. J., and Ober, W., *Ann. N.Y. Acad. Sci.* 336 (1980) 399.

Fro67   Fronteau, J., in *Focusing of Charged Particles*, ed. A. Septier (New York: Academic, 1967), p. 421.

Ful80   Fuller, G. M., Fowler, W. A., and Newman, M., *Ap. J. Suppl.* 42 (1980) 447 and 48 (1982) 279.

Gal78   Gallagher, J. S., and Starrfield, S., *Ann. Rev. Astr. Ap.* 16 (1978) 171.

Gam28   Gamow, G., *Z. Phys.* 51 (1928) 204.

Gam37   Gamow, G., *Structure of Atomic Nuclei and Nuclear Transformations* (Oxford: Clarendon, 1937).

Gam48   Gamow, G., *Nature* 162 (1948) 680.

Gam53   Gamow, G., *Danske Videnskab. Selskab Mat.-Fys. Medd.* 27, No. 10 (1953), and *Scient. Am.*, September 1956.

Geb79   Geballe, T. R., *Scient. Am.*, July 1979.

Ger31   Gerthsen, C., *Naturwissenschaften* 20 (1931) 743.

Gla80   Glashow, S. L., *Rev. Mod. Phys.* 52 (1980) 539.

Gob66   Gobert, G., Mami, G. S., and Sadeghi, M., *Nucl. Instr. Meth.* 42 (1966) 250.

Goe80   Goerres, J., Kettner, K. U., Kräwinkel, H., and Rolfs, C., *Nucl. Instr. Meth.* 177 (1980) 295.

Goe82   Goerres, J., Rolfs, C., Schmalbrock, P., Trautvetter, H. P., and Keinonen, J., *Nucl. Phys.* A385 (1982) 57.

Goe83   Goerres, J., Becker, H. W., Buchmann, L., Rolfs, C., Schmalbrock, P., Trautvetter, H. P., and Vlieks, A., *Nucl. Phys.* A408 (1983) 372.

Goe85   Goerres, J., and Rolfs, C., *Nucl. Instr. Meth.* A236 (1985) 203.

Goe85a   Goerres, J., Becker, H. W., Krauss, A., Redder, A., Rolfs, C., and Trautvetter, H. P., *Nucl. Instr. Meth.* A241 (1985) 334.

Gol37   Goldschmidt, V. M., *Norske Vidensk. Skrifter, Mat. Nat. Kl.* 4 (1937).

Gol68   Gold, T., *Nature* 218 (1968) 731 and 221 (1968) 25.

Gol80   Goldhaber, M., Langacker, P., and Slansky, R., *Science* 210 (1980) 851.

Got74   Gott, J. R., Gunn, J. E., Schramm, D. N., and Tinsley, B. M., *Ap. J.* 194 (1974) 543 and *Scient. Am.*, March 1976.

Gou74   Gouldring, F. S., and Pehl, R. H., in *Nuclear Spectroscopy and Reactions*, Part A, ed. J. Cerny (New York: Academic, 1974), p. 289.

Gou74a   Gouldring, F. S., and Landis, D. A., in *Nuclear Spectroscopy and Reactions*, Part A, ed. J. Cerny (New York: Academic, 1974), p. 413.

Gov60   Gove, H. E., and Litherland, A. E., in *Nuclear Spectroscopy*, ed. F. Ajzenberg-Selove (New York: Academic, 1960), p. 260.

Gre21   Greinacher, H., *Z. Phys.* 4 (1921) 195.

Gre54   Greenstein, J. L., *Modern Physics for the Engineer* (New York: McGraw-Hill, 1954).

Gre64   Greenstein, J. L., and Schmidt, M., *Ap. J.* 140 (1964) 1.

Gri63   Griffiths, G. M., Lal, M., and Scarfe, C. D., *Can. J. Phys.* 41 (1963) 724.

Gry65   Gryzinski, M., *Phys. Rev.* 138 (1965) 336.

Gud65   Gudden, F., and Stehl, P., *Z. Phys.* 185 (1965) 111.

Gur75   Gursky, H., and Ruffini, R., *Neutron Stars, Black Holes and Binary X-Ray Stars* (Dordrecht: Reidel, 1975).

Gur76   Gursky, H., in *Frontiers of Astrophysics*, ed. E. H. Avrett (Cambridge: Harvard University Press, 1976) p. 147.

Gut81   Guth, A. H., *Phys. Rev.* D23 (1981) 347 and 876.

Gut84   Guth, A. H., and Steinhardt, P. J., *Scient. Am.*, May 1984.

Gut85   Guth, A. H., and Pi, S. Y., *Phys. Rev.* D32 (1985) 1899.

Haa73   Haas, F. X., and Bair, J. K., *Phys. Rev.* C7 (1973) 2432.

Haa74   Haas, F. X., and Bair, J. K., *Phys. Rev.* C10 (1974) 961.

Hab79   Habing, H. J., and Israel, F. P., *Ann. Rev. Astr. Ap.* 17 (1979) 345.

Hab84   Habs, D., Jaeschke, E., Kienle, P., Koerner, H. J., Lynen, U., Povh, B., and
        Schuch, R., in *Proc. Workshop on Physics with Heavy-Ion Cooler Rings*
        (Heidelberg: Max-Planck-Institut, 1984).

Hag57   Hagedorn, F. B., *Phys. Rev.* 108 (1957) 735.

Hah70   Hahn, R. L., Stone, R. L., Tarrant, J. R., and Hunt, L. D., *Nucl. Instr. Meth.*
        86 (1970) 331 and 96 (1971) 481.

Hai83   Haight, R. C., Mathews, G. J., White, R. M., Aviles, L. A., and Woodard, S.
        E., *Nucl. Instr. Meth.* 212 (1983) 245.

Hal50   Hall, R. N., and Fowler, W. A., *Phys. Rev.* 77 (1950) 197.

Hal64   Hall, I., and Tanner, N. W., *Nucl. Phys.* 53 (1964) 673.

Hal73   Halbert, M. L., Hensley, D. C., and Bingham, H. G., *Phys. Rev.* C8 (1973)
        1226.

Ham75   Hammer, J. W., Schüpferling, H. M., Bergandt, E., and Pflaum, T., *Nucl.
        Instr. Meth.* 128 (1975) 409.

Ham75a   Hammer, J. W., and Niessner, W., *Kerntechnik* 11 (1975) 477.

Ham79   Hammer, J. W., Fischer, B., Hollick, H., Trautvetter, H. P., Kettner, K. U.,
        Rolfs, C., and Wiescher, M., *Nucl. Instr. Meth.* 161 (1979) 189.

Han47   Hanson, A. O., and McKibben, J. L., *Phys. Rev.* 72 (1947) 673.

Han67   Hanley, P. R., Haberl, A. W., and Taylor, A., *IEEE Trans. Nucl. Sci.* NS-14,
        No. 3 (1967) 933.

Han74   Hanson, D. L., Stokstad, R. G., Erb, K. A., Olmer, C., Sachs, M. W., and
        Bromley, D. A., *Phys. Rev.* C9 (1974) 1760.

Har17   Harkins, W. D., *J. Am. Chem. Soc.* 39 (1917) 856.

Har67   Hara, E., *Nucl. Instr. Meth.* 54 (1967) 91.

Har67a   Harrison, W. D., Stephens, W. E., Tombrello, T. A., and Winker, H., *Phys.
        Rev.* 160 (1967) 752.

Har73   Harwitt, M., *Astrophysical Concepts* (New York: Wiley, 1973).

Har74   Harrison, E. R., *Physics Today*, February 1974, p. 30.

Har83   Harris, M. J., Fowler, W. A., Caughlan, G. R., and Zimmerman, B. A., *Ann.
        Rev. Astr. Ap.* 21 (1983) 165.

Hau52   Hauser, W., and Feshbach, H., *Phys. Rev.* 87 (1952) 366.

Haw74   Hawking, S. W., *Nature* 248 (1974) 30.

Hay62   Hayashi, C., Hoshi, R., and Sugimoto, D., *Prog. Theoret. Phys.* 22 (1962) 1.

Hay66   Hayashi, C., *Ann. Rev. Astr. Ap.* 4 (1966) 171.

Heb60   Hebbard, D. F., and Vogl, J. L., *Nucl. Phys.* 21 (1960) 652.

Heb60a   Hebbard, D. F., *Nucl. Phys.* 15 (1960) 289.

Hei55   Heilpern, W., *Helvetica Phys. Acta* 28 (1955) 485.

Hel83   Helfand, D. J., Ruderman, M. A., and Shaham, J., *Nature* 304 (1983) 423.

Hel84   Helfand, D. J., and Becker, R. H., *Nature* 307 (1984) 215.

Hen67   Hensley, D. C., *Ap. J.* 147 (1967) 818.

Hen74   Hendrie, D. L., in *Nuclear Spectroscopy and Reactions*, Part A, ed. J. Cerny
        (New York: Academic, 1974), p. 365.

Her35   Herb, R. G., Parkinson, D. B., and Kerst, D. W., *Rev. Sci. Instr.* 6 (1935) 261.

Her37   Herb, R. G., Parkinson, D. B., and Kerst, D. W., *Rev. Sci. Instr.* 8 (1937) 75.

Her59   Herb, R. G., *Handbuch der Physik* 44 (1959) 64.

Her72   Herb, R. G., *Rev. Brasileira Fis.* 2 (1972) 17.

Her74   Herb, R. G., *Nucl. Instr. Meth.* 122 (1974) 267.

Her79   Herbst, W., and Assousa, G. E., *Scient. Am.*, August 1979.

Hes58   Hester, R. E., Pixley, R. E., and Lamb, W. A. S., *Phys. Rev.* 111 (1958) 1604.

Hes61   Hester, R. E., and Lamb, W. A. S., *Phys. Rev.* 121 (1961) 584.

Hew67   Hewish, A., Bell, S. J., Pilkington, J. P. H., Scott, P. F., and Collins, R. A., *Nature* 217 (1967) 709.

Hew70   Hewish, A., *Ann. Rev. Astr. Ap.* 8 (1970) 265.

Hie75   Hietzke, W. H., Ph.D. thesis, California State University, Los Angeles (1975).

Hig77   High, M. D., and Cujek, B., *Nucl. Phys.* A282 (1977) 181.

Hil76   Hilf, E. R., Groote, H., von, and Takahashi, K., CERN 76-13 (1976) 142.

Hil82   Hillebrandt, W., *Astr. Ap.* 110 (1982) L3.

Hil84   Hilgemeier, M., Jahresbericht, Institut für Kernphysik, Universität Münster (1984).

Hil87   Hilgemeier, M., Ph.D. thesis, Universität Münster (1987).

Hin62   Hintenberger, H., *Ann. Rev. Nucl. Sci.* 12 (1962) 435.

Hod81   Hodge, P. W., *Ann. Rev. Astr. Ap.* 19 (1981) 357.

Hol56   Holland, L., *Vacuum Deposition of Thin Films* (London: Chapman & Hall, 1956).

Hol76   Holms, J. A., Woosley, S. E., Fowler, W. A., and Zimmerman, B. A., *At. Nucl. Data Tables* 18 (1976) 305.

Hol79   Holt, S. S., White, N. E., Becker, R. H., Boldt, E. A., Mushotzky, R. F., and Smith, B. W., *Ap. J. (Lett.)* 234 (1979) L65.

Hor64   Hortig, G., *Nucl. Instr. Meth.* 30 (1964) 355.

Hor68   Hortig, G., Mokler, P., and Müller, M., *Z. Phys.* 210 (1968) 312.

How71   Howard, W. M., Arnett, W. D., and Clayton, D. D., *Ap. J.* 165 (1971) 495 and 175 (1972) 201.

How74   Howard, A. J., Jensen, H. B., Fowler, W. A., and Zimmerman, B. A., *Ap. J.* 188 (1974) 131.

How74*a*   Howard, W. M., and Nix, J. R., *Nature* 247 (1974) 17.

Hoy46   Hoyle, F., *M.N.R.A.S.* 106 (1946) 343.

Hoy49   Hoyle, F., *M.N.R.A.S.* 109 (1949) 365.

Hoy53   Hoyle, F., Dunbar, D. N. F., Wenzel, W. A., and Whaling, W., *Phys. Rev.* 92 (1953) 1095.

Hoy54   Hoyle, F., *Ap. J. Suppl.* 1 (1954) 121.

Hoy60   Hoyle, F., and Fowler, W. A., *Ap. J.* 132 (1960) 565.

Hoy64   Hoyle, F., and Taylor, R. J., *Nature* 203 (1964) 1108.

Hoy65   Hoyle, F., and Fowler, W. A., in *Quasi-Stellar Sources and Gravitational Collapse*, ed. I. Robinson, A. Schild, and E. L. Schucking (Chicago: University of Chicago Press, 1965), p. 62.

Hub20   Hubble, E. P., *Pub. Yerkes Observatory* 4 (1920) 69.

Hub25   Hubble, E. P., *Ap. J.* 62 (1925) 409.

Hub29   Hubble, E. P., *Proc. Nat. Acad. Sci.* 15 (1929) 168.

Hub36   Hubble, E. P., *The Realm of the Nebulae* (New Haven: Yale University Press, 1936).

Hul80   Hulke, G., Rolfs, C., and Trautvetter, H. P., *Z. Phys.* A297 (1980) 161.

Hul82   Hulke, G., Rolfs, C., and Trautvetter, H. P., *Nucl. Instr. Meth.* 198 (1982) 299.

Hum76   Humblet, J., Dyer, P., and Zimmerman, B. A., *Nucl. Phys.* A271 (1976) 210.

Hur76   Hurst, G. J., Payne, M. G., Kramer, S. D., and Young, J. P., *Rev. Mod. Phys.* 51 (1979) 767.

Hut67   Hutter, R., in *Focusing of Charged Particles*, ed. A. Septier (New York: Academic, 1967), p. 3.

Ibe63   Iben, I., *Ap. J.* 138 (1961) 1090.

Ibe65   Iben, I., *Ap. J.* 141 (1965) 993.

Ibe67   Iben, I., *Ann. Rev. Astr. Ap.* 5 (1967) 571 and *Ap. J.* 147 (1967) 650.

Ibe67a Iben, I., Kalata, K., and Schwartz, J., *Ap. J.* 150 (1967) 1001.

Ibe70   Iben, I., *Scient. Am.*, July 1970.

Ibe74   Iben, I., *Ann. Rev. Astr. Ap.* 12 (1974) 215.

Ibe75   Iben, I., *Ap. J.* 196 (1975) 525 and 549.

Ibe82   Iben, I., and Renzini, A., *Ap. J. (Lett.)* 263 (1982) L188.

Ibe83   Iben, I., and Renzini, A., *Ann. Rev. Astr. Ap.* 21 (1983) 271.

Ibe84   Iben, I., and Tutukov, A. V., *Ap. J. Suppl.* 55 (1984) 335.

Ibe84a Iben, I., and Renzini, A., *Phys. Rept.* 105 (1984) 330.

Ibe84b Iben, I., *Ap. J. (Lett.)* 275 (1984) L65.

Ich82   Ichimaru, S., *Rev. Mod. Phys.* 54 (1982) 1017.

Ich84   Ichimaru, S., and Utsumi, K., *Ap. J.* 278 (1984) 382.

Ima68   Imanishi, B., *Phys. Lett.* B27 (1968) 267 and *Nucl. Phys.* A125 (1968) 33.

Jam83   James, K., and Demarque, P., *Ap. J.* 264 (1983) 206.

Jas70   Jaszczak, R. J., Gibbons, J. H., and Macklin, R. L., *Phys. Rev.* C2 (1970) 63 and 2452.

Joh70   Johlige, H. W., Aumann, D. C., and Born, H. J., *Phys. Rev.* C2 (1970) 1616.

Jol74   Jolivette, P. L., Goss, J. D., Rollefson, A. A., and Browne, C. P., *Phys. Rev.* C10 (1974) 2629.

Jon62   Jones, C. M., Phillips, G. C., Harris, R. W., and Beckner, E. H., *Nucl. Phys.* 37 (1962) 1.

Jud50   Judd, D. L., *Rev. Sci. Instr.* 21 (1950) 213.

Käp82   Käppeler, F., Beer, H., Wisshak, K., Clayton, D. D., Macklin, R. L., and Ward, R. A., *Ap. J.* 257 (1982) 821.

Kat69   Kato, S., *Nucl. Instr. Meth.* 75 (1969) 293.

Kav60   Kavanagh, R. W., *Nucl. Phys.* 15 (1960) 411.

Kav69   Kavanagh, R. W., Tombrello, T. A., Mosher, J. M., and Goosman, D. R., *Bull. Am. Phys. Soc.* 14 (1969) 1209.

Kav72   Kavanagh, R. W., in *Cosmology, Fusion and Other Matters*, ed. F. Reines (Boulder: University of Colorado Press, 1972), p. 169.

Kav82   Kavanagh, R. W., in *Essays in Nuclear Astrophysics*, ed. C. A. Barnes, D. D. Clayton, and D. N. Schramm (Cambridge: Cambridge University Press, 1982), p. 159.

Kei79   Keinonen, J., and Antilla, A., *Nucl. Instr. Meth.* 160 (1979) 211.

Ket77   Kettner, K. U., Lorenz-Wirzba, H., Rolfs, C., and Winkler, H., *Phys. Rev. Lett.* 38 (1977) 337.

Ket80   Kettner, K. U., Lorenz-Wirzba, H., and Rolfs, C., *Z. Phys.* A298 (1980) 65.

Ket82   Kettner, K. U., Becker, H. W., Buchmann, L., Goerres, J., Kräwinkel, H., Rolfs, C., Schmalbrock, P., Trautvetter, H. P., and Vlieks, A., *Z. Phys.* A308 (1982) 73.

Kie79   Kieser, W. E., Azuma, R. E., and Jackson, K. P., *Nucl. Phys.* A331 (1979) 155.

Kin69   Kinbara, S., and Kumahara, T., *Nucl. Instr. Meth.* 70 (1969) 173.

Kir73   Kirshner, R. P., Oke, J. B., Penston, M. V., and Searle, L., *Ap. J.* 185 (1973) 303.

Kir76   Kirshner, R. P., *Scient. Am.*, December 1976.

Kir78   Kirsten, T., in *The Origin of the Solar System*, ed. S. F. Dermott (New York: Wiley, 1978), p. 267.

Kir80   Kirshner, R. P., and Blair, W. P., *Ap. J.* 236 (1980) 135.

Kla83   Klapdor, H. V., *Progr. Part. Nucl. Phys.* 10 (1983) 131.

Kno79   Knoll, G. F., *Radiation Detection and Measurement* (New York: Wiley, 1979).

Kod75   Kodama, T., and Takahashi, K., *Nucl. Phys.* A239 (1975) 489.

Kol83   Kolb, E. W., and Turner, M. S., *Ann. Rev. Nucl. Part. Sci.* 33 (1983) 645.

Koo74   Koonin, S. E., Tombrello, T. A., and Fox, G., *Nucl. Phys.* A220 (1974) 221.

Kra59   Kraft, R. P., *Scient. Am.*, July 1959.

Kra68   Kraner, H. W., Chasman, C., and Jones, K. W., *Nucl. Instr. Meth.* 62 (1968) 173.

Kra86   Kratz, K. L., et al., *Z. Phys.* A325 (1986) 489.

Kra87   Krauss, A., Becker, H. W., Trautvetter, H. P., and Rolfs, C., *Nucl. Phys.* A467 (1987) 273.

Krä82   Kräwinkel, H., Becker, H. W., Buchmann, L., Goerres, J., Kettner, K. U., Kieser, W. E., Santo, R., Schmalbrock, P., Trautvetter, H. P., Vlieks, A., Rolfs, C., Hammer, J. W., Azuma, R. E., and Rodney, W. S., *Z. Phys.* A304 (1982) 307.

Kuc76   Kuchowicz, B., *Rept. Progr. Phys.* 39 (1976) 291.

Kui60   Kuiper, G. P., and Middlehurst, B. M. (eds.), *Stars and Stellar Systems*, Vol. 1 *Telescopes* (Chicago: University of Chicago Press, 1960).

Kun83   Kunth, D., and Sargent, W. L. W., *Ap. J.* 273 (1983) 81.

Lak84   Lake, G., *Science* 224 (1984) 675.

Lam39   Lamb, W. E., *Phys. Rev.* 55 (1939) 190.

Lam57   Lamb, W. A. S., and Hester, R. E., *Phys. Rev.* 107 (1957) 550 and 108 (1957) 1304.

Lam77   Lamb, S. A., Howard, W. M., Truran, J. W., and Iben, I., *Ap. J.* 217 (1977) 213.

Lan32   Landau, L. D., *Phys. Z. Sowjetunion* 1 (1932) 285.

Lan79   Lande, K., *Ann. Rev. Nucl. Part. Sci.* 29 (1979) 395.

Lan81   Langacker, P., *Phys. Rept.* C72 (1981) 185.

Lan85   Langanke, K., and Koonin, S. E., *Nucl. Phys.* A439 (1985) 384.

Lap70   Lapostolle, P. M., and Septier, A. L., *Linear Accelerators* (Amsterdam: North-Holland, 1970).

Lar71   Larson, R. B., and Starrfield, S., *Astr. Ap.* 13 (1971) 190.

Lau51   Lauritsen, C. C., quoted by W. A. Fowler, J. L. Greenstein, and F. Hoyle, *Am. J. Phys.* 29 (1961) 393.

Lau51a   Laubenstein, R. A., and Laubenstein, M. J. W., *Phys. Rev.* 84 (1951) 18.

Law30   Lawrence, E. O., and Edlefson, N. E., *Science* 72 (1930) 376.

Law32   Lawrence, E. O., and Livingston, M. S., *Phys. Rev.* 40 (1932) 19.

Lea81   Learned, J. G., and Eichler, D., *Scient. Am.*, February 1981.

Led78   Lederer, C. M., and Shirley, V. S., *Table of Isotopes* (New York: Wiley, 1978).

Lee69   Lee-Whiting, G. E., and Bezic, N., *Nucl. Instr. Meth.* 71 (1969) 61.

Lee77   Lee, T., Papanastassiou, D., and Wasserburg, G. J., *Ap. J.* 211 (1977) L107.

Lee79   Lee, T., *Rev. Geophys. Space Sci.* 17 (1979) 1591.

Lej80   Lejeune, C., and Aubert, J., in *Adv. Electronics Electron Phys.*, ed. A. Septier (New York: Academic, 1980), p. 159.

Lem27   Lemaître, G., *Ann. Soc. Sci. Bruxelles* A47 (1927) 49.

Lew62   Lewis, H. W., *Phys. Rev.* 125 (1962) 937.

Lew81   Lewin, W. H. G., *Scient. Am.*, May 1981.

Lin82   Linde, A. D., *Phys. Lett.*, B108 (1982) 389.

Lit67   Litherland, A. E., Ollerhead, R. W., Smulders, P. J. M., Alexander, T. K., Broude, C., Ferguson, A. J., and Kuehner, J. A., *Can. J. Phys.* 45 (1967) 1901.

Lit80   Litherland, A. E., *Ann. Rev. Nucl. Part. Sci.* 30 (1980) 437.

Liv62   Livingston, M. S., and Blewett, J. P., *Particle Accelerators* (New York: McGraw-Hill, 1962).

Liv69   Livinghood, J. J., *The Optics of Dipole Magnets* (New York: Academic, 1969).

Loe67   Loebenstein, H. M., Mingay, D. W., Winkler, H., and Zaidins, C. S., *Nucl. Phys.* A91 (1967) 481.

Loe73   Loeffler, M., Scheerer, H. J., and Vonach, H., *Nucl. Instr. Meth.* 111 (1973) 1.

Lor57   Lorrain, P., Beigne, R., Gilmore, P., Girard, P. E., Breton, A., and Picke, P., *Can. J. Phys.* 35 (1957) 299.

Lor79   Lorenz-Wirzba, H., Schmalbrock, P., Trautvetter, H. P., Wiescher, M., and Rolfs, C., *Nucl. Phys.* A313 (1979) 346.

Low63   Lowenheim, F. A., *Modern Electroplating* (New York: Wiley, 1963).

Mac63   Macklin, R. L., Gibbons, J. H., and Inada, T., *Nucl. Phys.* 43 (1963) 353 and *Nature* 197 (1963) 369.

Mac65   Macklin, R. L., and Gibbons, J. H., *Rev. Mod. Phys.* 37 (1965) 166 and *Ap. J.* 149 (1967) 577.

Mac71   Macklin, R. L., Hill, N. W., and Allen, B. J., *Nucl. Instr. Meth.* 96 (1971) 509.

Mah82   Mahoney, W. A., Ling, J. C., Jacobson, A. S., and Lingenfelter, R. E., *Ap. J.* 262 (1982) 742.

Mak72   Mak, H. B., Mann, E. M., and Kavanagh, R. W., *Bull. Am. Phys. Soc.* 17 (1972) 1180.

Mak74   Mak, H. B., Ashery, D., and Barnes, C. A., *Nucl. Phys.* A226 (1974) 493.

Mak75   Mak, H. B., Evans, H. C., Ewan, G. T., McDonald, A. B., and Alexander, T. K., *Phys. Rev.* C12 (1975) 1158.

Man75   Mann, F. M., Ph.D. thesis, California Institute of Technology (1975).

Man75a   Mann, F. M., Dayras, R. A., and Switkowski, Z. E., *Phys. Lett.* B58 (1975) 420.

Mar57   Marion, J. B., and Fowler, W. A., *Ap. J.* 125 (1957) 221.

Mar66   Marion, J. B., *Rev. Mod. Phys.* 38 (1966) 660.

Mar68   Marion, J. B., and Young, F. C., *Nuclear Reaction Analysis* (Amsterdam: North-Holland, 1968).

Mar76   Markham, R. G., Austin, S. M., and Shababuddin, M. A. M., *Nucl. Phys.* A270 (1976) 489.

Mar80   Margon, G., *Scient. Am.*, October 1980.
Mat83   Mathews, G. J., Mengoni, A., Thielemann, F. K., and Fowler, W. A., *Ap. J.*, 270 (1983) 740.
Mat84   Mathews, G. J., and Käppeler, F., *Ap. J.* 286 (1984) 810.
Max67   Maxman, S. H., *Nucl. Instr. Meth.* 50 (1967) 53.
May68   May, R., and Clayton, D. D., *Ap. J.* 153 (1968) 855.
Maz73   Mazarakis, M. G., and Stephens, W. E., *Phys. Rev.* C7 (1973) 1280.
McC73   McCaslin, S. J., Mann, F. M., and Kavanagh, R. W., *Phys. Rev.* C7 (1973) 489.
McD77   McDonald, A. B., Alexander, T. K., Beene, J. R., and Mak, H. B., *Nucl. Phys.* A288 (1977) 529.
McK41   McKellar, A., *Pub. Dom. Ap. Obs. Victoria* 7 (1941) 251.
Mee72   Meer, S. van der, CERN ISR-PO 72-31.
Mei79   Meier, D. L., and Sunyaev, R. L., *Scient. Am.*, November 1979.
Mer52   Merrill, P. W., *Science* 115 (1952) 484 and *Ap. J.* 116 (1952) 21.
Mey74   Meyer, P., Ramaty, R., and Weber, W. R., *Physics Today* 27 (1974) 23.
Mic69   Michaud, G., and Vogt, E., *Phys. Lett.* B30 (1969) 85 and *Phys. Rev.* C1 (1970) 864 and C5 (1972) 350.
Mic70   Michaud, G., and Fowler, W. A., *Phys. Rev.* C2 (1970) 2041 and *Ap. J.* 173 (1972) 157.
Mic73   Michaud, G., *Phys. Rev.* C8 (1973) 525 and *Ap. J.* 175 (1972) 751.
Mid74   Middleton, R., and Adams, C. T., *Nucl. Instr. Meth.* 118 (1974) 229 and 122 (1974) 35.
Mid83   Middleton, R., *Nucl. Instr. Meth.* 214 (1983) 139.
Mil67   Miller, R. G., and Kavanagh, R. W., *Nucl. Instr. Meth.* 48 (1967) 13.
Mil79   Miller, G. E., and Scalo, J. M., *Ap. J. Suppl.* 41 (1979) 513.
Mol71   Molnar, M. R., *Ap. J.* 163 (1971) 203.
Moo70   Moon Issue, Age Measurements, *Science* 167 (1970) 461.
Mor69   Morric, J. M., and Ophel, T. R., *Nucl. Instr. Meth.* 68 (1969) 344.
Mox63   Moxon, M. C., and Rae, E. R., *Nucl. Instr. Meth.* 24 (1963) 445.
Mug61   Muggleton, A. H., and Howe, T. A., *Nucl. Instr. Meth.* 13 (1961) 211.
Mye66   Myers, W., and Swiatecki, W. J., *Nucl. Phys.* 81 (1966) 1.
Mye77   Myers, W., *Droplet Model of Atomic Nuclei* (New York: Plenum, 1977).
Nag69   Nagatani, K., Dwarakanath, M. R., and Ashery, D., *Nucl. Phys.* A128 (1969) 325.
Nei74   Neijs, de E. O., Meyer, M. A., Reinecke, J. P. L., and Reitmann, D., *Nucl. Phys.* A230 (1974) 490.
Nit84   Nitschke, J. M., in *Proc. Workshop on Prospects for Research with Radioactive Beams from Heavy Ion Accelerators* (LBL-18187, 1984).
Nol76   Nolen, J. A., and Austin, S. M., *Phys. Rev.* C13 (1976) 1773.
Nom76   Nomoto, K., Sugimoto, D., and Neo, S., *Ap. Space Sci.* 39 (1976) L37.
Nom84   Nomoto, K., Thielemann, F. K., and Yokoi, K., *Ap. J.* 286 (1984) 644.
Noy76   Noyes, R. W., in *Frontiers of Astrophysics*, ed. E. H. Avrett (Cambridge: Harvard University Press, 1976), p. 41.
Obs72   Obst, A. W., Grandy, T. B., and Weil, J., *Phys. Rev.* C5 (1972) 738.
Obs76   Obst, A. W., and Braithwaite, W. J., *Phys. Rev.* C13 (1976) 2033.
Odi56   Odiot, S. O., and Dandel, R., *J. de Phys. et le Radium* 17 (1956) 60.
Oes75   Oeschger, H., and Wahlen, M., *Ann. Rev. Nucl. Sci.* 25 (1975) 423.

O'Ne63   O'Neill, G. K., *Science* 141 (1963) 679.
O'Ne66   O'Neill, G. K., *Scient. Am.*, November 1966.
Öpi51   Öpik, E. J., *Proc. Roy. Irish Acad.* A54 (1951) 49.
Opp39   Oppenheimer, R. J., and Volkoff, G. M., *Phys. Rev.* 55 (1939) 374.
Opp39a   Oppenheimer, R. J., and Snyder, R., *Phys. Rev.* 56 (1939) 455.
Osb82   Osborne, J. L., Barnes, C. A., Kavanagh, R. W., Kremer, R. M., Mathews, G.
         J., Zyskind, J. L., Parker, P. D., and Howard, A. J., *Phys. Rev. Lett.* 48
         (1982) 1664 and *Nucl. Phys.* A419 (1984) 115.
Osm82   Osmer, P. S., *Scient. Am.*, February 1982.
Ost63   Ostic, R. G., Russell, R. D., and Reynolds, P. H., *Nature* 199 (1963) 1160.
Pac68   Pacini, F., *Nature* 219 (1968) 145.
Pal83   Palla, F., Salpeter, E. E., and Stahler, S. W., *Ap. J.* 271 (1983) 632.
Par63   Parker, P. D., and Kavanagh, R. W., *Phys. Rev.* 131 (1963) 2578.
Par63a   Park, J. T., and Zimmermans, E. J., *Phys. Rev.* 131 (1963) 202.
Par64   Parks, P. B., Beard, P. M., Bilpuch, E. G., and Newson, H. W., *Rev. Sci.
         Instr.* 35 (1964) 549.
Par64a   Parker, P. D., Bahcall, J. N., and Fowler, W. A., *Ap. J.* 139 (1964) 602.
Par66   Parker, P. D., *Phys. Rev.* 150 (1966) 851 and *Ap. J.* 145 (1966) 960.
Par68   Parker, P. D., *Ap. J.* 153 (1968) L85.
Par73   Parker, P. D., Pisano, D. J., Cobern, M. E., and Marks, G. H., *Nature* 241
         (1973) 106.
Par85   Parker, P. D., in *Physics of the Sun*, ed. P. A. Sturrock (Dordrecht: Reidel,
         1985), p. 15.
Par85a   Park, H. S., et al., *Phys. Rev. Lett.* 54 (1985) 22.
Pat56   Patterson, D., *Geoch. Cosmoch. Acta* 10 (1956) 230.
Pat69   Patterson, J. R., Winkler, H., and Zaidins, C. S., *Ap. J.* 157 (1969) 367.
Pat71   Patterson, J. R. Nagorcka, B. N., Symons, G. D., and Zuk, W. M., *Nucl.
         Phys.* A165 (1971) 545.
Pau74   Paul, P., in *Nuclear Spectroscopy and Reactions*, Part A, ed. J. Cerny (New
         York: Academic, 1974), p. 345.
Pee66   Peebles, P. J. E., *Ap. J.* 146 (1966) 542 and *Phys. Rev. Lett.* 16 (1966) 410.
Pee80   Peebles, P. J. E., *The Large Scale Structure of the Universe* (Princeton:
         Princeton University Press, 1980).
Pee84   Peebles, P. J. E., *Science* 224 (1984) 1385.
Pel60   Peel, E. M., *J. Appl. Phys.* 31 (1960) 291.
Pen37   Penning, F. M., and Moubis, J. K. A., *Physica* 4 (1937) 71 and 1190
Pen61   Penner, S., *Rev. Sci. Instr.* 32 (1961) 150.
Pen65   Penzias, A. A., and Wilson, R. W., *Ap. J.* 142 (1965) 419.
Per68   Persico, E., Ferrari, E., and Segrè, S. E., *Principles of Particle Accelerators*
         (New York: Benjamin, 1968).
Per84   Perkins, D. H., *Ann. Rev. Nucl. Part. Sci.* 34 (1984) 1.
Pie54   Pierce, J. R., *Theory and Design of Electron Beams* (New York: Van Nos-
         trand, 1954).
Pix57   Pixley, R. E., Ph.D. thesis, California Institute of Technology (1957).
Pla87   Plaga, R., Becker, H. W., Redder, A., Rolfs, C., Trautvetter, H. P., and
         Langanke, K., *Nucl. Phys.* A465 (1987) 291.
Pod78   Podosek, F. A., *Ann. Rev. Astr. Ap.* 16 (1978) 293.
Pon46   Pontecorvo, B., Chalk River Lab. Rept. PD-205 (1946).

Pon67   Pontecorvo, B., *Zh. Eksper. Teoret. Fiz.* 53 (1967) 1717 and *Phys. Lett.* B28 (1969) 493.

Pow73   Powers, D., Lodhi, A. S., Lin, W. K., and Cox, H. L., *Thin Solid Films* 19 (1973) 202.

Pue70   Puehlhofer, F., Ritter, H. G., Bock, R., Brommundt, G., Schmidt, H., and Bethge, K., *Nucl. Phys.* A147 (1970) 258.

Ram77   Ramstroem, F., and Wiedling, T., *Ap. J.* 211 (1977) 223.

Red82   Redder, A., Becker, H. W., Lorenz-Wirzba, H., Rolfs, C., Schmalbrock, P., and Trautvetter, H. P., *Z. Phys.* A305 (1982) 325.

Red85   Redder, A., Becker, H. W., Goerres, J., Hilgemeier, M., Krauss, A., Rolfs, C., Schröder, U., Trautvetter, H. P., Wolke, K., Donoghue, T. R., Rinckel, T. C., and Hammer, J. W., *Phys. Rev. Lett.* 55 (1985) 1262 and *Nucl. Phys.* A462 (1987) 385.

Ree59   Reeves, H., and Salpeter, E. E., *Phys. Rev.* 116 (1959) 1505.

Ree66   Reeves, H., in *Stellar Evolution*, ed. R. F. Stein and A. G. W. Cameron (New York: Plenum, 1966), p. 83.

Ree70   Rees, M., and Silk, J., *Scient. Am.*, June 1970.

Ree70a   Reeves, H., Fowler, W. A., and Hoyle, F., *Nature* 226 (1970) 727.

Ree74   Rees, M., Ruffini, R., and Wheeler, J. A., *Black Holes, Gravitational Waves and Cosmology* (New York: Gordon & Breach, 1974).

Ree79   Reeves, H., *Ap. J.* 231 (1979) 229.

Reg67   Regenstreif, E., in *Focusing of Charged Particles*, ed. A. Septier (New York: Academic, 1967), p. 353.

Rey60   Reynolds, J. H., *Phys. Rev. Lett.* 4 (1960) 8.

Ric60   Richards, H. T., in *Nuclear Spectroscopy*, Part A, ed. F. Ajzenberg-Selove (New York: Academic, 1960), p. 99.

Ric69   Richards, D. W., and Comella, J. W., *Nature* 222 (1969) 551.

Rin56   Rindler, W., *M.N.R.A.S.* 116 (1956) 663.

Rio75   Rios, M., Schweitzer, J. S., and Anderson, B. D., *Ap. J.* 199 (1975) 173.

Rob61   Robertson, L. P., White, B. L., and Erdman, K. L., *Rev. Sci. Instr.* 32 (1961) 1405.

Rob77   Robertson, R. H. G., Warner, R. A., and Austin, S. M., *Phys. Rev.* C15 (1977) 1072.

Rob81   Robertson, R. H. G., Dyer, P., Warner, R. A., Melin, R. C., Bowles, T. J., McDonald, A. B., Ball, G. C., Davies, W. G., and Earle, E. D., *Phys. Rev. Lett.* 47 (1981) 1867.

Rob83   Robertson, R. G. H., Dyer, P., Bowles, T. J., Brown, R. E., Jarmie, N., Maggiore, C. J., and Austin, S. M., *Phys. Rev.* C27 (1983) 11.

Rob84   Robertson, R. G. H., Dyer, P., Melin, R. C., Bowles, T. J., McDonald, A. B., Ball, G. C., Davies, W. G., and Earle, E. D., *Phys. Rev.* C29 (1984) 755.

Rol73   Rolfs, C., *Nucl. Phys.* A217 (1973) 29.

Rol74   Rolfs, C., and Winkler, H., *Phys. Lett.* B52 (1974) 317.

Rol74a   Rolfs, C., and Rodney, W. S., *Ap. J.* 194 (1974) L63.

Rol74b   Rolfs, C., and Azuma, R. E., *Nucl. Phys.* A227 (1974) 291.

Rol74c   Rolfs, C., and Rodney, W. S., *Nucl. Phys.* A235 (1974) 450.

Rol75   Rolfs, C., Rodney, W. S., Shapiro, M. H., and Winkler, H., *Nucl. Phys.* A241 (1975) 460.

Rol75a   Rolfs, C., and Rodney, W. S., *Nucl. Phys.* A250 (1975) 295.

Rol75b   Rolfs, C., Rodney, W. S., Durrance, S., and Winkler, H., *Nucl. Phys.* A240 (1975) 221.

Rol78   Rolfs, C., Goerres, J., Kettner, K. U., Lorenz-Wirzba, H., Schmalbrock, P., Trautvetter, H. P., and Verhoeven, W., *Nucl. Instr. Meth.* 157 (1978) 19.

Rol79   Rolfs, C., *Proc. Hirschegg Workshop* (1979) (INKA-79-1-69), p. 183.

Rol86   Rolfs, C., and Kavanagh, R. W., *Nucl. Phys.* A455 (1986) 179.

Rol86a   Rolfs, C., and Kavanagh, R. W., *Nucl. Instr. Meth.* A247 (1986) 507.

Ros50   Rosenblum, R. S., *Rev. Sci. Instr.* 21 (1950) 586.

Ros64   Rose, P. H., Bastide, B. P., Brooks, N. B., Airey, J., and Wittkower, A. B., *Rev. Sci. Instr.* 35 (1964) 1283.

Rou70   Roush, M. L., West, L. A., and Marion, J. B., *Nucl. Phys.* A147 (1970) 235.

Rub83   Rubin, V. C., *Scient. Am.*, June 1983.

Ruf71   Ruffini, R., and Wheeler, J. A., *Physics Today* 24 (1971) 30.

Ruf75   Ruffini, R., *Neutron Stars, Black Holes and Binary X-Ray Sources* (New York: Gordon & Breach, 1975), p. 59.

Rus14   Russell, H. N., *Pop. Astr.* 22 (1914) 275.

Rut13   Rutherford, E., and Robinson, H., *Phil. Mag.* 26 (1913) 717.

Rut29   Rutherford, E., *Nature* 123 (1929) 313.

Sal52   Salpeter, E. E., *Phys. Rev.* 88 (1952) 547 and *Ap. J.* 115 (1952) 326.

Sal53   Salpeter, E. E., *Ann. Rev. Nucl. Sci.* 2 (1953) 41.

Sal54   Salpeter, E. E., *Australian J. Phys.* 7 (1954) 373.

Sal57   Salpeter, E. E., *Phys. Rev.* 107 (1957) 516.

Sal59   Salpeter, E. E., *Ap. J.* 129 (1959) 608.

San61   Sandage, A., *The Hubble Atlas of Galaxies* (Washington, D.C.: Carnegie Institution of Washington, 1961).

San67   Sanders, R. H., *Ap. J.* 150 (1967) 971.

San82   Sandage, A., and Tammann, G. A., *Ap. J.* 256 (1982) 339.

San84   Sandage, A., and Tammann, G. A., *Nature* 307 (1984) 326.

Saw53   Sawyer, G. A., and Phillips, J. A., LASL Rept. 1578 (1953).

Sca77   Scalo, J. M., Caltech preprint OAP-498 (1977).

Sch19   Schenkel, O., *Elektrotech. Z.* 40 (1919) 333.

Sch25   Schottky, W., *Z. Phys.* 31 (1925) 163.

Sch41   Schein, M., Jesse, W. P., and Wollan, E. O., *Phys. Rev.* 59 (1941) 615.

Sch51   Schatzmann, E., *Comptes Rendus* 232 (1951) 1740.

Sch52   Schardt, A., Fowler, W. A., and Lauritsen, C. C., *Phys. Rev.* 86 (1952) 527.

Sch53   Schatzmann, E., *Ann. d'Ap.* 16 (1953) 162.

Sch58   Schwarzschild, M., *The Structure and Evolution of the Stars* (Princeton: Princeton University Press, 1958).

Sch59   Schmidt, M., *Ap. J.*, 129 (1959) 232.

Sch62   Schwarzschild, M., and Härm, R., *Ap. J.* 136 (1962) 158.

Sch63   Schmidt, M., *Nature* 197 (1963) 1040 and *Ap. J.* 137 (1963) 758.

Sch67   Schwarzschild, M., and Härm, R., *Ap. J.* 150 (1967) 961.

Sch70   Schramm, D. N., and Wasserburg, G. J., *Ap. J.* 163 (1970) 57.

Sch71   Schmidt, M., and Bello, F., *Scient. Am.*, May 1971.

Sch71a   Schramm, D. N., and Fowler, W. A., *Nature* 231 (1971) 103.

Sch73   Schramm, D. N., and Fiset, E., *Ap. J.* 180 (1973) 551.

Sch76   Schmalbrock, P., Staatsexamensarbeit, Universität Münster (1976).

Sch77   Schramm, D. N., and Wagoner, R. V., *Ann. Rev. Nucl. Part. Sci.* 27 (1977) 37.

Sch78   Schramm, D. N., and Clayton, R. N., *Scient. Am.*, October 1978.

Sch82   Schramm, D. N., in *Essays in Nuclear Astrophysics*, ed. C. A. Barnes, D. D. Clayton, and D. N. Schramm (Cambridge: Cambridge University Press, 1982), p. 325.

Sch83   Schmalbrock, P., Becker, H. W., Buchmann, L., Goerres, J., Kettner, K. U., Kieser, W. E., Kräwinkel, H., Rolfs, C., Trautvetter, H. P., Hammer, J. W., and Azuma, R. E., *Nucl. Phys.* A398 (1983) 279.

Sch84   Schulte, W. H., and Cleff, B., Universität Münster, private communication (1984).

Sch87   Schröder, U., Ph.D. Thesis, Universtität Münster (1987) and Schröder, U., *et al.*, *Nucl. Phys.* A467 (1987) 240.

Sea52   Seagrave, J. D., *Phys. Rev.* 85 (1952) 197.

See63   Seeger, P. A., and Kavanagh, R. W., *Ap. J.* 137 (1963) 704 and *Nucl. Phys.* 46 (1963) 577.

See65   Seeger, P. A., Fowler, W. A., and Clayton, D. D., *Ap. J. Suppl.* 11 (1965) 121.

Seg47   Segrè, E., *Phys. Rev.* 71 (1947) 274.

Seg49   Segrè, E., and Wiegand, C. E., *Phys. Rev.* 75 (1949) 39.

Seg77   Segrè, E., *Nuclei and Particles* (New York: Benjamin, 1977).

Sei67   Seiler, R. F., Cleland, M. R., and Wegner, H. E., *Rev. Sci. Instr.* 38 (1967) 972.

Sep67   Septier, A., *Focusing of Charged Particles* (New York: Academic, 1967).

Ses73   Seshadri, S. R., *Fundamentals of Plasma Physics* (New York: American Elsevier, 1973).

Seu85   Seuthe, S., Diplomarbeit, Universität Münster (1985) and Seuthe, S., *et al.*, *Nucl. Instr. Meth.* A260 (1987) 33.

Sev79   Sevier, K. D., *At. Nucl. Data Tables* 24 (1979) 323.

Sey43   Seyfert, C. K., *Ap. J.* 97 (1943) 28.

Sha71   Shapiro, S. L., *Ap. J.* 76 (1971) 291.

Sho49   Shoupp, W. E., Jennings, B., and Jones, W., *Phys. Rev.* 76 (1949) 502.

Shu82   Shu, F. H., *The Physical Universe* (Mill Valley: Universal Science Books, 1982).

Sid65   Sidenius, G., *Nucl. Instr. Meth.* 38 (1965) 19.

Sie68   Siegbahn, K., α-β-γ *Ray Spectroscopy* (Amsterdam: North-Holland, 1968).

Sil61   Silverstein, E. A., Salisbury, S. R., Hardie, G., and Oppliger, L. D., *Phys. Rev.* 124 (1961) 868.

Sil70   Silber, R. R., *Nucl. Instr. Meth.* 87 (1970) 221.

Sil83   Silk, J., Szalay, A., and Zel'dovich, Y. B., *Scient. Am.*, October 1983.

Sim83   Simpson, J. A., *Ann. Rev. Nucl. Part. Sci.* 33 (1983) 323.

Sin71   Singh, B. P., and Evans, H. C., *Nucl. Instr. Meth.* 97 (1971) 475.

Slo31   Sloan, D. H., and Lawrence, E. O., *Phys. Rev.* 38 (1931) 2021.

Smi59   Smith, L., *Handbuch der Physik* 44 (1959) 341.

Smi79   Smit, J. J. A., Meyer, M. A., Reinecke, J. P. L., and Reitmann, D., *Nucl. Phys.* A318 (1979) 111.

Smo77   Smoot, G. F., Gorenstein, M. V., and Muller, R. A., *Phys. Rev. Lett.* 39 (1977) 898.

Spe63   Spear, R. H., Larson, J. D., and Pearson, J. D., *Nucl. Phys.* 41 (1963) 353.

Spe67   Spencer, J. E., and Enge, H. A., *Nucl. Instr. Meth.* 49 (1967) 181.

Spi74   Spinka, H., and Winkler, H., *Nucl. Phys.* A233 (1974) 456.

Sta74   Starrfield, S., Sparks, W. M., and Truran, J. W., *Ap. J. Suppl.* 28 (1974) 247.

Ste64   Steffen, K. G., *High Energy Beam Optics* (New York: Wiley, 1964).

Ste76   Steigman, G., *Ann. Rev. Astr. Ap.* 14 (1976) 339.

Ste79   Steigman, G., *Ann. Rev. Nucl. Part. Sci.* 29 (1979) 313.

Ste81   Stegmann, W., and Begemann, I., *Earth Planetary Sci. Lett.* 55 (1981) 266.

Ste84   Steinhardt, P. J., *Comm. Nucl. Part. Phys.* A12 (1984) 273.

Ste85   Steigman, G., in *Nucleosynthesis*, ed. W. D. Arnett and J. W. Truran (Chicago: University of Chicago Press, 1985), p. 48.

Sto70   Storm, E., and Israel, H. I., *Nucl. Data Tables* A7 (1970) 565.

Sto71   Stocker, H., Rollefson, A. A., and Browne, C. P., *Phys. Rev.* C4 (1971) 1028.

Sto76   Stokstad, R. G., Switkowski, Z. E., Dayras, R. A., and Wieland, R. M., *Phys. Rev. Lett.* 37 (1976) 888.

Sto78   Stokstad, R. G., Eisen, Y., Kaplanis, S., Pelte, D., Smilansky, U., and Tserruya, L., *Phys. Rev. Lett.* 41 (1978) 465.

Sto79   Stokes, R. H., Crandall, K. R., Stovall, J. E., and Swenson, D. A., *IEEE Trans. Nucl. Sci.* NS-26 (1979) 3469.

Str58   Strominger, D., Hollander, J. M., and Seaborg, G. T., *Rev. Mod. Phys.* 30 (1958) 585.

Str68   Strehl, P., and Schucan, T. H., *Phys. Lett.* B27 (1968) 641 and *Z. Phys.* 234 (1968) 416.

Str75   Strom, S. E., Strom, K. M., and Grasdalem, G. L., *Ann. Rev. Astr. Ap.* 13 (1975) 187.

Str76   Strom, S. E., in *Frontiers of Astrophysics*, ed. E. H. Avrett (Cambridge: Harvard University Press, 1976), p. 95.

Stu39   Stueckelberg, E. C. G., *Helvetica Phys. Acta* 11 (1939) 225.

Sue56   Suess, H. E., and Urey, H. C., *Rev. Mod. Phys.* 28 (1956) 53.

Sut84   Sutherland, P., and Wheeler, J. C., *Ap. J.* 280 (1984) 282.

Sva47   Svartholm, N., and Siegbahn, K., *Ark. Mat. Astr. Fys.* A33 (1947) 1.

Swi75   Switkowski, Z. E., and Parker, P. D., *Nucl. Instr. Mech.* 131 (1975) 263.

Swi76   Switkowski, Z. E., Stokstad, R. G., and Wieland, R. M., *Nucl. Phys.* A274 (1976) 202 and A279 (1977) 502.

Swi77   Switkowski, Z. E., Wu, S. C., Overlay, J. C., and Barnes, C. A., *Nucl. Phys.* A289 (1977) 236.

Sym81   Symbalisty, E. M. D., and Schramm, D. N., *Rep. Progr. Phys.* 44 (1981) 293.

Taa78   Taam, R. E., and Picklum, R. E., *Ap. J.* 224 (1978) 210 and 233 (1979) 327.

Tak73   Takahashi, K., Yamada, M., and Kondoh, T., *At. Nucl. Data Tables* 12 (1973) 101.

Tak78   Takahashi, K., El Eid, M. F., and Hillebrandt, W., *Astr. Ap.* 67 (1978) 185.

Tak83   Takahashi, K., and Yokoi, K., *Nucl. Phys.* A404 (1983) 578.

Tam74   Tammann, G. A., in *Supernovae and Supernova Remnants*, ed. C. B. Cosmovici (Dordrecht: Reidel, 1974), p. 155.

Tam77   Tammann, G. A., in *Supernovae*, ed. D. N. Schramm (Dordrecht: Reidel, 1977), p. 95.

Tan59   Tanner, N. W., *Phys. Rev.* 114 (1959) 1060.

Tan65   Tanner, N. W., *Nucl. Phys.* 61 (1965) 297.

Tat76   Tatsumoto, M., Unruh, D. M., and Desborough, G. A., *Geoch. Cosmoch. Acta* 40 (1976) 617.

Tay80   Taylor, R. J., *Rept. Progr. Phys.* 43 (1980) 253.

Thi80   Thielemann, F. K., Ph.D. thesis, Technische Hochschule Darmstadt (1980).

Thi83   Thielemann, F. K., Metzinger, J., and Klapdor, V., *Z. Phys.* A309 (1983) 301 and *Astr. Ap.* 123 (1983) 162.

Thi83a   Thielemann, F. K., in *Proc. Workshop on Stellar Nucleosynthesis* (Erice, Italy) (Dordrecht: Reidel, 1983), p. 389.

Thi85   Thielemann, F. K., and Arnett, W. D., in *Nucleosynthesis*, ed. W. D. Arnett and J. W. Truran (Chicago: University of Chicago Press, 1985), p. 151.

Tho46   Thonemann, P. C., *Nature* 158 (1946) 61.

Tho76   Thomann, C., and Benn, J. E., *Nucl. Instr. Meth.* 138 (1976) 293.

Tim87   Timmermann, R., Diplomarbeit, Universität Münster (1987).

Tin75   Tinsley, B. M., *Ap. J.* 198 (1975) 145.

Tin80   Tinsley, B. M., *Fundamentals Cosmic Phys.* 5 (1980) 287.

Toe71   Toeves, J. W., Fowler, W. A., Barnes, C. A., and Lyons, P. B., *Ap. J.* 168 (1971) 421.

Toe71a   Toeves, J. W., *Nucl. Phys.* A172 (1971) 589.

Tom63   Tombrello, T. A., and Parker, P. D., *Phys. Rev.* 131 (1963) 2582.

Tom65   Tombrello, T. A., *Nucl. Phys.* 71 (1965) 459.

Tom82   Tombrello, T. A., Koonin, S. E., and Flanders, B. A., in *Essays in Nuclear Astrophysics*, ed. C. A. Barnes, D. D. Clayton, and D. N. Schramm (Cambridge: Cambridge University Press, 1982), p. 233.

Tou79   Toussaint, B., Treiman, S. B., Wilczek, F., and Zee, A., *Phys. Rev.* D19 (1979) 1036.

Tra75   Trautvetter, H. P., and Rolfs, C., *Nucl. Phys.* A242 (1975) 519.

Tra78   Trautvetter, H. P., Wiescher, M., Kettner, K. U., Rolfs, C., and Hammer, J. W., *Nucl. Phys.* A297 (1978) 489.

Tra79   Trautvetter, H. P., Elix, K., Rolfs, C., and Brand, K., *Nucl. Instr. Meth.* 161 (1979) 173.

Tra84   Trautvetter, H. P., Habilitationsarbeit, Universität Münster (1984).

Tri82   Trimble, V., *Rev. Mod. Phys.* 54 (1982) 1183.

Tru66   Truran, J. W., Hansen, C. J., Cameron, A. G. W., and Gilbert, A., *Can. J. Phys.* 44 (1966) 151 and 563.

Tru69   Truran, J. W., and Arnett, W. D., *Ap. J.* 157 (1969) 339 and 160 (1970) 181.

Tru72   Truran, J. W., *Ap. Space Sci.* 18 (1972) 308.

Tru77   Truran, J. W., and Iben, I., *Ap. J.* 216 (1977) 797.

Tru82   Truran, J. W., in *Essays in Nuclear Astrophysics*, ed. C. A. Barnes D. D. Clayton, and D. N. Schramm (Cambridge: Cambridge University Press, 1982), p. 467.

Tru83   Truran, J. W., in *The Neutron and Its Applications*, ed. P. Schofield (Inst. of Phys. Conf. Ser. No. 64, Bristol, 1983), p. 95.

Tru84   Truran, J. W., *Ann. Rev. Nucl. Part. Sci.* 34 (1984) 53.

Ulr82   Ulrich, R. K., in *Essays in Nuclear Astrophysics*, ed. C. A. Barnes, D. D. Clayton, and D. N. Schramm (Cambridge: Cambridge University Press, 1982), p. 301.

Ure55   Urey, H. C., *Proc. Nat. Acad. Sci.* 41 (1955) 127.

Van31   Van de Graaff, R., *Phys. Rev.* 38 (1931) 1919A.

Van33   Van de Graaff, R., Compton, K. T., and van Atta, L. C., *Phys. Rev.* 43 (1933) 149.

Van83   VandenBerg, D. A., *Ap. J. Suppl.* 51 (1983) 29.

Vau70   Vaughn, F. J., Chalmers, R. A., Kohler, D. A., and Chase, L. F., *Phys. Rev.*
         C2 (1970) 1657.
Vie75   Viefers, W., von Witsch, W., and Richter, A., *Nucl. Phys.* A248 (1975) 518.
Vli83   Vlieks, A. E., Hilgemeier, M., and Rolfs, C., *Nucl. Instr. Meth.* 213 (1983)
         291.
Vog60   Vogt, E., and McManus, H., *Phys. Rev. Lett.* 4 (1960) 518.
Vog64   Vogt, E., McPherson, D., Kuehner, J., and Almqvist, E., *Phys. Rev.* 136
         (1964) B99.
Vog68   Vogt, E., *Adv. Nucl. Phys.* 1 (1968) 261.
Voi72   Voit, H., Ischenko, G., Siller, F., and Helb, H. D., *Nucl. Phys.* A179 (1972)
         23.
Vol83   Volk, H., Kräwinkel, H., Santo, R., and Wallek, L., *Z. Phys.* A310 (1983) 91.
Wad75   Wade, N., *Science* 189 (1975) 358.
Wad77   Waddington, C. J., *Fundamentals Cosmic Phys.* 3 (1977) 1.
Wag67   Wagoner, R. V., Fowler, W. A., and Hoyle, F., *Ap. J.* 148 (1967) 3.
Wag73   Wagoner, R. V., *Ap. J.* 179 (1973) 343.
Wal62   Walters, W. L., Costello, D. G., Skofronick, J. G., Palmer, D. W., Kane, W.
         E., and Herb, R. G., *Phys. Rev.* 125 (1962) 2012.
Wal81   Wallace, R. K., and Woosley, S. E., *Ap. J. Suppl.* 45 (1981) 389.
Wam73   Wampler, E. J., Robinson, L. B., Baldwin, J. A., and Burbidge, E. M., *Nature*
         243 (1973) 336.
Wan66   Wang, N. M., Novatskii, V. M., Osetinskii, G. M., Chien, N. K., and Che-
         purchenko, I. A., *Soviet. Nucl. Phys.* 3 (1966) 777.
Wap85   Wapstra, A. H., and Audi, G., *Nucl. Phys.* A432 (1985) 1 and 55.
War47   Warren, R. E., Powell, J. L., and Herb, R. G., *Rev. Sci. Instr.* 18 (1947) 559.
War54   Warren, J. B., Laurie, K. A., James, D. B., and Erdman, K. L., *Can. J. Phys.*
         32 (1954) 563.
War75   Ward, W. R., *Rev. Geophys. Space Phys.* 13 (1975) 422.
War76   Ward, R. A., Newman, M. J., and Clayton, D. D., *Ap. J. Suppl.* 31 (1976) 33.
War76a   Warner, B., in *Structure and Evolution of Close Binaries,* ed. P. Eggleton, J.
         Whelan, and S. Mitton (Dordrecht: Reidel, 1976), p. 85.
War77   Ward, R. A., *Ap. J.* 216 (1977) 540.
War78   Ward, R. A., and Newman, M. J., *Ap. J.* 219 (1978) 195.
War80   Ward, R. A., and Fowler, W. A., *Ap. J.* 238 (1980) 266.
Was82   Wasserburg, G. J., and Papanastassiou, D. A., in *Essays in Nuclear Astro-
         physics,* ed. C. A. Barnes, D. D. Clayton, and D. N. Schramm (Cambridge:
         Cambridge University Press, 1982), p. 77.
Wea78   Weaver, T. A., Zimmerman, G. B., and Woosley, S. E., *Ap. J.* 225 (1978)
         1021.
Wea80   Weaver, T. A., and Woosley, S. E., *Ann. N.Y. Acad. Sci.* 336 (1980) 335.
Wea80a   Weaver, T. A., Wilson, J. R., and Bowers, R. L., *Energy and Technology
         Review, Lawrence Livermore Laboratory* (February 1980).
Web70   Weber, J., *Phys. Rev. Lett.* 25 (1970) 180 and *Scient. Am.,* May 1971.
Wei37   Weizsäcker, C. F. von, *Phys. Z.* 38 (1937) 176.
Wei38   Weizsäcker, C. F. von, *Phys. Z.* 39 (1938) 633.
Wei72   Weinberg, S., *Gravitation and Cosmology* (New York: Wiley, 1972).
Wei74   Weisser, D. C., Morgan, J. F., and Thompson, D. R., *Nucl. Phys.* A235
         (1974) 460.

Wei77   Weinberg, S., *The First Three Minutes* (New York: Basic Books, 1977).
Wei79   Weinberg, S., *Phys. Rev. Lett.* 42 (1979) 850.
Wei80   Weinberg, S., *Rev. Mod. Phys.* 52 (1980) 515.
Wei81   Weinberg, S., *Scient. Am.*, June 1981.
Wer71   Werntz, C., *Phys. Rev.* C4 (1971) 1591.
Wet81   Wetherill, G. W., *Scient. Am.*, June 1981.
Wey29   Weyl, H., *Z. Phys.* 56 (1929) 330.
Whe81   Wheeler, J. C., *Rept. Progr. Phys.* 44 (1981) 85.
Whe82   Wheeler, J. C., in *Supernovae*, ed. M. J. Rees and R. J. Stoneham (Dordrecht: Reidel, 1982), p. 167.
Whe82a   Wheeler, J. C., *Type I Supernovae* (Austin: University of Texas Press, 1982).
Wid28   Wideroe, R., *Arch. Elektrotech.* 21 (1928) 387.
Wie77   Wiezorek, C., Kräwinkel, H., Santo, R., and Wallek, L., *Z. Phys.* A282 (1977) 121.
Wie80   Wiescher, M., Becker, H. W., Goerres, J., Kettner, K. U., Trautvetter, H. P., Kieser, W. E., Rolfs, C., Azuma, R. E., Jackson, K. P., and Hammer, J. W., *Nucl. Phys.* A349 (1980) 165.
Wie82   Wiescher, M., and Kettner, K. U., *Ap. J.* 263 (1982) 891.
Wig49   Wigner, E. P., *Proc. Am. Phil. Soc.* 93 (1949) 521.
Wil73   Wilson, R. G., and Brewer, G. R., *Ion Beams* (Malabar: Krieger, 1973).
Wil80   Wilczek, F., *Scient. Am.*, December 1980.
Wil80a   Wilson, R. R., *Scient. Am.*, January 1980.
Wil81   Williams, R. E., *Scient. Am.*, April 1981.
Wil82   Williams, R. E., *Ap. J. (Lett.)* 261 (1982) L77.
Wil83   Wilson, J. R., in *Numerical Astrophysics*, ed. R. Bowers et al. (Portola: Science Books, 1983).
Win67   Winkler, H., and Dwarakanath, M. R., *Bull. Am. Phys. Soc.* 12 (1967) 16.
Wis78   Wisshak, K., and Käppeler, F., *Nucl. Sci. Engineering* 66 (1978) 363.
Wol67   Wollnick, H., in *Focusing of Charged Particles*, ed. A. Septier (New York: Academic, 1967), p. 163.
Wol83   Wolke, K., Diplomarbeit, Universität Münster (1983).
Won73   Wong, C., Haight, R. C., Grimes, S. M., Andersen, J. D., Davis, J. D., McClure, J. W., and Pol, B. A., *Bull. Am. Phys. Soc.* 18 (1973) 652 and *Phys. Rev. Lett.* 31 (1973) 766.
Woo49   Woodbury, E. J., Hall, R. N., and Fowler, W. A., *Phys. Rev.* 75 (1949) 1462.
Woo52   Woodbury, E. J., and Fowler, W. A., *Phys. Rev.* 85 (1952) 51.
Woo71   Woods, R., Jennings, H. K., Collins, M. W., Harder, D. G., and McMillan, D. E., *J. Vacuum Sci. Technol.* 8 (1971) 352.
Woo72   Woosley, S. E., Arnett, W. D., and Clayton, D. D., *Ap. J.* 175 (1972) 731.
Woo73   Woosley, S. E., Arnett, W. D., and Clayton, D. D., *Ap. J. Suppl.* 23 (1973) 231.
Woo78   Woosley, S. E., Fowler, W. A., Holmes, J. A., and Zimmerman, B. A., *At. Nucl. Data Tables* 22 (1978) 371.
Woo78a   Woosley, S. E., and Howard, W. M., *Ap. J. Suppl.* 36 (1978) 285.
Woo78b   Woodward, P. R., *Ann. Rev. Astr. Ap.* 16 (1978) 555.
Woo79   Woosley, S. E., private communication (1979).
Woo81   Woosley, S. E., and Weaver, T. A., *Ann. N.Y. Acad. Sci.* 375 (1981) 357.
Woo82   Woosley, S. E., and Weaver, T. A., in *Essays in Nuclear Astrophysics*, ed. C.

A. Barnes, D. D. Clayton, and D. N. Schramm (Cambridge: Cambridge University Press, 1982), p. 377.

Woo82a   Woolf, N. J., *Ann. Rev. Astr. Ap.* 20 (1982) 367.

Woo84   Woosley, S. E., Axelrod, T. S., and Weaver, T. A., in *Stellar Nucleosynthesis*, ed. C. Chiosi and A. Renzini (Dordrecht: Reidel, 1984), p. 263.

Woo85   Woosley, S. E., private communication (1985).

Wu78   Wu, S., Ph.D. thesis, California Institute of Technology (1978).

Wyn81   Wynn-Williams, G., *Scient. Am.*, August 1981.

Yaf62   Yaffe, L., *Ann. Rev. Nucl. Sci.* 12 (1962) 153.

Yag64   Yagi, K., *Nucl. Instr. Meth.* 25 (1964) 371.

Yan68   Yang, Y. T., *Nucl. Instr. Meth.* 66 (1968) 341.

Yan79   Yang, J., Schramm, D. N., Steigman, G., and Rood, R. T., *Ap. J.* 227 (1979) 697.

Yan84   Yang, J., Turner, M. S., Steigman, G., Schramm, D. N., and Olive, K. A., *Ap. J.* 281 (1984) 493.

Ynt62   Yntema, J. L., and Ostrander, H. W., *Nucl. Instr. Meth.* 16 (1962) 69.

Yok83   Yokoi, K., Takahashi, K., and Arnould, M., *Astr. Ap.* 117 (1983) 65.

Yok85   Yokoi, K., and Takahashi, K., Kernforschungszentrum Karlsruhe, Rept. 3849 (1985).

Yok85a   Yokoi, K., Takahashi, K., and Arnould, M., *Astr. Ap.* 145 (1985) 339.

Yos80   Yoshizawa, Y., Iwata, Y., and Iinuma, Y., *Nucl. Instr. Meth.* 174 (1980) 133.

You68   Young, H. D., *Fundamentals in Optics and Modern Physics* (New York: McGraw-Hill, 1968).

Zei78   Zeilik, M., *Scient. Am.*, April 1978.

Zel65   Zel'dovich, Y. B., *Adv. Astr. Ap.* 3 (1965) 241.

Zel70   Zel'dovich, Y. B., *Comm. Ap. Space Phys.* 2 (1970) 12.

Zie77   Ziegler, J. F., *Helium Stopping Powers and Ranges in All Elemental Matter* (London: Pergamon, 1977).

Zwi74   Zwicky, F., in *Supernovae and Supernova Remnants*, ed. C. B. Cosmovici (Dordrecht: Reidel, 1974), p. 1.

Zys79   Zyskind, J., and Parker, P. D., *Nucl. Phys.* A320 (1979) 404.

Zys81   Zyskind, J., Rios, M., and Rolfs, C., *Ap. J. (Lett.)* 243 (1981) L53.

Index